British Red Data Books
1 Vascular plants

3rd edition

Compiled and edited by
M.J. Wigginton

Published by
Joint Nature Conservation Committee
Monkstone House, City Road
Peterborough PE1 1JY
United Kingdom

Contents

3

Foreword

It is now 20 years since the first edition of the vascular plants Red Data Book was published, and much has since changed, both in our natural heritage and our documentation of it. That book broke new ground in being the first document to draw attention in a systematic way to the likelihood of extinction of a substantial part of our native flora. Five years later a second edition was produced, with updated information. Yet both editions were very thin books, not because few plants were threatened but because we had so little information to draw on. The average species account was a mere six lines long! Now, 20 years on, with a wealth of autecological studies, targeted surveys, initiatives such as Plantlife's Back from the Brink project, Biodiversity Action Plans and so on, we are in a much better position to make informed scientific statements about the status of our most threatened plants. This has meant that the species accounts in the present book are not just longer but also packed with information distilled from the expertise of many British botanists over the past 20 years and more.

Indeed, readers will notice many differences between this book and its progenitors. As well as the much more detailed introductory section and species accounts, each species is also accompanied by a distribution map, with some species having an additional, more detailed, map. In respect of both text and maps the present book has much more in common with Stewart, Pearman & Preston's *Scarce plants in Britain*, published by the JNCC in 1994, than with the previous Red Data Books, and could in fact be regarded as a companion volume to that publication.

This is the second British Red Data Book to use the revised IUCN threat categories (the first covered lichens),

which are based on more scientific criteria than the original categories, and I believe this has resulted in a much more accurate assessment of the degree of threat to each species. Although it is not always possible to apply the new categories to plants with the degree of precision one would expect for large mammals, because of the general lack of population viability analyses and the problem of determining just what constitutes an individual plant, the new system is undoubtedly a substantial improvement on the old one.

I would like to stress that this volume has been a collaborative effort between botanists in the statutory agencies, those in the voluntary societies and other non-governmental bodies and, principally, the host of individual botanists throughout Britain whose enthusiasm is so important for sustaining interest in our native plantlife and ensuring its conservation for future generations. The Editor, Martin Wigginton, is to be congratulated on synthesising such a disparate body of information into a coherent and valuable portrait of the current state of our most endangered plants; this work will continue to be useful to conservationists and botanists for many years to come.

Sir Angus Stirling
Chairman, Joint Nature Conservation Committee
December 1998

1 Introduction

Red Data Books play a crucial role in focusing attention on the plants and animals most in need of conservation action. Prior to this volume, there were four published Red Data Books (RDB) concerning the British flora. The first, in two editions (Perring & Farrell 1977, 1983), were of vascular plants, and these were the first national Red Data Books published for any European country. The other two are of stoneworts (Stewart & Church 1992) and lichens (Church *et al.* 1996). A Red Data Book of bryophytes is shortly to be published (Stewart & Church in prep.). The main objectives of the present volume are to provide a definitive statement on the current distribution and status of rare plants in Britain and to describe the recent changes and trends in the distribution and status of those species and the conservation measures taken to safeguard them.

The first and second editions of the British Red Data Book for vascular plants included those species, and a few subspecies, that occurred in Great Britain in one to fifteen 10 km x 10 km squares (hectads) of the Ordnance Survey national grid. This fifteen-grid-square threshold was established as the main selection criterion for defining a rare plant on the basis of a detailed investigation of the distribution of taxa in the British flora. This threshold was considered to provide an appropriate measure of rarity and was subsequently adopted by the Nature Conservancy Council (NCC) and non-government organisations as a national standard for plants and invertebrates. In the selection of species for inclusion in this edition, we have adopted the revised IUCN Criteria (described in Chapter 3), in which other aspects of rarity are considered, including decline, threat and population dynamics, in addition to distributional data. We have placed a greater emphasis on the status of species in a European or

worldwide context, highlighting in the introductory chapters those for which Britain has a special responsibility. The following groups of species and subspecies are treated in this edition as 'Red Data Book' species:

- *Critically Endangered, Endangered, Vulnerable, Extinct* and *Extinct in the Wild* species and subspecies as defined by the revised IUCN criteria (i.e. nationally threatened = Red List species);
- non-Red List species in the revised IUCN categories *Lower Risk (Conservation Dependent), Lower Risk (Lower Risk (near threatened))* and *Data Deficient*;
- species and subspecies endemic to Britain;
- species included in Schedule 8 of the Wildlife & Countryside Act;
- species included in Annexes IIb and IVb of the European Community Habitats & Species Directive;
- other species threatened internationally, or for which Britain has a special responsibility.

The selection of taxa for inclusion is explained more fully in the chapters that follow.

A national Red Data Book is mainly concerned with species that are nationally threatened, and accounts of such species comprise the main bulk of this volume. They range from those that are endemic to Britain to those that are geographically widespread and common on a worldwide scale. There are a few endemic taxa that are not nationally rare but whose international importance is recognised by their inclusion in this volume. The international context is further explored in Chapter 4.

2   Production of the Red Data Book

2.1 Data sources and data gathering

The Biological Records Centre (BRC) at the Institute of Terrestrial Ecology, Monks Wood, holds the national database of single species records of all vascular plants, including the most complete historical record of Red Data Book species. This database is managed jointly by the Institute for Terrestrial Ecology (ITE) and the Joint Nature Conservation Committee (JNCC). In the mid-1980s the Nature Conservancy Council (NCC) became responsible for maintaining a computer database of rare plants, and this database came to hold the most complete set of post-1980 records. At the start of the current project in January 1993, the NCC database was transferred to the JNCC and was held on RECORDER, a data management system that is also widely used by voluntary bodies and individuals to hold locational species data. Because the NCC/JNCC database was incomplete at the time of transfer, a program of data verification and enhancement was begun. Mr R. Smith, Mr J. Farthing, Miss T. Sykes and Miss C. Minto were employed on short contracts over the next three years to verify existing data, to add new records and to document the occurrence of Red Data Book species in protected sites. The JNCC and BRC databases provided the initial bases for assessing the status of species and selecting those for inclusion in this volume. In 1996, following a wide-ranging review of its remit, the JNCC relinquished the responsibility for maintaining a rare plants database. All records from the JNCC were transferred to BRC at Monks Wood and will be incorporated into the national plants database held there.

Much of the data gathering was organised through the Botanical Society of the British Isles (BSBI), whose vice-county recorders represent one of the main sources of detailed information on the local distribution of the British flora. Print-outs from the JNCC database were sent to BSBI vice-county recorders, who checked the accuracy of existing data, amending where necessary, and provided additional and later records of RDB species at known sites from their own files. Some recorders and other BSBI members and local botanists undertook or organised fieldwork to obtain up-to-date information on the species in their particular areas. Nearly all BSBI vice-county recorders responded to requests for information, and many new records were received. The BSBI further organised a number of field meetings with the principal aim of obtaining information on particular species whose current status was uncertain. Known localities of *Polygala amarella* and some *Alchemilla* species were surveyed during a meeting in Yorkshire, and the very local species *Juncus mutabilis* and *J. capitatus* were specially sought during the meeting held on the Lizard peninsula, Cornwall. More general requests for records of rare species were published in issues of *BSBI News*, together with provisional lists of 'candidate' RDB species considered likely to be included in this volume. Some regional biological records centres, notably the Cornwall Biological Records Unit, provided information on their particular areas.

The JNCC commissioned surveys of a number of species for which up-to-date information was required. These were *Gentiana verna* and *Potentilla fruticosa* (G.G. & P.S. Graham), *Lotus angustissimus* (S.J. Leach), *Phleum phleoides, Silene otites* and *Thymus serpyllum* (Y. Leonard) and *Carex ornithopoda* (M. Porter & F.J. Roberts). Oxford University was commissioned to carry out an investigation of putative *Apium repens* populations at Port Meadow, and the Botanical Society of Scotland to survey some localities for Scottish montane species. Survey, monitoring and commissioned research carried out by and for English Nature (EN), Scottish Natural Heritage (SNH) and the Countryside Council for Wales (CCW) were sources of many new data on rare species, as also were their recovery programmes (or the equivalent) for threatened species. Staff of the country agencies also sent data on RDB species extracted from files held in local offices.

Data were culled from a wide range of published and unpublished sources, most of which are cited in the general bibliography or in individual species accounts. The most comprehensive collated information on the locations, habitats and populations of rare species in England and Wales is to be found in the series of county rare plant reports produced by the NCC and EN under the direction of L. Farrell between 1978 and 1993. The reports are cited in full in the bibliography: G. Beckett (1993), J. Blakemore (1979, 1980, 1981), G. Crompton (1974-1986), S. Everett (1988), R. FitzGerald (1988a-e, 1990a-d), Y. & D. Leonard (1991), V. Morgan (1987a-b, 1988a-b, 1989a-g) and I. Taylor (1987a-d, 1990a-f). The dissolution of the NCC into three country agencies intervened before such detailed surveys and reports could be produced for Scottish counties or regions, but the status of rare Scottish plants is being investigated and documented under the direction of C. Sydes, and species action plans compiled. Likewise, in Wales, similar investigations continue on rare species, much of which is carried out under the direction of A. Jones. Individual species files, which were maintained by L. Farrell in NCC and EN, and now held by JNCC, were valuable sources of locational and ecological data on species and sites in all three countries.

Plantlife reviews and monitoring reports of threatened species have provided much useful up-to-date information on populations, and also details of current and proposed management aimed at enhancing them. The National Trust for England & Wales, the National Trust for Scotland, and the Royal Society for the Protection of Birds provided information on rare plants occurring in their designated sites and areas. The main source of data on the occurrence and status of British RDB species outside Britain was the World Conservation Monitoring Centre (WCMC), which maintains a database of threatened species, compiled from a very wide range of sources, including national Red Data Books and lists.

2.2 Geographical scope

We follow the first and second editions of the British Red Data Book for vascular plants in confining the geographical coverage to Great Britain and the Isle of Man. As pointed out by Perring & Farrell (1983), it would be difficult to justify on phytogeographical grounds the preparation of a single list of threatened vascular plants for the United Kingdom as a whole, or for the United Kingdom and Ireland together. The vascular plant flora of the island of Ireland is considerably smaller than that of mainland Britain and differs substantially in the occurrence, distribution and status of species. For instance, many of Ireland's rare species are common in Britain and, conversely, some species occur in Ireland which are unknown in mainland Britain (Perring & Walters 1982; Perring 1996). These differences are reflected in the different priorities for the conservation of species adopted in the Republic of Ireland and in Northern Ireland, and, furthermore, different legislative frameworks provide for species protection in Ireland. An Irish Red Data Book for vascular plants (including Northern Ireland) was published a few years ago (Curtis & McGough 1988), and revisionary work covering both geopolitical parts of the island is currently underway. The Channel Islands are also excluded on phytogeographical as well as political grounds, their flora having a greater affinity with that of the nearby European mainland.

3   Species selection

3.1 Taxonomic scope

The aim has been to include all native or probably native threatened and near-threatened species and subspecies of vascular plants that occur in Britain. The distribution and status of these taxa is, for the most part, very well known. Plants of lower taxonomic rank, including varieties, forms, races, ecotypes and genetic variants, have been excluded (though a few are mentioned in some accounts of species) because in most cases their geographical distribution and ecology are inadequately known, or their taxonomy disputed. Thus, at the present time, the status of many of such taxa of lower rank cannot be determined.

Some apomictic groups are included, but, because our knowledge of them is uneven, different selection criteria have been applied to different genera. The distribution and status of species of *Alchemilla* and *Sorbus* are well known, and all rare species were described in the second edition of the vascular plants Red Data Book (Perring & Farrell 1983). The standard selection criteria have been applied to species in those two genera. Since the publication of the second edition, the occurrence and status of the larger, more 'difficult' apomictic genera, *Hieracium*, *Rubus* and *Taraxacum* have become better known, and despite some of their microspecies being rather unstable or ill-defined, it was nonetheless considered worthwhile highlighting some of their rarest taxa, if only to emphasise that their conservation value should not be disregarded. Because these groups are large, a different approach has been necessary. We list species of *Hieracium* that are thought to occur in five or fewer hectads, and provide a short list of species of *Rubus* and *Taraxacum* which specialists in these groups consider likely to be among the rarest in Britain. Of course, a severely restricted distribution does not necessarily imply a severe threat to a species. For example, many microspecies of *Rubus* have very restricted distributions (perhaps occurring in only a single hectad), but may be locally abundant in vigorous colonies and under little apparent threat in their particular localities. The recent revision of the *Limonium binervosum* aggregate (Stace & Ingrouille 1986) resulted in a large number of species and subspecies being newly described. Many of these appear to be very local and rare, although it is also probable that the status of many are inadequately known. However, since most of these new segregates are thought to be endemic to Britain, we considered that mention should made of all the species and subspecies currently thought to be rare.

Though it has not been possible to select taxa at a lower rank than subspecies for inclusion in this volume, their potential importance should not be overlooked. The loss of genetic variants or ecotypes might be detrimental to the long-term survival of a species, and represent genetic diversity and potential for evolution. Genetic aspects of conservation are discussed in section 5.4.

Taxonomy and nomenclature follows Kent (1992), except in very few instances where names used by other authorities are adopted, including *Asparagus prostratus* (Kay 1997), *Cochlearia atlantica* (Rich 1991) and *Diphasiastrum issleri* (Jermy *in litt.*).

3.2 Assessment of native status

In setting national priorities for species conservation, the greatest emphasis has always rightly been placed on those species which are native to Britain. In this book we treat as native those species which are presumed to have colonised Britain by natural means independent of human activity. Helpful discussions on the criteria for assessing native status are found, for example, in Webb (1985) and Preston (1986), and the status and distribution of plants deemed to be alien or probably so are given in Clement & Foster (1994) and Ryves *et al.* (1996).

Whilst the status of most species in the British flora is clear and undisputed, there has been much debate and disagreement over the status of others, and because the weighting of available evidence (sometimes conflicting) is a matter of personal judgement, there remains a divergence of opinion. For example, Webb (1985) gives a list of species accepted as native by Clapham *et al.* (1962) but which he (Webb) considered to be the clearest cases of non-native origin. Ten species on Webb's list of probable aliens are nationally rare, but of these ten, six are listed in Stace (1991) as native (*Centaurea cyanus, Echium plantagineum, Lavatera cretica, Muscari neglectum, Valerianella rimosa, Veronica triphyllos*), one as native in a single locality (*Eryngium campestre*), two as possibly native (*Melampyrum arvense* and *Lonicera xylosteum*), and only a single species (*Cynodon dactylon*) as probably introduced. Likewise, Clement & Foster (1994) and Ryves *et al.* (1996) differ from other authorities in their assessment of species: in the former, for example, *Iris spuria, Leucojum vernum* and *Matthiola incana* are accepted, with reservations, as native.

In selecting species for inclusion in the present volume, of the approximately 331 rare species first considered, 258 have been accepted as probably or certainly native, with thirteen subsequently re-assessed as nationally scarce and two as *Data Deficient*. A further 30 species, which were included in the second edition RDB, have been rejected as certain or almost certain introductions (Table 1). Their alien status is variously indicated - for example, by their distribution outside Britain, by their recent appearance in Britain or history of recording, or by the habitats which they occupy. Degrees of difficulty were presented by 30 other rare species whose claim to native status is doubtful, or at least has not been entirely unquestioned. For some of these (e.g. *Draba aizoides*), recent evidence points strongly to native status, whilst for others (e.g. *Cerastium brachypetalum*), non-native status seems very possible. It was particularly difficult to judge the status of species that could lay claim to be native in some of their sites, though certainly alien in others, examples of which are *Anisantha madritensis, Centaurea cyanus, Eryngium campestre* and *Lonicera xylosteum*. A special case is *Senecio cambrensis*, which is non-native in terms of its introduced parent, *S. squalidus*, but arose naturally through hybridisation with the native *S. vulgaris*. A list of questionably native species is shown in Table 2, and includes those which have been variously described as 'probably native', 'possibly native' and 'probably introduced'. Accounts of all of those species have been included in this volume. It is, of course, possible that some of our weeds of cultivation were introduced in ancient times (perhaps as far back as the Neolithic period), and whether some are truly native is probably unknowable. Where there is no clear opposing evidence, such species have been treated as native in this volume.

Table 1 Introduced species included in the 2nd ed. RDB but excluded from the 3rd ed. RDB	
Alyssum alyssoides	Leucojum vernum
Anisantha tectorum	Linaria supina
Anthoxanthum aristatum	Matthiola incana
Campanula rapunculus	Narcissus obvallaris
Caucalis platycarpos	Oenothera stricta
Crocus vernus	Paeonia mascula
Cyclamen hederifolium	Rorippa austriaca
Equisetum ramosissimum	Sagittaria rigida
Galium spurium	Schoenoplectus pungens
Iris spuria	Silene italica
Iris versicolor	Sisymbrium irio
Isatis tinctoria	Spartina alterniflora
Juncus subulatus	Tetragonolobus maritimus
J. tenuis var. dudleyi	Trifolium stellatum
Ledum groenlandicum	Veronica praecox

Table 2 Species whose native status has been considered doubtful by some authorities	
Adonis annua	Echium plantagineum
Alchemilla acutiloba	Eryngium campestre
A. subcrenata	Homogyne alpina
Althaea hirsuta	Lavatera cretica
Anisantha madritensis	Limosella australis
Bupleurum falcatum	Lonicera xylosteum
Centaurea calcitrapa	Melampyrum arvense
C. cyanus	Muscari neglectum
Cerastium brachypetalum	Petrorhagia prolifera
Chenopodium vulvaria	Pulmonaria obscura
Cotoneaster integerrimus	Spergularia bocconei
Crassula aquatica	Tordylium maximum
Cynodon dactylon	Valerianella rimosa
Draba aizoides	Veronica triphyllos
A. monticola	Galium tricornutum

3.3 The revised IUCN threat categories and selection criteria

The IUCN[1] Criteria for assigning threat status and category have been used in various forms in Red Data Books and Red Lists for thirty years. They are recognised internationally, provide a simple and readily understood method for highlighting species under threat, and are a means by which conservation priorities may be determined. In the original system, the categories *Endangered, Vulnerable,* and *Rare* were defined rather loosely and without quantitative qualifiers (Perring & Farrell 1983). Their application was, therefore, to a large extent a matter of subjective judgement, and it was not easy to apply them consistently within a taxonomic group or to make comparisons between groups of different organisms. The deficiencies of the old system had been recognised for some time, and in the mid-1980s proposals were made to replace it with one which could be more objectively and consistently applied. Finally in 1989, the IUCN's Species Survival Commission Steering Committee requested that a new set of criteria be developed to provide an objective framework for the classification of species according to their extinction risk. The aims of the revised criteria would be:

- to provide a system that could be applied consistently;

- to improve objectivity by providing clear guidance on how to evaluate the different factors which affect the risk of extinction

- to provide a system which will facilitate comparisons across widely different taxa

- to provide users of threatened species lists a better understanding of how individual species were classified.

The first, provisional, outline of the new system was published in Mace & Lande (1991). This was followed by a series of revisions, and the final version was adopted as the global standard by the IUCN Council in December 1994. The guidelines were recommended for use also at the national level. In 1995, the JNCC endorsed their use as the new national standard for Great Britain.

A brief outline of the revised IUCN criteria and their application is given below, but it is important that users of the new system refer to the published document (IUCN 1994), which fully explains it and contains many qualifying remarks. The definitions of the categories are given in Figure 1 and the hierarchical relationship of the categories in Figure 2.

Figure 1 Definitions of IUCN threat categories (IUCN 1994)

Threat category	Definition
Extinct (EX)	A taxon is *Extinct* when there is no reasonable doubt that the last individual has died.
Extinct in the wild (EW)	A taxon is *Extinct* in the wild when it is known to survive only in cultivation, in captivity or as a naturalised population (or populations) well outside the past range. A taxon is presumed extinct in the wild when exhaustive surveys in known and/or expected habitat, at appropriate times (diurnal, seasonal, annual) throughout its range have failed to record an individual. Surveys should be over a time frame appropriate to the taxon's life cycle and life form.
Critically Endangered (CR)	A taxon is critically endangered when it is facing an extremely high risk of extinction in the wild in the immediate future, as detailed by any of the criteria A to E*
Endangered (EN)	A taxon is *Endangered* when it is not *Critically Endangered* but is facing a very high risk of extinction in the wild in the near future, as defined by any of the criteria A to E*
Vulnerable (VU)	A taxon is *Vulnerable* when it is not *Critically Endangered* or *Endangered* but is facing a high risk of extinction in the wild in the medium term future, as defined by any of the criteria A to D*
Lower Risk (LR)	A taxon is Lower Risk when it has been evaluated but does not satisfy the criteria for any of the categories *Critically Endangered, Endangered* or *Vulnerable.* Taxa included in the Lower Risk category can be separated into three sub-categories.
Conservation Dependent (cd)	Taxa which are the focus of a continuing taxon-specific or habitat-specific conservation programme targeted towards the taxon in question, the cessation of which would result in the taxon qualifying for one of the threatened categories above within a period of five years.
Near Threatened (nt)	Taxa which do not qualify for *Lower Risk (conservation dependent),* but which are close to qualifying for *Vulnerable.*
Least Concern (lc)	Taxa which do not qualify for *Lower Risk (conservation dependent)* or *Lower Risk (near threatened).*
Data Deficient (DD)	A taxon is *Data Deficient* when there is inadequate information to make a direct or indirect assessment of its risk of extinction based on its distribution and/or population status. A taxon in this category may be well studied, and its biology well known, but appropriate data on abundance and/or distribution are lacking. *Data Deficient* is therefore not a category of threat or Lower Risk. Listing of taxa in this category indicates that more information is required and acknowledges the possibility that future research will show that a threatened category is appropriate.
Not Evaluated (NE)	A taxon is Not Evaluated when it has not been assessed against the criteria*.

Key: *see Appendix 1.

---

[1] now The World Conservation Union (WCU)

Figure 2 Hierarchical relationships of the categories

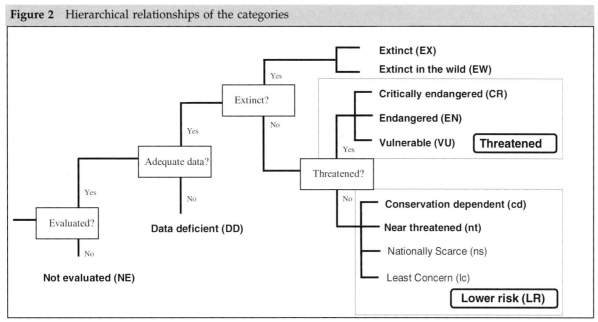

Source: adapted from IUCN (1994) *Red List Categories.*

Newly established categories are *Extinct in the wild* (EW), and *Critically Endangered* (CR). Whilst the names *Endangered* (EN) and *Vulnerable* (VU) have been maintained, they are now differently defined, and species in one of these threat categories in the old system will not necessarily be in the same category in the new. Most species deemed to be 'Rare' in the old system have been assigned to the *Lower Risk (near threatened)* (LR(nt)) category in the new system, though on the basis of the new criteria some are now regarded as *Vulnerable*. The *Lower Risk (least concern)* (lc) subdivision of the *Lower Risk* category represents all other species, including the most widespread and ubiquitous.

At the national level, countries are permitted to refine the definitions for the Lower Risk categories and to define additional ones of their own. JNCC has established one extra category and two definitions as a national standard. The *Lower Risk (near threatened)* category is defined as - species occurring in 15 or fewer hectads, but which are not threatened (i.e. not qualifying as *Critically Endangered, Endangered* or *Vulnerable*). The *Nationally Scarce* category is defined as - species occurring in 16-100 hectads, but which are not *Threatened, Lower Risk (near threatened)* or *Lower Risk (conservation dependent)*. Comments on the last category are included in the following section.

The revised criteria can be applied to any taxon at or below species level, and within any specified geographical

area. However, the IUCN guidelines suggest that their application at regional or national levels (as opposed to global) are best used with two key pieces of information: the global status of the taxon, and the proportion of the global population or range that occurs within the region or nation. Thus, species endemic to Britain are treated as Red Data Book species, even though they may not qualify by the strict application of the criteria (see also chapter 4). More detailed guidelines for the use of national Red List categories are still being developed by IUCN to take account of the movement of organisms across national boundaries, especially highly mobile species such as birds.

Taxa listed as *Critically Endangered, Endangered* or *Vulnerable* are defined as Threatened (Red List) species. For each of these threat categories there is a set of five main criteria A-E (an additional sub-criterion for the *Vulnerable* category), any one of which qualifies a taxon for listing at that level of threat. The qualifying thresholds within the criteria A-E differ between threat categories. They are summarised in Table 3, and given in full in Appendix 1.

Species have been assigned to a threat category solely on the basis of their status in Great Britain, and without reference to their status outside this country. A full list of these species is given in Appendix 2, arranged in order of IUCN threat category, together with their qualifying criteria. Endemic species are discussed in 4.3. below.

Table 3 Summary of the thresholds of the IUCN Criteria

Criterion		Critically Endangered	Main thresholds Endangered	Vulnerable
A.	Rapid decline	>80% over 10 yrs or 3 generations in past or future	>50% over 10 yrs or 3 generations in past or future	>20% over 10 yrs or 3 generations in past or future
B.	Small range – fragmented, declining or fluctuating	extent of occurrence <100 km² or area of occupancy <10 km²	extent of occurrence <5,000 km² or area of occupancy <500 km²	extent of occurrence <20,000 km² or area of occupancy <2,000 km²
C.	Small population and declining	<250 mature individuals, population declining	<2,500 mature individuals, population declining	<10,000 mature individuals, population declining
D1.	Very small population	<50 mature individuals	<250 mature individuals	<1,000 mature individuals
D2.	Very small range			<100 km² or <5 locations
E.	Probability of extinction	>50% within 5 years	>20% within 20 years	>10% within 100 years

3.4 The application of the revised IUCN criteria

Most of the criteria are quantitative, at least in part, so a degree of objectivity can be applied. Nonetheless, subjective assessments are still required as, for example, in predicting future trends and judging the quality of the habitat. Since the criteria have been designed for global application and for a wide range of organisms, it is hardly to be expected that every one will always be appropriate to every taxonomic group or taxon. Thus, a taxon need not meet all the criteria A-E, but is allowed to qualify for a particular threat category on any single criterion. The guidelines emphasise that a precautionary principle should be adopted when assigning a taxon to a threat category, and this should be the arbiter in borderline cases. The threat assessment should be made on the basis of reasonable judgement, and it should be particularly noted that it is not the worst-case scenario which will determine the threat category to which the taxon will be assigned. Generally speaking, these new criteria have been relatively straightforward to apply, though some difficulties of interpretation remain; some of the problems encountered are mentioned below.

Criterion A. Rapid decline of population

A decline in the population of a species over a particular period may be readily determined in terms of numbers of individuals or range, and therefore assigning a species to a threat category should be relatively straightforward. However, because of the lack of consistency of baseline data, some subjective judgement has been required. Determining the causal factors for any future decline may be much less easy, particularly, for example, in determining the decline in the quality of the habitat, and the effects of hybridisation, of competitors or pollutants. Estimating past and future changes in our arable weed flora presented problems. Though it is clear that several species are in sharp decline, the lack of comprehensive data covering the past ten years (the period specified in criterion A) makes it uncertain whether they qualify as *Threatened*. *Scarce Plants in Britain* (Stewart *et al.* 1994) highlight some arable species, such as *Fumaria parviflora*, *Galeopsis angustifolia*, *Ranunculus arvensis*, *Scandix pecten-veneris* and *Torilis arvensis*, which have declined markedly. It is possible that some have declined at least

20% over the past ten years in terms of population size or geographical occurrence, and so qualify as *Threatened*, but in the absence of reliable data they have not been classified as such in this book.

Criterion B. Small range combined with fragmentation, decline or fluctuation

The criterion 'extent of occurrence' (definition in Appendix 1) has not been used in assessing the status of rare species in this book, as such estimates are meaningless in the present context. The 'area of occupancy' also presents problems of interpretation. It should be measured, according to the Guidelines, on grid squares "which are sufficiently small", and its size should be appropriate to the biological aspects of the taxon. This criterion is perhaps more appropriate to a mobile animal holding a territory or a 'home range' whose area can be measured. For a plant it is either much less easy to determine the area needed for its survival, or the area might be tiny in comparison. For simplicity, the 'area of occupancy' of the rare plants considered in this book has been interpreted in terms of hectads, which appears to be appropriate for a country the size of Britain. Thus, for example, under criterion B, a *Vulnerable* plant is one that occurs in twenty or fewer 10 km x 10 km grid squares and satisfies two of the three sub-criteria.

Criterion C. Small and declining population

This criterion includes thresholds of numbers of mature individuals (i.e. those that are capable of reproduction), and is generally readily applied where numbers are known. The definition of an 'individual' has been interpreted in this book to include clumps where single plants are difficult to define (for example, species with an underground rhizome which may interconnect what appear to be separate plants above ground, such as *Homogyne alpina*, *Lithospermum purpureocaeruleum*). For species whose populations naturally fluctuate, the definitions of the criteria state that the minimum population number should be used. However, this is a severe ruling for a plant species which may occur in vast numbers when the habitat is in an ideal condition, yet fall to near zero (though retaining a good seed-bank) when it is not (e.g. *Filago pyramidata*). Much also depends on the interpretation of what is a 'natural' fluctuation. Therefore, reasonable judgement has been applied, rather than taking the minimum number of plants as the qualifying figure.

Criterion D. Very small or restricted population

No qualifiers are given in the *Critically Endangered* and *Endangered* categories; the population size alone will admit the taxon to criterion D irrespective of any decline or threat. In the *Vulnerable* category, D1 relates to the number of mature individuals, and D2 is applicable to a taxon with a very small population and a very restricted area of occupancy. As in criterion B, the area of occupancy is difficult to apply, and in D2 the threshold of the number of locations (normally fewer than five) was the key criterion used in this book. The IUCN guidelines defines a 'location' as "a geographically or ecologically distinct area in which a single event (e.g. pollution) will soon affect all individuals of the taxon present . . . and usually contains all or part of a subpopulation". The main difficulty was in defining locations for species with a scattered distribution (as, for example, along a length of coastline or across a range of hills), and again, some subjectivity was applied.

Criterion E. Population viability analysis (PVA)

This criterion invites a quantitative estimate of the percentage probability of extinction in the wild within a specified number of years. Such analysis should be explicit and based on the known life history and specified management options. Clearly this criterion can be applied only to well-studied organisms whose population dynamics and life histories are adequately known, and has not been applied to any plant in this book.

Conservation Dependent (cd) species

Many *Threatened* taxa are dependent upon regular and continuing conservation management, but this category is reserved for *Lower Risk (near threatened)* taxa that might very rapidly (within five years) become *Threatened* if such conservation action ceases. None of our *Lower Risk (near threatened)* species appears to qualify, though some of these species may, of course, be heading towards *Threatened* status through habitat change or neglect.

Threshold for 'Near Threatened' and 'Nationally Scarce' species

In assigning species to the revised IUCN categories, it should be emphasised that a species may qualify as *Threatened* (i.e. *Critically Endangered, Endangered or Vulnerable*) on the basis of its decline or other criteria, irrespective of its geographical occurrence (in grid squares, in the interpretation of the criteria adopted in this volume). The differentiation of *Lower Risk (near threatened)* species from those that are *Nationally Scarce* and *Lower Risk (least concern)* is more contentious. In this book, we follow established practice in defining species that occur in 16-100 hectads as *Nationally Scarce* unless they qualify as *Threatened*. Whilst this threshold has proved workable over many years, it is admittedly a coarse 'filter', and with the ever increasing availability of detailed locational data, a more refined system is now required. Pearman (1997) suggests many ways by which the definitions of a nationally rare and scarce species can be more precisely defined, based on their occurrence and frequency in smaller Ordnance Grid squares and ideally also on measures of population size. This points a useful way forward, and it is hoped that future Red Data Books will adopt a more refined system.

4 The international context

4.1 Legislation

An obligation to conserve certain species and habitats is laid upon the UK government by a number of international nature conservation conventions and directives. The most important ones that explicitly concern the protection of internationally important species that occur in the UK are the Bern Convention, the EC Habitats & Species Directive, the Ramsar Convention, the CITES Convention and the UN Convention on Biological Diversity.

The Bern Convention (*Council of Europe Convention No. 104 - convention on the conservation of European wildlife and natural habitats*) aims to conserve wild flora and fauna and their natural habitats. It incorporates the principle of sustainable development, and particular emphasis is given to endangered and vulnerable species, especially endemic ones. Appendix 1 of the Bern Convention lists plants requiring strict protection in the signatory states. Nine of the listed vascular plants occur in the wild in Britain (Table 4).

The main aims of the EC Habitats & Species Directive (*European Communities Directive 92/43/EEC, on the conservation of natural habitats and of wild fauna and flora*) were to establish, by the year 2000 at the latest, a network of protected areas in the member states, designed to maintain both the distribution and abundance of threatened species and habitats, and to confer overall protection for the most threatened European species. Annex IVb lists the plants requiring special protection, and Annex IIb lists the species for which protected areas must be designated. Annexes IIb and IVb list nine vascular plant species which occur in the wild in Britain. The Directive also requires, where it is deemed necessary, management of the exploitation of certain species (listed in Annex V). Under the Directive, member states will establish measures, including statutory measures, for the conservation of threatened habitats and species listed in the Annexes. The EU will designate a series of Special Areas of Conservation (SACs) to protect habitat types and species (listed in Annex II) considered to be of European Community interest and requiring conservation. Annex I includes 83 habitat types that occur in the UK, of which 22 are 'priority habitats' because of their Europe-wide rarity. The national network of SACs will contribute in a major way to the conservation of the British flora, including rare species. The protection of habitats in SACs may, of course, conserve nationally rare species other than those listed in the Annexes. For example, those occurring on wet heaths (*Erica ciliaris*), dry calcareous grassland (*Ophrys fuciflora, Orchis simia*), 'Mediterranean' temporary pools (*Cyperus fuscus, Mentha pulegium*), and alpine pioneer formations (*Carex microglochin*). The *Conservation (Natural Habitats etc.) Regulations 1994* came into force in response to the Directive and implements the Directive in GB.

The Ramsar Convention (*Convention on wetlands of international importance, especially as wildfowl habitat*) aims to stem the progressive encroachment on and loss of wetlands, which are broadly defined to include marsh, fen, peatland and water. Sites for designation may, through Article 2, be selected on the basis of their botany as well as other factors, which thus provides opportunity to conserve threatened or endemic wetland plants in designated sites.

The CITES Convention (*Convention on international trade in endangered species of wild fauna and flora*) regulates the international trade in species which are endangered or may become so unless their exploitation is controlled. The EC Regulation 3626/82 (as amended) implements CITES directly in the UK and other member states. Appendix I of CITES lists species whose trade is permitted only in exceptional circumstances (no UK vascular plant species are listed), and Appendix II lists species whose trade is subject to licensing (all orchids, and *Galanthus nivalis*). However, the EC Regulation 3626/82 treats all species of orchid as if they were listed on Appendix I (category C1).

The Convention on Biological Diversity, ratified by the UK government, was an important component of the 1992 United Nations Conference on Environment and Development (the 'Earth Summit') held in Rio de Janeiro. This commits the UK government, *inter alia*, to "develop national strategies, plans or programmes for the conservation and sustainable use of biological diversity". The first product of this commitment was the UK Biodiversity Action Plan, published in 1994, which contains a wide ranging review of our biological resources, and indicates the means by which they may be sustained and enhanced. The Plan outlines programmes and targets for the conservation of both habitats and species; for the latter, these include the preparation of Action Plans for globally threatened and endemic species, the development of a strategy for the *ex situ* conservation of genetic resources, and the preparation of guidelines on species translocations and reintroductions. Programmes for the conservation of habitats are also of key importance for the conservation of our rare and threatened plants.

Other international legislation, although not directly aimed at conserving threatened species, may nonetheless benefit rare plants indirectly by the protection of habitats. Examples include the World Heritage Convention and the EC Birds Directive, some of whose designated sites contain rare UK plants.

Vascular plants listed in international directives and conventions are shown in Table 4 and Appendix 3. Sixteen of the species covered by these international directives and conventions are *Threatened* in Britain (this includes ten orchids additional to *Cypripedium calceolus*). Recent research casts doubt on whether the snowdrop *Galanthus nivalis* is native anywhere in Britain.

14

Table 4 International obligations for the protection of UK vascular plant species

Species	EC Habitats & Species Directive Annex	Bern Convention Appendix	CITES Appendix	Wildlife & Countryside Act (GB) Schedule	National IUCN threat category
Apium repens	IIb, IVb	I		8	*Critically Endangered*
Cypripedium calceolus	IIb, IVb	I	II	8	*Critically Endangered*
Galanthus nivalis	Vb		II		*Least Concern*
Gentianella anglica	IIb, IVb	I		8	*Nationally Scarce*
Liparis loeselii	IIb, IVb	I	II	8	*Endangered*
Luronium natans	IIb, IVb	I		8	*Nationally Scarce*
Lycopodium sensu lato - all species	Vb				*Near Threatened/Nationally Scarce/ Least Concern*, according to species
Najas flexilis	IIb, IVb	I		8	*Nationally Scarce*
Orchidaceae (all orchids)	-		II	8 (11 species)	*Critically Endangered* to *Lower Risk* according to species
Rumex rupestris	IIb, IVb	I		8	*Endangered*
Ruscus aculeatus	Vb	-			*Least Concern*
Saxifraga hirculus	IIb, IVb	I		8	*Vulnerable*
Trichomanes speciosum	IIb, IVb	I		8	*Vulnerable*
Bromus interruptus		I			*Extinct in the Wild*
Spiranthes aestivalis	IVb	I	II		*Extinct*

4.2 International importance of the British flora

In comparison with many other European countries, particularly those bordering the Mediterranean, Britain has a relatively impoverished flora, only Ireland, the Low Countries and Scandinavian countries having fewer species. Kent (1992) lists some 1,300-1,400 native species and subspecies of ferns and seed-bearing plants, not including taxa in the large apomictic genera. However, because of our geographical position, the British flora combines north-west European, Lusitanian and Mediterranean elements, together with oceanic, arctic and alpine species.

Many European species are at the limit of their range in Britain, and thus many of our rare plants occur more abundantly in other parts of Europe. Nonetheless, this need not detract from their value in Britain. Species at or near the limit of their geographical range are close to their tolerance for environmental factors. They develop locally adapted populations (ecotypes) which may be genetically or physiologically distinct. Such ecotypes often occupy different ecological niches to those nearer the centre of the range. The need to conserve such diversity is readily apparent, and the importance of conserving the full extent of a species' natural range is also reflected in the new IUCN threat criteria, and is identified in the UK Biological Action Plan as a key objective in conserving biodiversity.

4.3 Endemic and near-endemic taxa, and other species of global concern in the British flora

From an international standpoint, the conservation of endemic taxa is clearly of the greatest importance, and should be the focus of priority action. Since they have such restricted distributions on a global scale they are, by definition, more threatened internationally than more cosmopolitan taxa. In Britain, the relative poverty of our vascular plant flora is matched by the few endemic taxa occurring here. Only about 25 species and subspecies are endemic to the UK and the Isle of Man, apart from the numerous endemic microspecies in *Euphrasia*, *Hieracium*, *Limonium*, *Rubus*, *Sorbus* and *Taraxacum*. Of these 25, two are extinct in the wild (*Bromus interruptus*, *Sagina boydii*), and the taxonomy of four others (*Alchemilla minima*, *Athyrium flexile*, *Calamagrostis scotica*, *Cochlearia atlantica*) is

disputed. It is sobering to reflect that 71% of the UK endemic species are threatened or near-threatened.

Seventeen of the British endemic or near endemic species are nationally threatened or near-threatened, and accounts of such species are included in this book. The few other endemic species that are nationally scarce in Britain do not qualify for inclusion on the strict application of the IUCN criteria. However, the preamble to the Red List categories (IUCN 1994) advises that the global status category for a taxon should also be taken into consideration when applying the criteria at a national level. All taxa endemic to Britain are thus treated as 'Red Data Book' taxa. The following list of such taxa includes

Table 5 Species and subspecies endemic to the United Kingdom, Isle of Man and the Channel Islands*

Taxon	Threatened or Lower Risk (near threatened)	Nationally Scarce	Occurrence
Alchemilla minima	VU		**England**
Anthyllis vulneraria ssp. *corbierei*	VU		England, Channel Islands
Arenaria norvegica ssp. *anglica*	EN		England
Athyrium flexile	VU		Scotland
Bromus interruptus	EW		England
Calamagrostis scotica	VU		Scotland
Cerastium fontanum ssp. *scoticum*	VU		Scotland
Cerastium nigrescens	VU		Scotland
Cochlearia atlantica (?endemic)	DD		Scotland
Cochlearia micacea (?endemic)	LR-nt		Scotland
Coincya monensis ssp. *monensis*		LR-ns	England, Scotland, Wales, Isle of Man
Coincya wrightii	VU		England
Dactylorhiza majalis ssp. *cambrensis*		LR-ns	England, Scotland, Wales
Epipactis youngiana	EN		England, Scotland
Fumaria occidentalis	LR-nt		England
Gentianella amarella ssp. *septentrionalis*		LR-ns	England, Scotland
Gentianella anglica		LR-ns	England
Helianthemum canum ssp. *levigatum*	VU		England
Herniaria ciliolata ssp. *ciliolata*	LR-nt		England, Channel Islands
Herniaria ciliolata ssp. *subciliata*			Jersey
Linum perenne ssp. *anglicum*		LR-ns	England, Scotland
Pilosella flagellaris ssp. *bicapitata*	VU		Scotland
Primula scotica		LR-ns	Scotland
Sagina boydii	EW		Scotland
Scleranthus perennis ssp. *prostratus*	EN		England
Senecio cambrensis	LR-nt		England, Scotland, Wales
Tephroseris integrifolia ssp. *maritima*	VU		Wales
Ulmus plotii		LR-ns	England

Key: *not including the many microspecies of *Euphrasia*, *Hieracium*, *Limonium*, *Rubus*, *Sorbus* and *Taraxacum*. Note: *Cotoneaster integerrimus* from the Great Orme has recently been published as a new and endemic species, *C. cambricus* (Fryer & Hylmö 1994), but there are doubts as to its origin and whether it is distinct from continental stock.

seven endemics, and two non-endemics (*Luronium natans, Najas flexilis*) that are afforded special protection under the EC Habitats and Species Directive. The British endemic, *Gentianella anglica*, is also protected under that EC Directive, and under the Bern Convention.

British 'Red Data Book' taxa:

> *Coincya monensis* ssp. *monensis*
> *Dactylorhiza majalis* ssp. *cambrensis*
> *Gentianella amarella* ssp. *septentrionalis*
> *Gentianella anglica*
> *Linum perenne* ssp. *anglicum*
> *Luronium natans*
> *Najas flexilis*
> *Primula scotica*
> *Ulmus plotii*

Accounts of these species (except for *Dactylorhiza majalis* ssp. *cambrensis* and *Gentianella amarella* ssp. *septentrionalis*) are given in Stewart *et al.* (1994). Conservation programmes should also take account of near-endemic taxa which are not threatened, and those for which Britain has a special responsibility because a high proportion of the world population occurs here. A list of such species, whilst not exhaustive, is given in Table 6.

The percentage of the world population occurring in Britain is an estimate based on available evidence, though this is often difficult to assess as the quality of data from other European countries is variable. For some countries (mainly northern European) there is a detailed and up-to-date record of the flora, but for others (especially eastern and southern European), data are decidedly patchy. Sources of information on the distributions of European species include Jalas & Suominen (1972 *et seq.*), Hultén & Fries (1986), regional atlases and Floras and unpublished data from WCMC. Cook (1983) describes the status of aquatic plants endemic to Europe, and detailed information is available for some European countries in published Red Data Books.

Some of the species listed in Table 6 are not uncommon in parts of their British range, and where there are large populations, a watching brief is perhaps all that is required. However, all these species are represented in some localities by small populations, sometimes comprising only a few individuals. Such endangered populations require positive conservation action if they are to stand any chance of longer term survival. The bluebell *Hyacinthoides non-scripta* is, of course, a celebrated species in Britain and something of a national icon. A high proportion of the world population of this north-west European species occurs in Britain and, though it is still widespread and locally common, it is under increasing threat from commercial exploitation.

Table 6   Select list of other species for which Britain has special responsibility

Taxon	Threat category in Britain	Endemic to Europe	% of the world population[1]	Remarks
Alopecurus borealis	LR-ns	No	?	In Europe occurs only in Britain and Svalbard
Carum verticillatum		Yes	25-50	
Deschampsia setacea	LR-ns	Yes	25-50	Threatened throughout its range
Dryopteris aemula		Yes	25-50	
Fumaria purpurea	LR-ns	Yes	?	Endemic to Britain, Ireland and the Channel Isles, perhaps declining
Hammarbya paludosa	LR-ns	No	?	Threatened throughout Europe
Hyacinthoides non-scripta		Yes	>50?	Under threat through commercial exploitation
Hymenophyllum tunbrigense		?	25-50	If taxonomically distinct from similar taxa in North America and elsewhere
Hymenophyllum wilsonii		Yes	25-50	
Oenanthe fluviatilis		Yes	25-50	Decreasing in Europe
Petroselinum segetum		Yes	25-50?	
Pilularia globulifera	LR-ns	Yes	25-50	Decreasing in much of mainland Europe
Ranunculus hederaceus		Yes	25-50	Decreasing in Europe

Key: [1] main source: UK Biodiversity Action Plan

5 Conservation

5.1 GB legislation

The first legislation directly providing significant protection to all wild plants was the Conservation of Wild Creatures and Wild Plants Act 1975. This Act made it an offence, except in certain circumstances, for any unauthorised person to uproot any wild plant, without reasonable excuse. The Act included a list of specially protected plants (Schedule 2), which it was an offence to pick, uproot or destroy, except as an incidental result of an operation carried out in accordance with good agricultural or forestry practice, or in certain other circumstances.

The law providing protection for plants was strengthened by the passing of the *Wildlife and Countryside Act* in 1981, which includes many sections which can be used to further the conservation of plants and their habitats. Under this Act, it remains an offence for an unauthorised person intentionally to uproot any wild plant. However, under section 13(1)a, the picking, destruction or removal of *any part* (including seed) of a specially protected (Schedule 8) plant was made unlawful, except where it could be shown that it was the incidental result of a lawful action and could not reasonably have been avoided. There are also additional restrictions on the sale of plants.

The *Wildlife and Countryside Act* (WCA) also tightened the criteria used for the selection of species for special protection (listed in Schedule 8), so that a plant can be added to the Schedule if it is in danger of extinction or likely to become so unless conservation measures are taken. In addition, species may be added in order to comply with international obligations, such as that conferred by European Union legislation. There are currently 110 species of vascular plant listed on Schedule 8 (see Appendix 3). There is a requirement under the WCA for the Secretary of State for the Environment, with advice from the statutory conservation agencies, to review the Schedule every five years, at which times recommendations may be made for species to be removed or added. The results of these quinquennial reviews could be seen as one measure of the effectiveness of conservation action. The Secretary of State can, however, add species to Schedule 8 at any time, not only at quinquennial reviews.

Rare plants also benefit from legislation providing for site safeguard and habitat protection. Other sections of the WCA provide for statutory designation and management of Sites of Special Scientific Interest (SSSI) and National and Marine Nature Reserves (NNR, MNR).

Rare plants may also benefit indirectly from other environmental legislation not specifically directed towards their conservation. Examples are the Natural Heritage (Scotland) Act, 1991, which enables SNH to enter into management agreements, the Environmental Protection Act, which seeks, *inter alia*, to control pollution and the release of genetically altered organisms, and Town & Country Planning Acts which control development.

5.2 Site protection

Populations of some threatened plants are safeguarded in Britain through the network of protected sites. The principal site designations are: Sites of Special Scientific Interest (SSSI), in which the appropriate statutory conservation agency can negotiate management agreements for the benefit of the wildlife; National Nature Reserves (NNR), which may be owned by one of the statutory agencies and which are managed specifically for nature conservation; Local Nature Reserves (LNR), managed by local authorities; and Special Areas of Conservation (SACs). The last is a designation soon to be brought into operation under the EC Habitats and Species Directive, and will contribute to the Natura 2000 site series, a network of 'elite' sites throughout Europe. The National Park network also provides some protection for important wildlife sites.

The series of non-statutory protected sites is also important for the conservation of threatened plants. National organisations such as the Royal Society for the Protection of Birds (RSPB), the National Trust and the National Trust for Scotland have extensive land holdings, which are managed primarily for the conservation of the wildlife they support, including many sites for threatened plants. The Ministry of Defence is an major landholder, and conservation management of its property is one of its responsibilities, in liaison with the country agencies. Plantlife owns some small nature reserves, mainly herb-rich meadows. At a local level, the Wildlife Trusts also manage substantial areas as nature reserves and some private landowners do the same. Of course, many of the areas owned and managed by these bodies are also SSSIs or NNRs.

Internationally, the Planta Europa network was established as a consortium of statutory and non-statutory organisations involved in plant conservation across Europe, including Plantlife, following the first European conference on the conservation of wild plants at Hyères, France, in 1995. Planta Europa is currently working on a Europe-wide list of Important Plant Areas (IPA), which is intended to feed into statutory site designations under international law such as the Natura 2000 series. In Britain, most of these may already receive some degree of protection through a statutory designation.

Designation of protected sites may, however, be of little value for ephemeral species, such as some of the arable weeds. Some of these species may be highly mobile, appearing at any particular location only when appropriate conditions prevail, and perhaps not remaining for very long. Less intensive management of the wider countryside, such as that provided under the various agricultural incentive schemes, may be more effective in conserving these species. Examples of these schemes, also mentioned below, include Environmentally Sensitive Areas (ESA) and the Countryside Stewardship scheme, both of which are administered by the Ministry of Agriculture, Fisheries and Food (MAFF).

5.3 Conservation

The conservation of wild flora in Britain during this century has mirrored the development of wildlife conservation overall, with the first steps involving purchase of nature reserves followed by the realisation of the need for specific management regimes to suit the needs of plants with different requirements. This in turn has led to more sophisticated assessments of the responses of different species to alternative management approaches and more detailed ecological studies of threatened plants.

Threats to plants are many and diverse, and are described, as far as they are known, in the individual species accounts. Habitat destruction through the loss of semi-natural areas to agriculture, forestry and urban development continues to threaten our rare plants, as do inappropriate management and neglect. Pollution, of all kinds, is an insidious and ever-present threat. Other, more local, threats include the invasion of semi-natural habitats by alien species such as *Rhododendron ponticum* and *Carpobrotus edulis,* and the increasing pressures from the 'leisure industry' leading to the loss of habitat, for example, to golf courses and coastal caravan sites. The importance of appropriate and continuing conservation management for many habitats and species can hardly be over-emphasised.

Grasslands have been particularly badly affected in recent decades, with about 97% of our semi-natural lowland grassland having been destroyed since 1932, either by ploughing, or by the application of artificial fertilisers and herbicides (Fuller 1987). Fenlands and other wetlands, along with the Red List plants they support, are affected by the lowering of the water table, a problem that is becoming more serious in areas such as East Anglia, now officially classified as 'semi-arid'. Large areas of ancient semi-natural woodland are now protected as nature reserves, but other areas remain unprotected and are always vulnerable to new commercial fashions in woodland management and changes in government policy on subsidies etc., any of which are potentially threatening to Red List plants.

Pollution, particularly nitrate pollution, is a serious problem for some species, causing an increased growth of common, nutrient-demanding, 'weedy' species at the expense of those species that thrive in conditions of low nutrient availability. This can affect whole habitats and wide areas, gradually contributing to the degeneration of heathland, for example. Although sulphur dioxide pollution has been reduced over the past two decades, other pollutants, such as the nitrogen oxides, primarily from vehicle emissions, have assumed a greater importance. Cutting harmful emissions from industry is a costly procedure, although the Government has committed itself to reducing the 1980 levels of sulphur dioxide by 80% by 2010 (Critical Loads Advisory Group 1995). The challenge of cutting emissions from vehicle exhausts is even more daunting, but it is an issue that is gradually becoming recognised as crucial for the health of the environment, though usually for reasons other than the well-being of plants!

Collecting by botanists, a factor that has in the past caused a catastrophic decline in some species (such as Killarney fern *Trichomanes speciosum,* which was severely depleted during the Victorian 'fern craze'), is now at a very low level and is not now generally considered a major threat. Most active botanists in Britain are also conservationists and are content with a photographic record of their finds, or a note on paper or in a database. There is a need to collect a rare plant only if there is doubt over the identity of a specimen, and this can normally be accommodated through collecting a minimal amount of material in a responsible way. Having said that, it would take only a single unscrupulous collector to wipe out some of the species covered in this book, so vigilance continues to be necessary.

Global warming is a likely, but at present unquantifiable, threat. It has hitherto often been assumed that as the mean global temperature rises, Britain will experience a warmer climate, with serious consequences for its alpine and snow-bed plants. However, such warming is by no means certain. Another possible scenario, for example, according to some experts, is a movement of the Gulf Stream southwards so that our climate becomes cooler and wetter. A continuing rise in sea-level seems certain, however, and this will have an impact on low coastal habitats and any nationally rare plants they support, including *Corrigiola litoralis* (on shingle) and *Petrorhagia nanteuilii* (on shingle and sand dunes).

Many actions have been, and are being, taken to counter the threats to our native flora, most notably the establishment of a network of protected sites (see above). Apart from designating and managing sites for nature conservation, the statutory agencies have a wider role in advising government, landowners and the public on nature conservation, including the conservation of threatened plants. English Nature's Species Recovery Programme, a programme of action for bringing threatened species back from the brink of extinction, has provided resources for addressing the problems faced by many of our most threatened plants. Similarly, SNH's Rare Plants Programme has addressed the same issues in Scotland, and CCW have their own programme of threatened plant conservation. More recently, other statutory bodies, such as the Environment Agency, have been drawn into the conservation of threatened plants, with the advent of the Biodiversity Action Plans (see below). Local Authorities have an important part to play in rare plant conservation, through local planning, organising surveys and designation and management of Local Nature Reserves.

The botanical societies have a particularly vital role in the conservation of threatened plants, as it is through them that specialist expertise is mobilised and the good-will of members put into practice. The Botanical Society of the British Isles is particularly strong in the field of recording and mapping. This is due in no small part to its system of vice-county recorders, whereby a local botanist takes on responsibility for plant recording in a particular area. Local Floras are often the product of the efforts of vice-county recorders, usually supported by a small number of local enthusiasts, and remain the definitive sources of data for years or decades after their publication. Increasingly, vice-county recorders are keeping their botanical data on computerised databases.

The launch of Plantlife in 1989 heralded a new era in British plant conservation. Plantlife, with the aim of being to plants what the Royal Society for the Protection of Birds is to birds, has grown steadily and is now an important player in plant conservation both in the UK and abroad. Its main aims include carrying out ecological investigations of endangered species and undertaking practical conservation management at sites where they are declining or have recently disappeared. Emphasis is placed on four main groups of plants which are particularly threatened, namely:

- plants of ponds and commons, such as *Mentha pulegium* and *Pulicaria vulgaris;*
- Mediterranean' annuals - plants close to their northern limit, often of droughted soils in open habitats maintained by grazing, e.g. *Ajuga chamaepitys, Filago pyramidata;*
- plants of coppice, woodland rides and edges. Traditional coppice management is becoming rare, as are plants such as *Carex depauperata* and *Cynoglossum germanicum*, which are characteristic of coppiced woodland;
- other important plants under particular threat, such as the endemic *Gentianella anglica.*

Plantlife's 'Back from the Brink' initiative has directly addressed the conservation of certain species (for example *Damasonium alisma, Filago lutescens, Thlaspi perfoliatum*) that are in particularly grave danger of disappearing from our flora.

Other societies, such as the Wildflower Society and the RSPB, are also active in the conservation of rare species. The Wildlife Trusts and the National Trust are important in survey and protection of particular threatened species, as well as designating and managing nature reserves and other protected sites.

One of the messages highlighted by this Red Data Book is the importance of management for threatened plants *outside* nature reserves and other protected sites. For example, a number of threatened species are arable weeds or otherwise ephemeral species that are dependent on certain forms of management over a wide area rather than in a small specially protected area. It is clear, therefore, that land managers whose prime concern is other than with nature conservation may have a significant responsibility, in partnership with nature conservation bodies, to ensure the continuing survival of such species. MAFF has acknowledged this in the implementation of schemes such as ESA and Countryside Stewardship (inherited from the Countryside Commission), which enable land to be managed sympathetically for plants and other wildlife.

One particular international event has given species conservation a boost in recent years. At the Rio Earth Summit, the UK signed and ratified the Convention on Biodiversity. This has resulted in *Biodiversity: the UK Action Plan* (HMSO 1994) (UKBAP). New partnerships have developed between government departments and agencies, voluntary conservation organisations, business and the private sector, unlocking new sources of funds, with the new partnerships taking action under agreed plans for threatened species. One of the features of UKBAP is the production of lists of species 'of conservation concern'. These were subdivided into three: the 'short', 'middle' and 'long' lists. All short list species and many middle list species now have, or will have very soon, written action plans targeted specifically at the conservation of each species. The remainder of the species on the middle list and all those on the long list are targeted for lower level action or surveillance. The lists - essentially a means of determining conservation priorities - cover all taxonomic groups. Their composition was determined by considering, for each species, its status in Britain, its decline and its international importance. Action plans are not intended to be immutable. Both the individual plans and the composition of the lists will be reviewed at regular intervals and amended as necessary.

In addition to the species listed, thirty-seven habitats of conservation importance are listed in *The UK Action Plan*. Nearly all of these are of at least some importance for plants, including Red List species. Particularly important in the context of the conservation of threatened plants are broad-leaved and yew woodland, native pine woodland, calcareous grasslands, unimproved neutral grasslands, cereal field margins, fen, standing open water, rivers and streams, canals, montane habitats, maritime cliff and slope, sand dunes, coastal strandline and limestone pavement.

In response to the publication of *Biodiversity: the UK Action Plan*, a report (HMSO 1995) has been presented to Government from the Biodiversity Steering Group, a consortium of representatives from statutory conservation agencies, the voluntary conservation sector, Government Departments, industry and other bodies. This report contains costed action plans for 14 habitats, including fens, cereal field margins, native pine woods and limestone pavements, and also action plans for the conservation of 28 species of vascular plant. For example, the action plan for *Alisma gramineum* contains recommendations for, among other things, protecting it at its existing sites, restoration to its formerly occupied sites (either by appropriate habitat management or re-introduction), work on its water quality requirements and depositing seed in the National Seed Bank, run by Kew Gardens at Wakehurst Place.

Ex-situ techniques have an important part to play in the conservation of threatened plants. Whilst cultivation of wild stock or the cryogenic storage of seed is not a alternative to maintaining species *in situ*, these means of conserving genetic resources do provide important insurance against potential loss due to unforeseen events. Populations of rare and endangered plants are maintained in many botanical gardens and provide a valuable resource for research into propagation, breeding systems and genetic typing. Small collections of rare plants are also maintained in many private gardens, having been established from seed or other parts collected from the wild under licence. Such collections are often important in preserving genetic stock from populations of plants that have subsequently become extinct at a particular locality. A recent example is *Filago gallica*, which became extinct in the wild, stock of native origin subsequently being discovered in a private garden, enabling it to be re-introduced to one of its former native sites. Plants in collections can, however, be exposed to hybridisation with related taxa, and genetic drift, leading to a much diminished value.

Outside Britain, many other European countries have produced or are producing Red Data Books for vascular plants. Together, these form a valuable body of information on the status of threatened plants internationally. The World Conservation Monitoring Centre (WCMC), based in Cambridge, takes a lead in collating information on plants worldwide, and is an important source of information on rare and threatened species internationally.

A key aspect of all the conservation initiatives mentioned above is co-operation and partnership and many such initiatives require co-ordinated input from a number of organisations and individuals. A good example of this is the co-operation between Plantlife and English Nature in implementing Species Recovery Programmes. Current initiatives in data collection and handling and the proposals for a National Biodiversity Network will become increasingly important as more people become involved in threatened plant conservation. A prerequisite for the successful conservation of our most threatened plants is the continuing involvement of amateur and professional specialists, local monitoring and surveillance of Red List species, and co-ordination at the national level. Such co-ordinated action can ensure the long-term survival of Britain's most threatened plants.

N. G. Hodgetts

5.4 Genetic aspects of rare plant conservation

Introduction

Genetic aspects of rare plant conservation have received relatively little attention until recently. There is now a growing realisation that biodiversity extends below the level of species, to include heritable variation between and within populations of the same species. Intraspecific variation may be expressed in qualitative or quantitative characters of a structural, physiological or biochemical nature, and only a small fraction has been formally recognised in infra-specific taxonomies. Conservation biologists are concerned that species may be losing genetic variation, primarily as a consequence of habitat decline and fragmentation, and that this genetic depletion may compromise their ability to evolve and adapt to changing environmental conditions (Falk & Holsinger 1991; Ellstrand & Elam 1993; Loeschke *et al.* 1994). Genetics also has considerable relevance to rare plant conservation as a technique for investigating various aspects of taxonomy, status and population biology of plants. This fundamental distinction between genetics as an investigative tool and as a component of biodiversity is maintained in the following synoptic account of recent work on genetic conservation in rare British vascular plants.

Genetics and biodiversity

Recent case studies on rare British plants have mostly involved population surveys of isoenzyme variation, occasionally supplemented by molecular analyses of the genetic material itself. These studies illustrate the wide diversity of ways in which genetic variation is distributed in vascular plants, particularly in relation to the breeding system. For example, John (1992) contrasted the regional patterning and high levels of heterozygosity in *Mibora minima*, which is wind-pollinated and outbreeding, with the local population differentiation and greater homozygosity found in *Gastridium ventricosum* and *Ononis reclinata*, both of which are autogamous inbreeders. Such fundamental differences in genetic make-up have important implications for rare plant conservation strategies, and genetic survey data can help assessments of priorities for population persistence.

Although vascular plants are polymorphic at an average of 50% of their isoenzyme loci (Hamrick *et al.* 1991), genetic screening has occasionally failed to reveal variation within or between the sampled populations, as in the case of the Welsh populations of *Liparis loeselii*, *Gentianella uliginosa* and *Eleocharis parvula* studied by Kay & John (1995). Similarly, very little variation was detected by isoenzyme and DNA analysis in the British endemic *Primula scotica* (Glover & Abbot 1995). However, a subsequent study of *P. scotica* found high levels of variation in quantitative morphological characters, both between and within populations (Ennos *et al.* 1997). This raises important and as yet not fully resolved questions about the extent to which variation detected at the molecular level is representative of biometric traits of potentially greater adaptive significance.

Conservation of genetic diversity is likely to be a particular priority in species with global population concentrations in Britain, but may also be important in geographically peripheral populations of internationally widespread species. For example, the Snowdonia corrie populations of *Lloydia serotina* are genetically divergent from populations in the French Alps and North American Rockies (B.Jones in prep.). Similarly, the Gower populations of *Draba aizoides* are clearly differentiated from their continental counterparts (John 1992). Lesica & Allendorf (1995) concluded that geographically marginal populations are potentially important sites for future speciation events and therefore have high value for biodiversity conservation.

A central tenet of population genetic theory is that small, isolated populations tend to lose genetic variation and decline in fitness due to genetic drift and inbreeding depression. The extent to which these processes occur in natural populations has been investigated in Dutch material of *Salvia pratensis*. Small populations were found to contain lower levels of variation in isoenzyme and phenotypic characters than large populations, as predicted by genetic theory (Treuren *et al.* 1991; Ouborg *et al.* 1991). However, there was no significant correlation between population size and various fitness attributes, probably because population contraction had taken place relatively recently and inbreeding effects had yet to be fully realised (Ouborg & Treuren 1995). Further work has focused on modelling the effects of inbreeding depression on demographic characteristics and population viability (Ouborg & Treuren 1997).

Genetics as an investigative tool

Genetic studies have provided important insights into conservation-relevant aspects of the biology of rare species. In the context of plant taxonomy, DNA analysis was used recently to confirm the occurrence of putative *Apium repens* alongside a phenotypically similar variant of *A. nodiflorum* at the former's only extant British locality (Grassly *et al.* 1996). Similarly, genetic screening can aid assessments of the extent to which populations have been 'contaminated' by hybridisation with allied taxa, as in the case study of *Cirsium tuberosum*, which crosses with *C. acaule*, reported by Kay & John (1994). Assessments of native or introduced status can also be examined using genetic techniques, and in *Luronium natans* genetic inter-relationships corroborate the notion that populations in eutrophic lowland watercourses are recently derived from native colonies in upland oligotrophic lake systems (Kay *et al.* in prep.).

The gene pool of a species may also provide important information on its evolutionary origins or population structure. For example, studies of isoenzyme and chloroplast DNA variation in *Senecio cambrensis* have provided evidence for the independent origins and subsequent genetic diversification of this endemic polyploid hybrid derivative of *S. vulgaris* and *S. squalidus* in Wales and Scotland (Ashton & Abbott 1992; Harris & Ingram 1992). DNA analysis has similarly been valuable in elucidating the clonal structure of British populations of the self-incompatible *Pyrus cordata*, with important implications for its species recovery plan (Jackson *et al.* 1997).

Conclusions

Notwithstanding our inadequate understanding of the importance of hereditary factors in extinction, sufficient case studies have been compiled on the British flora to illustrate the relevance of genetics for nature conservation. Genetics has an important role to play in both traditional and progressive approaches to conserving rare vascular plants (Stevens & Blackstock 1997). Representation and maintenance of genetic variation are important considerations in conservation site selection and *ex situ* procedures, and Gray (1997) has emphasised the need to protect populations across their full range of ecological variation. Conversely, problems of crossing and competition need to be assessed in species recovery programmes, and Kay (1993; Kay & John 1997) has cautioned against translocations of inappropriate genotypes which may lead to disruption of locally adapted genotypes or distortion of long-established patterns of genetic variation. With an increasingly interventionist stance being adopted by conservationists towards the end of the 20th century, genetic considerations are likely to play a prominent role in the conservation strategies of the next millennium.

D. Stevens

6   Accounts of species

Many botanists, both amateur and professional, contributed to the species accounts. Offers to write accounts were received in response to general invitations issued in *BSBI News*, and also from specialists who were individually approached. All the edited draft accounts were subsequently circulated to specialists for comment, were then modified in the light of comments received, and edited into a standard format. Most of the final accounts are essentially those which were originally contributed, but all have been modified to some extent, and more recent information incorporated when this has become available.

6.1 Explanation of the text and maps

With some exceptions, the paragraphs in the text follow a more or less standard sequence. The first summarises the species' principal habitats and plant associates, and the second its reproductive biology. The third paragraph briefly describes its historic and present geographical range in Britain, together with some details of populations. Threats to the species and its conservation management are described in a fourth paragraph, sometimes with more general comment on other topics. The final paragraph summarises the species' distribution outside Britain. The amount of information provided inevitably varies between species, and sometimes there is little on a particular aspect. In particular, information on reproductive biology was sometimes difficult to come by, and time did not permit a thorough literature search. However, many species seem to have been little studied in this respect. At the head of the page, the species status in Europe refers to the continent as a whole: it should be noted that a species recorded as 'Not threatened' may nevertheless be threatened in some countries or geographical regions.

Gaelic and Welsh names are given for species which occur or have occurred in Scotland or Wales (or close to their borders), except for a few species for which there seems to be no such local name (Pankhurst & Preston 1996; Davies & Jones 1995).

Maps

The distribution of each species is summarised as a 10 km square distribution map, which distinguishes three date classes for native records: pre-1970 (o), 1970-1987 (•) and post-1987 (■); and two date classes for alien records: pre-1970 (+) and 1970 onwards (x). Where it has been considered useful, some species have an additional map not only showing presence in a square but also giving an indication of the frequency within that square, in terms of the number of 1 km squares, from one to nine: an 'x' on these maps means that the plant has been recorded from 10 or more 1 km squares within the 10 km square. The number of records of Critically Endangered, Endangered, Vulnerable and Near Threatened species mapped in 10 km squares are shown in Appendix 2.

- Re-introduction sites have not been differentiated on the distribution maps.
- No special effort has been made to collect alien records as part of the RDB exercise and therefore the records mapped may not accurately reflect recent change.

6.2 Accounts of the individual species

Adonis annua L. (Ranunculaceae)
Pheasant's eye, Llygad y Goediar
Status in Britain: VULNERABLE.
Status in Europe: Not threatened.

A. annua is a plant of cultivated ground, formerly widespread on the chalk and limestone of southern England, but now persisting in very few places. It occurs most frequently in cornfields, in the past sometimes throughout the crop, but in recent years progressively confined to field margins and corners that escape the most intense herbicide and fertiliser treatments. It typically occurs in a species-rich community and may sometimes accompany other uncommon species such as *Papaver hybridum, Petroselinum segetum, Torilis arvensis, Scandix pecten-veneris* and *Valerianella dentata* (Wilson 1990).

This species is generally a winter annual, flowering from early June to July, setting seed in July, with the seedlings germinating in late autumn. Germination can, however, take place in the spring, with later flowering and, after harvest, seeding into the stubble. Even under ideal conditions, *A. annua* produces relatively few seeds, and these are heavy and immobile, preventing it from colonising new areas rapidly even when suitable habitats exist nearby. Seeds may have considerable longevity in the soil; for instance, in Friston Wood, East Sussex, it has appeared when clearings are made or rides opened up, even in places that have not been open land for many years (P. Harmes pers. comm.).

Between 1930 and 1960 *A. annua* was known from about 40 hectads, although apparently casual in half of them. It occurred mainly in southern and south-east England, with scattered records north to Leicestershire and Lincolnshire. Since 1987 it has been recorded from only about eighteen sites, the main strongholds now being in central Hampshire and south Wiltshire. It is seldom seen in large quantity, and several recent records are of very small populations, although such is the nature of the habitat that a suitable change in management could result in increases in numbers.

This species was at one time so abundant that it was collected and sold in bunches at Covent Garden, under the name 'Red Marocco'. A steady decline since the 1880s has been attributed to climatic fluctuations and improved seed cleaning (Salisbury 1961), and more recently to arable intensification, including the increased use of nitrogen and herbicides. The earlier sowing of winter cereals in the autumn may prevent the effective establishment of seedlings. Wilson (1990) has shown that it performs best in crops sown between the middle of October and the beginning of November.

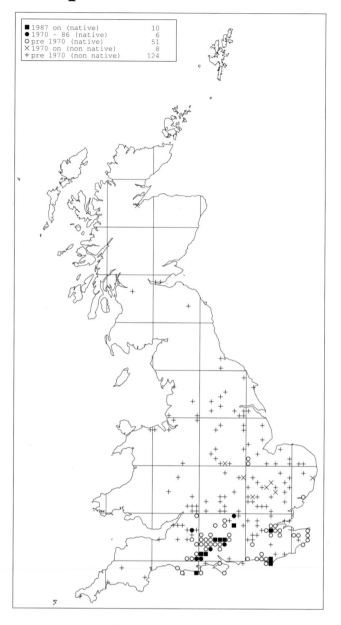

■ 1987 on (native)	10
● 1970 - 86 (native)	6
○ pre 1970 (native)	51
✕ 1970 on (non native)	8
+ pre 1970 (non native)	124

A. annua is at the edge of its range in southern England and is considered by many to be an introduction, albeit an ancient one. Its centre of distribution is around the Mediterranean, where it is increasingly threatened by arable intensification. It is rare or unknown in several central European countries, but its range extends eastwards to Iran and southwards to North Africa.

P.J. Wilson and M.J. Wigginton

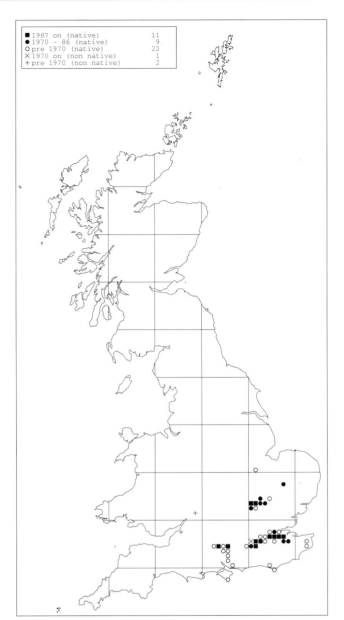

Ajuga chamaepitys (L.) Schreber (Lamiaceae)

Ground pine, Palf y Gath Bali
Status in Britain: VULNERABLE. WCA Schedule 8.
Status in Europe: Not threatened.

A. chamaepitys was familiar to the Tudor herbalists, who thought that it and *Teucrium botrys* were one and the same species, the former being male and the latter female. It is an attractive plant with intense yellow flowers and a strong resinous scent when crushed or trampled. Characteristic habitats include the upper edges of cultivated fields, crumbling banks and track sides, ground disturbed by scrub removal, road works or pipe-laying operations, and the edge of chalk or gravel pits. Dry sun-baked slopes are favoured, particularly on chalk escarpments. It is a poor competitor, confined to bare ground and the earliest seral stages. It used to be a characteristic member of the flora of temporary arable or fallow fields on the chalk downs of Surrey and Kent. Though it is no longer a typical arable weed, it sometimes

still occurs with *Filago pyramidata, T. botrys* and other rare species on ploughed field edges that have escaped herbicide spraying. *A. chamaepitys* also occurs in semi-natural calcareous grassland, generally on the steepest, hottest part of the slope, and on bare soil scratched by rabbits. It may grow on bare chalk or chalky clay, but more typically prefers a thin surface layer of sandy or gravelly drift. Commoner associates include *Arenaria serpyllifolia* ssp. *leptoclados, Euphorbia exigua, Kickxia spuria, Sherardia arvensis* and *Veronica persica*.

The flowering period is unusually long, extending from June to October, depending on whether the seedlings are autumn-, winter- or spring-germinated. Autumn-germinated plants survive the winter as rosettes. Although frequently thought of as an annual, some robust plants are plainly short-lived perennials with a woody tap-root. *A. chamaepitys* is said to be vulnerable to cold, wet, prolonged winters, which kill off the autumn-germinated seedlings. There is also evidence that seeds fail to ripen in cold summers (Grubb 1976). It may owe its survival here at the northern edge of its range to the flexible seed germination strategy, which may be genetically controlled.

A. chamaepitys can survive as dormant seed for many years - in places up to half a century. This helps to explain its erratic appearances, flowering prolifically after sudden disturbance, and then disappearing again as the vegetation closes. On nature reserves it can be induced to flower annually by shallow ploughing or rotavating. This species has become much less frequent during the past 50 years, partly because of the use of herbicides, but more because of the abandonment of fallow land on chalk slopes and the spread of coarse grass, scrub and secondary woodland on its downland localities. It is particularly vulnerable at the outlying parts of its British range in Hampshire and the

Chilterns, where only a few sites remain; it is less vulnerable on the North Downs in Surrey and Kent, where new sites are still being found. Most of the known populations are small, however, and the species now often depends on conservation management to survive. Exceptionally, a population may number more than 1,000 plants. The improved level of scrub control and the restocking of some old sites with sheep are hopeful signs, as is the possible prospect of hotter, drier summers.

Outside Britain, *A. chamaepitys* occurs across central and southern Europe, usually on light calcareous soils. Its distribution extends eastwards into the Levant and southwards to North Africa.

P.R. Marren

Alchemilla acutiloba Opiz (Rosaceae)
Lady's mantle
Status in Britain: LOWER RISK - Near Threatened.
Status in Europe: Not threatened.

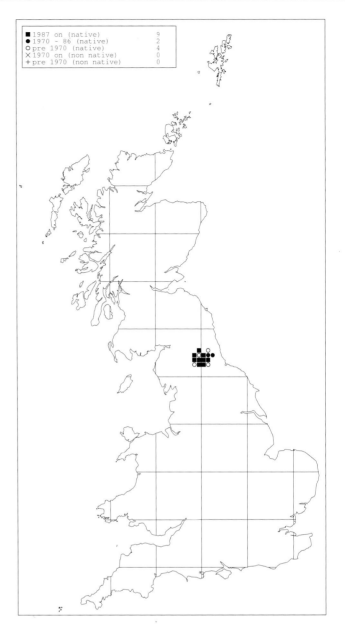

A. acutiloba is a plant of traditionally managed upland hay meadows which have not been markedly improved by re-seeding or the heavy application of fertilisers, and whose cutting is delayed until after seed- set. In the richer grassland communities, associated species may include *Anthoxanthum odoratum*, *Briza media*, *Conopodium majus*, *Festuca rubra*, *Geranium sylvaticum*, *Geum rivale*, *Leontodon hispidus*, *Primula veris*, *Rumex acetosa*, *Sanguisorba officinalis*, *Trifolium pratense* and other *Alchemilla* species including *A. monticola* and *A. xanthochlora*. It is also characteristic of roadside verges and railway banks, particularly where relict hay meadow communities have survived, sometimes occurring in large stands.

This species is a perennial apomict, flowering in June and July.

Apart from a very few populations in southernmost Northumberland, this species occurs only in Co. Durham. It is almost entirely confined to Teesdale, where it is local, and Weardale, where is much more widespread and locally abundant. This species was added to the British flora as recently as 1946, when A.J. Wilmott found a specimen in the Natural History Museum herbarium, collected in Teesdale in 1933 and labelled *A. pastoralis*. The very restricted distribution in Britain of this species (and also *A. monticola* and *A. subcrenata*) is remarkable, and their native status has been questioned. Bradshaw (1962) considers the possibility of their accidental or intentional introduction in historical time and concludes that the question of their origin remains an open one.

The survival of this species and other rare *Alchemilla* species in Dales hay meadows is largely dependent on the continuation of traditional management. They will not survive in intensively managed grass fields, especially not those cut for silage. Very many hay fields in Teesdale and Weardale have been considerably 'improved' by re-seeding and the heavy application of fertilisers, and recent evidence suggests that some populations of *A. acutiloba* (and other *Alchemilla* species) have been lost. Occasional cutting of roadside verges is likely to be beneficial, but not before seed-set.

This northern-montane species is widespread in central, northern and eastern Europe, extending from the Alps to Greece, and northwards to southern Scandinavia and Russia. It is often abundant by paths and in pastures in montane and subalpine regions.

M.J. Wigginton

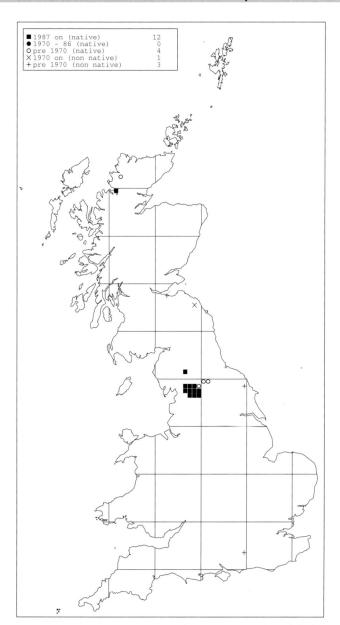

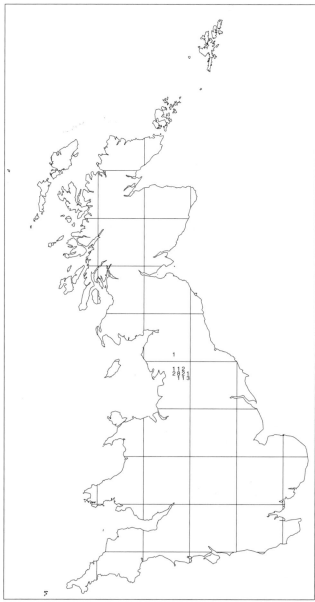

Alchemilla glaucescens Wallr. (Rosaceae)

Lady's mantle
Status in Britain: LOWER RISK - Near Threatened.
Status in Europe: Not threatened.

The stronghold of *A. glaucescens* is in the Carboniferous limestone country of north-west England. Most of its sites are in Yorkshire, but there is one site extant on Gragareth in West Lancashire and one in Cumbria. It occurs in limestone grassland on open hillsides or flat ground, on grassy banks by roads, rivers and streams, and in vegetated patches on limestone pavement. Typical associates include *Bellis perennis, Campanula rotundifolia, Centaurea nigra, Conopodium majus, Cynosurus cristatus, Festuca ovina, Lotus corniculatus, Plantago lanceolata, Ranunculus bulbosus, Sanguisorba minor* and *Trifolium pratense*. At some sites it is locally abundant as, for instance, on limestone pavement on the slopes of Ingleborough, but at others only a single clump may occur. Because its habitat is so widespread in the Pennines, it is likely that further sites will be found.

In common with most other species of the *A. vulgaris* agg., it is a perennial apomict, flowering in June and July. It is one of the smaller species of the aggregate, readily distinguished by its glaucous appearance and partial covering of dense, silky hairs.

In Scotland, only four populations have been recorded, in West Ross and West Inverness. One of them has been destroyed, one has not been seen recently and may be extinct, and a third is very small and imminently threatened by quarrying. The other population, of about 130 plants, is on a flushed rocky (limestone) path on a bracken-covered hillside near Ullapool. This species is sometimes grown in cultivation, and escapes have occasionally been recorded, for example, in Surrey, south-west Yorkshire and Midlothian, though some records have been errors of identification.

The limestone grasslands in which this species occurs are maintained by livestock grazing, particularly by sheep. Agricultural improvement of the grass sward, including the heavy application of fertilisers or herbicides, would be detrimental to the plant, as also to *A. minima*, with which it occurs at some sites.

A. glaucescens is widespread in central and north-eastern Europe, extending from the western Alps to southern Scandinavia and West Siberia, mainly on calcareous rock and in limestone grassland. Outside its main range, it is sparsely distributed in Britain, northern Fennoscandia, Italy, Corsica and the Crimea.

M.J. Wigginton

Alchemilla gracilis Opiz (Rosaceae)
Lady's mantle
Status in Britain: VULNERABLE.
Status in Europe: Not threatened.

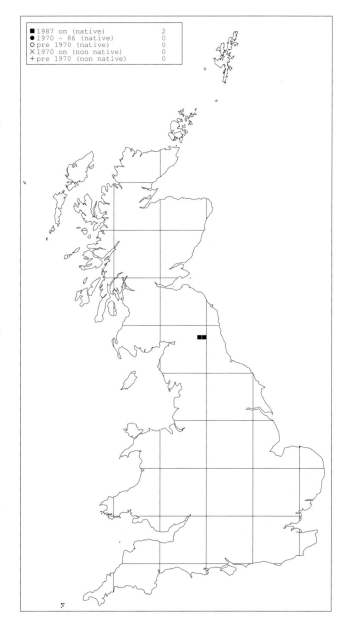

■ 1987 on (native)	2
● 1970 - 86 (native)	0
○ pre 1970 (native)	0
✕ 1970 on (non native)	0
+ pre 1970 (non native)	0

A. gracilis is one of the most recent additions to the British flora, discovered in 1976 in Northumberland by G.A. Swan. In its main site, it occurs in more or less closed grassland on Carboniferous limestone and on a small rocky limestone knoll nearby. The turf is species-rich, and associated species include *Bellis perennis*, *Briza media*, *Festuca ovina*, *Galium verum*, *Helictotrichon pratense*, *Koeleria macrantha*, *Lotus corniculatus*, *Ranunculus bulbosus*, *Sanguisorba minor*, *Thymus polytrichus* and *Trifolium pratense* (Swan & Walters 1988). Elsewhere, a few clumps have been recorded growing in a rough pasture, amongst tall herbage in an ungrazed hay meadow (associated with various coarse grasses and common herbs including *Plantago lanceolata*, *Ranunculus acris* and *Veronica chamaedrys*) and on a grassy roadside verge.

A. gracilis is a perennial apomict that usually flowers in May, earlier in the year than most of the related *Alchemilla* species. Seed is shed by late June or early July, at which time the inflorescence becomes relatively inconspicuous.

Despite searches in many other suitable localities, this species is known from only four sites in Northumberland, all in the same general area in the south of the county. At the main site there were about 800 flowering plants in 1984, but by 1994 the plant was said to be difficult to find (G.A. Swan, pers. comm. 1997) because of excessive grazing by rabbits. The population may have declined. At the other three sites populations are small, and consist of a few clumps. Its occurrence in this country was predicted prior to its discovery because of its common association in many parts of Europe with other members of the aggregate which also occur in the northern Pennines. An early specimen of *A. gracilis* had in fact lain unrecognised in the Natural History Museum, London, until re-determined following the Northumberland discovery.

The cessation of grazing and the improvement of limestone grassland and hay meadows are likely to be the main threats to its survival. One site is in an SSSI.

A. gracilis has a wide distribution in northern, eastern and central Europe, extending from southern France, Italy and Greece to south-west Norway and western Siberia.

M.J. Wigginton

Alchemilla minima Walters (Rosaceae)
Lady's mantle
Status in Britain: VULNERABLE. ENDEMIC.

This endemic taxon, restricted to a limited area of the limestones of northern England, rarely occurs below 300 m but can be found almost to the summits of two Yorkshire fells in the short turf of base-rich *Festuca ovina* grassland. Its habitats are generally moist. At lower altitudes it grows within and on the margins of periodically flushed areas, characterised by the presence of *Carex panicea*, with *Leontodon autumnalis* and *Prunella vulgaris* as typical associates. At higher altitudes it occurs in non-flushed areas amongst limestone boulders and debris where *C. panicea* is absent but *Achillea millefolium*, *Cerastium fontanum* ssp. *holosteoides* and *Sagina procumbens* are typical associates. Here the increased precipitation helps maintain a moist substrate.

A. minima is a small herbaceous apomictic perennial. It is at a selective advantage in sheep-grazed pasture, where the plant is able to set seed straggling at ground level in closely grazed turf, where other *Alchemilla* species are largely unable to flower (Walters 1970).

It is probably an overlooked and under-recorded plant, especially in the higher altitude habitats, and new localities are still being found within its limited range. Although its distribution is centred upon the two fells of Ingleborough and Whernside, there are other doubtful records from further afield, particularly from the northern Pennines. In its present upland habitats populations seem to be stable, the plant surviving vegetatively even where the habitat is periodically overgrazed. Quarrying for limestone occurs within a short distance of some of its localities, but other than from this and from major changes in farming practice, its habitat appears to be little threatened.

This taxon was first recognised as a separate species in 1947 (Walters 1949). Subsequent study by Bradshaw (1964) seemed to establish that its dwarf habit is genetically determined, and she confirmed its specific status. However, S.M. Walters (1986, and *in litt*. 1995), who originally described this species, has come to the view that it is best treated as a dwarf variant of the widespread and variable *A. filicaulis* that has been selected from the *A. filicaulis* population by centuries of intensive grazing by sheep. The extreme dwarf characters are to a large extent lost in cultivation over a long period, and no single distinctive character survives.

M.J.Y. Foley

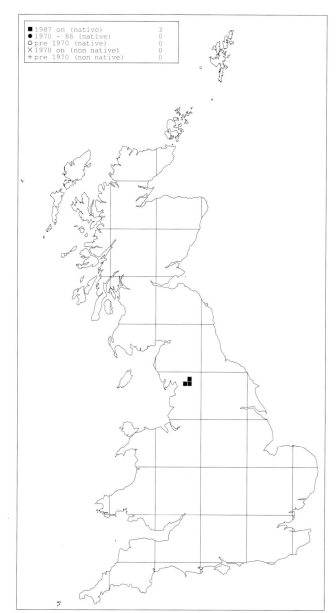

■ 1987 on (native)	3
● 1970 - 86 (native)	0
○ pre 1970 (native)	0
✕ 1970 on (non native)	0
+ pre 1970 (non native)	0

Alchemilla monticola Opiz (Rosaceae)

Lady's mantle
Status in Britain: LOWER RISK - Near Threatened.
Status in Europe: Not threatened.

A. monticola is a characteristic species of traditionally managed hay meadows, and of roadside verges and banks. It may occur throughout the meadow or, in meadows that have been more intensively managed, be largely confined to edges, paths or farm tracks. In hay meadows and on the more species-rich roadside verges, associated herbs include *Anthoxanthum odoratum, Crepis paludosa, Geranium sylvaticum, Geum rivale, Primula veris, Ranunculus bulbosus, Sanguisorba officinalis,* and other *Alchemilla* species including *A. acutiloba* and *A. xanthochlora. A. monticola* may also be found in species-poor swards on roadside verges.

Like most other *Alchemilla* species, it is a perennial apomict, flowering in June and July.

A. monticola was first detected in Britain in the early 1920s. It is almost entirely confined to Co. Durham, where it is strongly centred in Upper Teesdale. In the 1950s and 1960s it was frequent and locally abundant in Teesdale from Romaldkirk to Langdon Beck (Bradshaw 1962), but there has been no thorough survey since that time and its current status is unknown. However, a partial survey in 1996 seemed to indicate a possible decline. There are scattered populations south of Teesdale, and it just extends into the South Tyne valley in Northumberland. Oddly, it does not occur in Weardale. There have been casual occurrences in Surrey and Buckinghamshire and erroneous records elsewhere.

Like *A. acutiloba* and *A. subcrenata,* its survival in meadows depends on the continuation of traditional farming, and it will quickly succumb under intensive regimes. Populations on road verges may be threatened by inappropriate management, including mowing before seed-set.

A. monticola may be native in Britain, or a long-established introduction. In Europe it is widespread, extending from the Alps to Greece and north and east to southern Scandinavia and Russia, a distribution very similar to that of *A. acutiloba.* Outliers are in the Netherlands (where it is endangered) and in Britain. In Switzerland it is very common in montane and subalpine regions, and in Scandinavia it is frequently dominant in meadows.

M.J. Wigginton

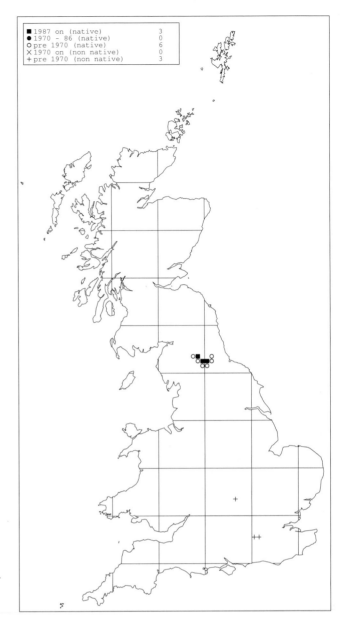

■ 1987 on (native)	3
● 1970 - 86 (native)	0
○ pre 1970 (native)	6
✕ 1970 on (non native)	0
+ pre 1970 (non native)	3

Alchemilla subcrenata Buser (Rosaceae)

Lady's mantle
Status in Britain: ENDANGERED.
Status in Europe: Not threatened.

A. subcrenata is a plant of species-rich hay meadows and pastures, occurring in only a very limited area of Co. Durham. Associates are those typical of upland hay meadows, including *Alchemilla monticola, A. xanthochlora, Anthoxanthum odoratum, Conopodium majus, Festuca rubra, Geranium sylvaticum, Leontodon hispidus, Luzula campestris, Plantago lanceolata, Ranunculus acris, Rumex acetosa, Sanguisorba officinalis* and *Trifolium pratense.*

This species is an obligate apomict, flowering in June and July.

A. subcrenata was not detected in Britain until 1951 (Walters 1952). Its British localities are restricted to two very limited areas of Teesdale and Weardale, where it has been recorded from about five sites, in one of which it was formerly locally abundant. It may now be extinct in Weardale, and it is not known whether it remains at all its Teesdale sites since there has been no comprehensive survey since the 1950s. It is certainly extant in one, but it is discouraging that a search of three of its Teesdale sites in 1996 failed to reveal it: one site, formerly a hay meadow, is now an improved pasture. However, because of its similarity to commoner species, new localities may yet be discovered and a thorough survey is required. As with other rare *Alchemilla* species, traditional methods of hay cropping hold the key to its survival, including cutting late in the year after it has set seed.

Its occurrence in Britain is a disjunction at the extreme western edge of its wide European range. In the Alps it is an abundant hay meadow plant and occurs in similar habitats in Scandinavia, where it is also known from woodland. On the continental mainland it has a wide altitudinal range, ascending to the sub-alpine zone.

M.J.Y. Foley

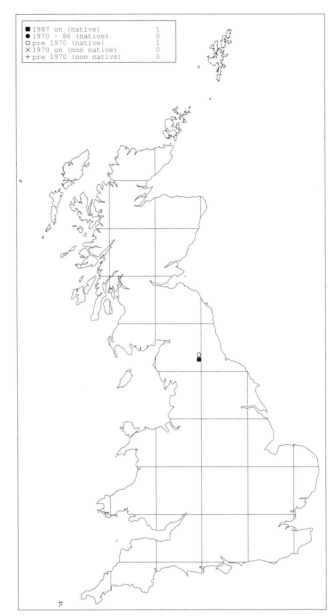

■ 1987 on (native)	1
● 1970 - 86 (native)	0
○ pre 1970 (native)	1
✕ 1970 on (non native)	0
+ pre 1970 (non native)	0

Alisma gramineum Lej. (Alismataceae)
Ribbon-leaved water plantain, Dër-Lyriad Hirfain
Status in Britain: CRITICALLY ENDANGERED. WCA
Schedule 8.
Status in Europe: Vulnerable.

A. gramineum is recorded from shallow, eutrophic water at the edge of lowland lakes and rivers and in fenland drains. At Westwood Great Pool it usually grows amongst stands of emergents in water less than 10 cm deep, although it sometimes grows in open water or on recently exposed mud. Lousley (1957) found it with abundant *Eleocharis palustris*, but in recent years most plants have grown amongst *Typha latifolia*. In Lincolnshire it formerly grew in water more than a metre deep over a soft, muddy substrate in the River Glen and its associated drains. It has been seen recently in only a single ditch in this area, where it appeared in 1991 and 1992 after clearance, but did not survive once *Phragmites australis* recolonised the site. The other British records are from a fenland ditch in Cambridgeshire (Libbey & Swann 1973) and Langmere in Breckland (Swann 1975).

In the shallow water in which it usually grows, *A. gramineum* forms rosettes of numerous submerged leaves. Plants with submerged leaves may flower, but mature plants may also develop floating and emergent leaves. The observation in Clapham *et al.* (1952) that the flowers open between 6 and 7.15 am is totally fictitious: in the wild British plants flower throughout the day and in cultivation Björkqvist (1967) found that the flowers of this species remained open for 13-15.5 hours, longer than those of the other species, which he studied. Cleistogamous flowers may develop and seed under water. Seed-set is good in the wild, and reproduction by seed occurs, although the number of seeds that germinate at Westwood varies greatly from year to year. Many plants in Britain are annuals, but some persist as short-lived perennials.

This species was first discovered in Britain in 1920 at an artificial lake, Westwood Great Pool, but it was initially confused with *A. lanceolatum*. It still persists at this site, though plants may fail to appear in some years; the population size is usually small but there may be at least 150 flowering plants in a favourable year. The reason for the fluctuations is unknown. It was discovered in the Glen and nearby ditches in 1955. At its other sites it does not appear to have persisted for long, although it may survive as buried seed. It is an inconspicuous species which could be overlooked elsewhere. It has been suggested that the East Anglian populations may be derived from seed carried by wildfowl arriving from Denmark or the Baltic (Libbey & Swann 1973), but *A. gramineum sensu stricto* is replaced by *A. wahlenbergii* in the Baltic.

A. gramineum is widespread in temperate latitudes in Europe, Asia (except the Far East) and North America. In

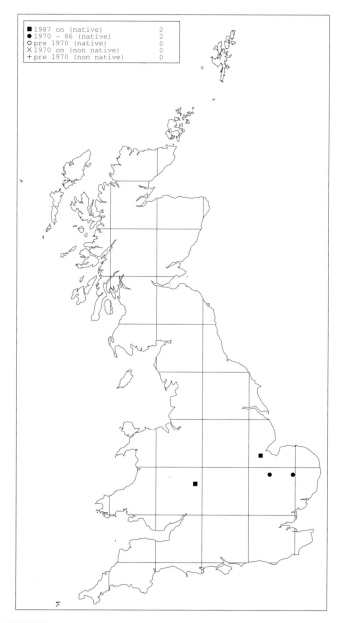

Europe it is absent from the Iberian peninsula and the Mediterranean region; it now reaches its northern limit in Denmark, having become extinct in Sweden. In mainland Europe it is uncommon throughout its range and is normally found as small populations (Björkqvist 1967). It is difficult to account for its rarity in western Europe. Its sporadic appearance and apparent dependence on disturbance make it a difficult plant to conserve. By contrast, it has spread along the St Lawrence River in North America, where it has been described as a 'pesky weed' (Raymond & Kucyniak 1948; Countryman 1968).

C.D. Preston, adapted from an account in Preston & Croft (1997).

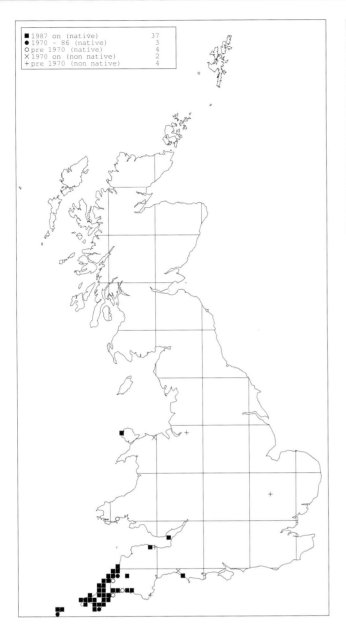

Allium ampeloprasum agg.

Allium ampeloprasum L. (Liliaceae)
Wild leek, Cenhinan Wyllt
Status in Britain: LOWER RISK - Nationally Scarce.
Status in Europe: Not threatened.

A. *ampeloprasum* is typically a plant of anthropogenic habitats, associated with ruined settlements and historic field systems and occurring, for example, amongst brambles and rough vegetation on stone-faced Cornish 'hedges', in old orchards, by footpaths, on roadsides, and on disturbed and waste ground (FitzGerald 1990c). Associates are mainly undistinguished species of rough places and walls, such as *Arrhenatherum elatius*, *Chaerophyllum temulentum*, *Galium aparine*, *Heracleum sphondylium*, *Iris foetidissima*, *Rubus fruticosus*, *Rumex acetosa*, *Silene dioica*, *Umbilicus rupestris* and *Urtica dioica*. It is also found occasionally in more natural habitats, including the tops and faces of rocky sea-cliffs and sandy places on the coast.

It is a robust perennial, flowering in July and August and reproducing by seed and vegetatively. Three morphologically distinct varieties are usually recognised; var. *ampeloprasum* with dense globose umbels lacking bulbils, var. *babingtonii* having rather loose globose umbels with many bulbils, and var. *bulbiferum* with fewer smaller bulbils in the umbel. Vegetative reproduction is by offsets (bulblets), and, except in var. *ampeloprasum*, by bulbils, and colonies of limited extent can arise by these means. Seed is dispersed much more widely.

Var. *babingtonii* has become abundant in the Isles of Scilly, where it occurs on most of the main islands. It has also spread and increased in numbers on the mainland, where it was recorded in eleven 1 km squares in the 1970s, 54 in the 1980s and 80 in the 1990s. Though there has undoubtedly been more recording in recent years, the increase appears to be real. FitzGerald (1990c) suggests that since the plant has strong associations with man and

A. ampeloprasum var. *babingtonii*

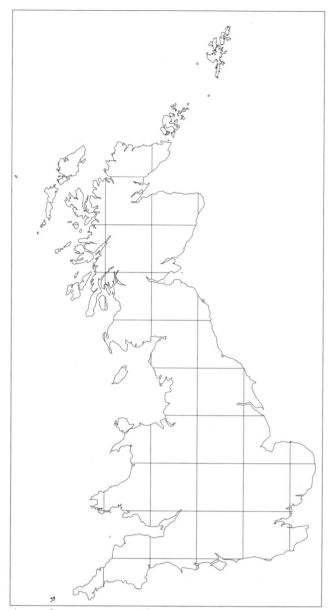

A. ampeloprasum var. *ampeloprasum*

may have been cultivated in ancient times, it may thrive on the increasing management of the countryside, and this may partly account for its increase. Its large size, easy reproduction via bulbils, and vigorous spring growth, make it well able to compete with the rank wayside and ruderal species which are its usual associates. Populations are generally small, but a few number in the hundreds or even thousands of plants. Var. *ampeloprasum* has been recorded in very few sites on the coasts of England and Wales and has been seen in only three in the past twenty years: on Steep Holm and Flat Holm in the Bristol Channel and near South Stack, Anglesey. When first recorded on Steep Holm in 1625, it was described as abundant, and in 1891 as plentiful, but it has since declined. In 1989, only 216 plants were found (Taylor 1990f). Neither is it thriving on Flat Holm, and only about 300 plants were counted recently (Morgan 1989f). The South Stack population consists of a few tens of plants.

There are differing views on the status of *A. ampeloprasum*. Some authorities (e.g. White 1912; Margetts & David 1981) treat it as a likely ancient introduction. Stearn (1987) noted that the bulbilliferous variants may owe their [British] localities to a now forgotten culinary use, and Roberts & Day (1987) thought that the Anglesey plants had no appearance of being indigenous. However, Stace (1991) and Sell & Murrell (1996) treat all varieties as native.

Var. *ampeloprasum* occurs throughout western and southern Europe, including the Mediterranean islands, with its stronghold in Iberia and the Balearic Islands. It ranges eastwards to Turkey, Iraq and the Caucasus, and is considered to be naturalised in the Azores. Var. *babingtonii* is endemic in Britain, Ireland and the Channel Islands, and var. *bulbiferum* is endemic in the Channel Islands.

M.J. Wigginton

Allium sphaerocephalon L. (Liliaceae)
Round-headed leek, Cenhinan Bengrwn
Status in Britain: ENDANGERED. WCA Schedule 8.
Status in Europe: Not threatened.

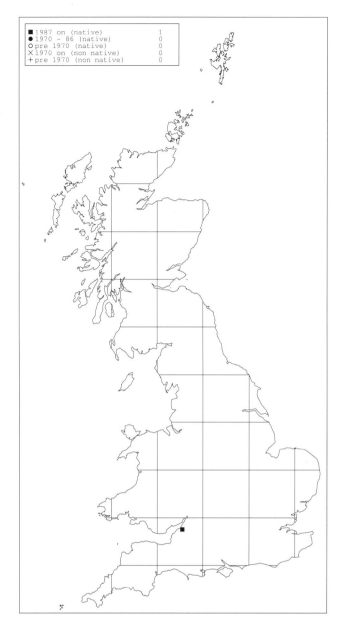

A. sphaerocephalon occurs on the wooded Carboniferous limestone cliffs of the Avon Gorge. It grows on shallow soil on ledges and slopes, most of which face south to west and are open to direct sunlight and consequently can become very hot in the summer (Lovatt 1982). It grows in a community where the plant cover is incomplete because of the exposure of bedrock or the presence of rock fragments in the soil. Associated species include *Bromopsis erecta, Crataegus monogyna, Dactylis glomerata, Festuca ovina, Geranium sanguineum, Hedera helix, Leucanthemum vulgare, Scabiosa columbaria, Sedum album*, and sometimes the other onions *A. roseum, A. vineale* and *A. carinatum*, the last locally becoming a troublesome competitor.

It is a perennial, reproducing vegetatively and by seed. Flowers appear in June to August and are pollinated by insects. The flowers are normally protandrous and the plants are normally out-breeders, but can be self-pollinated. Vegetative propagation is by offsets (bulblets) and by bulbils.

The earliest record of this species in Britain was made in 1847, when H.O. Stevens discovered it in the Avon Gorge near Bristol, and this remains its only mainland site. The first record in the British Isles, however, pre-dates the Avon one, the species having been recorded in Jersey in 1833, where it still just survives on sandy ground in St Aubyn's Bay. In the Avon Gorge, populations have apparently always been relatively small and localised, and the plant is now restricted mainly to two small areas about 1.5 km apart. Flowering is variable from year to year, ranging from a few hundred to several thousand flowering heads. There is some indication of a general population decline, and in 1989 only 79 plants flowered after an exceptionally dry autumn and wet spring. Plants in the two main populations show slight morphological differences and may be of different genetic stock. Both colonies lie within the Avon Gorge SSSI.

The Avon Gorge is heavily used for recreation and this undoubtedly has a major impact, particularly on the smaller of the two populations. In the past, large quantities of flowers were collected here, as mentioned by several early authors (e.g. White 1912). Picking still continues, despite legal protection, though to a much lesser extent. Works, aimed at reducing the risk from rock-falls to traffic on the road below, have in the past damaged the main colony (Taylor 1990f), but it seems to have recovered reasonably well. A threat is posed by encroaching scrub, particularly the alien species *Quercus ilex* and *Cotoneaster* spp., and regular cutting back is required. Some control of the alien *A. carinatum*, whose distribution overlaps that of *A. sphaerocephalon*, has been effected by the selective removal of whole plants (Frost *et al.* 1991). A severe threat to *A. sphaerocephalon* has recently arisen from the cleaning, in 1995, of the Clifton Suspension Bridge by shot-blasting with copper slag containing high levels of other heavy metals toxic to plants. Spent slag, which fell into the Avon Gorge below, formed in heavy deposits on ledges and slopes, and it was estimated that 80% of the *A. sphaerocephalon* population may be affected (Rich *et al.* 1996). Some clean-up works have been carried out.

In mainland Europe it is widespread and has a sub-Mediterranean distribution pattern, though it extends north to Belgium and eastwards to Poland, south-central Russia, Turkey and Israel. In the Mediterranean region it is typical of garrigue on rocky limestone and igneous mountainsides, but it also occurs in vineyards, on waste ground and roadsides.

A. sphaerocephalon is generally regarded as a native species, though its status has been questioned and the possibility of an ancient introduction cannot be entirely dismissed. Lovatt (1982) discusses its status in some detail.

M.J. Wigginton

Althaea hirsuta L. (Malvaceae)
Rough marsh-mallow, Hocysen Flewog
Status in Britain: ENDANGERED. WCA Schedule 8.
Status in Europe: Not threatened.

A. hirsuta occurs on open, dry, calcareous soils. In Somerset it grows on south-facing calcicolous grassland slopes over Rhaetic clays and Lias limestone with *Bellis perennis, Blackstonia perfoliata, Centaurium erythraea, Potentilla reptans* and *Sherardia arvensis*. In Sussex, its site is an area of bare chalk on the top of a covered reservoir, where it is found with other annuals including *Anagallis arvensis* and *Aphanes arvensis*. In Kent, it grows on chalky ground on a south-facing arable field headland with a rich weed flora including *Ajuga chamaepitys, Filago pyramidata* and *Teucrium botrys,* and also at the nearby grassy wood edge with *Brachypodium sylvaticum* and *Viola* species.

It is an annual or, rarely, biennial herb flowering mainly from May to early July. The hard, long-lived seed is shed in July and August, each seed head producing about 12-14 seeds. *A. hirsuta* behaves mainly as a winter annual, but in a wet season may also be a summer annual. In the mid-Somerset site in 1992, one- or two-leaved seedlings were present in August in the major pulse of germination, but in the following year most seedlings were still at the cotyledon stage in late October, indicating considerably later germination that year. It is a poor competitor and requires bare soil for germination and seedling establishment. Rabbit scrapes provide suitable conditions, as does soil disturbed in scrub clearance or in arable fields. Summer baking and cracking of soils can also be beneficial. Germination in such conditions can be so good as to cause density-dependent mortality. It does poorly in closed turf where a plant may produce only 1-3 flowers, but in open ground individual plants can have more than 50 flowers. Robust biennial plants with branches over a metre long have also been recorded (Rich 1993a).

Although considered by some (e.g. Stace 1991) as likely to have been introduced, its presence in semi-natural habitat in Somerset and Oxfordshire and its long history in Kent (where it has been known since 1792) suggest that it might be native in those places. It is, however, known to have been introduced at other sites, including one in North Lincolnshire where it has occurred sporadically for almost a century and was probably introduced with pheasant fodder. In the latter site, it appeared in an area of felled woodland, and it is likely that seed had lain dormant there for at least 35 years (Weston 1994). Population size varies considerably in response to climate, the management regime and the intensity of rabbit-grazing. For example, the largest population (in mid-Somerset) ranged between 450 and 770,000 plants during the period 1992 to 1996 (Ulf-Hansen 1994; Rich, Ulf-Hansen & Goddard 1996). At most sites, fewest numbers occurred in 1993-4, and this was attributed to the wet winter and spring (Rich & Ulf-Hansen 1994). The newly-discovered Oxfordshire site held a small colony of a few plants in 1996.

All established localities are SSSIs, which should afford a degree of protection. However, the immediate threat at three sites is lack of sufficient disturbance, which, in the

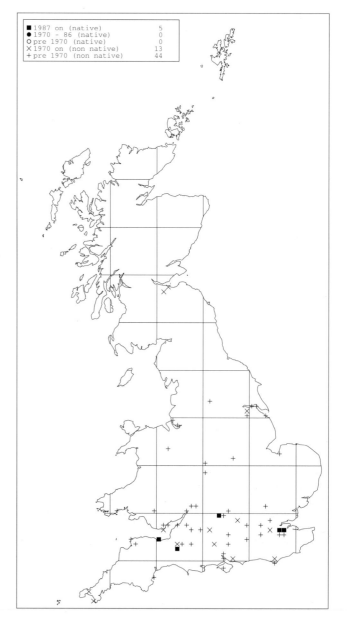

■ 1987 on (native)	5
● 1970 - 86 (native)	0
○ pre 1970 (native)	0
✕ 1970 on (non native)	13
+ pre 1970 (non native)	44

absence of grazing, could lead rapidly to a closed turf. At the main colony in mid-Somerset, scrub clearance has re-started and grazing been re-established. The West Somerset site is very vulnerable (only three plants in 1994), despite regular small-scale scrub clearance. Rabbit fencing of the arable field edge in Kent now prevents grazing, and coarse grasses colonise the now undisturbed woodland/field edge. Active management of most sites is necessary to prevent succession and create open areas. However, the very long-lived seed provides a buffer against periodic catastrophe.

In Europe, the species has its stronghold in dry calcareous turf or bare stony places in the Mediterranean region but occurs across central and southern Europe eastwards to Iran, and northwards to Germany, where it is vulnerable. The English populations are at the north-western limits of its distribution.

P.F. Ulf-Hansen

Anisantha madritensis (L.) Nevski (Poaceae)

Bromus madritensis L.
Compact brome, Pawrwellt Dwysedig
Status in Britain: VULNERABLE.
Status in Europe: Not threatened.

A. madritensis is a grass of dry, open habitats, occurring mainly in man-made sites including wall-tops, old buildings, open waste ground, roadsides and cracks in pavements. It occurs in a few semi-natural sites, such as in the Avon Gorge, where it grows on limestone outcrops, stony slopes and open grassland on warm, freely draining calcareous soils. Associated species are varied, as the grass grows as an intrusive species in the gaps in many communities. In the Avon Gorge, the most frequent associates are *Centranthus ruber* and *Crepis vesicaria*, but in its most natural sites the associates are *Bromus hordeaceus*, *Catapodium rigidum*, *Geranium robertianum*, *Hornungia petraea*, *Saxifraga tridactylites* and *Sedum acre*.

This is an annual species, flowering from May to July. It is tetraploid and may have arisen from the hybridisation of *A. sterilis* and *A. rubens*. There is some evidence of good seed longevity and germination being triggered by disturbance and light (Lovatt 1982). Hot dry summers may reduce competitors, allow light to the seed-bank, with larger populations of *A. madritensis* the following year.

In the Avon Gorge, where it was first recorded by Lightfoot in 1773, it is often regarded as a native (e.g. Ryves *et al.* 1996). There it grows on screes and on unstable limestone rock ledges, which regularly provide newly disturbed ground into which the plant can seed. The population is large, and several tens of thousands of plants may occur in good years. It has also recently been reported at a second site nearby, at Shirehampton (S. Parker pers. comm.). Another possibly native site is in Pembrokeshire, where it grows in limestone grassland and on rocky outcrops near Castle Carew. It is probably also native on the Channel Islands. Otherwise, *A. madritensis* is generally regarded as an alien, though an old and well-established one. For instance, it grows on ancient city-walls in Exeter, and has been known in Southampton since 1806. The plant is scattered throughout southern England, with a few locations in northern Britain and in southern Ireland. It appears to be under no special threat. Urban colonies persist despite herbicides, tarmac, re-turfing and the cleaning of buildings.

This species has its stronghold in southern Europe, where it is often a common weed, but its range extends eastwards to northern France, and to North Africa, south-west Asia and Macaronesia. It also is found in Afghanistan and on Atlantic Islands and is now naturalised in Australia and North America.

D.A. Pearman, R.M. Walls and M.J. Wigginton

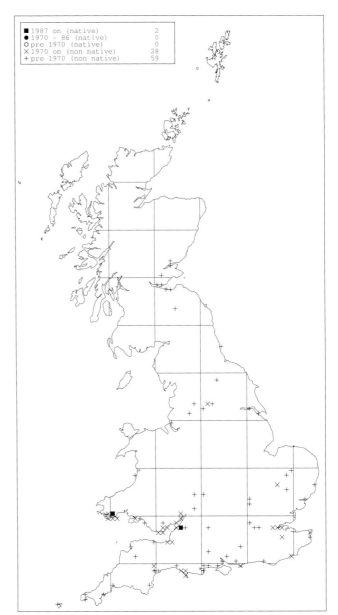

■ 1987 on (native)	2
● 1970 – 86 (native)	0
O pre 1970 (native)	0
X 1970 on (non native)	28
+ pre 1970 (non native)	59

Anthyllis vulneraria ssp. *corbierei* (Salmon & Travis) Cullen (Fabaceae)

Kidney vetch, Placen Felen

Status in Britain: LOWER RISK Near Threatened. Possibly ENDEMIC.

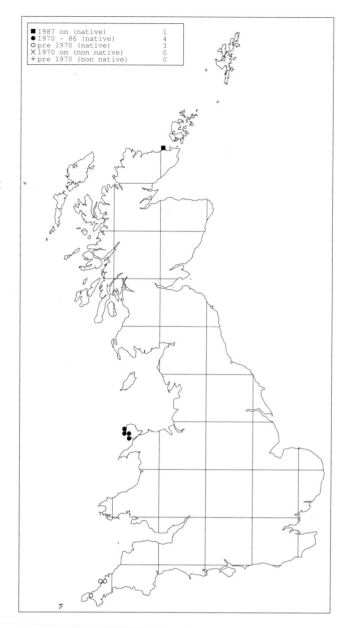

The genus *Anthyllis* is taxonomically complex, with many intergrading variants that are to some extent ecologically and geographically separated. Subspecies *corbierei* is much the rarest of the four subspecies native to Britain and is confined to coastal habitats. It is principally a plant of grassy sea-cliffs and rocky slopes, accompanying such species as *Cochlearia officinalis, Daucus carota, Festuca rubra, Plantago maritima, Scilla verna, Silene uniflora* and *Tripleurospermum maritimum*, but also occurs on dunes and dune grassland.

It is a perennial, flowering in midsummer. The reproductive biology of this subspecies has not been investigated, though it is likely to be similar to that of subspecies *vulneraria*: protandrous flowers pollinated by insects, and reproduction entirely by seed.

Subspecies *corbierei* has been recorded at few sites in Britain; apparently only two in Cornwall, several in Anglesey between Aberffraw and South Stack (Roberts 1982), and one in Caithness. It may be overlooked and more widespread (e.g. Stace 1991), though it has not been found during recent occasional investigations of *Anthyllis* populations at a number of sites in South Wales and on the Lleyn peninsula (Jones 1994). It is noteworthy that in Cornwall, a well-botanised county, there are only two records, made in 1933 and 1958 (Margetts & David 1981). The distribution of types and intermediates needs further investigation in western and northern Britain.

Elsewhere, subspecies *corbierei* is recorded only on Sark, though searches on the coast of France may reveal its presence there (Cullen 1986).

M.J. Wigginton

Apium repens (Jacq.) Lag. (Apiaceae)

Creeping marshwort
Status in Britain: CRITICALLY ENDANGERED. WCA Schedule 8. EC Habitats & Species Directive, Annexes II and IV.
Status in Europe: Vulnerable.

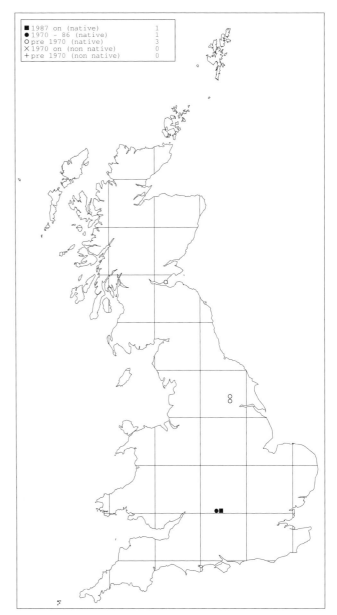

■ 1987 on (native)	1
● 1970 - 86 (native)	1
○ pre 1970 (native)	3
✕ 1970 on (non native)	0
+ pre 1970 (non native)	0

A. *repens* has been reported from about 10 locations in Britain during this century, always in damp places, usually by rivers and generally on sites subject to flooding in winter. Associates include *Alopecurus geniculatus*, *Galium palustre*, *Glyceria notata*, *Juncus articulatus*, *Mentha aquatica*, *Myosotis scorpioides* and *Ranunculus flammula*, as well as the ever-present *Apium nodiflorum* and, at its Oxford station, the only known extant site in Britain, large amounts of prostrate *Oenanthe fistulosa*.

A. *repens* is perennial, forming a small rosette of simply-pinnate leaves in which each lobe has one major cleft and several smaller teeth. Once established, a horizontal stem or runner sets out at or slightly below soil level. Nodes are a few centimetres apart, each giving rise to roots, one or more long-petiolate leaves and sometimes a long-pedunculate small umbel furnished with 3-6 bracts and bearing white flowers. Recent workers have remarked on the absence of fully-formed fruit but fruit *is* sometimes seen in untrampled spots. The mericarp is brown-black, 1 mm long and broad.

The status of A. *repens* has long been a topic of discussion due to the apparent mimicry by two or more named varieties of *Apium nodiflorum*. Further, this mimicry is not constant but varies from season to season at any one location. Many herbarium specimens determined as A. *repens* are more or less obviously A. *nodiflorum*, though others from, for instance, Witney and Binsey, Oxfordshire, Skipwith Common, Yorkshire, and Kinghorn, Fife, appear to be the true species. The plants found at Port Meadow, Oxford, have been at the centre of the discussion for many years, as reference to Botanical Exchange Club notes and reports from the late 19th and early 20th centuries will confirm. The long-standing uncertainty has recently led to a programme of DNA analysis being undertaken (Grassly *et al.* 1996). This work appears to indicate that there is a colony of unalloyed A. *repens* at Port Meadow, that there is a mimicking taxon of A. *nodiflorum* growing in close proximity, and that there are no hybrids between these two at this site. It remains to be seen what analysis will indicate about populations elsewhere, if any can be found, but the recent work at Port Meadow tends to refute the suggestion that A. *repens* may have been lost in Britain through hybridisation with A. *nodiflorum*. Now that the presence of A. *repens* seems to have been confirmed, the question arises as to whether it is extant at other sites, where it was reported in earlier decades. A site at Witney appears to have been lost due to habitat spoliation. Sites in East Yorkshire need renewed study, as does one at Kinghorn, Fife. It may be that several others exist, and the discovery of new sites is not entirely impossible.

Judging by the large numbers of horses and cattle in Port Meadow, grazing is not a threat unless the site becomes a cattle-pen. Disturbance of soil is, however, required, ideally by the hooves of grazing animals, to provide openings into which the advancing horizontal stems can progress. Complete closure of the turf would probably be fatal to A. *repens*. Drainage, inappropriate grazing levels, dumping and spraying are almost certainly the chief problems. Grazing by hoofed animals sometimes renders what might have been a discrete, organised plant into an ill-defined assortment of much-modified rooted pieces, but this may not be a matter for concern.

We do not know the extent of confusion between these two *Apium* species abroad. However, A. *repens* is attributed to a considerable number of countries in *Flora Europaea*, although the scarcity of records and specimens suggests that it is thinly scattered. The species has also been recently collected on the Atlas mountains in Morocco. The distribution abroad, as in Britain, is unclear and would certainly merit further investigation.

M.J. Southam

Arabis alpina L. (Brassicaceae)

Alpine rock-cress, Biolair na Creige Ailpeach
Status in Britain: ENDANGERED. WCA Schedule 8.
Status in Europe: Not threatened.

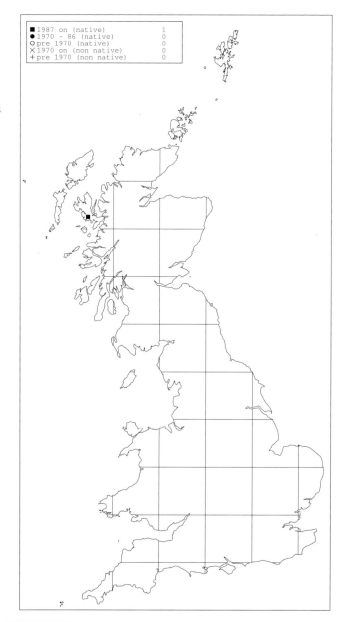

In view of its relatively remote location, it is hardly surprising that *A. alpina* was not discovered in Britain until 1887, when it was found by chance by H.C. Hart, an Irish mountaineer who had set out to check the height of Sgurr Alasdair. It occurs in only a single corrie in the Cuillin Hills and has yet to be found elsewhere, despite rumours of other stations. It grows on a few shady ledges of basic cliffs at about 820-850 m altitude, with such species as *Arabis petraea, Chrysosplenium oppositifolium, Cystopteris fragilis, Oxyria digyna, Sedum rosea, Trollius europaeus* and *Philonotis fontana* in close proximity. There are two main colonies, one lower and one higher on cliff ledges. In 1989, two small patches and four other small plants of one or two rosettes were reported on lower ledges. In 1993, three colonies were found, of 41, 35 and seven plants (L. Farrell pers. comm.), but information on the present size of the population seems not to be available.

A. alpina is a perennial mat-forming herb with slender branching stolons, which give rise to rosettes, then flowering and non-flowering shoots up to 40 cm high, the stems sometimes becoming decumbent in fruit. The flowers are homogamous and are usually produced in June or July, though climatic conditions strongly influence both the time of flowering and the numbers of flowers and flowering stems. Though the flowers are visited by insects, they are probably mostly self-pollinated. The degree of seed germination and of establishment of new plants in the wild is not known. Plants are, however, successful in cultivation, showing vigorous growth and abundant flowering and seed-set.

The relative inaccessibility of the plant does not necessarily confer full protection. Some plants are particularly vulnerable, and scrambling nearby or other attempts to reach the colonies could result in the loss of plants. Restocking of the existing populations could be beneficial. In 1975 some seedlings taken from a locally cultivated population (established from native stock) were transplanted on to cliff ledges higher up the corrie, but the success or otherwise of the transplant is not known. Further introductions into nearby sites from which no record is known may also be desirable.

Outside Britain, *A. alpina* is a widespread and often frequent plant of arctic and alpine regions of the northern hemisphere and on some East African mountains. In Europe it is not threatened, though it is rare in eastern Europe.

M.J. Wigginton

Arabis glabra (L.) Bernh. (Brassicaceae)

Tower mustard, Twrged Esmwyth
Status in Britain: VULNERABLE.
Status in Europe: Not threatened.

A. glabra occurs on light, freely draining sandy soils over chalk and limestone. In the Breckland of East Anglia it is a colonist of clear-felled conifer plantations and of waste ground such as the sites of old straw stacks, but always where the underlying soils are calcareous. When these sites become overgrown with a dense cover of grasses or are shaded by newly planted trees, individual plants appear able to hold their own for a while, but with no open space for seedlings the population may decline to extinction. It is often found in association with *Lactuca serriola* and *Verbascum thapsus*, both being species that can take advantage of temporary clearances and open soil, and it often behaves like an invasive weed. Other habitats include a sandy roadside bank in Hampshire, and sandy tracksides in Surrey, both on the Greensand.

A. glabra is a biennial or, rarely, a short-lived perennial in Britain. It germinates in spring and flowers from May to July in its second year. It seems that self-pollination occurs, and the flowers are self-compatible. Plants produce copious seed, which may remain viable for many years in the soil. Cultivation *ex situ* seems to indicate that the plant is susceptible to winter damp (Rich & Rose 1995).

This species has declined strongly across most of its British range (Stewart *et al.* 1994). It is now known from only six sites in southern England (North Wiltshire, North Hampshire and Surrey), about twelve in West Norfolk and perhaps a few in the West Midlands (Baldock & Rich 1996; Rich & Rose 1995). There has been no comprehensive survey of its historic sites in the Midlands and northern England, and it may yet survive in some of them. Although it occurs in small numbers and is threatened in southern England, numbers have increased in parts of the Norfolk Breckland in recent years. Trist (1979) describes it as usually occurring singly, but some colonies in the Breck now consist of thousands of plants, although they occupy very restricted areas. This may in part be due to more open habitat created after the 1987 storm, which devastated East Anglian plantations. *A. glabra* will presumably decline again with the closing over of its habitats.

Whilst some colonies have fallen to very low numbers, a persistent seed-bank may allow recovery, and at some sites it has reappeared after long intervals (Lousley 1976). Recovery is often rapid: for example, scarification of one Hampshire site in 1994 resulted in a doubling of the population (to 200) in 1995. Wire cages have been used locally to protect against rabbit-grazing. However, because

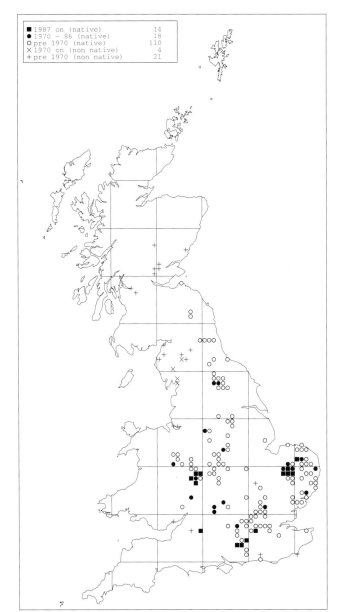

■ 1987 on (native)	14
● 1970 - 86 (native)	18
○ pre 1970 (native)	110
✕ 1970 on (non native)	4
+ pre 1970 (non native)	21

of its rather invasive tendencies in Breckland, special conservation measures are not required in that area at the present time.

It has a wide distribution throughout the world, being found in Europe up to 70°N, and in Africa, Asia, North America and Australia. It is believed to be native in Europe and western Asia and introduced elsewhere (Hultén & Fries 1986).

G. Beckett and M.J. Wigginton

Arabis scabra All. (Brassicaceae)

Arabis stricta Huds.

Bristol rock-cress

Status in Britain: VULNERABLE. WCA Schedule 8.

Status in Europe: Not threatened.

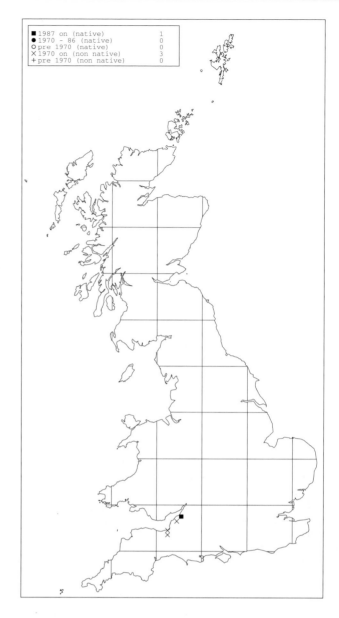

A. scabra is a plant of highly calcareous soils derived from Carboniferous limestone. It grows on very shallow soil in open turf, on loose rubble and scree, and directly on the limestone rock, where it roots into crevices. It is intolerant of competition, and cannot establish itself from seed in closed turf. *A. scabra* accumulates strontium in its leaves, and it may be favoured by strontium-rich soils (Bowen & Dymond 1955). In the Avon Gorge it is usually found on south-facing rocky slopes and crags, though some plants occur low down near tide level. Nearby at Shirehampton it grows in two disused quarries. Its capacity to withstand drought enables it to survive in open sites that become very dry in summer. In open communities, where it grow best, it is often associated with such species as *Bromopsis erecta, Catapodium rigidum, Festuca ovina, Helianthemum nummularium, Pilosella officinarum* and *Sanguisorba minor*. It is able to survive in more enclosed areas where scrub has encroached (*Cotoneaster* species, *Crataegus monogyna, Hedera helix, Quercus cerris*) but will succumb if too densely shaded or overgrown.

It is a perennial species, flowering from late March to May, the seeds maturing in midsummer and being shed from July onwards, even as late as October. Germination normally takes place in early autumn, the seedlings overwintering as small rosettes. The slightly protandrous flowers are highly self-fertile and a high proportion set seed. Though seeds are freely produced, they are not usually dispersed more than a few centimetres, and seedlings develop in a cluster around the parent plant. However, most seedlings do not survive, and mature plants typically occur singly or in very small groups. There is little or no vegetative spread.

This species has long been known in Britain, having been first recorded in the Avon Gorge by John Ray in 1686 "in rupe S. Vincentii prope Bristolium", where it is still found. The largest populations are in the Avon Gorge, where it occurs in at least twelve sub-sites, on both sides of the river, in Somerset and Gloucestershire. Numbers appear to vary from year to year, and counts of the whole population between 1977 and 1989 ranged from about 500 to 5,400 individuals (Taylor 1990f). There has been a serious decline in some areas because of scrub growth and rock stabilisation work, but several thousand individuals occur in the Gorge at the present time. The Shirehampton population is much smaller: in 1989, only four sub-populations with a total of 146 individuals were recorded (Taylor 1990f). Attempts in the 1950s to establish the plant in suitable habitats elsewhere in the district by sowing seed or planting rooted plants have been largely unsuccessful (Hope-Simpson 1987). An old record for Radnorshire is now regarded as unreliable.

Conservation management mainly entails the control of scrub, which has vigorously encroached at some sites and is threatening some populations. Grazing would be an ideal means of control, but may not be practical in a largely urban situation. It has been suggested that trampling may be a problem, but the threat from this activity appears to be minimal; indeed, it may increase the amount of bare ground and marginal habitat favoured by the plant. Of prime concern to the local authorities is the prevention of rock-falls from the cliffs of the Avon Gorge, which might endanger traffic below. This has involved the extensive netting of rock faces, which traps litter and leads to the accumulation of humus and the development of closed communities. This has adversely affected populations of *A. scabra*, which requires open ground for germination. Debris from shot-blasting the Clifton Suspension Bridge by copper smelter slag in 1995, leading to toxic levels of copper, zinc and lead in the Gorge, may have had some deleterious effect on the plant.

In continental Europe the species is rather rare, being recorded from mainly montane areas of eastern and southern France, Switzerland and Spain.

Further details are given in Pring (1961) and Thompson (1928).

M.J. Wigginton

Arenaria norvegica ssp. *anglica* Halliday (Caryophyllaceae)
English sandwort
Status in Britain: VULNERABLE. ENDEMIC. WCA
Schedule 8.

This plant is endemic to the Yorkshire Pennines. Most of its populations in natural habitats occur on freely draining substrates, normally thin peat or earthy rubble over limestone bedrock. Typical niches include thin peaty soil in slight depressions on flat limestone outcrops or surrounding them, peaty soil in cracks and hollows in limestone and open habitats alongside trackways. Most sites are subject to periodic flushing or waterlogging but may dry quickly. It appears to be a poor competitor, growing mainly in sparsely-vegetated ground. The more frequent associates include *Carex flacca*, *Festuca ovina*, *Minuartia verna*, *Sagina nodosa* and *Sesleria caerulea*. Two populations occur in continuously wet bryophyte-rich tufaceous flushes in a very open community containing *Primula farinosa*, *Sedum villosum* and *Palustriella commutata*. Some colonies have established themselves on man-made tracks of compacted limestone chippings (mimicking natural habitats), and one on a quarry floor where periodically there is standing water. Because of its preference for permeable, quickly-drying substrates, colonies can be severely reduced in dry spells, but natural recovery probably fully compensates in the longer term.

This subspecies is annual or biennial. It normally behaves as a winter annual, and though overwintering is common it is doubtful whether plants survive more than two winters. The long flowering season extends from May to October. Most recruitment is from spring-germinated seed, and most seedlings germinating in autumn are killed by winter frosts. Re-appearance of plants at sites where it has been absent for many years indicates long-term viability of seed (Walker 1995).

A. norvegica ssp. *anglica* occurs in only three hectads. A detailed census carried out in 1995 revealed a total of about 2,600 plants in about 27 discrete populations, most of which lay within a small area on the eastern slopes of Ingleborough, at altitudes between 300 and 400 m, and all but one within an SSSI. Populations ranged from three plants to more than 600, but most occurred in very small and scattered colonies, the four largest holding about 70% of the British population (Walker 1995). However, numbers fluctuate considerably from year to year in response to environmental conditions, with frost and drought limiting (Halliday 1960). For instance, in 1981 totals of between 1,000 and 1,500 plants were counted in eight sites, and in 1990 a 70% decrease was observed at two of them. There was a severe decline of 50-90% overall in 1992, with fewer than twenty individuals present at five of the eight monitored sites. The cause of this poor showing is likely to have been local drought conditions that year.

Provided land management and use remain suitable, this endemic plant appears to be in no immediate threat of

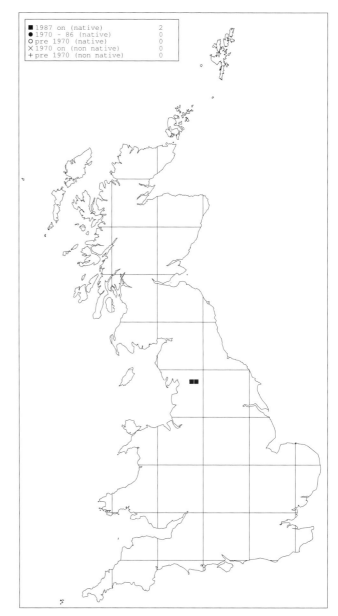

extinction. However, some populations, particularly those on tracksides, are at risk and many have been depleted or show signs of deterioration. Likely causes include the increased use of the area by walkers and bikers or by farm vehicles and the parking of cars. Re-routing of some tracks and the Pennine bridleway away from sensitive areas is desirable and there is now some control of car-parking. However, some disturbance is beneficial in keeping habitats open and grasses under control. The general area contains much suitable habitat, and it is possible that further sites may exist. Autecological studies (Walker 1995) show that low dispersal ability, poor reproductive success, slow growth and low genetic diversity all limit its abundance and distribution.

M.J. Wigginton

Arenaria norvegica Gunn. ssp. *norvegica*
(Caryophyllaceae)
Arctic sandwort
Status in Britain: LOWER RISK - Near Threatened. WCA
Schedule 8.
Status in Europe: Vulnerable. Endemic.

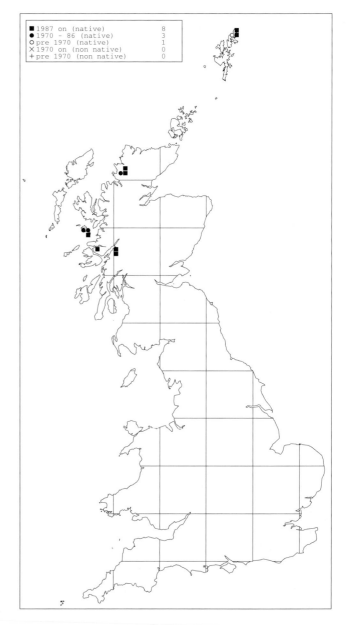

This calcicole species is confined to freely draining
base-rich substrates. Its habitats are varied: steep, unstable,
coarse or fine screes derived from dolomitic limestones or
basalts, limestone slabs or knolls (sometimes on summit
ridges), shallow gravel over bedrock and, in Shetland,
open serpentine debris weathering from rock *in situ*. At
one site it grows on partially stabilised river gravels. In
mountains it ascends to 610 m, but it also occurs on lower
ground as, for instance, on Rum (250 m) and in Shetland
(below 50 m). It is a poor competitor, growing in open
communities where the stony substrate limits competition.
In montane areas, plants recorded in close proximity
include *Alchemilla alpina, Carex capillaris, Draba incana,
Dryas octopetala* and *Saxifraga oppositifolia*. It has been
recorded in a flush in summit grasslands with *Sedum
villosum*. In Shetland, common associates on serpentine
debris include *Anthyllis vulneraria, Arabis petraea, Cerastium
nigrescens, Plantago maritima* and *Silene acaulis*.

At most of its sites, *A. norvegica* ssp. *norvegica* appears
to be perennial, but recent studies in Shetland (S.Kay pers.
comm.) have shown that most plants are annual or
biennial: only 5% of mature plants recorded in July 1994
were still alive in July 1996. It sets seed freely, often
producing a prolific crop of seedlings, but there is also
high mortality of plants, resulting in large fluctuations in
the size of populations. It flowers mainly in June and July,
but in Shetland, and perhaps elsewhere, flowers can be
seen from May to September.

It is very sparsely distributed in western and northern
Scotland, occurring on Rum and Eigg, on a few mountains
in Argyll (Coker 1969), Inverness and Sutherland (Slack &
Dickson 1959) and at several adjacent sites in Unst,
Shetland. In the past twenty years *A. norvegica* has been
recorded in seventeen sites in nine hectads, though it has
not been seen since 1978 in West Ross. In many of its sites,
there are healthy populations of several hundred plants,
and in Shetland, a fluctuating colony at the Keen of Hamar
has ranged from about 5,000 to 15,000 individuals between
1978 and 1993 (Slingsby *et al.* 1993).

The plant appears to be relatively safe, and there is no
evidence of an overall decline. Many localities are remote,
and populations scattered. Changes in populations do
occur as a result of natural processes such as scree falls,
but recolonisation generally follows and these processes
are unlikely to lead to a longer term decline. There is,
however, some concern that heavy grazing pressure, from
the increasing numbers of sheep over the past few years,
may have an adverse effect on some populations.
Recreational activities could pose a local threat.

This subspecies is endemic to north-west Europe,
occurring only in Scotland, Norway and Finland (where it
is rare and legally protected), and in Iceland, where it is
fairly common. A population reported in the Burren,
Ireland in the 1960s has not been re-found and may either
be extinct or an erroneous determination (Curtis &
McGough 1988).

M.J. Wigginton

Armeria maritima ssp. *elongata* (Hoffm.) Bonner (Plumbaginaceae)
Thrift
Status in Britain: VULNERABLE.
Status in Europe: Not threatened.

A. maritima ssp. *elongata* was not recognised in Britain until the 1950s, even though an inland *Armeria* had been observed at many sites in Lincolnshire and Leicestershire as early as 1726. Indeed, Pulteney (1757) described it as "very plentiful about Grantham and Sleaford". Gibbons & Lousley (1958) fully document the history of recording and its status.

Most colonies existing in the 1950s were characterised by good drainage and an absence of competition from taller vegetation (Gibbons & Lousley 1958). They were in areas of neutral grassland overlying post-glacial river gravels or alluvium, except for one which was in limestone grassland. The two extant populations are in neutral grassland. One is on well-drained sandy ground in a cemetery at Ancaster, where it grows with *Achillea millefolium*, *Festuca rubra*, *Galium verum*, *Leucanthemum vulgare* and *Trifolium pratense*. The other is in neutral to basic grassland in a nearby pasture, where the clumps of *A. maritima* ssp. *elongata* grow in tall-herb swards together with *Achillea millefolium*, *Anthoxanthum odoratum*, *Centaurea nigra*, *Cynosurus cristatus*, *Leontodon hispidus*, *Plantago lanceolata*, *Primula veris*, *Rhinanthus minor*, *Rumex acetosa*, *Sanguisorba minor* and *Trifolium pratense*.

A. maritima ssp. *elongata* is similar to the common subspecies (ssp. *maritima*) but is a rather larger plant with longer leaves and taller, glabrous flower-stalks. It is perennial with a stout rootstock. The flowers appear mainly in June and July, though some may be seen as late as October. They are self-incompatible and the plant is out-breeding.

Even as late as the mid-1950s, it was known in twelve localities between Ancaster and Wilsford in Lincolnshire, but ploughing, re-seeding and other agricultural improvement destroyed eight of these by the early 1980s, by which time it was almost extinct in two of the four remaining. Much the largest population occurs in the cemetery site (many thousands of plants), although its flowering there is dependent upon sympathetic management. For instance, in July 1994 there were hardly any flowers because almost the entire area had been mown early in the summer and was covered with rotting grass cuttings. Only three flowering plants were seen, in a patch of ox-eye that had been spared the mower. The nearby pasture holds several small populations: a total of about 100 plants were observed in 1994.

Both remaining sites are afforded protection within an SSSI and County Wildlife Trust Reserve, and provided suitable management continues, the plant should be reasonably safe. In the natural site, a management regime

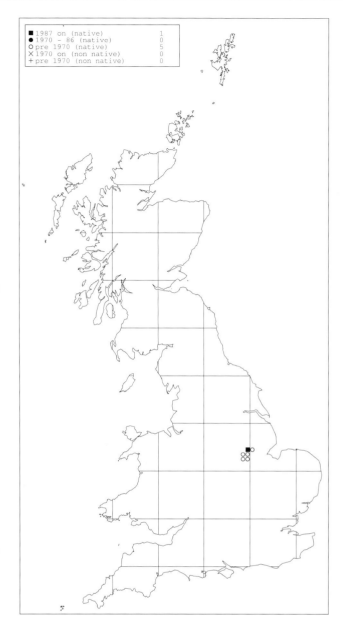

of light cattle-grazing and no application of artificial fertilisers has maintained the small population. Mowing or scything of the cemetery grassland should be carried out after seed is set, although a second, albeit modest, flowering in autumn has been observed in a sward cut too early. Consideration might be given to transplanting *A. maritima* ssp. *elongata* to other suitable sites within its historic area, if any can now be found.

Its European distribution is centred on the sandy heaths of the north German plain, where it is common, but it extends from the Netherlands to western Russia, Denmark, southern Finland and southern Sweden.

M.J. Wigginton and N.G. Hodgetts

Artemisia campestris L. (Asteraceae)

Field wormwood, Llysiau'r Corff
Status in Britain: ENDANGERED. WCA Schedule 8.
Status in Europe: Not threatened.

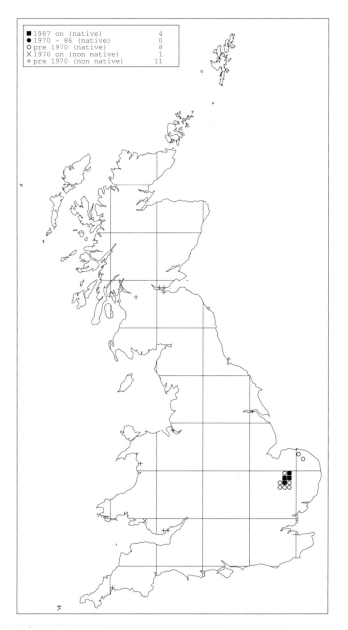

Native sites of *A. campestris* are restricted to Breckland, where, like other special plants of this area, it grows in short grassy swards with some ground disturbance. Its habitats include roadside verges, tracks, abandoned arable fields, forest rides and the barer areas in Breckland pine belts. It is found on drought-prone calcareous to slightly acidic sands. Associates include *Achillea millefolium*, *Arenaria serpyllifolia*, *Crepis capillaris*, *Erophila verna*, *Festuca rubra*, *Filago minima*, *Galium verum*, *Koeleria macrantha*, *Ornithopus perpusillus*, *Plantago lanceolata* and *Sedum acre*, together with some of the rarer Breckland species, *Medicago falcata*, *M. minima* and *Silene conica*. It does not survive for long in rank tussocky grassland, the plants becoming smothered, particularly when overwintering buds and shoots lie close to the ground.

A. campestris is a deep-rooting perennial with ascending stems to 60 cm, woody at the base. Flowering panicles can be seen from May to September. The flowers lack nectaries and are likely to be wind-pollinated (Birkinshaw 1990d). Once established, mature plants are long-lived, but plants in early stages of development are intolerant of competition. It is sensitive to grazing (Watt 1971) and no longer appears on rabbit-grazed heaths, except within rabbit exclosures, but will survive as rosettes in closely mown grass as, for instance, around factory units at Brandon.

Records of *A. campestris* date back to 1670, since which time it has been recorded in about 20 locations in 11 hectads (Crompton 1974-1986), principally in the area bounded by Mildenhall, Brandon and Thetford. Many sites have been lost to agriculture, forestry and building, and the plant is now found at only three native sites. By far the largest numbers - some 350 plants - occur on an industrial estate in Brandon, where a small undeveloped area containing 150 plants is specially managed for it. The other two native sites held 97 and about 40 plants in 1994. Because the plant is potentially threatened on sites lacking statutory protection, populations derived from planned translocations have been established on other Breckland SSSIs. It sometimes occurs as a casual outside Breckland but, with the exception of a small and declining population at Crymlyn Burrows near Swansea, rarely persists for long.

A. campestris is widespread in Europe, extending from Spain eastwards to Greece and eastern Europe, and north to Fennoscandia, even to Novaya Zemyla at 75°N. Outside Europe it is found from Turkey to central Asia, in North Africa and North America.

Y. Leonard

Artemisia norvegica Fries (Asteraceae)

Norwegian mugwort, Gròban Lochlannach
Status in Britain: VULNERABLE.
Status in Europe: Vulnerable.

This species is one of the most recent additions to the British flora, having remained undetected on a few remote Scottish mountains until 1950. Exclusively a plant of mountain tops, at 700-870 m, it occurs on or near summit ridges, usually in exposed situations and subject to a severe climate. Habitats include moist sandy or gritty debris in bared areas of sparse vegetation, closed but stony *Racomitrium lanuginosum*-heath, amongst boulders on flat summit plateaux, the more sheltered hollows between blocks, and exposed bouldery crests of small solifluction terracettes on shallow slopes just below the summit ridge. Associated species include *Alchemilla alpina*, *Carex bigelowii*, *Deschampsia flexuosa*, *Festuca ovina*, *Juncus trifidus*, *Minuartia sedoides*, *Salix herbacea*, *Racomitrium lanuginosum* and, where soils are base-enriched, such species as *Antennaria dioica*, *Gnaphalium supinum* and *Thymus polytrichus*.

A. norvegica is a rhizomatous perennial. Flowering is usually between July and September, the timing and degree of flowering likely to depend on climatic conditions. The numbers of flowers on each stem varies, but at one site where counts have been made, more than half bore two flower heads and most of the remainder were single. In some years, very few flowers occur, a possible reason being prolonged dry spells in the plant's early growing season. Little appears to be known about its reproductive biology and the relative importance of vegetative and sexual reproduction. It has, however, been suggested that conditions are unsuitable for seed production in normal seasons (Raven & Walters 1956).

It is known from only three mountains in West Ross, each supporting at least several hundred plants. The largest site holds many thousands of plants, while population counts at other sites have ranged between 150 and 'hundreds' of individuals. Some populations are restricted in extent, covering only a few square metres, but the largest comprises colonies scattered over about 9 hectares of a summit ridge. Counts of individual plants within a colony have provided some evidence that populations fluctuate in size, presumably in response to climatic or biotic factors. It has been suggested that at one site an apparent population decline could be because of the removal of sheep, resulting in reduced disturbance and decreased amounts of bare ground receptive to dispersed seed. In general, however, populations appear to be reasonably stable.

Two of its sites are safeguarded within SSSIs, one of which is also an NNR. Their relative remoteness affords further protection, and no special conservation measures seem to be required. Botanical collecting is not a significant threat, nor is the plant likely to be of interest to the horticultural trade.

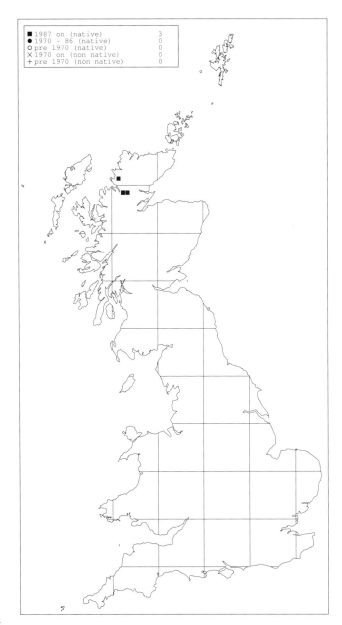

■ 1987 on (native)	3
● 1970 – 86 (native)	0
○ pre 1970 (native)	0
✕ 1970 on (non native)	0
+ pre 1970 (non native)	0

Elsewhere it occurs only in Norway and the northern Ural mountains and thus has one of the most disjunct and restricted distributions of any of our native plants. The Scottish taxon has been regarded as an endemic var. *scotica* Hultén (Hultén 1954) on the basis of its different leaf shape, fewer capitula and dwarf growth form. More recent chemical and morphological evidence (Øvstedal & Mjaavatten 1992) indicates some real, albeit small, differences between plants from Scotland and south and south-central Norway, but whether this warrants varietal status is debatable.

M.J. Wigginton

Asparagus prostratus Dumort. (Liliaceae)

A. officinalis ssp. *prostratus* (Dumort.) Corbière
Sea asparagus, Merllys Gorweddol
Status in Britain: VULNERABLE.
Status in Europe: Vulnerable. Endemic.

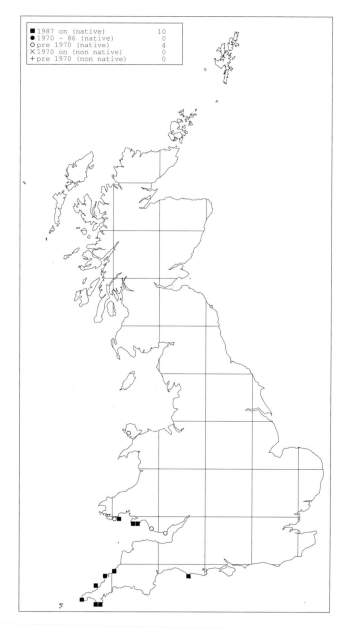

Most populations of *A. prostratus* occur on sea-cliffs on freely-draining, often rocky soils, with a few populations on sand-dunes. In Cornwall, they are often in relatively inaccessible sites on cliffs, the plants growing in grassy turf on rock ledges or slopes, sometimes on thin soils over 'clitter' from old mineral workings. Associated species include *Anthyllis vulneraria, Armeria maritima, Bromus hordeaceus, Crithmum maritimum, Daucus carota, Festuca ovina, F. rubra, Plantago coronopus, Scilla verna* and *Spergularia rupicola*. At some sites it grows through *Calluna vulgaris* or *Erica cinerea*. In Wales, populations on open, south-facing cliff-slopes are found mainly in wind-beaten *Festuca rubra* turf with such species as *Armeria maritima, Carex flacca, Dactylis glomerata, Ononis repens* and *Plantago lanceolata*. Dune populations occur in closed or fairly open turf, particularly by footpaths, with associates including *Ammophila arenaria, Anthyllis vulneraria, Carex arenaria, Elytrigia juncea, Festuca rubra, Galium verum, Plantago lanceolata* and *Senecio jacobaea*. The Dorset site, in an area where *A. prostratus* has been known at intervals over 200 years, is under *Atriplex portulacoides* on the edge of an old railway embankment.

A. prostratus is a dioecious long-lived perennial with stems arising from creeping rhizomes. Vegetative spread takes place only by the extension of the occasionally branching rhizomes, and plants become separated and independent from the parent as older parts of the rhizomes die. It is common, therefore, to find groups of plants growing in close proximity. However, rhizomes grow extremely slowly, and a local spread of a few metres in extent is likely to take several hundred years (Kay & John 1995). Flowers appear in June and July and are pollinated by insects. The red fruits, ripening from July to October, are perhaps taken mainly by birds. Some populations, though containing both sexes, regularly fail to produce fruit or seedlings, but the reasons for this are not clear. Natural dispersal is likely to be mainly or exclusively by seed, though no seedlings or young plants have been observed in Welsh populations.

This species was at one time more widespread along the coasts of Wales and south-west England but now appears to be extant in only nine hectads. In England, it is present at only three sites on the north coast of Cornwall, a few on the Lizard peninsula and one in Dorset. Populations are very localised and mostly small: for instance, counts made between 1976 and 1996 at the three north coast sites recorded totals of 15, 49, and seven plants respectively. In Wales, there are currently four known sites, one in Pembrokeshire and three on the Gower coast (Jones 1994; Wilkinson 1997). No Welsh colony exceeds 10 plants.

Several populations are in decline, the reasons for which include trampling, the erosion of its substrate and a lack of grazing leading to overgrowth by rank herbage. In Cornwall, the spread of *Carpobrotus edulis* might threaten some colonies on the Lizard, but populations on inaccessible cliff sites, though small, appear to be more or less stable in numbers. In Wales, it is severely threatened. All populations that have been studied appear to consist of old, long-established plants, often in decline, with no recruitment

(Kay & John 1995). Some Welsh populations have been reported as containing only male plants, but this is erroneous. Both sexes are currently present and flowering at two sites, and both were present at another site in 1990 and are probably still present at the fourth (Q.O.N. Kay *in litt.* 1997). The increasingly isolated populations are likely to suffer a loss of genetic variation. Collection of seed from Welsh populations and *ex situ* propagation are currently being undertaken with the aim of enhancing existing populations by reintroducing plants to former sites. A detailed study of populations in England is desirable.

A. prostratus is endemic to western Europe, occurring on the coasts of northern Spain, western France and Brittany, the Channel Islands, south-west England and Wales, south-east Ireland, the Netherlands and north-west Germany. It is thought to be still locally frequent in southern Brittany and northern Spain but is considered to be threatened over much, if not all, of its range.

A. officinalis and *A. prostratus* were both originally described as full species, but subsequently *A. prostratus* has often been treated as subspecies or a variety. However, they are morphologically and ecologically distinct and have different native geographical ranges. The prostrate habit of *A. prostratus* has been shown to remain constant in cultivation for more than 50 years. In addition, evidence suggests that the two taxa are genetically isolated because of a barrier to gene-flow caused by their difference in chromosome ploidy (2n=20 in *A. officinalis* and 2n=40 in *A. prostratus*). Supporting evidence for treating the two taxa as separate species is reviewed in Kay (1997).

L. Wilkinson

Asplenium trichomanes ssp. *pachyrachis*
(Christ) Lovis & Reichstein (Aspleniaceae)
Lobed maidenhair spleenwort, Duegredynen Gwallt y Forwyn
Status in Britain: DATA DEFICIENT.
Status in Europe: Not threatened.

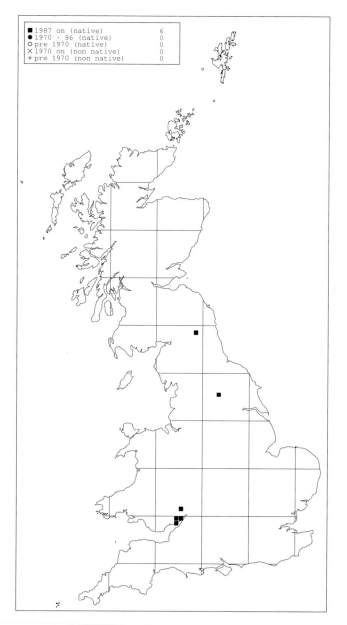

A. *trichomanes* ssp. *pachyrachis* grows directly on perpendicular calcareous rocks, in crevices or under overhangs, or on walls where non-calcareous rock is used with mortar. In such habitats it is likely to be sheltered from direct rain, being apparently intolerant of water retained in the crown. It normally grows in well-shaded sites, for instance on north-facing walls or rock outcrops in woodland, with relatively high humidity. At its British sites it is, in many instances, the sole vascular plant. It is sometimes found with A. *trichomanes* ssp. *quadrivalens*, with which it forms a vigorous hybrid, A. nothosubsp. *stauffleri* (Rickard 1989). Other associates include *Asplenium scolopendrium, Dryopteris affinis, Polystichum aculeatum* and *P. setiferum*, with *Ctenidium molluscum, Homalothecium sericeum* and other hypnoid mosses. In the southern Welsh Marches it is also found on old castle walls, for instance at Chepstow and Caldicot, often with *Asplenium ceterach, A. trichomanes* ssp. *quadrivalens* and *Polypodium cambricum*.

It reproduces readily from spores, and young plants have been observed at many sites. Nearly all British populations may be recognised by their distinct morphology (Jermy & Camus 1991), indicating that the individual populations may represent old colonisations from perhaps a single spore.

Eleven sites are currently known in Britain. Its stronghold is in the Wye valley, with the largest populations on castle walls: 50-100 plants on Chepstow and Caldicot Castles, and 50-70 on Cas Troggy. A population of 50-70 plants was recently found near Bellingham in Northumberland, in deep shade on a vertical sandstone cliff with other lime-loving plants. Three colonies of 14, three and one plant respectively are currently known in Knaresborough, on the castle and walls close to the Nidd. Herbarium specimens exist from Barmouth. It is likely to occur elsewhere in Britain on limestone or calcareous sandstone. In Britain this taxon has been known (and formally named) as an unusual form in the A. *trichomanes* species complex since the middle of the nineteenth century. Victorian fern collectors sought it for cultivation, and several slightly different forms were described by E.J. Lowe and others. Recent evidence suggests that British plants are conspecific with A. *csikii* Kümm. & Andr.

In natural habitats it is rare but not particularly threatened. However, plants on castle walls may be vulnerable when civic authorities seek to clear them of 'harmful' vegetation. Further sites may be found on medieval castles.

In mainland Europe, this taxon is confined to limestone massifs up to 1,000 m in the Alps. The centres of its distribution are found in southern France, southern Germany, the Alps and the mountains of south-east Europe. It is also reported to occur in the Atlas Mountains, Morocco and in the Himalayas.

A.C. Jermy, J.C. Vogel, F.J. Rumsey and M.H. Rickard

Aster linosyris (L.) Bernh. (Asteraceae)

Linosyris vulgaris DC., *Crinitaria linosyris* (L.) Less.
Goldilocks aster, Gold y Môr
Status in Britain: LOWER RISK - Near Threatened.
Status in Europe: Not threatened.

A. linosyris is restricted to a few coastal localities in western Britain, where it occurs on limestone sea-cliffs and rocky slopes and in cliff-top grassland and heath overlying limestone. It is confined to more or less open communities, typically in a discontinuous tussocky turf on shallow soil or as isolated clumps. A wide variety of associated maritime and limestone herbs include *Anthyllis vulneraria*, *Crithmum maritimum*, *Daucus carota*, *Festuca rubra*, *Plantago coronopus* and *Scabiosa columbaria*. At some sites it occurs with other rare species, including *Helianthemum apenninum* and *H. canum*. Many of its populations are on inaccessible cliff ledges. In Pembrokeshire, colonies occur in low-growing sheep-grazed grassland and maritime heath on cliff-tops, with *Calluna vulgaris*, *Danthonia decumbens*, *Erica cinerea* and *Succisa pratensis*.

It is a woody, long-lived perennial, and appears to be a sexually-outcrossing species. Flowers appear in late summer and are usually at their best in September. Seed is freely produced, but the production of fertile seed may sometimes be sparse or absent in some colonies (Jones 1994). The relative importance of sexual and vegetative reproduction, conditions of seedling establishment, and genetic exchange between colonies, are unknown. It is a poor competitor, unable to compete successfully in closed turf, and is intolerant of heavy grazing by sheep or rabbits.

There are about eight localities in Devon, Somerset, Glamorgan, Pembrokeshire, Caernarvonshire and Westmorland. Pembrokeshire holds much the largest populations, large stands several metres across occurring in two main areas of cliff-top between Stackpole and Linney Head. In 1988, the population there was estimated to be about 7,000 plants (Morgan 1989e). By contrast, some colonies are small and vulnerable: for example, at Humphrey Head only a single clump survives (I. Slater pers. comm. 1997). Casual occurrences, for instance on ballast tips and railways, have only rarely been reported.

At some sites, colonies are under threat from overgrowth of scrub, usually *Prunus spinosa* but locally also bramble or bracken. In these areas, some grazing is desirable to check the growth of these invasive species. Local disturbance of the vegetation may be beneficial in providing open areas in the turf; indeed at one locality the plant appears to grow preferentially along cliff-top sheep-walks. Excessive trampling is damaging, however, and because of it some populations have declined. Other potential local threats include quarrying.

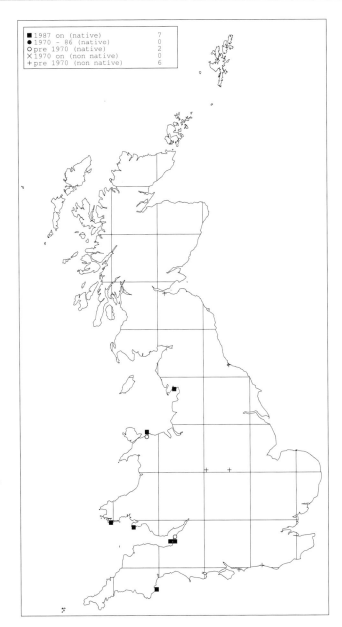

■ 1987 on (native)	7
● 1970 - 86 (native)	0
O pre 1970 (native)	2
X 1970 on (non native)	0
+ pre 1970 (non native)	6

The species has a continental distribution in Europe, being common in the south and east and becoming rarer in the north and west, but reaching southern Sweden. Though normally occurring on steep limestone slopes or outcrops, it is also found on the chalk in north-western France, and its absence from the English chalk is, therefore, noteworthy. It may be that it failed to find suitable habitats on this soft limestone during the post-glacial forest maximum (Pigott & Walters 1954).

M.J. Wigginton

Astragalus alpinus L. (Fabaceae)
Alpine milk-vetch, Bliochd-pheasair Ailpeach
Status in Britain: VULNERABLE.
Status in Europe: Not threatened.

The largest colonies of *A. alpinus* occur in species-rich, locally flushed calcicolous grassland that forms a patchy mosaic with communities of *Calluna vulgaris*, *Erica tetralix* and *Vaccinium myrtillus*. It also grows on the margins of flushes but is rarely present in areas dominated by the dwarf shrubs. A wide range of associates includes *Carex pulicaris*, *Deschampsia cespitosa*, *Festuca ovina*, *Galium verum*, *Persicaria vivipara*, *Saussurea alpina*, *Sibbaldia procumbens* and *Thymus polytrichus*. It also grows on base-rich ledges and rocky outcrops with such species as *Carex rupestris*, *Dryas octopetala*, *Erigeron borealis* and *Silene acaulis*.

It is a small herbaceous perennial. Flowers normally appear between mid-June and mid-July, but as for many mountain plants, flowering time is greatly influenced by climate. The number of inflorescences may also vary from year to year. For instance, in a monitored colony in Caenlochan between 1986 and 1992, numbers ranged from 29 to 120. It is not known, however, whether this is indicative of a natural cycle or merely reflects variable levels of grazing.

In Britain, *A. alpinus* is known from only four sites in the eastern Scottish Highlands, three of them holding several disjunct populations. In total, there are about twenty discrete colonies, most of which form patches of only 1-15 square metres. However, the largest (of more than 10,000 plants) covers some 120 x 30 m, in which *A. alpinus* may attain a cover of 50% or more.

At some sites, *A. alpinus* grows on very unstable rocks and soil which cannot withstand much trampling. Indeed significant damage to ledges and crags has already happened in certainly one station, and plants have been lost from the eroded rocks. This damage may be partly attributed to sheep but, in at least one site, visiting botanists have added to it. Plants are grazed by sheep and deer, and this may be the main reason for the lack of flowers and seed-pods that is often reported. The survival of the whole of the smallest population may be particularly uncertain: monitoring over the past few years has shown that very few if any flowers remain ungrazed, and no seed-pods are produced.

A. alpinus is widespread in northern Europe and on central European mountain ranges, including the Alps, Pyrenees and Carpathians. It ranges through temperate Asia and occurs in North America and Greenland.

M.J. Wigginton

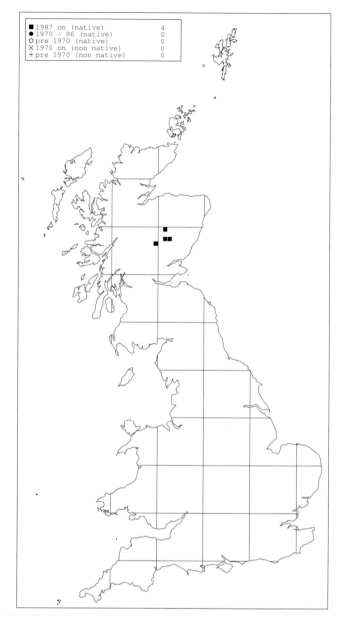

■ 1987 on (native)	4
● 1970 – 86 (native)	0
○ pre 1970 (native)	0
✕ 1970 on (non native)	0
+ pre 1970 (non native)	0

Athyrium flexile (Newman) Druce (Woodsiaceae)
Newman's lady-fern
Status in Britain: VULNERABLE. Probably ENDEMIC.

A. flexile is found in high alpine corries between 750 and 1,140 m in the Scottish Highlands in areas of late snow-lie. It grows in block screes, mainly on acidic rocks, but may spread on to more basic rocks where there is surface leaching. Its habitats are cool and shaded, and mostly north-east to north-west facing, though the Glen Prosen colony is on a south-facing slope beneath overhanging rocks. While generally growing with *A. distentifolium*, *A. flexile* is much less abundant, occurring in discrete patches within the extensive areas occupied by the former.

Occasional plants have been reported from up to twenty sites in the central Highlands of Scotland, though it has not been seen since 1950 on Caenlochan, Meall nan Tarmachan, Ben Nevis and Aonach Mor. The Ben Alder site has been known since 1867 and holds the largest population of several hundred plants. *A. flexile* was found on nearby Ben Eibhinn in 1957, where it has a population nearly as large. Three clumps were found on Creag Meagaidh in 1994. There are both older and more recent records from the Cairngorms, most notably Glen Einich. Although not recorded in Glen Prosen since the 1800s, nearly 100 plants were found there in one small area in 1995. *A. flexile* has also been reported from Glen Doll and Lochnagar. Another vigorous population was found in 1943 below Meall Buidh near Bridge of Orchy, and a few plants were seen in an adjacent corrie below Beinn a'Chreachain.

A. distentifolium seems to be a more attractive species to sheep and deer than most ferns, and is often nibbled. There are currently high numbers of deer in Glen Prosen, which may explain the almost complete absence of ferns from the valley floor. The rocks with *A. flexile* have large clumps of *A. distentifolium* in the centre, which were severely grazed twice in 1995. Several of the *A. flexile* fronds also lost their tips. *A. flexile* is usually smaller and less erect, which may make it less conspicuous, and the screes it frequents are less attractive to grazing animals. A more significant threat could come from global warming: both *A. distentifolium* and *A. flexile* are part of a specialised snowbed community exploiting a habitat that only a limited number of species can tolerate; the smaller *A. flexile* might be especially vulnerable to any increased competition.

Ever since *A. flexile* was first described in 1853 from Glen Prosen, there has been discussion about its status. *A. distentifolium*, with which *A. flexile* is usually found, occurs extensively throughout Europe and the northern hemisphere. Herbarium material has not revealed specimens with the characters of *A. flexile* from anywhere outside Scotland, suggesting that this form is endemic. It

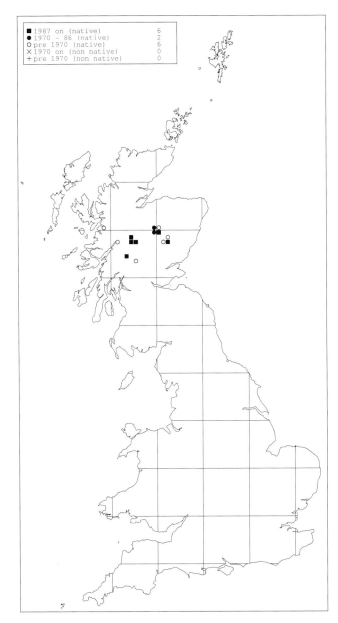

■ 1987 on (native)	6
● 1970 - 86 (native)	2
○ pre 1970 (native)	6
✕ 1970 on (non native)	0
+ pre 1970 (non native)	0

differs morphologically from *A. distentifolium* in having a narrow, congested frond with most of the sporangia borne on the lower pinnae, contrasting with the *A. distentifolium* pattern of densest fertility near the top of the frond. Its disjunct distribution and the occasional occurrence of only a single plant have suggested that *A. flexile* may have arisen from occasional mutations (Page 1982).

The status of this taxon thus remains uncertain. A research programme is currently underway to investigate its relationship to *A. distentifolium*, its molecular identity and the limits of its distribution in Scotland. This account is based on preliminary findings.

H. McHaffie

Atriplex pedunculata L. (Chenopodiaceae)

Halimione pedunculata (L.) Aellen
Pedunculate sea-purslane
Status in Britain: CRITICALLY ENDANGERED. WCA
Schedule 8.
Status in Europe: Vulnerable.

Until its rediscovery in 1987 (Leach 1988), it was
assumed that *A. pedunculata* had become extinct in Britain
in the 1930s. There are old records from at least sixteen
localities on the east coast of England from Kent to
Lincolnshire (Perring & Farrell 1983), but it had been
observed this century at only three of these: Pegwell Bay
in East Kent, the Wash at Freiston, Lincolnshire, and at
Walberswick, Suffolk. In 1987 it was found in Essex, on a
narrow strip of dryish saltmarsh behind a sea-wall near
Shoeburyness. It grows there in an open sward in which
Aster tripolium and *Puccinellia maritima* are co-dominant,
along with *Parapholis strigosa*, *Spergularia media*, *Suaeda
maritima* and *Triglochin maritimum*. This is similar to
vegetation in which *A. pedunculata* occurs on the European
mainland (Géhu & Meslin 1968; Géhu 1969; Westhoff &
Den Held 1969). *Hordeum marinum* is also found at the
Essex site, growing along the cattle-poached margins of an
adjoining ditch. The saltmarsh is ungrazed, and the *A.
pedunculata* colony in 1987 was hemmed in on all sides by
dense stands of *Elytrigia atherica* and *Festuca rubra*.

A. pedunculata is a spring-germinating annual,
flowering in August and fruiting in September and
October. It is uncertain to what extent it is capable of
developing a persistent seed-bank. Bare ground is essential
for germination and successful seedling establishment, and
many of its old localities were in disturbed areas of upper
saltmarsh or grazing marsh, on tracks, ditch banks and
roadsides. It is a notoriously erratic species, with numbers
of plants varying dramatically from year to year (Bennett
1905; Géhu 1969; R. Beyersburgen pers. comm.). In Essex,
annual counts between 1987 and 1994 ranged from fewer
than 400 to more than 4,100 plants. The plant is
surprisingly immobile, having failed to colonise nearby
patches of apparently suitable habitat, and by 1994 it was
in danger of being ousted by *Elytrigia atherica*.

The plant was brought into cultivation at Cambridge
Botanic Garden, and in 1989 a second population was
successfully established in the wild, on Foulness Island,
using seed from Cambridge (Gibson 1991). From an initial
sowing of 3,600 seeds, over 300 plants matured and fruited
in 1990 and more than 800 in 1991. Unfortunately, the
colony was later destroyed when a scrape for wading
birds was excavated at the site. More recently, new
colonies have been established on patches of open
saltmarsh near the original site at Shoeburyness, and
attempts made to limit the spread of *Elytrigia atherica* and
Festuca rubra. Further introductions may be necessary if the
species is to be prevented from becoming extinct,
especially as in 1995 and 1996 no plants were found at the
original site but only at nearby introduction sites (C.
Gibson pers. comm.).

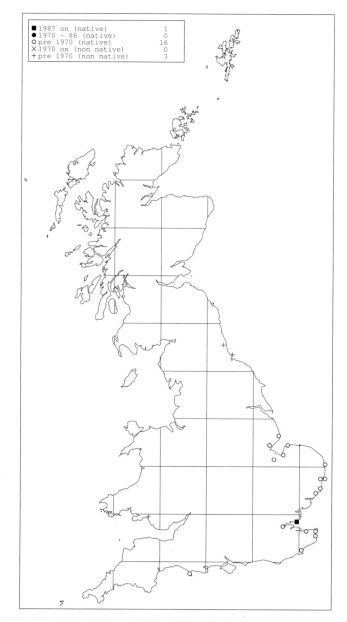

■ 1987 on (native)	1
● 1970 – 86 (native)	0
O pre 1970 (native)	16
X 1970 on (non native)	0
+ pre 1970 (non native)	3

It is an open question as to whether *A. pedunculata* has
always been present at the Essex locality, or is a recent
arrival. Its centre of distribution is the Danish coast, which is
an important feeding area for brent geese during their annual
migration (Owen *et al.* 1986). Large flocks pass through in late
September just as the seeds of *A. pedunculata* are ripening,
arriving on the Essex coast in October. It is therefore quite
possible that seeds might have been transported to this
country by the geese, or indeed by other wildfowl.

In mainland Europe *A. pedunculata* occurs in coastal
saltmarshes and tidally inundated dune-slacks from
northern France to Estonia. It is also found around the
Black Sea, and inland on saline soils in parts of western
Asia (Turkestan and Siberia). It is declining throughout its
range and is extinct in some regions.

S.J. Leach

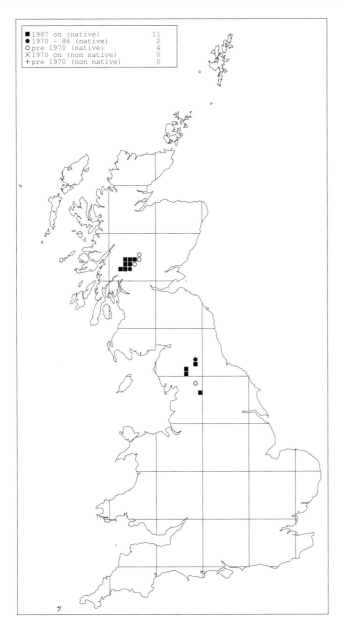

Bartsia alpina L. (Scrophulariaceae)

Alpine bartsia, Bairtsia Ailpeach
Status in Britain: LOWER RISK - Near Threatened.
Status in Europe: Not threatened.

B. alpina has a long history in Britain, having been first noted by John Ray in 1668 "prope Orton in Westmorlandia", where it still occurs. It is a plant of moist basic soils in upland meadows and pastures, and of montane rock ledges, at altitudes ranging from 380 to 950 m, most sites being between 600 and 800 m. Most of its sites are in the Scottish Breadalbane mountains, where nearly all its known colonies are in herb-rich swards on periodically-inundated ledges of calc-schist crags. A wide range of associated species may include *Carex panicea, Dryas octopetala, Festuca vivipara, Geranium sylvaticum, Saussurea alpina, Saxifraga aizoides, Thalictrum alpinum* and *Trollius europaeus*. In the largest Westmorland site, *B. alpina* grows in clumps on the drier hummocks within base-rich flushes and marshes, and in damp pasture, with such

species as *Festuca ovina, Molinia caerulea, Parnassia palustris, Potentilla erecta* and *Primula farinosa*. Elsewhere, it occurs on species-rich flushed banks as, for instance, by the Tees.

It is a perennial hemi-parasite, with a wide range of hosts, including *Andromeda polifolia, Astragalus alpinus, Betula pubescens, Pinguicula vulgaris* and *Vaccinium vitis-idaea*. Flowers are produced between June and August, with pollination mainly by bumble-bees. Propagation seems generally to be by vegetative spread, and it is unclear how important seed is for maintaining populations. There can be substantial pre-dispersal loss of seed because of predation by the larvae of a species of micro-lepidoptera, and this predation also seems to reduce the germination of apparently undamaged seed (Lusby & Wright 1996). However, seedlings have been noted occasionally and, although their survival has not been monitored, their presence indicates that recruitment sometimes takes place by this means.

B. alpina has been recorded in about thirty localities in Yorkshire, Co. Durham, Westmorland, Perthshire and Argyll. Although there are only six sites for *B. alpina* in England, two of them hold large populations. At one site in Westmorland, in which the plant has shown an overall increase, more than 1,000 plants were counted in 1993. Further large populations occur in Co. Durham. However, two of the English sites are very vulnerable, and in the most southerly, near Malham, apparently only a few plants remain (10-12 noted in 1992 and 3-4 during an incomplete search in 1994). The two largest populations in Scotland comprise some 2,000-3,000 plants, but most number between 100 and 500 individuals. In the Ben Lawers area, searches in the 1980s and in 1992 of previously known or likely sites in flush-pasture failed to detect any plants, and it is possible that heavy sheep-grazing combined with drainage has caused the demise of most or all colonies in that area.

Light grazing and trampling by cattle or sheep are crucial for the maintenance of populations in hummocky flush-pasture, by keeping the lusher vegetation down and the habitat open (Pigott 1956), though stock should be removed during the flowering and fruiting period. Heavy poaching by cattle is, however, detrimental (Taylor 1987a). Probably the greatest threats to the English sites are an unsympathetic grazing regime and excessive trampling of colonies by well-used paths. In Scotland, all Breadalbane sites are on ledges largely out of reach of grazing animals, and there seems little doubt that grazing restricts its occurrence on sites in Scotland.

B. alpina occurs in Greenland, around Hudson Bay (Canada), Iceland, the Faeroe Islands, Scandinavia and eastwards to the Urals. It is present in almost all alpine areas of Europe, reaching 3,000 m in the Alps. In Lapland, *B. alpina* occurs widely in forest mires with only very slight base enrichment.

M.J. Wigginton and G.P. Rothero

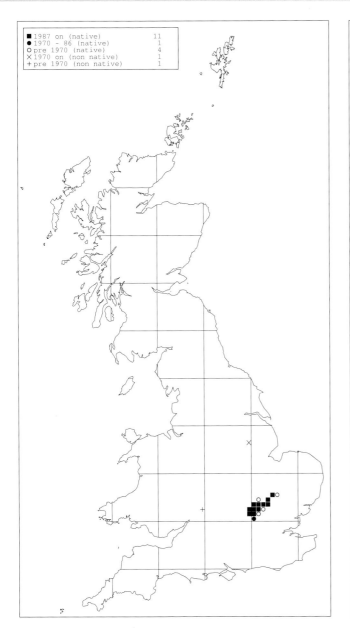

■ 1987 on (native)   11
● 1970 - 86 (native)   1
○ pre 1970 (native)   4
✕ 1970 on (non native)   1
+ pre 1970 (non native)   1

Bunium bulbocastanum L. (Apiaceae)

Great pignut
Status in Britain: LOWER RISK - Near Threatened.
Status in Europe: Not threatened.

This species is a plant of calcareous soils, and is one of the few that, in Britain, are confined to the chalk (Ratcliffe 1977). It is mainly a plant of arable fields, especially where cultivation has ceased, but is also frequent in rough or disturbed downland and open scrub. Road verges, quarries, tracksides and hedgerows are other habitats. It is very rare in established downland, but even there, past disturbance is often suspected (Dony 1953). Associated species in these communities include *Arrhenatherum elatius, Bromopsis erecta, Centaurea nigra, Chaerophyllum temulentum, Crataegus monogyna, Dactylis glomerata, Festuca rubra, Galium mollugo, Heracleum sphondylium, Knautia arvensis* and *Medicago sativa.* On arable land it occurs most frequently with *Sinapis alba,* and on rough or broken downland with *Anthyllis vulneraria* (Dony 1953).

B. bulbocastanum is a glabrous perennial up to a metre tall, flowering in June and July at about the same time that the leaves wither. It grows from a globose tuber which is said to be edible (Tutin 1980). The plant requires patches of bare ground into which to seed and is often abundant in the early seral stages of grassland development. However, once established, it can persist for many years in more mature and closed swards. It is probably intolerant of regular heavy grazing, being favoured by sites where there is only intermittent disturbance.

First described in Britain in 1839, *B. bulbocastanum* is now known from more than 50 localities, most of them in Bedfordshire and Hertfordshire. There are a few outliers in neighbouring Cambridgeshire, but in Buckinghamshire it may now occur in just one place (Ivinghoe Hills). However, despite its apparent local decline, there has been no significant change in its population nationally since the early days of recording. There are some colonies of several thousand plants, but most populations are smaller, some consisting of only a few plants.

In general, it does not appear to be particularly threatened, though individual populations are known to be at risk from residential development and road building, as well as from agricultural intensification. Neglect is also a factor at some sites, leading to dense scrub development. However, suitable habitat is still sufficiently frequent throughout its range such that it is likely to persist, particularly where there is a degree of disturbance.

The range of *B. bulbocastanum* extends from England southwards to north-west Africa and the Balearic Islands, and eastwards to central Germany and Croatia (Tutin 1980). It is rare in several countries and legally protected in Germany.

D. Soden

Bupleurum baldense Turra (Apiaceae)

Small hare's-ear
Status in Britain: ENDANGERED. WCA Schedule 8.
Status in Europe: Not threatened.

The habitat of *B. baldense* in both of its remaining
British sites is short, rabbit-grazed maritime grassland with
some open soil, over calcareous bedrock (Devonian
limestone and chalk). At Brixham, one population occurs
on made ground, the other in cliff-top turf, together with
*Anthyllis vulneraria, Erodium cicutarium, Koeleria macrantha,
Sanguisorba minor, Scilla autumnalis* and *Thymus polytrichus.*
Near Beachy Head, it is almost entirely restricted to a zone
of very short, rather sparse vegetation, no more than 60 cm
back from the bare chalk of the overhung cliff edge, with
almost the whole colony now confined to a 25 m strip of
ground. Characteristic associates there include *Asperula
cynanchica, Bromus hordeaceus, Carlina vulgaris, Diplotaxis
muralis, Echium vulgare, Euphorbia exigua, Festuca ovina* and
Koeleria macrantha.

B. baldense is a slender glabrous annual, seldom taller
than 3 cm in Britain and sometimes with only a single
inflorescence. In continental Europe, however, the plant
can attain a height of 25 cm. It normally flowers in June
and July, but in years with warm and wet springs it can
flower in May and the whole plant vanish by June, as
occurred in Sussex in 1992. However, seed is freely
produced, and there seems to be a good seed-bank.

The population in Devon fluctuates considerably from
year to year, but there may be up to 2,000 plants in a good
season. In Sussex also, populations are variable and
dependent on weather conditions during the growing
season. Numbers range from fewer than 10 to a few
hundred individuals. The plant used to be found at other
sites in both counties close to the existing ones. It has been
reduced in Devon by quarrying and urban development,
and in Sussex perhaps by a change of land use or by
extensive cliff falls.

The Sussex site might appear to be severely threatened,
as the plants grow where the turf breaks to bare chalk on
the crumbling, overhung cliff-edge, and since the cliffs are
retreating by about 25 cm a year, there have been several
predictions that cliff erosion would carry all the plants
away. However, scattered plants occur in the closed turf up
to 3 m from the edge, and it is likely that seed blows back
every year, supplementing the seed-bank. In Devon the
most significant threat may be from excessive trampling by
visitors, the vulnerable habitat being subject to heavy use
during the summer. It is possible that *B. baldense* still
occurs in other places near the known sites. However, the
depauperate plants are difficult to see from a standing
position, unless the eye is practised. Furthermore,
Euphorbia exigua is a frequent associate, and tiny plants of
both can look remarkably similar.

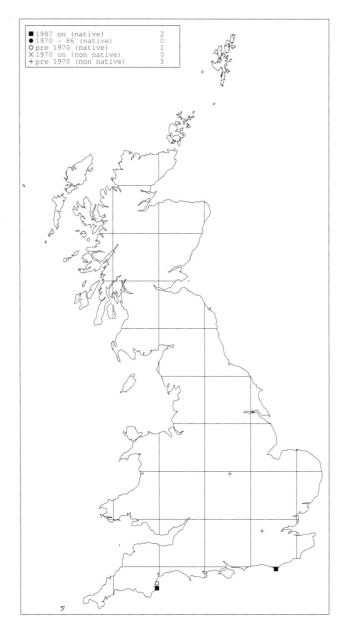

■ 1987 on (native)	2
● 1970 - 86 (native)	0
○ pre 1970 (native)	1
✕ 1970 on (non native)	0
+ pre 1970 (non native)	3

In Europe, *B. baldense* is a plant of dry stony places,
open grassland and grey dunes, having its stronghold in
Italy (from where, Monte Baldo, the species is named),
central and eastern Spain, southern France and Corsica. It
extends eastward to Romania and northwards to the south
coast of Britain. It is persistent in the Channel Islands, and
is locally abundant on the west coast of France north to the
River Seine.

M. Briggs and P.A. Harmes, based on an account by R.
FitzGerald.

Bupleurum falcatum L. (Apiaceae)
Sickle-leaved hare's-ear
Status in Britain: CRITICALLY ENDANGERED.
Status in Europe: Not threatened.

In 1831, when first found at Norton Heath, Essex (Sowerby 1934), *B. falcatum* was well-established and in great abundance in hedges between fields to a considerable distance on both sides of the road. It remained abundant until at least 1883 (Crompton 1974-86), but road improvements and changes in land use destroyed most of the population by 1943, when only about 50 plants remained. Despite low numbers, the population survived for the next twenty years. In 1962, during hedge cutting and ditch clearance, a fire was lit on the site (Jermyn 1974), which seems to have quickly led to the plant's extinction there. In 1979, a small colony was discovered near to the last. However, bearing in mind the short-term viability of the seed, it seems unlikely that seed or plants had survived since 1962 without being reported. Therefore, these new plants almost certainly arose from seed deliberately sown in 1978. Ironically, shortly afterwards, scrub was burnt on the very spot and, as if to make certain of its demise, in 1980 the verge was sprayed with herbicide all the way from Epping to Ongar.

The segment of road along which the plant formerly grew was finally by-passed in a road-straightening operation in 1985/6. A site between the old and new roads was landscaped out of the chalky boulder clay, and this was sown in 1988 (Birkinshaw 1990e) with large quantities of *B. falcatum* seed from plants in the author's garden and from the Cambridge Botanic Garden (derived from pre-1962 Norton Heath plants). Although still extant, the colony is small and vulnerable. There were 32 plants in 1991, but just 25 in 1994. In July 1995, 51 plants were present, but browsing by rabbits had reduced the number to 24 by early August. The latter plants survived because they were protected from grazing by the abundant *Cirsium arvense*, *C. vulgare* and *Picris echioides* that surrounded them. Numerous small plants and seedlings occurred in uncountable clusters around the larger plants. By July 1996, however, the population had been reduced to three flowering plants and a small number of seedlings. Unless rabbit-proof fencing is installed around the boulder clay mounds, it is unlikely that *B. falcatum* will persist at Norton Heath. It has been reported in the British Isles, as a casual only, from a very few other places.

B. falcatum is a biennial or short-lived perennial with umbels of conspicuous bright-yellow flowers. It is quite widely grown in botanic and private gardens throughout the British Isles, where it will grow luxuriantly unless grazed by rabbits. Only some plants have the asymmetric leaves that give the plant its name. That feature seems likely to be caused by unequal growth of the two half-laminas, either side of the midrib, owing to damage inflicted at the bud stage by frog-hoppers, which often infest the plants. Seed is copiously produced, but is viable for only about a year. Seedlings are produced in large numbers around parent plants but are unable to complete in a closed sward, being favoured by dry, bare, calcareous mineral-soils with minimal humus.

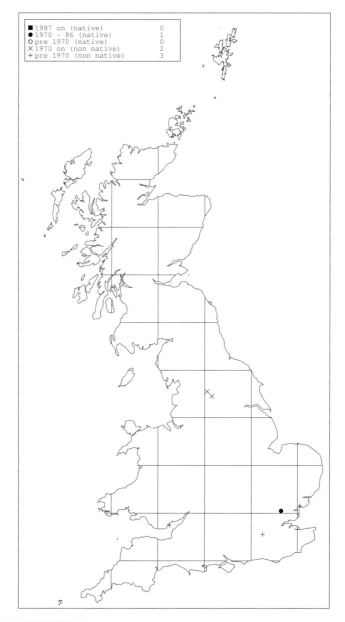

■ 1987 on (native)	0
● 1970 – 86 (native)	1
○ pre 1970 (native)	0
✕ 1970 on (non native)	2
+ pre 1970 (non native)	3

Although recently found in Britain as a sub-fossil in deposits from the Flandrian interglacial (Field 1994), it is extremely unlikely that the Norton Heath population was a relict from a more widespread distribution in the Flandrian. Such a showy plant is unlikely to have appeared at Norton Heath much before 1831 without someone reporting it. Its persistence there must also have gone hand-in-hand with rigorous control of the local rabbit population, a practice only relaxed in the area following the myxomatosis epidemic.

On the continent *B. falcatum* frequents dry sunny locations on slopes and ridges, and the scrubby southern fringes of woods on calcareous clayey and sandy soils. It is widely distributed in southern, central and eastern Europe, extending northwards to Belgium, Germany, Poland and Central Russia. Although plentiful in north-west France, its distribution elsewhere in northern Europe is patchy.

K. J. Adams

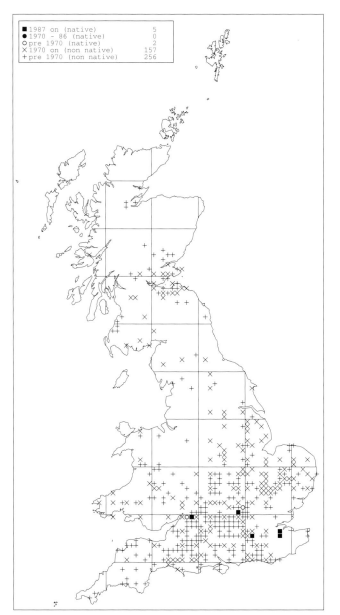

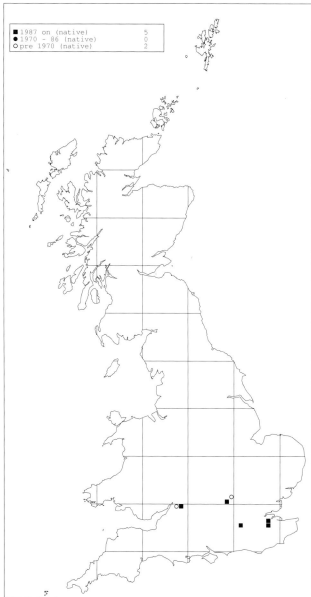

Buxus sempervirens L. (Buxaceae)

Box, Pren Bocs
Status in Britain: LOWER RISK - Near Threatened.
Status in Europe: Not threatened.

B. sempervirens occurs as an understory tree or shrub in woodland (usually beech), or as a major component in the canopy, occurring on chalk in southern England and on oolitic limestone in Gloucestershire. It is generally most dominant on steep slopes, presumably because the larger forest trees, such as beech, cannot maintain themselves on the steepest slopes and the shallowest soil (Pigott 1987). In Surrey, *Taxus baccata* is a frequent associate, and because of the dense canopy of the woodland, the ground is largely bare except for abundant box seedlings. At Ellesborough, it also occurs as scrub on chalk downland, dominant in places or mixed with *Cornus sanguinea, Crataegus monogyna, Euonymus europaeus, Ligustrum vulgare* and *Sambucus nigra*.

This species is a slow-growing evergreen attaining over 13 m where unrestricted. It flowers in April, producing copious seed, and in many sites seedlings are abundant. In the larger Gloucestershire site, seedlings have been noted as patchily present in the wood at densities of up to 100 per square metre. Old fallen trees may fall and sprout along their length in what appears to be a natural form of layering (Taylor 1990a).

Native sites are very localised in Britain, and there are now perhaps only about ten. The three largest are at Box Hill (and nearby) in Surrey, at Ellesborough in Buckinghamshire and at Boxwell in Gloucestershire. At Boxley in Kent, where it was once abundant, only about five trees remain. The status of *B. sempervirens* in its current Sussex sites is unclear, and local opinion tends to the view that it is not native at Arundel Park because of the centuries of management there and the likelihood of a deliberate introduction. Nonetheless, sites and habitats are similar to those of undisputed native sites. Certainly box was at one time native in Sussex, since at least two pre-historical records of box charcoal exist (Ross-Williamson 1930; Curwen & Ross-Williamson 1931). Although native sites are few, it has been widely planted in woods and often becomes naturalised. Garden escapes probably account for other occurrences.

Its ability to colonise steep and unstable chalk slopes that are not easily exploited by other trees has favoured its survival. However, while it is locally abundant and its best sites have statutory protection, it may still be vulnerable. At Box Hill, for instance, increased human activity in recent years (climbing up and down steep slopes) has been detrimental to this site, arguably the most important for box in Britain. There, seedlings and saplings of both box and yew were abundant up to the 1970s (Pigott 1987), and when gaps occurred in the canopy they grew up to fill them. But surveys in 1987 revealed that there was extensive exposure of box roots, leading to the selective loss of trees (since they are shallower-rooted than yew) and the loss of all box seedlings from the main walking areas. Box leaves may persist for several years on the tree and provide opportunity for epiphyllous species. They support the only epiphyllous lichen known in Britain, *Fellhanera (Catillaria) bouteillei*, and a liverwort, *Metzgeria fruticulosa*, has been recorded recently growing on box leaves (Porley 1996).

The status of box has been questioned since the earliest writings about trees, an introduction in Roman times sometimes having been assumed. However, the evidence for its native status has been reviewed by Pigott & Walters (1953) and Staples (1971), and it is now generally acknowledged to be native. Though never widespread in Britain, such was its importance that it has given rise to many historic place names, including Box Hill in Surrey, Boxwell in Gloucestershire and Buxted in Kent, and it was also referred to in Anglo-Saxon and Welsh charters.

In Europe, it is an important component of sub-alpine woodland in the Pyrenees, Alps and Apennines. Woodlands similar to those at Box Hill occur in France and are characteristic of the limestone regions of the Massif Central. It also occurs in the mountains of North Africa.

J. Stokes

Calamagrostis purpurea (Trin.) Trin. ssp. *phragmitoides* (Hartman) Tzvelev (Poaceae)

Scandinavian small-reed
Status in Britain: LOWER RISK - Near Threatened.
Status in Europe: Not threatened. Endemic subspecies.

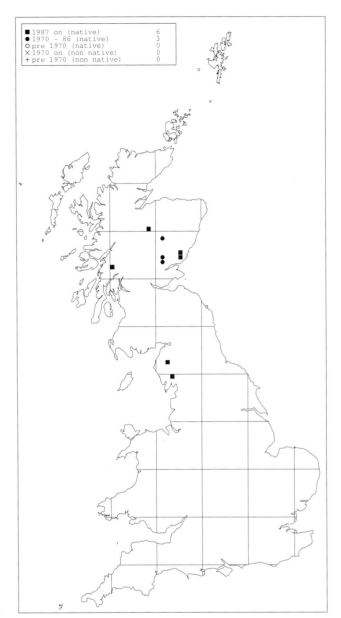

C. *purpurea* ssp. *phragmitoides* appears to be chiefly a plant of wet willow carr, often where there is standing water in winter, though seemingly never in the 'quaking' parts of mires. It also occurs in open marsh, wet ditches, in old peat diggings and more rarely on rather drier banks. In most sites it is locally abundant or even dominant in small areas, but elsewhere it may be in a more diverse community. A very wide range of associates include *Angelica sylvestris, Carex aquatilis, C. curta, C. rostrata, Epilobium palustre, Equisetum fluviatile, Filipendula ulmaria, Galium palustre, G. uliginosa, Juncus effusus, Pedicularis palustris, Ranunculus repens, Scutellaria galericulata* and *Valeriana officinalis.*

It is a rhizomatous perennial apomict, which has probably arisen through hybridisation of *C. epigejos* and *C. canescens* (Stace 1991). Little appears to be known about its reproductive biology, though most recruitment is likely to be by vegetative spread.

This taxon was not confirmed in Britain until 1980, though had been collected prior to that date. For example, R. Mackechnie and E.C.Wallace collected it from near Braemar in 1941, but, though they recognised it as different from other *Calamagrostis* species, they did not name it. Collections made by other botanists prior to 1980 had been named as either *C. epigejos* or *C. canescens*. It has now been confirmed at about eight sites, five of which are in eastern Scotland, and one in each of Argyll, Westmorland and Cumberland. At most sites it is vigorous, flourishing and present in good quantity. At its Braemar site, it forms extensive stands in willow carr. The most extensive stands are, however, in the Insh marshes near Kingussie, where it occurs in large swathes. Plants at this locality differ slightly in some respects from populations elsewhere, some morphological characters being akin to those of subspecies *langsdorfii* (O.M. Stewart pers. comm.). The taxonomic status of these plants is being investigated.

It seems secure in all its known sites, the only serious threat likely to be from drainage or direct habitat destruction.

This subspecies apparently has a restricted distribution in Europe, having been recorded only in Britain, Sweden, Norway and Finland. Other taxa treated as subspecies occur locally in Germany and Russia.

O.M. Stewart and M.J. Wigginton

Calamagrostis scotica (Druce) Druce (Poaceae)
Scottish small-reed, Cuilc-fheur Albannach
Status in Britain: VULNERABLE. ENDEMIC.

In 1863, Robert Dick found *C. scotica* in a single site in Caithness (Smiles 1878). It remains the only known site. Schoolbred and Druce re-found it during their tour of northern Scotland in 1902. The site is a drained loch that is moist in summer and covered to a depth of 50 cm by water during the winter months. Deep drainage channels cross the lowest part of the site and *C. scotica* is most abundant by the edges of these channels. The site is grazed by cattle, sheep and roe deer. The vegetation is chiefly *Juncus effusus - J. acutiflorus* rush pasture, the grass often growing in dense clumps of *J. effusus,* which probably limits grazing. The slender creeping rhizomes survive very readily in the dense root mat of the dominant rushes. In wetter, less grazed parts of the site it grows in fen dominated by *Filipendula ulmaria*, and there is some growing in willow carr interspersed with tall-herb swamp. *Agrostis canina* is common in the community.

C. scotica grows to 90 cm tall with leaves 2-5 mm wide and a flower panicle 7-16 mm long, consisting of erect closely-compact branches. It is very similar to *C. stricta,* which is less rare and grows in the same site and in a few other sites in Caithness. *C. scotica* is distinguished from *C. stricta* by its larger flower spikelets and more acuminate glumes. These differences are slight but distinctive, though intermediate forms occur. *C. scotica* was at one time considered identical to the Scandinavian *C.* x *strigosa*, but is currently considered to merit specific status (Stace 1991; Hubbard 1984).

The population size is difficult to assess because of flooding, grazing and probable hybridisation or introgression. In 1995 the north side of the loch was estimated to have 700 plants (Lusby & MacDonald 1995), while in 1993 the south side was estimated to have 300 plants (pers. obs.). Both counts are likely to be underestimates, and both probably include introgressed plants of *C. stricta*. Taking all aspects into account, the total population is probably in the range 500-1,500 plants.

The critical factors for this endemic species are probably the balance of wet and dry in the summer months, the level of grazing, and drainage of the site. Four farms have the grazing rights and there is no positive management of the water levels. There are no other known short-term threats to the population.

Reports of the occurrence of *C. scotica* in other locations in Caithness have not been substantiated and perhaps arise from confusion with *C. stricta*.

J.K. Butler

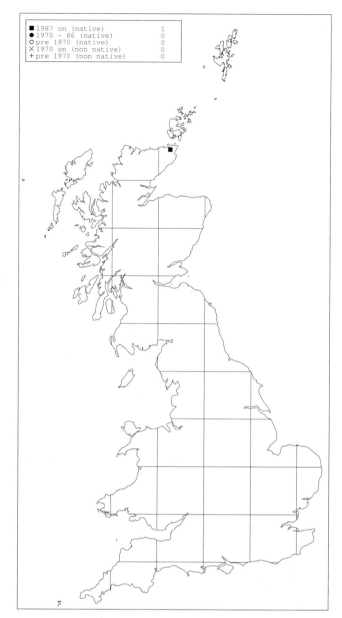

■ 1987 on (native)	1
● 1970 - 86 (native)	0
○ pre 1970 (native)	0
✕ 1970 on (non native)	0
+ pre 1970 (non native)	0

Calamagrostis stricta (Timm) Koeler (Poaceae)
Narrow small-reed, Cuilc-fheur Beag, Mawnwellt Cal
Status in Britain: LOWER RISK - Near Threatened.
Status in Europe: Not threatened.

C. stricta is a plant of near-neutral bogs and marshes, and the margins of lakes. At a site near Dalmellington it grows with *Deschampsia cespitosa, Filipendula ulmaria, Juncus effusus* and *Phalaris arundinacea*. In Yorkshire it is found as an emergent at the edge of the Leven Canal, and in Cheshire, in abundance around the margins of Oak Mere. It is primarily a lowland species, but ascends to 340 m at Kingside Loch, Selkirkshire.

C. stricta is a tufted perennial, with slender creeping rhizomes. It flowers from June to August, but little is known of its reproductive biology.

It is difficult to assess trends in the distribution of *C. stricta* as it has previously been confused with *C. scotica, C. purpurea* and hybrids with *C. canescens* (including *C. x gracilescens*), and perhaps with other species. It has certainly been lost from some sites through drainage, but may still be extant in some of the localities for which only a pre-1987 record is mapped.

British populations of *C. stricta* are variable, perhaps because of past hybridisation and introgression with other species (Stace 1975), and the presence of hybrids in some areas complicates the picture. Crackles (1994, 1995) describes in detail populations of *C. stricta* and *C. canescens* and their hybrids in south-east Yorkshire. Some plants in populations of *C. stricta* in Selkirkshire show some similarity to *C. scotica*, although they are well outside the southern limit of that species, and their identity remains uncertain.

C. stricta has a circumpolar distribution. It is widespread in the boreal zone of Europe, Asia and North America and occurs very locally in mountains in central Asia, Japan and South America.

Adapted from O.M. Stewart, in Stewart *et al.* (1994)

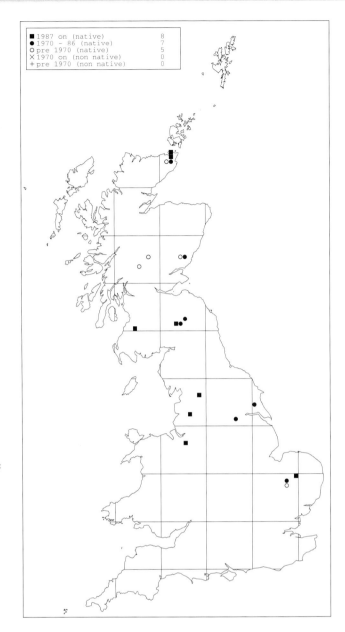

■ 1987 on (native)	8
● 1970 - 86 (native)	7
○ pre 1970 (native)	5
× 1970 on (non native)	0
+ pre 1970 (non native)	0

Carex atrofusca Schkuhr (Cyperaceae)
Scorched alpine sedge, Seisg Ailpeach Dhòthach
Status in Britain: LOWER RISK - Near Threatened.
Status in Europe: Not threatened.

C. atrofusca is a sedge of alpine calcareous, usually
micaceous, flushes on Scottish mountains, ranging from
800 to 1100 m in altitude, with one old record (1892) from
below 600 m. It typically occurs in open, stony flushes
where plant cover is patchy and is frequently protected by
a rock or tuft of *Nardus stricta* or *Festuca ovina*; its
associates include *Carex viridula* ssp. *oedocarpa*, *Eleocharis
quinqueflora*, *Juncus triglumis*, *Pinguicula vulgaris*, *Saxifraga
aizoides*, *Thalictrum alpinum*, *Tofieldia pusilla*, *Blindia acuta*
and *Campylium stellatum*. In particular sites it occurs with
such rare and local species as *Carex microglochin*, *Cerastium
cerastoides*, *Cochlearia micacea*, *Equisetum hyemale*, *Juncus
alpinoarticulatus*, *J. biglumis* and *J. castaneus* (McVean &
Ratcliffe 1962). It also grows in closed flushes with a range
of sedges including *Carex dioica* and *C. saxatilis*. Specially
favoured are sites where spring water first emerges on
hillsides. It is not associated with, and so ought not to be
confused with, *Carex atrata*, which is a plant of rock ledges,
though the two species can be quite similar
morphologically.

This shortly-rhizomatous perennial flowers in mid- to
late summer, producing ripe fruit between July and
September. It is readily recognised by its large rounded
purple- or red-black nodding female spikes.

It is known in five localities: two in the Ben Lawers
range, one each on Ben Heasgarnich and on Aonach Beag,
and one in the Bridge of Orchy Hills, Argyll. A record
from Rum is now discounted. There is no information on
the Argyll locality since the original find in 1976. In the
four other localities, colonies are scattered over large areas
of ground with populations numbered in hundreds (and at
two localities, thousands) of flowering plants. There are
very likely more colonies and even localities to be found.
For example, in 1994 a new colony with 300 plants was
found 500 m west of a previously known one. Like some
other montane plants, *C. atrofusca* has good years when it
is more likely to be found. One monitored colony, for
example, held five flowering plants in 1992, yet more than
900 in 1993.

Despite some collecting, some trampling and much
grazing, the species does not appear to be under any
immediate threat, and the total population is likely to have
changed little over many years.

Outside Scotland it also occurs in Scandinavia, the
Pyrenees and Alps, Greenland and arctic North America.

R.E. Thomas

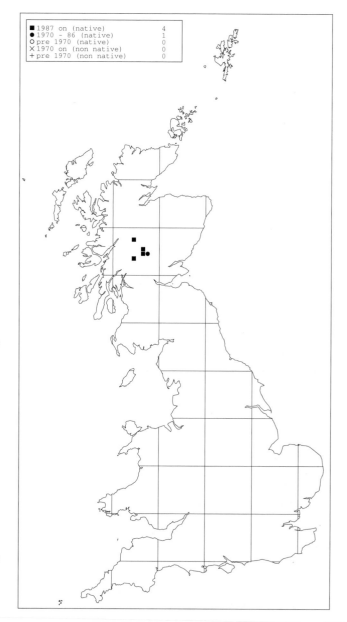

■ 1987 on (native)	4
● 1970 – 86 (native)	1
○ pre 1970 (native)	0
✕ 1970 on (non native)	0
+ pre 1970 (non native)	0

Carex buxbaumii Wahlenb. (Cyperaceae)
Club sedge, Seisg Chuailleach
Status in Britain: VULNERABLE.
Status in Europe: Not threatened.

C. buxbaumii is a sedge of mesotrophic fens that forms
extensive patches on the margins of lochs, inshore from *C.
lasiocarpa*, growing with *C. panicea* and *C. nigra*. Beyond the
loch edge, it also grows in fens in which *Molinia caerulea* is
the most frequent dominant, with other associates
including *Carex hostiana, Erica tetralix, Myrica gale, Potentilla
erecta* and, away from grazing, *Caltha palustris, Filipendula
ulmaria* and *Senecio aquaticus*. Some colonies border the
outflowing burns of lochs, and all are subject to periodic
flooding and may be under 60 cm of water after heavy
rain.

The plant is shortly rhizomatous, sending up stiffly
erect shoots and flowering stems to 80 cm tall. Plants can
form loose tussocks and in some places become the
dominant vegetation. Vegetative shoots are similar to those
of *Carex nigra* and, since the red fibrillose leaf sheaths are
noticeable only on close inspection, the plant is easily
overlooked when not in flower. Ripe fruit is usually
formed in July and August, but fruiting spikes have been
seen as late as the beginning of November.

In Britain, it was formerly known from just two sites,
one in East Inverness and one near Arisaig in West
Inverness, but four are now known following its discovery
in Argyll in 1986 and at a second site near Arisaig in 1989.
It was found on a small island in Loch Neagh, Northern
Ireland, in 1839, but became extinct there in about 1886
through drainage and grazing.

Populations appear stable and, though the number of
plants may vary little from year to year, monitoring has
shown that flowering can be very variable, in some years
with less than 10% of plants flowering. A lack of appreciation
of this variability may have given rise to erroneous views
about the true population sizes in known colonies. The largest
population is in East Inverness, the numbers of flowering
spikes there ranging from 4,000 in good years down to a few
hundreds in poor. Other sites usually have flowering
populations of between 500 and 1,500 plants. All colonies
appear to be thriving, though *C. buxbaumii* grows less well in
Myrica gale-dominated vegetation.

Most sites are lightly grazed by cattle and sheep, but
this does not seem adversely to affect the colonies of *C.
buxbaumii*. The main threat is likely to be through drainage,
which might also occur with any afforestation. Two of the
sites are within SSSIs, which afford a degree of protection.

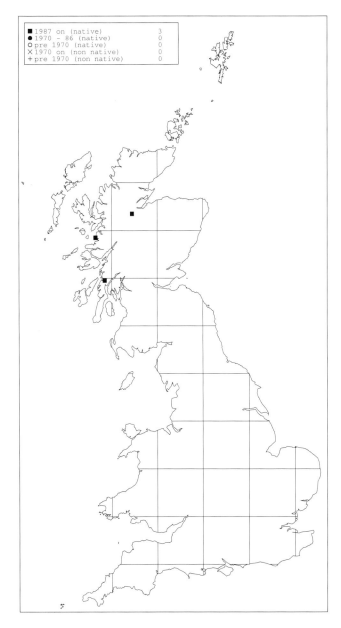

C. *buxbaumii* occurs widely in eastern and western
Europe, from western Norway southwards to the Alps,
and eastwards to the Urals. It is more disjunct further east
in Asia and in western Europe, and the plant ranges across
North America from Newfoundland to Alaska, mostly
south of the Arctic Circle. Though not threatened
throughout Europe, it is vulnerable or endangered in
several countries.

P.M. Batty

Carex chordorrhiza L. (Cyperaceae)

String sedge, Seisg Shreangach
Status in Britain: VULNERABLE.
Status in Europe: Not threatened.

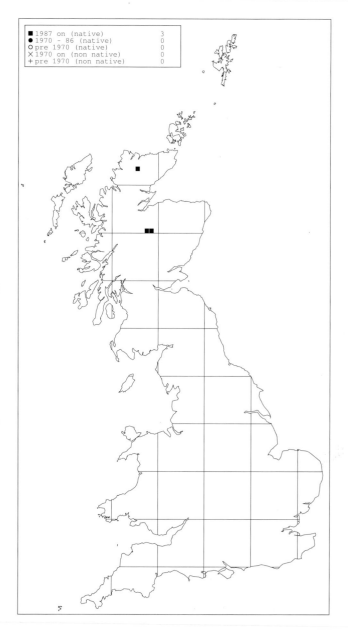

C. *chordorrhiza* is known from only two sites in Britain, both in the Highlands of Scotland. On the Insh Marshes in Strathspey, it occurs in fens dominated by *Carex rostrata*, *Equisetum fluviatile, Menyanthes trifoliata, Potentilla palustris* and, more rarely, in mires dominated by *C. rostrata* containing *Sphagnum squarrosum*. At Altnaharra, Sutherland, it is found in similar mire communities dominated by *Sphagnum recurvum*. These types of community are frequent in Scotland, and *C. chordorrhiza* may be a glacial relict (Matthews 1955). The pH of water samples from the two Scottish sites ranges from 5.4 to 6.5 and represents the upper range for this species in Europe. This species tolerates temporary submergence as, for instance, during winter flooding of the Insh Marshes (Legg *et al.* 1995). Many of the populations of *C. chordorrhiza* at Altnaharra may, however, avoid submergence by floating to the surface together with the substrate.

C. *chordorrhiza* is an extensively creeping perennial often producing very long, trailing stems or runners between short upright tufts of fine, bright green leaves. Plants flower during May, and the few seeds that are set are shed from July onwards. Regeneration is mainly clonal, with runners up to 100 cm long produced from early August. Fragments of plant dislodged by disturbance may be the main means of dispersal, and propagation by seed relatively unimportant. There is, however, considerable genetic variation in both populations (A. Hamilton pers. comm.), indicating that at least some reproduction by seed takes place.

The smaller population at Altnaharra was first discovered in 1897 and since then appears to have remained relatively unchanged in size. The species was discovered at Insh in 1978 (Page & Rieley 1985) and the large populations there were extensively mapped in 1989 (Wood 1989). A repeat sample survey during 1994 indicated that the known area covered by this species had increased five-fold (Legg *et al.* 1995), with evidence of both population expansion and under-recording. The presence of many old drains that are now derelict implies that the marsh may have been drier, but reverted to a wetter condition under current management, and this could explain the observed expansion of this sedge.

Populations in both of its localities are large and vigorous and are likely to remain so under current management. The main threat comes from proposals to drain the Insh marshes to control flooding in the Spey and, conversely, to flood the Altnaharra site to regulate water flow in the Naver. Experimental work has shown that *C. chordorrhiza* does not survive prolonged inundation; the few plants that survived grew very little and suffered high levels of shoot mortality (Legg *et al.* 1995). Clearly, a reduction in water table is also likely to result in

significant changes to the fen and mire communities in which this species is found.

C. *chordorrhiza* has a circumpolar distribution and is abundant in Fennoscandia, Iceland, north-west Russia, Poland and northern Germany, especially around the Baltic. Elsewhere in Europe it is uncommon, extending south to the Pyrenees and central Ukraine. It also occurs sporadically in arctic-boreal zones of northern Russia, and in North America from Greenland to Alaska southwards to Indiana (Hultén & Fries 1986). In Europe it is a constituent of small-sedge or transition mires (Page & Rieley 1985). It will colonise into *Sphagnum* nuclei developed in this vegetation, or denser vegetation of taller *Carex* species or sparse *Phragmites australis*.

N.R. Cowie

Carex depauperata Curtis ex With. (Cyperaceae)

Starved sedge, Hesgen Lom
Status in Britain: CRITICALLY ENDANGERED. WCA
Schedule 8.
Status in Europe: Not threatened. Endemic.

C. depauperata is a plant of dry, basic brown-earth soils in semi-shade. In particular, it is found in gaps and along tracks in deciduous woodland and among shrubs on rock outcrops. In both of its extant native sites in Britain it grows along the banks of lanes at the woodland edge. Associated species include *Allium ursinum, Arum maculatum, Brachypodium sylvaticum, Carex divulsa, Crataegus monogyna, Hedera helix, Ligustrum vulgare, Melica uniflora* and *Rubus fruticosus* (FitzGerald 1990a; Birkinshaw 1990c).

It is a tussock-forming perennial, flowering in April and May and carrying ripe seed from October to March. Unlike other British sedges, its seeds are few and very large. Inflorescences are borne at the end of long stems up to 1.2 m tall, which may be an adaptation to disperse seeds away from the parent. Seeds can remain dormant in the soil for long periods and are stimulated to germinate by changes brought about by the opening of the canopy when old trees die (Lousley 1976). Within newly created gaps, plants may flower and set seed. As the canopy closes over, plants slowly decline and eventually disappear, leaving dormant seed in the ground awaiting the formation of a new gap. The Godalming plants apparently resulted from the germination of seeds that had lain dormant for at least twenty years (Marren & Rich 1993).

Since plant recording began, this species has always been extremely rare in Britain, having been recorded from only twelve sites. These were located in Dorset, Kent, Surrey, Somerset and Anglesey and presumably represent relics of a once widespread species. Now, populations survive in only two locations: near Godalming, Surrey, and near Axbridge, Somerset. In 1995, near Godalming, there was one large plant which bore 72 flowering stems, and two smaller plants nearby (Rich & Fairbrother 1995). In 1997 four flowering plants occurred. Near Axbridge, there were 14 plants in 1995, but the population has been reinforced several times with cultivated native stock, and it may be that none of these plants is of natural origin. In addition, a 'back-up' population of 55 plants, originating from 100 plants translocated from Cambridge University Botanic Garden in 1988, grows close to the Axbridge population (Birkinshaw 1990c). All Axbridge plants are protected within an SSSI.

Populations of *C. depauperata* have declined or become extinct through woodland destruction, the cessation of traditional woodland management leading to the development of a closed canopy or a dense understorey of shrubs or bramble, and possibly the excessive collection of herbarium specimens by 19th century botanists. The habitats of the two extant populations are now being managed to maintain the favoured conditions of semi-

shade. It is possible that former populations that became extinct may be resurrected by the reinstitution of suitable management (Birkinshaw 1991). The plant's occurrence by tracks indicates that it may be tolerant of trampling, or perhaps survive because other less tolerant species are suppressed.

Outside Britain, the natural distribution of *C. depauperata* includes southern Europe, the Crimea, the Caucasus, south-west Germany, Belgium, and southern Ireland. It seems to be rare throughout its range, and in Germany may be extinct (Blab *et al.* 1984). In Belgium and Ireland only single populations are known (Delvosalle *et al.* 1969; O'Mahony 1976).

C.R. Birkinshaw

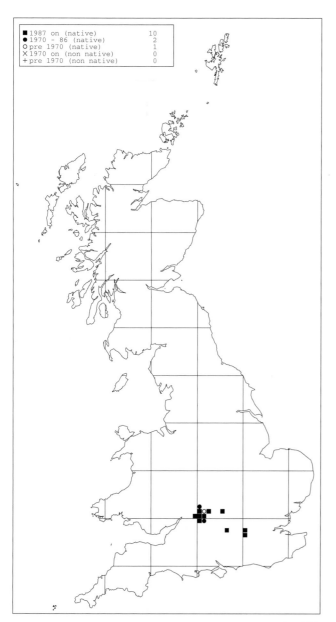

Carex filiformis L. (Cyperaceae)
Downy-fruited sedge, Hesgen Feindwf
Status in Britain: LOWER RISK - Near Threatened.
Status in Europe: Not threatened.

C. filiformis occurs in a wide range of habitats. Damp meadows appear to be the most characteristic, especially on gravel where the subsoil is continuously irrigated (David 1993), but it also occurs in woodland rides, on roadside verges and in dry grassland. (Riddelsdell *et al.* 1948; David 1983; Holland *et al.* 1986; Taylor 1990d). The common feature of its sites is the high availability of calcium, which in most is provided by the underlying bed-rock of oolitic limestone. At others, such as Otmoor, Oxfordshire, where the substrate is clay, the habitat is inundated with calcium-rich water for part of the year. Associates are as varied as its habitats. In damp meadows, accompanying species include *Carex flacca, C. nigra, C. panicea, Filipendula ulmaria, Molinia caerulea, Potentilla erecta, Ranunculus acris, Sanguisorba officinalis* and, in drier areas,

Bromopsis erecta, Centaurea nigra, Galium verum, Silaum silaus and *Trifolium pratense*.

It is a creeping perennial, producing flowering shoots in mid- to late May. Fruit ripens in June and July, and by late July the utricles have fallen, with subsequent disintegration of the flowering stem (David 1993). It is severely affected by late frosts, in which case flowering may be nil or very poor. The plant is undoubtedly difficult to find in the vegetative form and, moreover, it has a seeming propensity to disappear from a site, only to reappear in later years. This suggests that the plant may have a persistent seed-bank, or perhaps that it fails to produce many shoots in some years and thus remains undetected.

The species was first recorded in Britain by Robert Teesdale in 1799 at Marston Meysey, Wiltshire. This site is now lost, although two other Wiltshire populations remain. East Gloucestershire is its stronghold in Britain, with populations at seven sites. Oxfordshire holds populations at three sites and Surrey at two. Those formerly in Middlesex were destroyed by gravel extraction in the 1960s (Kent 1976). Other colonies were lost in the 1960s and early 1970s through drainage, fertilisers, conversion to arable land, the use of herbicides on road verges, and general development (e.g. David 1993; Taylor 1990d). However the current picture is a little more encouraging. Of the thirteen sites still holding *C. filiformis*, six have some statutory protection and eight sites are managed by local Wildlife Trusts or are under conservation agreements.

Numbers at the current sites vary from merely a few shoots to several thousand flowering spikes each year. With suitable management the number of flowering spikes can increase markedly at a site within a few years, as occurred at Westwell Gorse, Oxfordshire, where a five-fold increase occurred from 1989 to 1993. Ideally the sward should be left undisturbed from about late March to late July, followed by light mowing or grazing during autumn and winter to remove rank growth. Heavy grazing or mowing, especially at inappropriate times of year, may eliminate the plant, and since it is intolerant of shade in woodland situations, glades need to be kept open (Taylor 1990d).

C. filiformis is a mainly European species but ranges eastwards to western Siberia. In the south, it ranges from eastern Spain to central Italy and Greece and in the north to southern Sweden and the Baltic. The British populations are at the north-western limit of its range.

P.A. Ashton and D.A. Callaghan

Carex flava L. (Cyperaceae)
Large yellow-sedge
Status in Britain: VULNERABLE.
Status in Europe: Not threatened.

Typically in Europe, *C. flava* is a plant of base-rich fens, but in Britain, populations occupy a transition zone between the base of a shallow slope of limestone supporting ash woodland, and the edge of a lowland raised mire supporting *Betula pubescens-Molinia caerulea* woodland. It grows on peaty mineral soils kept moist by seepages and flushes from the neighbouring limestone outcrops, and does best where there are gaps in the canopy and in light shade. The canopy is of alder, ash and birch, with shrubs including *Frangula alnus* and *Prunus spinosa*. Associated species in the ground flora reflect the transitional nature of the habitat: *Ajuga reptans*, *Brachypodium sylvaticum*, *Carex remota*, *Festuca gigantea*, *Filipendula ulmaria*, *Lysimachia nemorum*, *Mentha aquatica*, *Mercurialis perennis*, *Oxalis acetosella*, *Ranunculus repens*, *Valeriana officinalis*, *Calliergonella cuspidata* and *Plagiomnium undulatum*.

It is a tufted perennial with a compact inflorescence, usually carrying its ripe fruits in July, when it is most readily distinguished from its near relatives. Seed usually germinates in early summer (mean day temperatures of about 15C are required). Populations of *C. flava* often have 'persistent seedlings', which may grow very slowly for a number of years but rapidly mature when released from near-neighbour competition. Individual plants can be long-lived under field conditions - at least six years (Schmid 1986) and up to twenty years (Davies 1956).

In Britain, *C. flava* is known at a single locality: Roudsea Wood, Cumbria. Numbers at Roudsea have increased over the past twenty years, mainly due to selective felling and canopy thinning, but there have been local declines over the same time span where shading has increased. Between 1967 and 1988, the numbers ranged between 1,300 and 2,300 plants, and there is no evidence that the population has changed significantly since 1988. If a light canopy is maintained, then populations are likely to remain relatively stable. Recent observations at Roudsea Wood reveal that the hybrid *C. x alsatica* (*C. flava* x *C. viridula* var. *oedocarpa*) is rare, comprising less than 1% of the *C. flava* population. It is also sterile, so that hybridisation does not pose a significant threat at the present time. *C. x alsatica* also occurs at Malham Tarn, in the absence of *C. flava*. There is an old record of *C. flava* from north-west Cumbria, and of presumed hybrids between *C. flava* and *C. viridula* ssp. *brachyrrhyncha* from Hampshire, Yorkshire and Ireland.

C. flava occurs throughout Europe, though it is generally rare and, in the south, confined to montane regions. In Finland, it occurs in a range of habitats including rich fens, spring feeds, marshy forests, and ditches in agricultural land (Pykälä & Toivenen 1994). It is found in eastern and western North America at roughly the same latitudes as in Europe and there are isolated

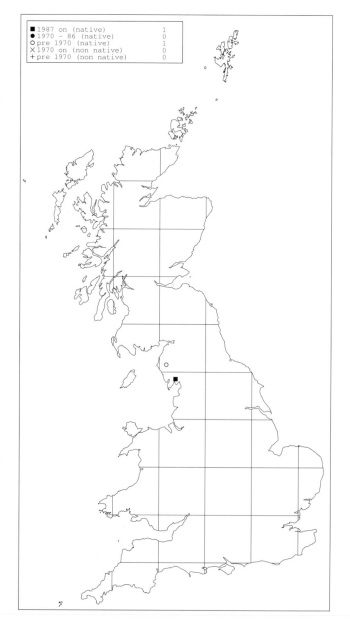

■ 1987 on (native)	1
● 1970 - 86 (native)	0
○ pre 1970 (native)	1
✕ 1970 on (non native)	0
+ pre 1970 (non native)	0

relict populations in the Caucasus, Iran, Siberia and the western Himalayas. In studies of *C. flava* populations throughout Europe, Schmid (1982) and Bruederle & Jensen (1991) found very little evidence of *C. x alsatica*, and hybrid plants that were found were highly sterile.

C. flava forms part of a species-complex that has long challenged the minds of taxonomists, but it is only in the last decade that a clearer picture of this group of sedges has emerged. As many as nine species were formerly recognised in the complex, but Schmid's work in Europe (e.g. Schmid 1983) and that in North America (Crins & Ball 1989) strongly supports the acceptance of just two - the relatively invariable *Carex flava* L. and the highly differentiated *C. viridula* Michaux.

I. Taylor

Carex lachenalii Schkuhr (Cyperaceae)

Hare's-foot sedge, Seisg Cas Maighiche
Status in Britain: LOWER RISK - Near Threatened.
Status in Europe: Not threatened.

C. lachenalii occurs in montane oligotrophic flushes, on rock ledges, and in *Salix herbacea* snow-beds in the Scottish Highlands between 750 and 1,140 m, growing principally in places where snow lies late (McVean & Ratcliffe 1962). On ledges, it grows in herb-rich ungrazed vegetation on base-poor rock with such widespread species as *Alchemilla alpina, Carex bigelowii, Cochlearia officinalis, Gnaphalium supinum, Juncus trifidus, Nardus stricta, Ranunculus acris, Sibbaldia procumbens* and *Racomitrium lanuginosum,* and locally with more distinguished species such as *Arabis petraea, Cerastium arcticum* and *Poa alpina*. It also occurs at the edge of *Caltha*-dominated flushes and in *Carex aquatilis-C. rariflora* mire.

It is a shortly-rhizomatous perennial tufted sedge that flowers in mid- to late summer, with fruit ripening between July and September.

C. lachenalii has been recorded on about 11 different hills since 1970, some of which support more than one population. The main cluster of sites is in the Cairngorms and eastern Highlands, and the plant is almost certainly extant in hectads where there is only a post-1970 record; unless in fruit, *C. lachenalii* is very difficult to identify. There are outliers on Aonach Beag, West Inverness, and in Glen Coe, though *C. lachenalii* has not been seen in the latter since 1954. A few colonies are large, producing several hundred flowering shoots (the largest with more than 1,000), but most are small. *C. lachenalii* can be confused with its hybrid with *C. curta (C. x helvola)*, which occurs with its parents in some abundance on Lochnagar and Cairntoul, and perhaps also on the Clova mountains (Jermy *et al.* 1982).

Skiing development is a potential threat to at least one colony in the Cairngorms, and erosion caused by increasing numbers of visitors is also of concern.

C. lachenalii has a circumpolar distribution, occurring on mountains of the northern hemisphere throughout the Arctic and sub-Arctic regions. It occurs widely in montane Europe from Iceland and Scandinavia, with southern outposts in Britain, the Alps and Pyrenees. There is a highly disjunct locality in the Southern Alps of New Zealand (Hultén & Fries 1986).

M. J. Wigginton and G. P. Rothero

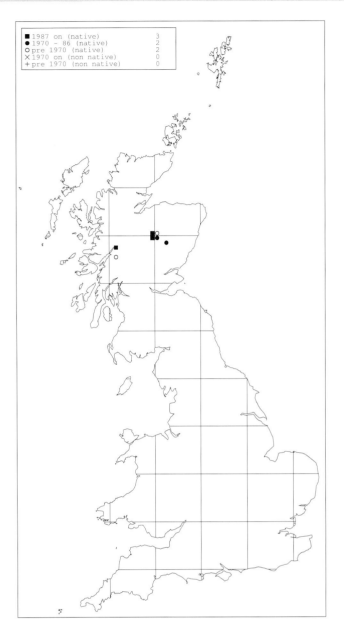

■ 1987 on (native)	3
● 1970 - 86 (native)	2
○ pre 1970 (native)	2
✕ 1970 on (non native)	0
+ pre 1970 (non native)	0

Carex microglochin Wahlenb. (Cyperaceae)
Bristle sedge, Seisg Chalgach
Status in Britain: VULNERABLE.
Status in Europe: Not threatened.

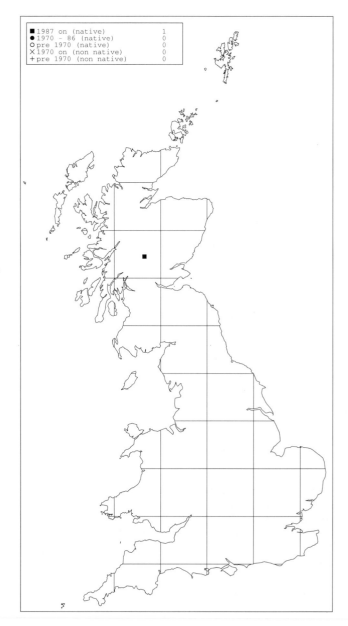

■ 1987 on (native)   1
● 1970 - 86 (native)   0
○ pre 1970 (native)   0
✕ 1970 on (non native)   0
+ pre 1970 (non native)   0

Before *C. microglochin* was first found in Britain, the possibility that it grew here had long haunted the botanist G.C.Druce. He searched for it in vain for many years in the Highlands of Scotland and also as far north as Shetland. It was eventually found in 1923, not by Druce, but by two ladies who had become separated from the main body of an expedition paying its respects to *Carex atrofusca* (Payne & Geddes 1980). It is a plant of moderately high-level base-enriched flushes ranging from 580 to 975 m altitude, but usually between 700 and 900 m. The flushes are mostly gently sloping, sometimes with steeper 'steps', and with a small to moderate water flow over the weakly acidic to weakly alkaline micaceous silt or gravelly substrate (Raven & Walters 1956; Evans 1982; Boddington 1995). The sites are rich in calcium, and the plant can tolerate high levels of magnesium, though is not dependent on it (Boddington 1995). The cover of vascular plants varies from less than 50% to nearly 100%, and bryophytes are usually abundant. Associates usually include *Carex dioica, C. panicea, C. viridula* ssp. *brachyrrhyncha, Eleocharis quinqueflora, Juncus alpinus, Pinguicula vulgaris, Saxifraga aizoides, Blindia acuta, Calliergon trifarium* and *Scorpidium scorpioides*. The flushes are mostly surrounded by *Nardus stricta*-dominated grassland. It has also been found in small quantities on steep ground beside burns downstream of large colonies, and in marshes with much *Equisetum palustre*.

This species is a perennial with a short, creeping, slender rhizome and, normally, single shoots (Jermy *et al.* 1982). Flowering is usually in early July but may often be delayed by late snow-lie until late July or early August. Utricles mature and become reflexed by late July to mid-August and then are gradually shed. Large fluctuations occur in the numbers of stems grown from year to year.

C. microglochin is recorded only from the Ben Lawers range, Perthshire, the whole population encompassed within a single hectad. It is distributed over a sizeable complex of flush systems on the slopes of two mountains, mostly in three corries and on an adjacent col, but with smaller, scattered, outlying colonies. It is very abundant in some flush systems. A reliable estimate of population size has not been possible, since only flowering or fruiting stems are conspicuous and distinctive enough to be counted, and these may be grazed. However, regular monitoring between the 1980s and 1996 has shown little change in its distribution.

The whole known range is subject to intensive sheep-grazing and a much lower intensity of deer-grazing. *C. microglochin* is grazed by sheep, and colonies are damaged by poaching of the soft substrate. There is some indication of a population increase where sheep-grazing is prevented (D. Mardon pers. obs.).

Despite the whole population being geographically remote and lying within an NNR, *C. microglochin* is threatened in a number of ways. The most serious stems from proposed major changes in grazing regimes, including the erection of permanent stock fencing, which is considered likely to have an adverse effect on both the plant and its habitat. Walkers are a threat locally, and all-terrain vehicles and mountain bikes (increasingly used) have also caused habitat damage in the same area.

C. microglochin has an arctic-alpine circumpolar distribution, including Iceland, Scandinavia, the mountains of central and eastern Europe to the Caucasus, the mountains of Asia and Siberia, in North America from Alaska and British Columbia to Labrador and Newfoundland, and in Greenland (Benum 1958). In Scandinavia it also grows within submontane woodland.

D.K. Mardon

Carex muricata L. ssp. *muricata* (Cyperaceae)
Prickly sedge, Hesgen Bigog Gynnar
Status in Britain: CRITICALLY ENDANGERED.
Status in Europe: Not threatened.

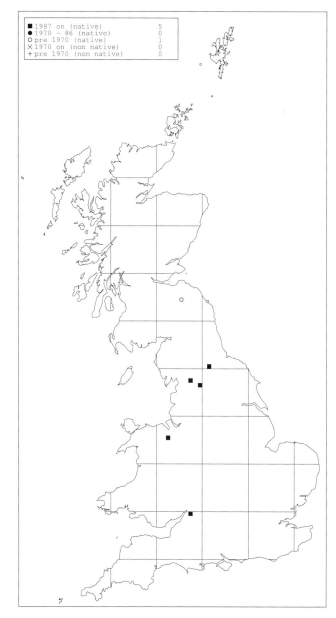

```
■ 1987 on   (native)          5
● 1970 - 86 (native)          0
O pre 1970  (native)          1
✗ 1970 on   (non native)      0
+ pre 1970  (non native)      0
```

C. muricata ssp. *muricata* is certainly known from five sites in Britain and has been recorded from four of these since 1990. It grows on rendzinas (pH 7.6-7.1) on dry slopes, screes and ledges on limestone crags, and on limestone pavement, often in places shaded by shrubs and trees. Except for *C. depauperata*, it appears to be the rarest of the British sedges, with only about 44 plants remaining in 1992, and perhaps fewer now.

The population near Wrexham, Denbighshire, was the largest, though there is evidence of a recent decline, and it may no longer be so. Two small colonies occurred about 25 m apart at a single site, growing on loose earth and limestone scree under a canopy of *Corylus avellana*. In 1987, one colony held about 12 plants (Morgan 1987a), but in 1992 there were only four. The plants grew in an almost complete ground cover of *Hedera helix*, with abundant *Rubus idaeus* above. The other colony held about 30 plants in 1987 and 23 in 1992. Here the most prominent herbs in the varied ground flora are *Carex flacca* and *Sanicula europaea*, together with occasional *Arrhenatherum elatius*, *Geranium robertianum*, *Hedera helix*, *Oxalis acetosella*, *Poa pratensis* and *Tamus communis*. Near Richmond, Yorkshire, about 15 plants grow on an open grassy limestone scree slope, where recorded associates include *Anthoxanthum odoratum*, *Cruciata laevipes*, *Galium sterneri*, *Geranium robertianum*, *Leontodon saxatilis*, *Teucrium scorodonia*, *Thymus polytrichus* and *Viola riviniana*. Near Ribblehead (David 1979), it occurs on wooded limestone pavement at an altitude of 340 m. Plants grow in dappled shade in a mossy carpet under stunted *Fraxinus excelsior*, one of them on the top of a rock in the centre of a grike. Four plants were found in 1977, but two are currently known (P.Corkhill pers. comm.). In Gordale, four plants (one flowering) were seen in 1985, but excessive trampling and erosion have destroyed the site. However, another population was discovered nearby on a steep grassy ledge, where an estimated 20 plants with 30-40 spikes were seen in 1997 (P. Jepson pers. comm.). Near Woodchester, West Gloucestershire, a sizeable colony of about 21 plants appeared after woodland clearance (David & Kelcey 1975), but as the site became overgrown it progressively declined to only four plants in 1982 and to extinction in 1983. There is a herbarium specimen from Lauder Castle, Berwickshire, collected in 1878.

C. muricata ssp. *muricata* is a tufted, short-rhizomatous perennial, producing ripe fruit between June and August. It is reported to be a month earlier than ssp. *lamprocarpa* in all stages of flowering and fruiting (David & Kelcey 1975). Evidence suggests that seed can remain viable in the soil for fairly long periods.

This species is intolerant both of excessive grazing and, conversely, of shading by scrub or overgrowth of ground cover where grazing is excluded. The population in Swaledale has been enclosed against rabbit grazing, which had prevented fruiting in previous years, but this has encouraged the growth of scrub, and the population of *C. muricata* ssp. *muricata* is still threatened there (D. Pearman pers. comm.). At the Ribblehead site, one plant is caged against rabbits, but the other is inaccessible to rabbit grazing. One of the Denbighshire colonies is threatened by rampant *Hedera helix* (95% cover in 1992), and the other by the closing of the canopy of *Corylus avellana*. The extant Gordale population is threatened by erosion caused by rock-climbers seeking access to the cliffs above. Plants derived from most of the British populations are in cultivation and seed held in seed-banks, and there have been attempts to bolster existing wild populations from this stock, which germinates easily.

The rarity of this sedge is puzzling in view of the large number of apparently suitable sites, particularly in the limestone districts of the Pennines, and in the Welsh Borders. Whilst it is possible that other sites remain to be found (as envisaged, for instance, by R.W. David), it is clear that the plant cannot be other than rare in Britain. On present evidence, it is critically endangered, and its survival in the wild seems precarious. Conservation management should be an urgent priority.

C. muricata ssp. *muricata* occurs throughout most of Europe, but is not found on the Mediterranean islands, or in Iceland, on the Faeroes or Svalbard. Its stronghold is in Scandinavia and eastern Europe. It is reported to have been introduced to New Brunswick and Pennsylvania (David & Kelcey 1975).

F.J. Roberts

Carex norvegica Retzius (Cyperaceae)
Close-headed alpine sedge, Seisg Lochlannach
Status in Britain: VULNERABLE.
Status in Europe: Not threatened.

C. norvegica is a sedge of wet rocky ledges, rocky slopes and grassy turf on, usually, the north or east faces of a very few mountains in Scotland, where snow lies late. It is confined to basic rock, often Dalradian mica-schist, at altitudes from 750 m in Angus to 960 m in mid-Perth. Its associated species are those typical of mica-schist; thus *Alchemilla alpina, Festuca rubra, Persicaria vivipara, Saxifraga oppositifolia, Selaginella selaginoides* and *Thalictrum alpinum* are constant associates, and *Carex capillaris* is usually present. In Angus it grows around rocks below montane willow scrub, and on rock ledges near *Alopecurus borealis* and *Juncus castaneus*. By contrast, in one of its mid-Perth sites it grows as a diminutive plant in a rich turf composed mainly of *Armeria maritima, Minuartia sedoides, Saxifraga oppositifolia* and *Silene acaulis*, and close to many other species including such local rarities as *Carex atrofusca, C. saxatilis, Galium sterneri, Juncus biglumis, Poa alpina* and *Saxifraga nivalis*.

C. norvegica is a shortly-rhizomatous perennial, flowering in June and July, and with ripe fruit in July and August.

It was first discovered in Angus in 1830 and is reliably known, for many years now, from two localities in mid-Perth, two in South Aberdeenshire and two neighbouring ones in Angus. Old records from Rum and the Uists are considered erroneous, though on northern Scandinavian coasts it does occur on blown shell-sand. Despite much hill botanising, there have been no recent confirmed additions. Some records in the last few years from the granite crags and summit plateaux of Lochnagar are undoubtedly misidentifications (perhaps chewed *Carex bigelowii*). A 1987 record from a mica-schist site near Ben Lui by an untraced recorder requires confirmation. A striking feature of the current populations is their restricted ground area; excluding the more spread-out Angus sites, areas are estimated at 25 and 200 square metres (South Aberdeenshire), and 20 and 120 square metres (mid-Perth). Monitoring of some populations has shown that numbers of flowering stems vary considerably from year to year, with typical maxima of 400 in Angus, 60 and 30 in south Aberdeenshire, and 150 in mid-Perth. Many plants are tiny. Whilst Angus specimens usually attain 10-20 cm in height, mid-Perth ones are usually under 5 cm, and a considerable proportion are merely 2-3 cm high. Notwithstanding, utricles always seem well developed.

The few sites, small area and stature and low flower numbers might suggest a very threatened species.

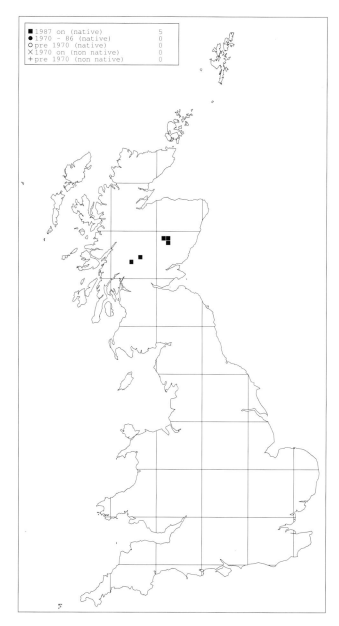

■ 1987 on (native)	5
● 1970 - 86 (native)	0
○ pre 1970 (native)	0
✕ 1970 on (non native)	0
+ pre 1970 (non native)	0

However, there is no documented evidence of any overall decline. Current concerns centre more on the increasingly severe grazing by deer and sheep, and the disturbance of unstable or easily damaged habitats by visiting botanists.

Outside Scotland this arctic-alpine species occurs throughout the Scandinavian mountains and Iceland, in the central and eastern Alps, in arctic Siberia and North America, on the south coast of Greenland and ascending to 2,500 m.

R.E. Thomas and J. Wright

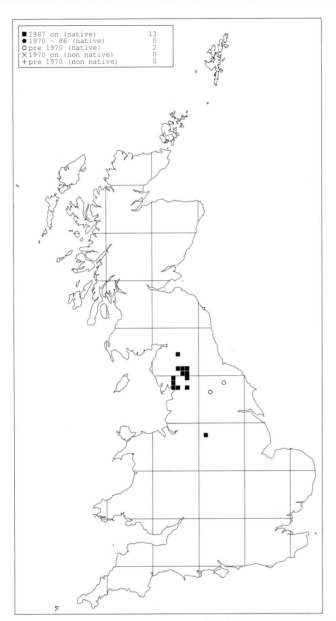

Carex ornithopoda Wild. (Cyperaceae)
Bird's-foot sedge
Status in Britain: LOWER RISK - Near Threatened.
Status in Europe: Not threatened.

C. ornithopoda is a plant of open calcareous grasslands on Carboniferous limestone in Cumbria, Yorkshire and Derbyshire. It often grows close to exposed rocks on south-facing slopes in full sun where the soils are thin and have a tendency to become parched during the summer months. Such situations enable the species to avoid competition from bulkier associates. In Cumbria it is typically found in swards dominated by *Sesleria caerulea*, many of the most extensive populations being closely associated with areas of shattered limestone pavement. However, at a number of Cumbrian sites the dominant grass is *Festuca ovina*, and this is the case at both of the Derbyshire localities. Typical associates in the characteristically species-rich swards include *Carex flacca*, *Helianthemum nummularium*, *Linum catharticum*, *Scabiosa columbaria* and *Thymus polytrichus*.

C. ornithopoda is a perennial that flowers on lateral shoots. It is best identified when in fruit, during May and June. Fertile seed is freely produced.

In Cumbria about 25 populations are shared between two main areas, one on the limestone scars to the north of Morecambe Bay and a second on cooler and higher exposures to the east of Shap. An outlying colony has also been detected further north in the Eden valley (Corner & Roberts 1989). Over 100 km to the south-east there are two substantial populations in Derbyshire's White Peak, and a colony of more than a hundred plants was discovered in 1992 at Twisleton Glen in Yorkshire. Confusion between *C. ornithopoda* and *Carex digitata* undoubtedly occurs, and a number of hitherto accepted records may well refer to the latter species. This is a particular problem where identification has been based on immature or non-fruiting specimens (David 1980a). Two historical records, from Mackershaw, near Ripon, and Hawnby on the North Yorks Moors, are discussed in David (1981). The former is confirmed from a specimen collected in 1887, and the latter record may also be correct. However, *C. ornithopoda* has not been seen east of the Pennines this century.

Many populations of *C. ornithopoda* lie within SSSIs because of their close association with species-rich, calcareous grasslands, which are of high nature conservation value. Communities on the thin, parched soils which favour *C. ornithopoda* generally require only light grazing to maintain their open character, and for most populations the current management regimes are serving the plants well. Some of the larger populations, such as those in Derbyshire, contain many thousands of plants.

C. ornithopoda is widespread in montane areas in Europe, with outlying populations in north-east Anatolia.

I. Taylor

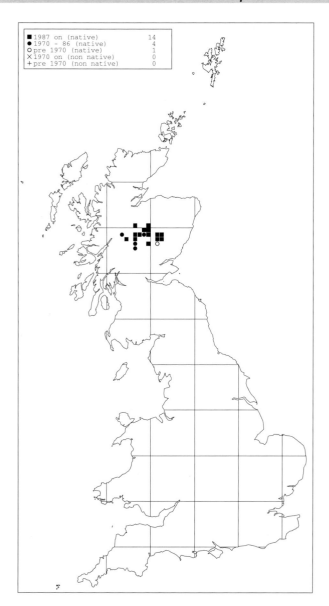

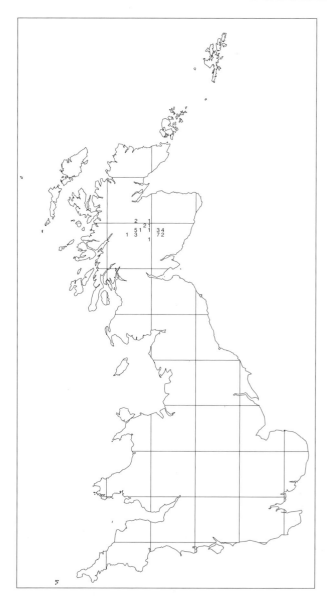

Carex rariflora (Wahlenb.) Sm. (Cyperaceae)

Mountain bog-sedge, Seig Ailpeach Thearc
Status in Britain: LOWER RISK - Near Threatened.
Status in Europe: Not threatened.

This small montane sedge occurs at altitudes between 750 and 1,050 m in the Scottish Highlands. It occurs mainly in acidic flush bogs with *Sphagnum fallax, S. papillosum, S. russowii, Carex bigelowii, C. curta, C. nigra, Rubus chamaemorus* and, less frequently, *Carex aquatilis, Polytrichum commune* or *Sphagnum lindbergii*. Sites are associated with late snow or with springs derived from such areas. On Glas Maol, at least one very large population grows on open peat by a pool, looking very reminiscent of *Carex limosa*. Populations on the Drumochter Hills have a different ecology. They occur on flattish terraces by the side of incised burns, often in fairly open vegetation on a peaty soil with *Carex bigelowii, C. curta, C. echinata, Eriophorum angustifolium, Nardus stricta* and *Sphagnum papillosum*. A characteristic habitat is a fixed bank of silt in the headwaters of a burn before the gradient steepens (David 1980b).

C. rariflora is a very shy flowerer in some seasons (David 1980b), and in a population there are often 8-15 times as many non-flowering as flowering plants. Because of this, it is likely that its abundance has often been under-estimated.

More than 30 localities are known. It occurs quite widely in the east-central Scottish Highlands, with clusters of sites around Drumochter and in the Clova Mountains, but is very rare in the Breadalbanes. It seems to be reasonably plentiful in many of its localities. Above Glen Feshie and in the Clova Hills, most colonies number in the thousands, some exceed 10,000, and the largest known population holds more 100,000 plants. The populations on the Drumochter Hills are much smaller and mostly number fewer than 1,000 plants. This species is likely to be threatened only by major changes of land use or, like other montane species, by climatic change.

C. rariflora has a circumpolar distribution and is widely distributed in this zone across northern Asia to Kamchatka and throughout boreal North America.

G.P. Rothero and M.J. Wigginton

Carex recta Boott (Cyperaceae)
Estuarine sedge, Seisg an Inbhir
Status in Britain: VULNERABLE.
Status in Europe: Not threatened.

C. *recta* is found on the banks of tidal rivers and on marshy flats, normally where silt is periodically deposited or where the water table fluctuates seasonally. In some places it forms extensive pure stands and in others may be abundant in sedge-dominated communities with *C. aquatilis* or *C. nigra*. It may also be dominant in tall swamp vegetation with *Phalaris arundinacea* or *Phragmites australis* and less frequently in species-rich marginal fen with such species as *Crepis paludosa*, *Deschampsia cespitosa*, *Filipendula ulmaria* and *Mentha aquatica*.

This species is partially fertile, with fruit ripening in August and September. Hybrids between *C. recta* and *C. aquatilis* occur in at least one site with *C. recta* and perhaps in another in the absence of *C. aquatilis*.

C. *recta* occurs in the lower reaches of the Wick River, the Kyle of Sutherland and the Beauly River. In the first two sites it forms extensive stands along 1-2 km of river, but by the Beauly River it appears to have declined markedly in the last ten years. It is rare in Europe as a whole, occurring only in Scotland and Finland, and perhaps more doubtfully from Norway and the Faeroe Islands. It also occurs on the Atlantic coast of North America.

Faulkner (1972), in a study of populations of *C. recta* from Scotland and Scandinavia, considered that on the basis of morphological and genetic studies, this taxon was likely to have arisen through hybridisation of *C. aquatilis* and *C. paleacea*, and might be better referred to as *Carex* x *recta*. Scottish plants differ in morphology and behaviour from those in Scandinavia.

M.J. Wigginton

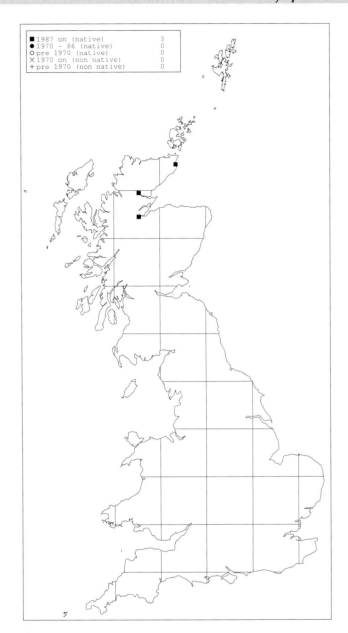

■ 1987 on (native)	3
● 1970 - 86 (native)	0
○ pre 1970 (native)	0
✕ 1970 on (non native)	0
+ pre 1970 (non native)	0

Carex vulpina L. (Cyperaceae)

True fox-sedge, Hesgen Dywysennog Fwyaf
Status in Britain: VULNERABLE.
Status in Europe: Not threatened.

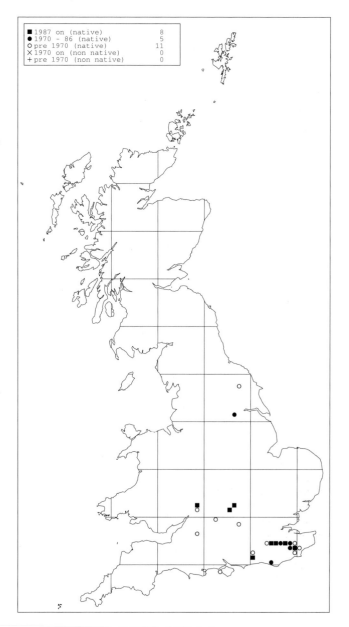

■ 1987 on (native)	8	
● 1970 – 86 (native)	5	
○ pre 1970 (native)	11	
✕ 1970 on (non native)	0	
+ pre 1970 (non native)	0	

C. vulpina is a lowland plant of wet ditches and pond-sides, for the most part on chalk or limestone. It requires more moisture than does the closely related *C. otrubae*; where the two grow together, as they often do, *C. otrubae* is likely to be on the bank of the ditch and *C. vulpina* in the standing water. Other species in the community may include *Alisma plantago-aquatica, Carex acuta, Glyceria fluitans, Iris pseudacorus, Juncus effusus, Oenanthe crocata, Phalaris arundinacea* and *Rorippa amphibia*.

In unshaded sites this sedge may form a large clump, but spreads by abundant seed rather than vegetatively. It is sensitive to changing conditions, and in shade soon ceases to flower and dwindles to extinction.

In Britain it was not recognised as distinct from *C. otrubae* until 1939, and its full distribution may still be masked by confusion between the two. Unconfirmed reports of *C. vulpina* should therefore be treated with caution, examination of utricles microscopically (Stace 1991) being desirable for correct identification. Most sites for *C. vulpina* are in Kent, but it also occurs in Sussex, Gloucestershire and Oxfordshire and perhaps still survives in South Yorkshire. In the last twenty years many colonies have disappeared because of drought or water-extraction, the clearing of ditches, infilling of ponds or shading by invasive scrub. Its survival seems extremely precarious at several sites where only single isolated plants may remain. In Oxfordshire, it is thought to persist at only one site, in very small numbers, in a neglected ditch between arable land and an improved pasture, where it appears to be threatened by dense shade cast by rampant growth of other plants, and perhaps by frequent lack of water (R. Porley pers. comm.). A review of the status of *C. vulpina* in Britain is urgently required.

C. vulpina occurs in suitable situations throughout Europe south to 60°N, though becoming much scarcer in the west and south. Its distribution in Asia is not certainly known.

M.J. Wigginton, adapted from R.W. David, in Stewart *et al.* (1994).

84

Centaurea calcitrapa L. (Asteraceae)
Red star-thistle
Status in Britain: VULNERABLE.
Status in Europe: Not threatened.

In Britain, *C. calcitrapa* is a plant of well-drained sandy, gravelly or light chalky soils. It occurs on waste ground and tracksides, in dry grasslands and on banks, usually in broken or disturbed ground. At one time it was a plentiful, though local, weed of arable fields, though it is now hardly known in this habitat, mainly because of improved seed-screening.

It is a biennial. The main flowering period is between late July and early August, but it may extend into September. The flowers are visited by bees and flies. Seed is freely produced, but dispersal is poor since the seed is heavy and falls close to the parent plant. It appears to have good viability in the seed-bank.

C. calcitrapa has declined markedly and now occurs regularly in few sites, most of which are on the South Downs in Sussex between Brighton and Eastbourne. At the present time, it appears to be established in just seven sites, five in Sussex and two in Kent. Populations are generally small, though one of the Sussex sites regularly holds several hundred plants. Elsewhere, records are generally of casual occurrences, originating mainly from wool shoddy or bird seed.

Its native status is disputed, with some authorities (e.g. Philp 1982; Stace 1991) regarding it as an introduction. Its occurrence in arable fields was dependent to a large extent on continued reintroductions as a seed impurity of lucerne from southern France or Italy. However, the fact that it was present on waste land before the introduction of lucerne around 1650 suggests that it might be native. F.Rose (*in litt.*) considers it to be probably native in Sussex on dry banks on the chalk, its habitats there being similar to those near the Somme estuary where also he considers it is probably native.

The two populations in Kent appeared, at least at one time, to be largely dependent on tethered or enclosed horses for maintaining open ground. At sites in Sussex, it benefits from erosion of slopes and, on tracksides, from regular mowing of surrounding vegetation and trampling by cattle (FitzGerald 1988c). Seed has been dispersed by hand in at least one Sussex site.

C. calcitrapa is widespread in southern and south-central Europe (including Mediterranean islands), from Iberia eastwards to Russia and northwards to southern England, and northern France and Germany. It also occurs in North Africa, the Canary Islands, and West Asia. It seems to be in general decline in western and northern Europe, probably mainly because of agricultural intensification and the destruction of marginal habitats.

M.J. Wigginton

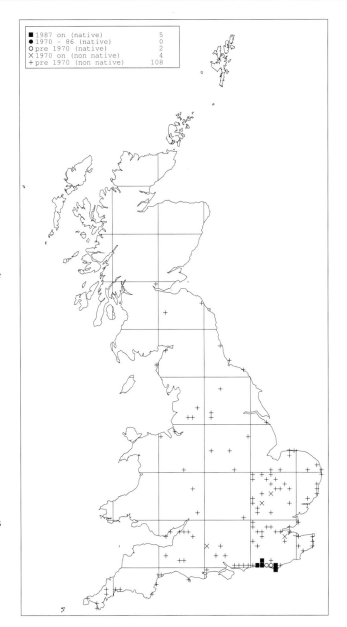

■ 1987 on (native)	5
● 1970 – 86 (native)	0
○ pre 1970 (native)	2
✕ 1970 on (non native)	4
+ pre 1970 (non native)	108

Centaurea cyanus L. (Asteraceae)
Cornflower, Penlas Yr ìd
Status in Britain: ENDANGERED.
Status in Europe: Not threatened.

C. cyanus is a plant of arable habitats, chiefly but not exclusively on sandy loam soils, and in the past it was a troublesome weed. It was formerly widespread in Britain, especially in the south and east, but in recent decades it has shown one of the most rapid declines of any plant in the British flora. It was known from about 264 hectads between 1930 and 1960, from fewer than a hundred between 1960 and 1975 and fewer than 50 between 1976 and 1985 (Smith 1986). Since 1986, *C. cyanus* has been recorded in very few arable fields, although it has occasionally appeared as a non-persistent casual on excavated soil in former arable land. Many of the recent occurrences on road verges can be attributed to the increasing use of 'wild flower' seed, probably of continental origin, and most of the mapped records may be of such origin.

C. cyanus is an annual, flowering between June and August. It is self-incompatible and is cross-pollinated especially by flies and bees. It is thought to be largely autumn-germinating and is associated with autumn-sown crops, although some seed can also germinate in the spring. There is evidence that most seed remains viable in the soil for up to four years (Svensson & Wigren 1986), though the occasional appearance of plants on old sites suggests that a small proportion may remain viable for much longer. Seed can be produced by early July by autumn-germinated plants. The seeds are relatively heavy for an annual species and tend to remain close to where the parent plant grew. In common with many other species of arable plant, *C. cyanus* is not a highly mobile, ephemeral species, but rather is restricted to areas where there has been long continuity of arable farming.

There are recent records from arable fields in Wiltshire, the Isle of Wight, Buckinghamshire, Lincolnshire and Suffolk, but it is believed that the last county supports the sole remaining arable field population in Britain that is persistent and self-sustaining. It is associated there with a typical flora of sandy arable fields, including *Anchusa arvensis*, *Anthriscus caucalis* and *Papaver argemone*.

Changes in arable practice have caused the decline of this once well-known and common species. It is possible that improved methods of seed cleaning introduced at the end of the 19th century eliminated the seeds from seed-corn. It is thought to be susceptible to a wide range of herbicides, and this combined with the mainly rather short-lived seed are probably the main reasons for its rapid decline. The Suffolk site is currently part of a conventionally farmed arable field, and research continues

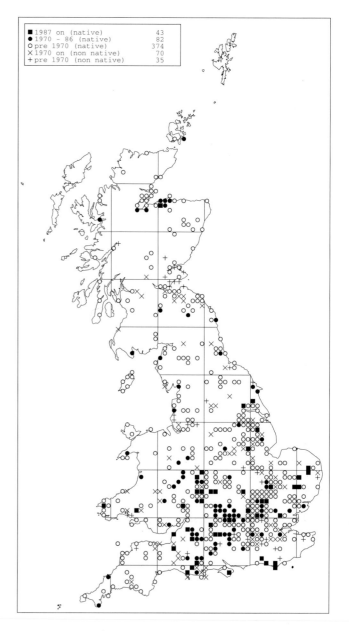

■ 1987 on (native)	43
● 1970 - 86 (native)	82
○ pre 1970 (native)	374
✕ 1970 on (non native)	70
+ pre 1970 (non native)	35

there to determine the ideal conservation management. This species is regarded as critically endangered in Britain because of the extreme restriction of persistent populations.

It is distributed throughout Europe and many of the other cereal-growing parts of the world. In common with many other species it has become much less frequent in parts of northern Europe where arable farming has become more intensive, and in a few other areas it is as threatened as in Britain.

P.J. Wilson

Centaurium scilloides (L. f.) Samp. (Gentianaceae)

C. portense (Brot.) Butcher
Perennial centaury, Canri Dryflwyddol
Status in Britain: VULNERABLE.
Status in Europe: Not threatened. Near Endemic.

The remaining undisputed native British populations of
C. scilloides are restricted to freely draining soils of variable
depth on coastal slopes with western and southern aspects
near Newport, Pembrokeshire. It is intolerant of shade and
is most abundant in less exposed maritime grassland
dominated by *Agrostis capillaris, Dactylis glomerata, Festuca
rubra* and *Holcus lanatus*. Other populations occur in
maritime heath with *Calluna vulgaris* and *Erica cinerea*. Other
frequent associates include *Aira caryophyllea, A. praecox,
Anthyllis vulneraria, Armeria maritima, Festuca ovina, Koeleria
macrantha, Lotus corniculatus, Plantago coronopus, P. maritima,
Scilla verna, Sedum anglicum* and *Thymus polytrichus*.
Although mainly found on the bevelled slopes of cliffs at
altitudes between 10 m and 85 m, populations extend into
dune grassland at Newport Sands. *C. scilloides* is better able
to withstand drought than many of the grasses with which
it grows. It also appears readily to colonise bare areas, and
extensive seedling establishment has been observed
following burning of *Calluna vulgaris* and *Ulex europaeus*.

This perennial species can form vigorous pink starry
flowering clumps up to 15 cm in height and 75 cm in
diameter. Its main flowering season extends from late June
to August, although flowers are sometimes found as late
as October. Seed production is prolific and the plant can be
easily established in gardens from seed. Outside the
flowering season it forms patches of bright green leaves,
which are usually visible throughout the winter.

The main British population holds several hundred
plants and appears to be stable. It is scattered along 3 km
of undeveloped coast north of the small estuary at
Newport, Pembrokeshire (Morgan 1989e), where it was
first discovered in Britain by T.B. Rhys in 1918 (Wilmott
1918). In 1952 it was found on cliffs in West Cornwall at
Porthgwarra, and in 1956 in East Cornwall at Sandymouth,
north of Bude. But it has not been seen at Porthgwarra
since 1967 and is thought to be extinct at Sandymouth,
since extensive searches there in the 1980s and
subsequently have failed to detect it.

The puzzling populations from West Kent and East
Sussex are on banks and lawns of early 20th century houses
and were first recorded in 1974 and 1982 respectively
(FitzGerald 1988a, 1988d). Four of the six sites have since
been lost to re-development and changes in management,
but plants have been transferred to a secure site at a nearby
golf course. Although it is known that this species has been
stocked by local nurseries (Briggs 1983), its presence in
species-rich turf including *Hypericum humifusum, Orchis
morio, Spiranthes spiralis* and formerly *Calluna vulgaris*, raises
the possibility of it being native there.

It seems that the decline of grazing on coastal slopes,
and the consequent reduction in open swards and the
growth of *Ulex europaeus* scrub, is likely to be responsible
for the disappearance of *C. scilloides* from Cornwall. In
Pembrokeshire many plants occur along the trampled and
eroded edges of the coastal path, on the cut margins of

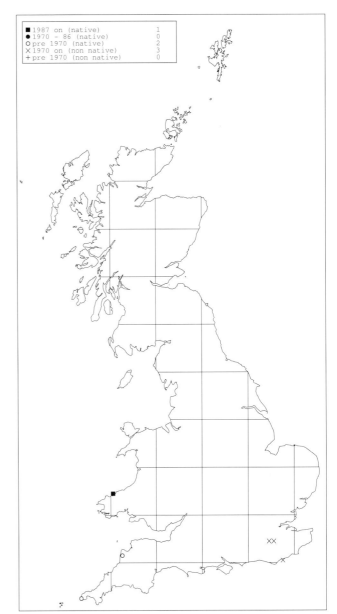

■ 1987 on (native)	1
● 1970 - 86 (native)	0
○ pre 1970 (native)	2
✕ 1970 on (non native)	3
+ pre 1970 (non native)	0

Newport golf course and on the steeper slopes of
unimproved but sheep- and cattle-grazed coastal fields.
Without such artificially maintained open conditions the
Welsh population would be much more fragmented and
confined to unstable cliff edges and the main salt-exposed
headlands. Disturbed areas around rabbit burrows may
also have benefited the species when rabbits were more
abundant than they are now. Restoration of appropriate
coastal-slope grazing combined with the control of bracken
and gorse would further boost the population. Virtually all
the colonies lie within SSSIs, and key populations are
further safeguarded by management agreements.

C. scilloides is an oceanic species reaching its most
northerly station in Wales. Other populations occur in
scattered localities along the western seaboard of Europe
from France to Portugal, and it also occurs in the Azores.

S.B. Evans

Centaurium tenuiflorum (Hoffsgg. & Link) Fritsch (Gentianaceae)

Slender centaury
Status in Britain: VULNERABLE. WCA Schedule 8.
Status in Europe: Not threatened.

Most floras describe *C. tenuiflorum* as a plant of damp grassy places. However, all extant colonies in Britain are on bare and open, poorly-draining sandy or clayey soils of the Middle and Lower Lias. Here *C. tenuiflorum* occurs in stands, often dense, either on its own or with a scattering of some of the following: *Agrostis stolonifera, Blackstonia perfoliata, Centaurium erythraea, Phragmites australis, Samolus valerandi, Tussilago farfara* and occasionally *Epipactis palustris, Isolepis cernua* and *Vicia bithynica*. The Dorset undercliffs where it grows are a mosaic of habitats from bare soil to dense thickets, and *C. tenuiflorum* tends to appear three or four years after the opening up of the habitat, and disappear after ten years unless it is opened up again. Nothing seems to be known of the detailed ecology at its former sites in the Isle of Wight and the Channel Islands.

C. tenuiflorum is an annual, germinating in spring, and in flower with a succession of plants, from late June until October. Abundant seed is set, and up to twenty seedlings may germinate from a well-grown plant. The seed-bank must be long-lived to enable it to survive in its undercliff habitats until the next opening.

The plant in Britain is currently confined to seven sites on the undercliff over a distance of 4 km to the east of Golden Cap in Dorset. The cliffs above are very unstable, and indeed after the heavy rains of early 1994 there were many landslips, at least two affecting sites of *C. tenuiflorum*. Nevertheless those sites contained in excess of 10,000 plants in 1994. If suitable conditions occur, the populations fluctuate little from year to year. The stretch of coast is almost all owned by the National Trust, but no specific conservation measures are feasible or necessary and access is possible only from the beach. There are old records that suggest occurrences to the west of Golden Cap, but in 1994 the whole area was covered with fresh falls, with no vegetation cover at all. It remains to be seen whether the plant can re-establish itself there. There are also old records from the Channel Islands, and from the Isle of Wight, where it was last seen at Newtown in 1953. It was reported in 1991 from near the old site at Newtown, but this plant was surely *C. pulchellum*.

This species is at the edge of its range in Britain, and it extends southwards to Spain, Portugal and North Africa, and then eastwards throughout the Mediterranean to the Crimea and Iraq. In all these areas it is predominantly a coastal plant but in Turkey, at least, it ascends to 1,150 m, where it is associated with inland saline deposits.

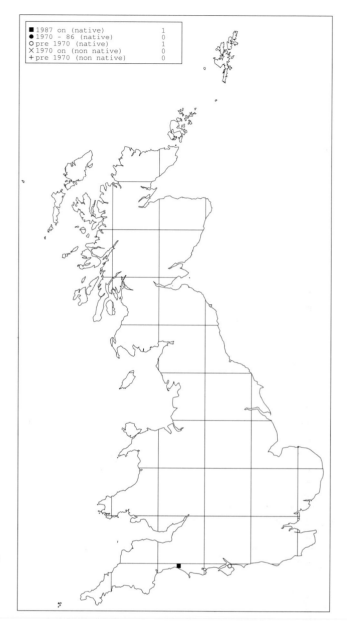

■ 1987 on (native)	1
● 1970 - 86 (native)	0
○ pre 1970 (native)	1
✕ 1970 on (non native)	0
+ pre 1970 (non native)	0

C. tenuiflorum is taxonomically close to *C. pulchellum*, and some authorities including Stace (1991) query its distinctness. In the field a well-grown specimen with its flat-topped, dense crown is very distinctive indeed. All the Dorset plants have white flowers, with only a very occasional pale pink one, but predominantly white-flowered populations do not seem to occur elsewhere in Europe.

D.A. Pearman

Cephalanthera rubra (L.) L.C.M.Rich. (Orchidaceae)
Red helleborine, Caldrist Coch
Status in Britain: CRITICALLY ENDANGERED. WCA
Schedule 8.
Status in Europe: Vulnerable.

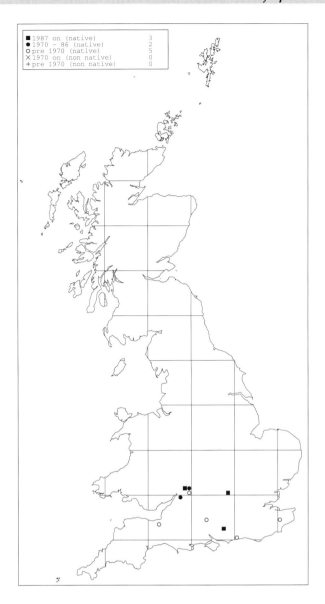

This species occurs in deciduous woods, principally of beech, on calcareous soils where the ground flora is sparse. Other trees and shrubs may include ash, elm, hazel, privet, sycamore, whitebeam and yew Three other orchid species - *Cephalanthera damasonium, C. longifolia* and *Epipactis helleborine* - typical of this habitat are present in one or other of its sites. It is intolerant of competition from other plants (Bateman 1980), but other species occurring sparsely in the general community include *Anemone nemorosa, Brachypodium sylvaticum, Campanula trachelium, Euphorbia amygdaloides, Hedera helix, Hieracium* species, *Hyacinthoides non-scripta, Lonicera periclymenum, Mercurialis perennis, Sanicula europaea* and *Viola riviniana*. All colonies of *C. rubra* are on well-drained sloping ground, but not with any common orientation.

According to Irmisch (1863) the plant has a long life cycle, with the first foliage leaves appearing in the sixth year, and the first flowers in the tenth. Flowers, when produced, appear in June and July, but only a few shoots may produce flowers. Seed-set is very low in Britain, presumably because of the limited opportunity for natural pollination, though bees and hoverflies have occasionally been observed visiting the flowers. The flowers lack nectar, but in studies on the island of Gotland, Nilsson (1983) suggested that it uses 'floral colour mimesis' to attract bee pollinators. He linked certain *Campanula* species to the pollination biology of *Cephalanthera rubra* and showed that in the bee-spectrum their colours are almost identical. He also showed that seed-set in *Cephalanthera rubra* was lower in the absence of the bees *Chelostoma campanularum* and *C. rapunculi*, which visit both *Campanula* and *Cephalanthera rubra*. However, though *C. campanularum* has been seen visiting *C. rubra* plants in Hampshire, this species is thought to be too small to effect the removal of pollinia, and *C. rapunculi* is not known in Britain (G.R. Else *in litt.*). If a similar link between *Campanula* and *C. rubra* exists in Britain, then other pollinators will be involved.

During the nineteenth century, *C. rubra* was found at several sites in Gloucestershire, and there have also been unconfirmed records from Somerset, Sussex and Kent. *C. rubra* now occurs only at single sites on chalk in Buckinghamshire and Hampshire (Rose & Brewis 1988), and one site on oolitic limestone in Gloucestershire. Its appearance, and especially flowering, has been sporadic at all three sites. For example, between 1955 and 1986 at the Buckinghamshire site, flowering occurred in only five of the 26 years when monitoring took place (Everett 1988). The number of individual plants is always small, the largest recent count being twenty at one site in 1990 (though 34 were noted in Buckinghamshire in 1958). The British population currently totals only about 30 plants.

Recent management has included the removal of invasive shrubs such as *Ligustrum vulgare* and *Rubus fruticosus* and the selective felling of trees to allow more light to the ground. This seems to have promoted flowering.

Leaf litter has been removed from one site and slug pellets have been scattered around the plants to control predation. Hand-pollination has been carried out to a limited extent, and research into germination and propagation is being undertaken at Kew, with a view to supplementing wild populations with plants grown *ex situ*. Plants may become small if the habitat is overgrown, and even disappear, but it is suggested that the mycorrhizal rhizome system may remain alive for a long time (Sell & Murrell 1996).

Though widespread in Europe, reaching northwards to southern England and Finland, it also occurs eastwards to Turkey and Iran. *C. rubra* is vulnerable over much of its range, and is legally protected in several central and northern countries. It is mainly found in beech and other deciduous woodland, principally in dappled shade or in open areas that receive direct sunlight for part of the day. It is also encountered on steep, barish roadside verges. It seems never to occur in large colonies, but always as small groups or scattered individuals.

L. Farrell

Cerastium brachypetalum Pers. (Caryophyllaceae)

Grey mouse-ear
Status in Britain: ENDANGERED.
Status in Europe: Not threatened.

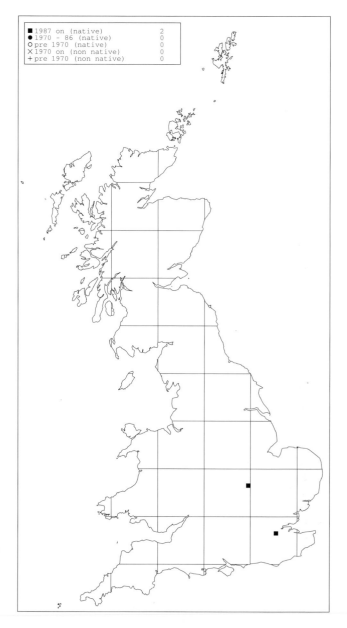

C. brachypetalum is a late-comer to the British flora, having been first discovered in 1947 in a railway cutting at Wymington in Bedfordshire. When first seen there, it was described as being "in large quantity over a considerable distance". Plants were mostly found on anthills in *Brachypodium pinnatum*-dominated grassland. The population, mainly in two colonies on west-facing banks, remained fairly constant for years, but in the 1980s a decline set in, and only small numbers have been seen in recent years, fewer than 20 plants in 1993 (Horn 1994). In 1973, the colony was found to extend just into Northamptonshire, but no plants have occurred in that county since 1990. A single plant was detected in 1984 in Wymington Fields Nature Reserve, which lies close to the railway.

The other area where *C. brachypetalum* is known is in West Kent. Its discovery there was of particular interest since it occurs not only on railway embankments but also in old grassland on chalk nearby. It has been known from about eight locations near Longfield over a distance of about 1 km, mostly in the more open habitats associated with old grassland, such as the edges of paths or on eroding earth banks. The species also occurs in former old grassland sites now altered by disturbance. In most sites, *Bromopsis erecta* is generally the dominant grassland species, but a wide range of associated species include those characteristic of chalk grassland, together with common early-flowerers of ruderal habitats, such as *Arabidopsis thaliana*, *Cerastium glomeratum* and *Erophila verna*. A survey of all six of the extant Kent sites in 1994 revealed a total of about 2,500 plants (Rich & Palmer 1994).

C. brachypetalum is a winter annual, varying in abundance from year to year, probably depending on prevailing weather conditions. Seed has been shown to remain viable in storage for at least three years (Brett 1955), and germinates rapidly after cold treatment, but viability in natural conditions is not known Flowering is in April and May (perhaps into June in a cool wet spring), and flowers are probably self-pollinated. Fruit is set from May onwards. Palmer (1994) has observed that many colonies appear in exactly the same areas each year (to within a few centimetres), thus suggesting very poor powers of dispersal.

The status of *C. brachypetalum* is uncertain, not least because of its occurrence in both areas near railways. However, its presence in semi-natural grassland in Kent has led some to suppose it might be a relict native species, more abundant in the past, but now able to survive only in the tiny areas which have escaped the plough and development. Conversely, it could have colonised this grassland from railway banks. The proposals for line widening for the Channel Tunnel Link pose a serious threat to the species and, if implemented, would result in the loss of five of its six extant sites and 85% of the British population. However, options for conserving *C. brachypetalum* in the face of such development are under discussion. In Bedfordshire, some controlled burning in 1994 and 1995 to open up the sward seems to have been beneficial, and more than 60 plants were counted there in 1995.

This species occurs right up to the Channel coast of France and is widespread throughout western and central Europe. It also extends eastwards to Bulgaria and the Crimea and northwards to southern Sweden and Norway.

M.J. Wigginton

Cerastium fontanum ssp. *scoticum* Jalas & Sell (Caryophyllaceae)

Scottish mouse-ear
Status in Britain: VULNERABLE. ENDEMIC.

C. fontanum ssp. *scoticum* is known only from Meikle Kilrannoch in Angus. There it is found at an altitude of 860 m, growing on soils largely derived from ultramafic (serpentine) rocks. It occurs in several different vegetation communities, most frequently those occurring in areas of thin, skeletal soils overlain by stones. These debris areas support open vegetation and are botanically rich, with such locally or nationally rare species as *Armeria maritima, Cochlearia officinalis* (cf. ssp. *alpina*), *Lychnis alpina* and *Minuartia sedoides* (Proctor *et al.* 1991). It is also found in *Empetrum nigrum* ssp. *hermaphroditum* and *Vaccinium myrtillus* heath, and more rarely in *Nardus stricta* grass-heath.

This taxon is distinguished from other subspecies of *C. fontanum* by its longer petals, larger seeds with larger tubercles, shorter capsules, sepals, flowering stems and leaves, and few-flowered inflorescences (Jalas & Sell 1967). It has a short-lived perennial life-history and flowers from June to September (and perhaps beyond these dates). Reproduction is by seed, though some vegetative reproduction through rooting of non-flowering shoots may also occur.

It is restricted to three small sites, situated within 1 km of each other and covering an area of approximately 6.5 hectares. There have been no direct counts of its population, but it is estimated as comprising around 100,000 individuals. It was formerly thought to occur at Strathy Point, West Sutherland, but plants from there have since been shown to be referable to ssp. *vulgare* (Wyse Jackson 1992).

There are no immediate or obvious threats to its survival. It may, however, be vulnerable to excessive grazing, and this should be monitored. Long-term management should ensure the maintenance of some debris areas, in the event of natural succession of these areas to closed vegetation. All three sites for ssp. *scoticum* lie within Caenlochan NNR.

M.B. Wyse Jackson

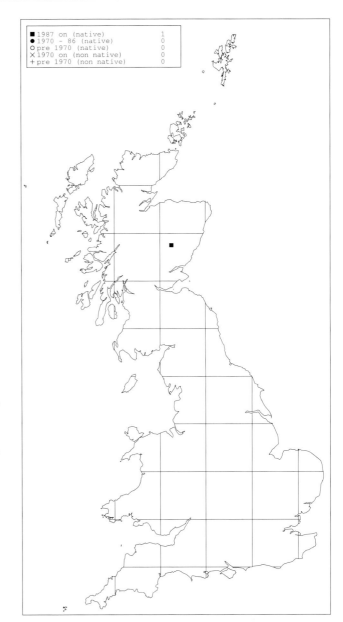

■ 1987 on (native)		1
● 1970 – 86 (native)		0
O pre 1970 (native)		0
X 1970 on (non native)		0
+ pre 1970 (non native)		0

Cerastium nigrescens (H.C.Watson) Edmondston ex H.C.Watson (Caryophyllaceae)

Cerastium arcticum ssp. *edmondstonii* (Edmondston) A. & D. Löve

Shetland mouse-ear

Status in Britain: VULNERABLE. ENDEMIC.

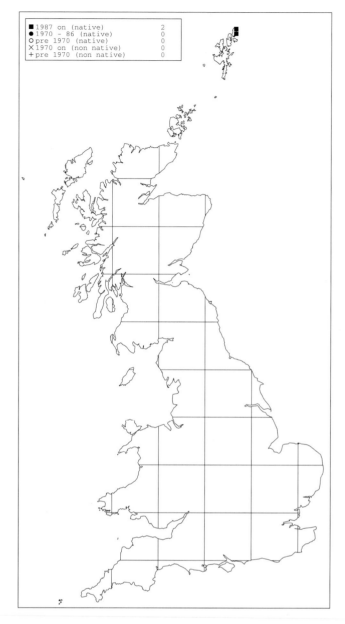

■ 1987 on (native)	2	
● 1970 – 86 (native)	0	
○ pre 1970 (native)	0	
✕ 1970 on (non native)	0	
+ pre 1970 (non native)	0	

This plant is endemic to Unst, Shetland Isles, discovered there by Edmondston in 1837. It is now to be found in only two places - the Keen of Hamar, and near Muckle Heog. It occurs on bare or sparsely to moderately vegetated serpentine debris ('fell-field'), growing between the shattered angular rocks and stones. At the Keen of Hamar, it is chiefly on well-drained ground, which can be droughted in summer, but on Muckle Heog it is also on more constantly moist debris (Scott & Palmer 1987). In winter plants can become dislodged by frost-heave and strong winds. Associated species on the serpentine debris are *Arabis petraea*, *Arenaria norvegica* ssp. *norvegica* and *Silene acaulis*, along with more widespread species including *Agrostis vinealis*, *Antennaria dioica*, *Festuca vivipara*, *Plantago maritima*, *Selaginella selaginoides*, *Silene uniflora* and *Thymus polytrichus* in areas of more closed vegetation.

C. nigrescens is a dwarf perennial, with more or less orbicular fleshy leaves, whose dark, purplish colour may be a response to high illumination or have a genetic basis (Scott & Palmer 1987). Most seedlings germinate in autumn, and they mature slowly, with only a small proportion flowering within two years of germination. Flowering is usually in June, but in some years flowers may be found up to late August. There is a good seed-bank. A recent study has shown a high survival rate of mature plants: 71% recorded in 1994 were alive in July 1996 (S. Kay pers. comm.).

The Keen of Hamar population is much the larger of the two, though numbers vary from year to year. Counts in 1978, 1985 and 1993 revealed about 4,500, 9,200 and 6,900 plants. The Muckle Heog population is much smaller, fluctuating between 300 and 800 plants (Slingsby *et al.* 1993). Plants from the Heogs differ in several characters from those at the Keen of Hamar, including narrower subacute leaves, a more straggly habit, smaller flowers and longer capsules. This form is probably the same as Edmonston's *C. latifolium* var. *acutifolium*.

The Keen of Hamar supports one of the finest examples of serpentine debris vegetation in Europe and is currently managed as an NNR. Unfortunately, a significant proportion of the habitat was lost in the 1960s to agricultural improvement. In the 1970s, the Lower Keen was grazed by cattle and suffered eutrophication owing to the provision of winter feed and dunging, with a consequent loss of the more interesting plant species. Parts of the area were later 'scalped' of vegetation, which

proved a successful strategy, since typical plants of the serpentine then moved in, including both *C. nigrescens* and *Arenaria norvegica*. However, it has become clear that the vegetation is now recovering naturally in the area, and 'scalping' has been discontinued (P.Harvey pers. comm.).

C. nigrescens is morphologically close to *C. arcticum* (of montane and arctic regions of Europe and North America), but its taxonomic status has long been disputed, and it has in the past been variously treated as a form, variety, or subspecies of it. Currently, the Shetland plant is considered to be a full species (Stace 1991).

M.J. Wigginton

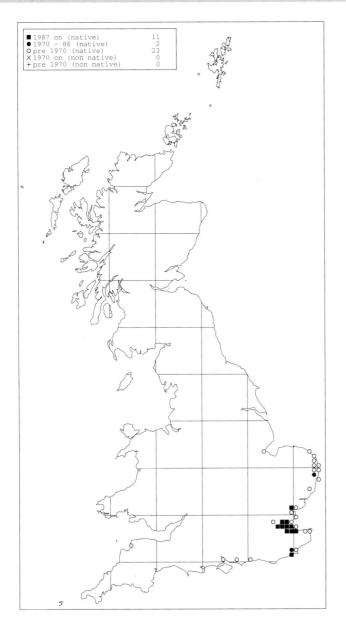

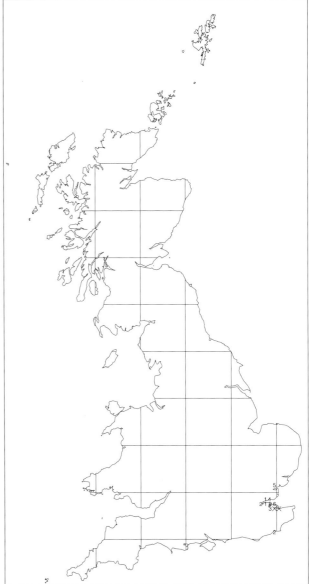

Chenopodium chenopodioides (L.) Aellen (Chenopodiaceae)

Chenopodium botryodes Sm.
Saltmarsh goosefoot
Status in Britain: LOWER RISK - Near Threatened.
Status in Europe: Not threatened. Endemic.

This little-known annual is characteristic of a restricted estuarine habitat, growing on dry brackish mud seasonally exposed on ditch sides and in the shallow winter inundations (or 'fleets') of saltings and grazing marshes. Associated species are mostly annual, often including *Chenopodium rubrum, Parapholis strigosa, Salicornia* species and *Suaeda maritima,* with the nationally scarce *Hordeum marinum, Polypogon monspeliensis* (and sometimes x *Agropogon kuttirakus,* its rare hybrid with *Agrostis stolonifera*), *Puccinellia fasciculata* and, occasionally, *P. rupestris.* Perennial associates are usually *Glaux maritima, Juncus gerardii* and *Spergularia media.*

Like most annuals of exposed mud, *C. chenopodioides* can appear in great quantity in suitable years, when the mud dries and warms early, but is scarce or even absent when water levels remain high. It appears to germinate from dormant seed only when the mud is exposed to the air, and flowering is normally only just beginning in July, but can build up to spectacular displays by mid-September, when fruiting plants turn bright red. Plant sizes vary, but (like its close relative *C. rubrum*) *C. chenopodioides* is often present as minute plants producing good seed in late years, when found on mud or in ephemeral habitats such as hoofprints.

Traditional management with fluctuating water levels and livestock trampling of the drying ditch edges was ideal, but grazing marshes have been much reclaimed for arable use, or drained with stabilised levels in the ditches. It is, however, still very frequent on the Isle of Sheppey and adjacent marshes south of the Swale, North Kent, though in contrast is very local on the nearby Cliffe, Stoke and All Hallows marshes (Williams 1996). The few colonies recorded outside the core populations of the

93

Thames estuary, including a wildfowl scrape in a coastal reserve and damp hollows on a sandy golf course (both in Sussex), are probably extinct.

This species is endemic to central and southern Europe, extending to Denmark and south-west Russia, but excluding the central and southern Mediterranean (Jalas & Suominen 1980).

Some past records may have been mis-determinations of the dwarf ephemeral form of *C. rubrum*, but the taxonomy is now clearly understood. Recorders no longer have to rely on the subjective character given by Fitt (1844), that "when fresh-gathered, the smell is like that of the pods of green peas".

Adapted from R. FitzGerald, in Stewart *et al.* (1994).

Chenopodium vulvaria L. (Chenopodiaceae)
Stinking goosefoot
Status in Britain: VULNERABLE. WCA Schedule 8.
Status in Europe: Not threatened.

C. vulvaria was characteristic of a wide variety of bare, nitrogen-enriched habitats, particularly in coastal areas, from arable fields and rubbish dumps to sandy or gravelly waste ground or tracks. In all these habitats it would be part of a typical open weed flora, with a vegetation cover of 25% or less. It is now found at only three regular sites, all of which are coastal. In Suffolk, its habitat is sandy shingle in an open community in which *Glaucium flavum*, *Lepidium latifolium, Ononis spinosa, Reseda lutea* and *Senecio viscosus* are prominent. In Dorset it has an atypical habitat on bare ground at the top of 40 m-high cliffs where erosion from spray and wind and disturbance by gulls keep open a strip up to a metre wide. Here the associates include *Beta vulgaris, Chenopodium album, Erodium cicutarium, Malva sylvestris* and *Phleum arenarium,* in addition to some of those listed above. In Kent its associates in open communities on sandy tracks in dune grassland included *C. album, E. cicutarium, Honkenya peploides, Polygonum aviculare* and *Solanum nigrum*. It can tolerate very high levels of nitrogen.

This annual species germinates in spring, and flowers from July to September. A second germination and flowering may occur in August and September, if a dry summer has been followed by rain. In its cliff site it germinates only where the majority of the substrate is soil, and not where there is rock or rock with soil pockets.

The recent decline of *C. vulvaria* in Britain has been remarkable. Once known from over 100 hectads, it declined to 15 by 1960, and is restricted to only three regular sites today. Sprays in arable fields, tidier rubbish dumps, housing developments and, above all, less open habitat in its coastal niches have accounted for this loss. At the Suffolk site, the colonies had been severely threatened by rabbit-grazing, and by soil compaction from vehicles. However, they are now fenced against both. Numbers have fluctuated greatly, with maxima in recent years of 125 plants in 1990 and 83 in 1986. Conservation management now involves opening up the enclosure during winter and spring to allow rabbits to graze down robust plants (especially *Glaucium flavum* and *Lepidium latifolium*), and rotavating the plot in late April to provide open ground for germination (Odin 1990; P. Holmes pers. comm. 1996). The Dorset site is kept open by the elements and, although only a few metres from the coast path, seems quite secure. Numbers fluctuate considerably: a seven year period has seen two years with fewer than ten plants, two years of over 2,000, and the other three with about 100 plants. There are rabbits on site, with burrows right by the plants,

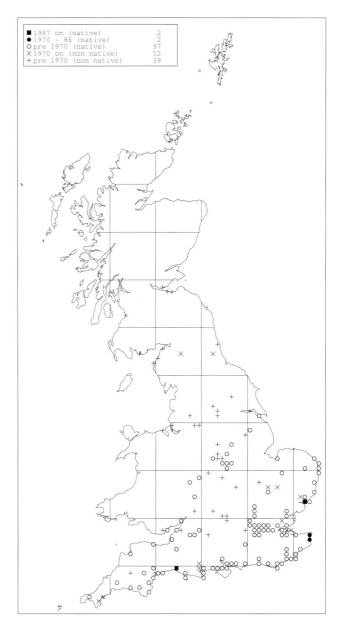

and this may be one of the reasons for the fluctuations. In one year a few plants were found on the beach below, and in two years some appeared by the coast path. In 1996 only a few plants were found on the cliff edge, but there were over 500 plants in bunkers on the adjoining golf course.

C. vulvaria is widespread throughout Europe north to Denmark, and is found also in North Africa and South-West Asia. Little is known of its current European status.

D.A. Pearman

Cicerbita alpina (L.) Wallroth (Asteraceae)
Alpine Blue sow-thistle, Bliochdan Gorm Ailpeach
Status in Britain: VULNERABLE. WCA Schedule 8.
Status in Europe: Not threatened.

C. alpina is one of the largest and leafiest mountain
plants in Britain. When in fresh flower it is a stately,
attractive plant, but despite predominantly occurring in
humid sites sheltered from strong winds, it often soon
succumbs to the mountain weather and wears a scorched
and battered look. Most of its colonies lie on broad ledges
or sloping rocky gullies with a good depth of soil at
altitudes between 530 and 1,090 m. The parent rock is
always acidic, but there is probably some base-enrichment
from irrigation which brings the soil pH to 4.8-5.9 (McVean
& Ratcliffe 1962). These ledges often carry snow well into
early summer. A characteristic associate is *Athyrium
distentifolium*, and other frequent associates in what is
usually a fairly lush ledge flora are *Alchemilla glabra,
Deschampsia cespitosa, Dryopteris* species, *Luzula sylvatica,
Rumex acetosa* and *Solidago virgaurea.*

This tall, rhizomatous perennial usually grows in
clumps of several square metres, with hollow reddish,
erect flowering stems about a metre high rising above a
'mattress' of succulent, broadly sagittate leaves. Because of
its general appearance, one colony was known to climbers
as 'the potato patch'. The deep blue-violet racemes of
flowers appear in late July and last until early September,
with bumble-bees the chief pollinators. Seed-set appears to
be very irregular and seed often not viable. Some, possibly
all, populations may be clones, with outbreeding
prevented by the isolation of one population from the next.

This species was first discovered by George Don in
1801, on one of his many ascents of Lochnagar. Later in
the century it was found in Glen Callater, Caenlochan,
Glen Canness and Glen Clova, and this portion of the
eastern Highlands in Aberdeenshire and Angus remains its
known range. About fifteen locations have been identified
since first recorded, but it has been lost from many,
leaving only four extant. The four colonies produced a
total of about 600 flowering spikes in 1995, compared with
about 500 in 1979-1983.

In living memory, *C. alpina* has been confined to steep
corries and cliffs out of reach of sheep and deer. Early
Floras indicate that the plant once grew by the side of
burns on more accessible ground, which suggests that its
present isolation was caused by overgrazing. Suitably
large, moist, sheltered ledges are few and far between, and
this limitation of habitat may alone explain the rarity of
the plant in the eastern Highlands. It seems poorly
adapted to mountain conditions and may be an example of
a former sub-montane species surviving only in sub-
optimal habitats at atypical elevations.

The Scottish populations appear to have long-term
stability, though they undergo short-term change. They
also sometimes suffer temporary damage from rock falls,
drought and occasional deliberate cutting. The seldom
visited and declining population in Glen Canness
succumbed to drought and overgrazing in 1976, but a

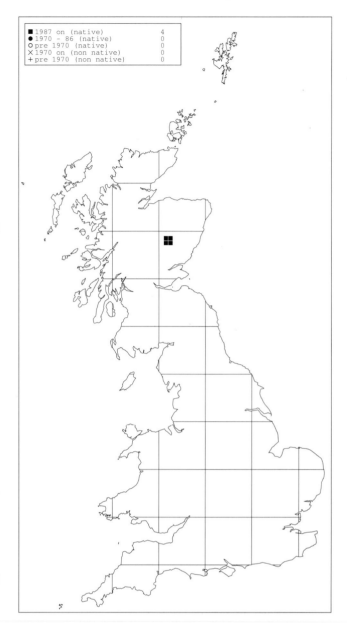

healthy population was rediscovered in Glen Clova by
A.G. Payne in 1979 after a 30-year interval. It has been
suggested that 'lost vitality' may threaten the survival of
the species, but this is not certain.

Outside Britain, *C. alpina* is a common submontane
plant in Scandinavia and in the main central European
mountain ranges. Here it occurs most typically in moist
conditions near the tree line as a member of a tall-herb
flora with *Trollius europaeus* and *Geranium sylvaticum* in
birch and pine forest on the more basic soils. In
Scandinavia it also occurs in submontane meadows and on
roadsides.

Marren *et al.* (1986) provide further details of the
ecology and distribution of *C. alpina*.

B.G. Hogarth and P.R. Marren.

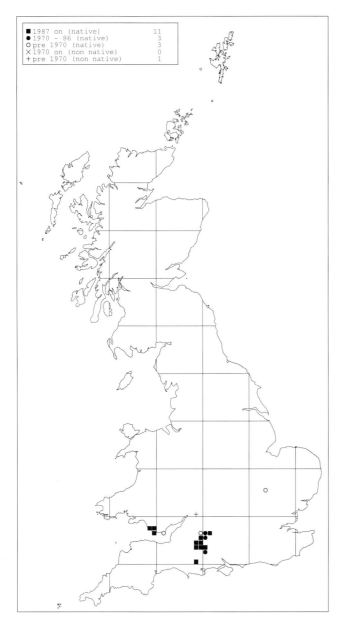

■ 1987 on (native)   11
● 1970 - 86 (native)   3
○ pre 1970 (native)   3
✕ 1970 on (non native)   0
＋ pre 1970 (non native)   1

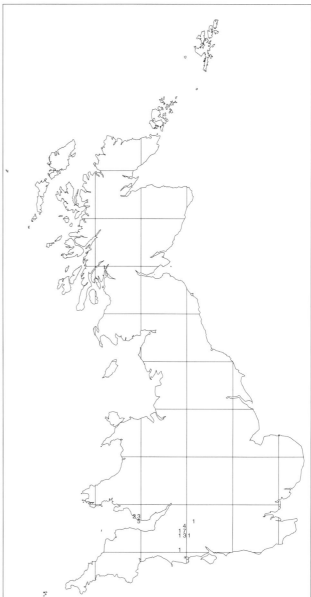

Cirsium tuberosum (L.) All. (Asteraceae)
Tuberous thistle, Ysgallen Oddfynog
Status in Britain: VULNERABLE.
Status in Europe: Not threatened. Endemic.

In Britain, the largest surviving populations of *C. tuberosum* grow in species-rich ancient chalk and limestone grassland, often on slopes with a northerly or north-westerly aspect. Typical associates include *Carex flacca, Centaurea scabiosa, Dactylis glomerata, Festuca rubra, Helictotrichon pubescens, Sanguisorba minor, Serratula tinctoria* and *Succisa pratensis*, with *Bromopsis erecta, Filipendula vulgaris* and *Genista tinctoria* in its Wiltshire sites.

It is a long-lived herbaceous perennial that spreads slowly by the production of axillary basal rosettes to form densely-leaved clonal patches, sometimes a metre or more in diameter. When vigorous and ungrazed, it may dominate or even exclude other vegetation. Flowering begins in late June and peaks in July, with a few heads continuing to appear through August and into September. In Welsh populations the erect flowering stems are typically 50-60 cm high with a single capitulum, but in Wiltshire they grow to 100 cm and bear two or three capitula. Flowers are produced freely by some plants, but rather sparsely or not at all by others, even when vegetative growth is vigorous. The flowers are visited by bumble-bees (*Bombus* species) and butterflies. Pollen fertility is high in apparently 'pure' populations in Wales and Wiltshire, and isolated plants are self-compatible, although seed-fertility is variable and sometimes low.

C. tuberosum has a disjunct distribution in Britain, now growing only in a group of about 8-10 limestone grassland sites along 5 km of the coast of south Glamorgan and in a series of increasingly scattered downland or old pasture localities in Wiltshire (Grose 1957), where there are at least seven extant colonies (Everett 1993), and Dorset, where one small population is known (Mahon & Pearman 1993). It also grew on an ancient grassy trackway in

97

Cambridgeshire, but became extinct there in 1974, although plants from the original stock were reintroduced to a nearby site in 1987, with initial success (Pigott 1988).

Threats come from habitat destruction by ploughing, coastal erosion and changes in land management and, especially, from hybridisation with *Cirsium acaule* and perhaps other *Cirsium* species. A local threat from rabbits, and perhaps other small mammals, which dig for the edible tubers of *C. tuberosum*, was successfully countered in a small Glamorgan population by wire netting laid on the soil surface (Kay & John 1995). In Wiltshire, many populations appear to have been severely affected or swamped by hybridisation with *C. acaule*. Here, a few refuges for *C. tuberosum* are found on unploughed north-facing slopes with tall swards unsuitable for the low-growing, thermophilous *C. acaule*. Genetic surveys indicate that the Glamorgan populations of *C. tuberosum* are little, if at all, affected by hybridisation with *C. acaule*, which is at the limit of its British range there.

C. tuberosum has a wide but scattered distribution in western and west-central Europe, normally growing in rather damp calcareous grassland. It is closely related to *Cirsium dissectum* and the European species *C. filipendulum* and *C. rivulare*, which show varying degrees of ecological, morphological and geographical distinction.

Q.O.N. Kay

Clinopodium menthifolium (Host) Stace (Lamiaceae)
Calamintha sylvatica Bromf.
Wood calamint
Status in Britain: ENDANGERED. WCA Schedule 8.
Status in Europe: Not threatened.

C. *menthifolium* is a plant of woodland edges and
scrubby thickets on chalk. It has only ever been known as
a native species in this country from a single location, a
dry chalk valley on the Isle of Wight, where it was first
described by William Bromfield as occurring "in the
greatest profusion and luxuriance" (Bromfield 1843). The
only current site is a scrubby woodland edge where there
is good light and the soil contains little humus. Associated
plants are those characteristic of chalk woods in this part
of the Isle of Wight, including *Allium ursinum,*
Brachypodium sylvaticum, Campanula trachelium, Rubus
caesius, Stachys sylvatica and *Tamus communis.* More
recently, the community has tended to become dominated
by more nutrient-demanding tall herbs such as *Dipsacus*
fullonum, Eupatorium cannabinum and *Urtica dioica.*

C. *menthifolium* spreads vegetatively by short creeping
rhizomes. Flowering is prolific, beginning in July and
generally continuing until October. The flowers are visited
by bumble-bees. Large quantities of fertile seed are set, but
recruitment from seed is very low. It grows well under
horticultural conditions and spreads freely by seed in the
absence of competition.

At its greatest extent in the nineteenth century, the
plant occupied a few hectares of a single valley, but the
population is now reduced to a few square metres. The
historic landscape was a mixture of woodlands, downland
and scrub that was maintained by free-roaming grazing
animals and coppicing of hazel. Bromfield described it as
"growing amongst the long herbage and under the shade
of bushes, in vast quantity, for a part of the way towards
the head of the vale, scattered over hill-side copses
wherever there is shade and shelter sufficient." Since 1940,
coppicing has declined and the hazel canopy has become
dense, eliminating all the lightly shaded areas where the
plant was known to occur. At this time, the adjoining lane
was improved and a newly created lay-by exposed fresh
chalk onto which the plant was able to spread. At the
present time, it occurs along an eastward-facing woodland
edge and is largely confined to two lay-bys alongside a
single-track road, with a few scattered plants growing
elsewhere along the road verge, over a distance of some
40 m. By 1960, fewer than five plants remained, but
following conservation management, the current
population is estimated to be in the low hundreds.

Conservation management since 1962 has included the
coppicing of hazel and the clearance of invasive ground
cover. Though this has been successful in maintaining

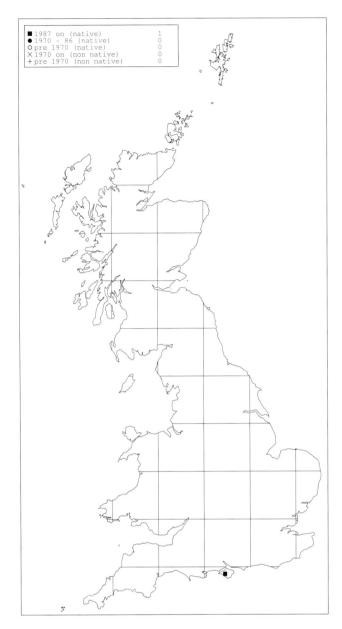

populations (largely through vegetative spread), it has not
resulted in any significant increase. As its current sites are
becoming eutrophic, competition from more aggressive
species is proving to be a problem. *C. menthifolium* would
probably benefit from the creation of freshly exposed chalk
rubble slopes into which it can seed. Despite protection
within an SSSI and intensive conservation effort, the plant
remains endangered (Winship 1994c).

C. *menthifolium* occurs in most European countries but
does not extend to Scandinavia.

C. Pope

Cochlearia atlantica Pobed. (Brassicaceae)
Atlantic scurvy-grass
Status in Britain: DATA DEFICIENT. Probably ENDEMIC.

The definition of the constituent taxa within the very complex *Cochlearia officinalis* group is notoriously problematic and has been debated for many years. The main difficulty lies in the considerable phenotypic plasticity, induced by such factors as water and nutrient stress, combined with genetic differentiation (Rich 1991). On the basis of its distinct morphology and favoured habitats (beach-head saltmarshes and other rocky-shore grasslands), Rich (1991) accepts *C. atlantica* as a good species, and it seems to maintain its characters in cultivation (T.C.G. Rich pers. comm. 1997). Other authorities have considered that these plants are merely extreme forms of *C. officinalis sensu stricto*, or a segregate of it that might merit subspecies rank or be of hybrid origin.

It grows on sandy, silty or stony sea shores, often with *Puccinellia maritima*. It has been recorded from Arran, Loch Linnhe, Rum, Lewis and South Uist and has also reported from Coll and Tiree (S.J. Leach pers. comm.), but probably occurs more widely along the west coast of Scotland. Further investigations are needed on its status, taxonomic relationships and occurrence, and for this reason no map of its currently-known distribution is included.

M.J. Wigginton, from an account by T.C.G. Rich

Cochlearia micacea Marshall (Brassicaceae)

Mountain scurvy-grass, Carran an t-Slèibhe
Status in Britain: LOWER RISK - Nationally Scarce.
Perhaps ENDEMIC.

C. micacea is an arctic-alpine species, with recent records from altitudes of 610-1,120 m, though there are older records from lower altitudes. It is found in four main types of habitat, generally rooting into a quite basic substrate. Most typically it is a plant of flushes, springs and stream-sides. It occurs in springs of various types, including those characterised by *Anthelia julacea*, *Philonotis fontana* or *Pohlia wahlenbergii* var. *glacialis*. On flatter streamsides, common associates include *Deschampsia cespitosa*, *Epilobium anagallidifolium* and *Saxifraga stellaris*. Other habitats are short calcicolous turf, cliffs and ledges (on which it grows to its maximum size), and stony gullies or ravines in cliffs (Dalby & Rich 1995).

This species is a perennial, but details of its reproductive biology are not well known. On Ben Lawers, Clarke (1993) recorded flowers from mid-May until September, with fruit first appearing at the end of June. Watson (1994) noted that flowers did not appear until June. In 1994, the setting of fruit was poor - in flushes as low as 5%, but rather higher in populations on cliffs. The reason for this is unknown, but self-incompatibility may be the most likely (Dalby & Rich 1995). There is some evidence that plants can reproduce vegetatively, but reproduction by seed is probably the usual method.

Since historical records from England and Wales are now rejected following a re-examination of herbarium specimens (Dalby & Rich 1995), *C. micacea* is now known to be confined to Scotland. It is currently present in about 19 hectads, most of which are in the Breadalbane Mountains. Records from Unst, Shetland, were errors (Stace 1997). It is undoubtedly to be found elsewhere. Populations range in size from many thousands down to a few tens of plants. This species appears to be quite safe, and no immediate threats are apparent apart from local development related to mountain sports. It seems to tolerate grazing, and no special conservation measures are needed.

C. micacea is included in the taxonomically difficult *C. officinalis* group. Plants now assigned by Dalby & Rich (1995) to *C. micacea* have in the past been variously placed under that name or under *C. groenlandica*, or treated as a mere variant of *C. officinalis*. The main morphological features differentiating this taxon from its close relatives are in the shape of mature fruits when they are dry and just before dehiscence. The other main diagnostic character is the chromosome number of 2n=26, which is unique

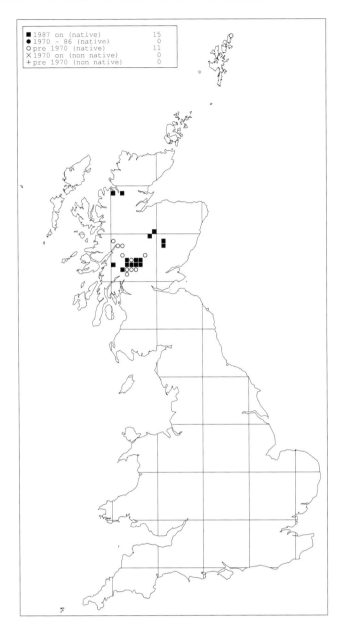

■ 1987 on (native)	15
● 1970 - 86 (native)	0
○ pre 1970 (native)	11
× 1970 on (non native)	0
+ pre 1970 (non native)	0

among British *Cochlearia* taxa. On the basis of these characters, Rich (1991) and Dalby & Rich (1995) consider that the specific status of *C. micacea* is fully justified. Modern methods of genetic fingerprinting, hitherto unattempted, would greatly assist in elucidating the relationship of this plant to other segregates within *C. officinalis* agg.

Further details are given in Rich & Dalby (1996).

M.J. Wigginton

Coincya wrightii (O.E. Schulz) Stace (Brassicaceae)
Rhynchosinapis wrightii (O.E. Schulz) Dandy
Lundy cabbage
Status in Britain: VULNERABLE. ENDEMIC. WCA Schedule 8.

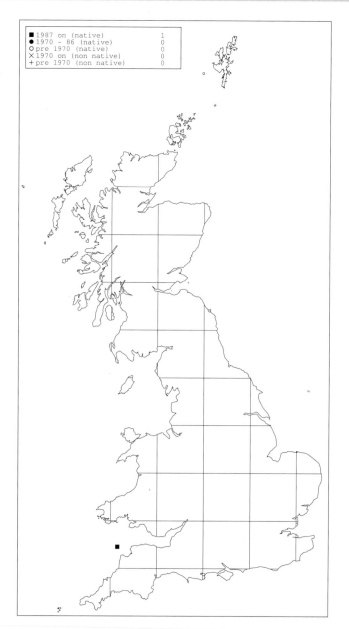

C. wrightii occurs only on Lundy, the main centre of distribution being the south-eastern part of the island (Irving 1984; L. Farrell pers. obs.). It grows in exposed, open communities subject to high atmospheric humidity during the cooler months and great heat coupled with a high light intensity during the summer (Marren 1971). Habitats are mainly the steep or vertical cliffs of granite and slate, but it also occurs on flat ground above the cliff-slopes where protected from grazing animals. It grows mainly in rather open communities where competitors are few. Species that occur in the same habitat include *Arrhenatherum elatius, Digitalis purpurea, Holcus lanatus, Lonicera periclymenum, Pteridium aquilinum, Rubus fruticosus, Sedum anglicum, Teucrium scorodonia* and *Umbilicus rupestris.* However, both seedlings and more mature plants can survive in quite closed swards, in deep bracken litter, and under heavy canopies of bramble, gorse or bracken (R. Key *in litt.*).

C. wrightii is a biennial or perennial of variable habit, sometimes tall and straggly, sometimes large and bushy, and dwarf in very dry soil. Populations fluctuate markedly, and in recent years numbers have ranged from about 320 in 1979 (Cassidi 1980) to 4,500 in 1994. The reasons for the recent increase are not clear, though it is quite possible that a series of warm and dry summers and intervening mild winters in the 1990s favoured the plant.

However, both overgrazing and spread of invasive shrubs are threatening some colonies. Young plants in particular are grazed by domestic animals (sheep and ponies), as well as by rabbits, Sika deer and the free-roaming Soay sheep. *Pteridium aquilinum* and *Rhododendron ponticum* locally threaten some colonies. Although there has been substantial control of the invasive shrubs under a Countryside Stewardship agreement, clearance has not so far been concentrated in areas where *C. wrightii* chiefly occurs.

Invertebrates associated with *C. wrightii* are of interest, particularly an endemic flea beetle *Psylloides luridipennis* which is specific to the plant. The biology of the genus *Coincya* is described in Leadley & Heywood (1990).

L. Farrell and M.J. Wigginton

Corrigiola litoralis L. ssp. *litoralis* (Caryophyllaceae)
Strapwort
Status in Britain: CRITICALLY ENDANGERED. WCA Schedule 8.
Status in Europe: Not threatened.

In Britain, *C. litoralis* has been recorded as a native plant only in Devon, Cornwall and perhaps Dorset, but it now occurs only around Slapton Ley in Devon. There it grows on areas of muddy shingle that is seasonally inundated by the rising water levels in the Ley and subject to poaching by cattle. It is a poor competitor, intolerant of competing vegetation taller than about 4 cm and growing best on open ground. Its usual associates are *Chenopodium album*, *C. rubrum*, *Persicaria hydropiper*, *P. maculosa*, *Poa annua*, *Polygonum aviculare*, *Potentilla anserina* and *Sisymbrium officinale*. Recently several of the open areas where it formerly grew have been invaded by *Phragmites australis* and *Urtica dioica*. Habitats at the former site at Loe Pool, Cornwall, are considered no longer suitable for the plant. There are casual records northwards to Yorkshire and a few colonies that are more persistent. Common associates, for instance on railway ballast, include ruderal species such as *Cerastium fontanum*, *Chamerion angustifolium*, *Linaria vulgaris* and *Senecio viscosus*.

This species is an annual, rarely overwintering, forming a flat rosette of several decumbent and radiating shoots. The tiny whitish-green axillary flowers appear in late July or August, though some may be found in September in some seasons. They are self-pollinated and often cleistogamous, and reproduction is entirely by seed (Coker 1962).

Populations have declined at Slapton in recent years. In 1989, about 50 plants occurred at about five locations around the Ley. In the early 1990s, water levels were exceptionally high because of high rainfall, and very few plants were seen. The causes of its decline include not only the high water levels but also changes in water quality and salinity, competition from vigorous species and lack of cattle poaching. However, cattle have recently been given access to more of the shoreline to see whether the resulting soil disturbance will lead to the germination of seed hitherto buried. 'Recovery' programmes aim to establish several strong populations at Slapton by special protection of existing plants, and by introducing plants propagated *ex situ*. Lower water levels during the growing season would also be beneficial.

C. litoralis is a subatlantic-mediterranean species of western and southern Europe, extending northwards to Britain and Denmark, it but is declining in many areas. It occurs in Russia, Turkey and quite widely in Africa. In western Europe is rarely found above 100 m but ascends to 2,500 m in Kenya. In France and Germany it occurs in a range of open habitats including sandy banks of rivers, cart-tracks and along footpaths (Coker 1962). In view of this catholic behaviour, it is difficult to account for the extreme restriction of native sites in Britain.

L. Farrell

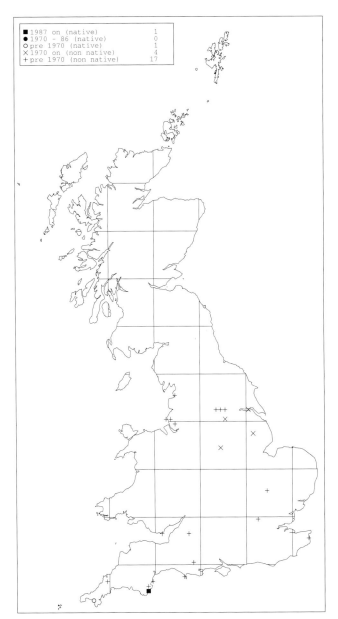

■ 1987 on (native)	1
● 1970 – 86 (native)	0
○ pre 1970 (native)	1
✕ 1970 on (non native)	4
+ pre 1970 (non native)	17

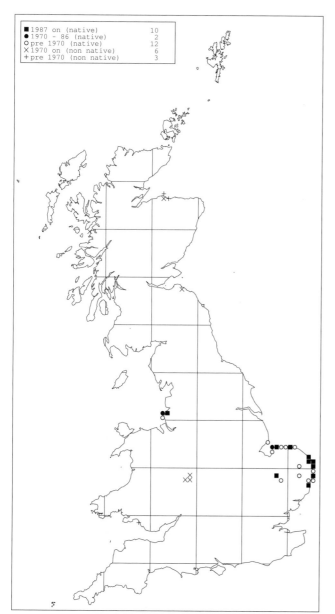

■ 1987 on (native)	10
● 1970 - 86 (native)	2
○ pre 1970 (native)	12
✗ 1970 on (non native)	6
+ pre 1970 (non native)	3

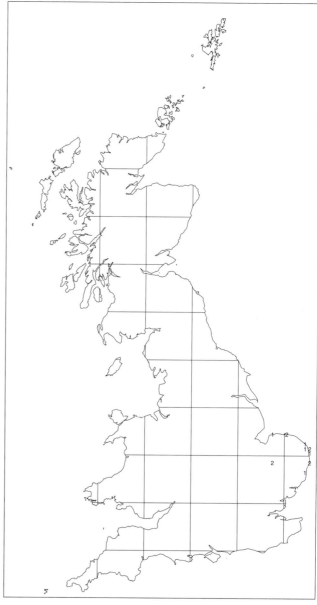

Corynephorus canescens (L.) P. Beauv. (Poaceae)
Grey hair-grass
Status in Britain: LOWER RISK - Near Threatened.
Status in Europe: Not threatened.

This is a mainly coastal species found in short sandy turf on beaches and on dunes to the landward side of shingle ridges. It also occurs on inland areas of mobile sand. It survives only where an annual accretion of sand around the base of the plants allows bud growth (and hence the development of roots and new shoots) low down on the culm. Dispersal of plants over a wide area is unusual and most are found in confined stretches of open vegetation. In areas with low rainfall, plants are well-spaced, and elsewhere it rarely forms a continuous sward. Its common associates are *Agrostis vinealis, Ammophila arenaria, Carex arenaria* and *Hypochaeris radicata*. In very dry sand it is also found with *Cladonia gracilis, C. impexa, C. pyxidata* agg. and *Cornicularia aculeata*.

C. canescens is a perennial reproducing by seed. There is no vegetative spread beyond individual plants arising from the parent position. Germination is erratic, and this species does not colonise new sites readily.

The distribution is probably stable, although populations may vary considerably in size from year to year. Prolonged gales will unseat plants in loose sand, and beach colonies are occasionally in danger from tidal encroachment and from the public. Management for this species must ensure that colonies remain exposed to wind and hence subject to sand movement. An adjacent conifer plantation will shelter plants from wind, as will the encroachment of bracken and gorse. The presence of some rabbits will help to control *Carex arenaria* by grazing, and their liking for *C. canescens* stems is offset by vegetative reproduction. For details of the introduced populations in Scotland, see Trist (1993).

C. canescens is found in Europe from the southern Baltic to southern Portugal, northern Italy and central Ukraine. It is local in the eastern part of its range. It is also recorded from North Africa, and occurs as an introduction in North America.

Further details of the ecology of this species are given in Marshall (1967).

P.J.O. Trist, in Stewart *et al.* (1994).

Cotoneaster integerrimus Medikus (Rosaceae)

Cotoneaster integerrimus L. auct., *C. cambricus* Fryer & Hylmö
Cotoneaster, Cotoneaster y Gogarth
Status in Britain: ENDANGERED. WCA Schedule 8.
Status in Europe: Not threatened.

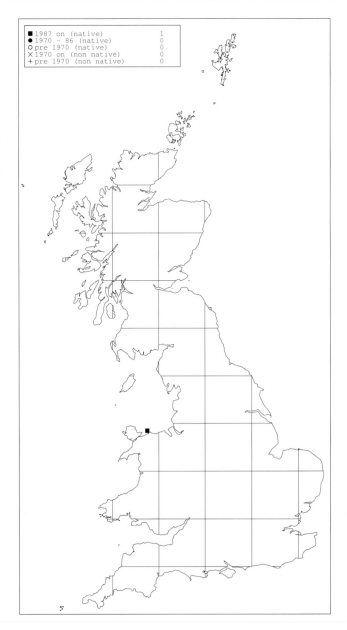

In Britain, *C. integerrimus* is found only on Great Orme's Head near Llandudno. It appears to have been first discovered in 1783 by John Wynn Griffith of Garn, but this record was not published until 1830. The earliest published report was in J.E. Smith's English Flora (1828) of a record made in 1825 of plants "on the limestone cliffs ... in various places", but in fact in a limited area above an old copper mine and near to houses, gardens and cultivated land. Lees (1850) described it as being "in some quantity" but, by the beginning of this century, Phillips (1905) thought it "exceedingly scarce ... on the brink of extinction". At the present time, only six plants from the historic population are extant. Other plants, propagated after much trial and error from original stock (Morris 1980), have been planted out in suitable locations on the Great Orme, and the whole population is regularly monitored (Gravett 1994).

The six original plants are widely dispersed on the limestone ledges of the Great Orme and differ markedly in size. Four are in sites that are easily accessible to sheep and, possibly a greater threat, the herd of feral Kashmir goats that live on the Great Orme. These four plants each consist of just a few branches, none longer than 60 cm. A fifth plant grows on a less accessible ledge, and although its spread is up to a metre it is no more than 25 cm in height, probably owing to its exposed western aspect. The sixth plant is in a sheltered east-facing location amongst other shrubs, which include *Crataegus monogyna, Euonymus europaeus* and *Rosa canina*. This is much the biggest specimen with a height and spread of up to 1.5 m, presumably because of the exclusion of grazing. Other associated species include *Carex flacca, Festuca ovina, Geranium sanguineum, Helianthemum canum* and *Thymus polytrichus*.

C. integerrimus is a deciduous spreading shrub, flowering in April to June, with ripe fruit in August. The flowers are visited by various insects, especially Hymenoptera. The species does not appear to regenerate naturally at Great Orme, either vegetatively or from seed. Probably the greatest threats to the continued survival of the remaining plants are excessive browsing by sheep and goats.

This plant has been widely accepted as a native on the Great Orme (e.g. Clapham *et al.* 1987; Stace 1991). However, in a recent review of its recorded history and localities, Kay & John (1995) consider that "there is every reason to suppose that [it] originated there as a garden

escape in the late eighteenth or early nineteenth century from plants cultivated locally". Further, they conclude that there is insufficient evidence to treat the taxon as a species (*C. cambricus*) distinct from *C. integerrimus* (*cf.* Fryer & Hylmö 1994). Further investigation is indicated, and it is important in the meanwhile to conserve the plants on the Great Orme for continued taxonomic and molecular studies. *C. integerrimus* occurs throughout much of Europe, except in the extreme north and much of the Mediterranean region, extending eastwards to Turkey and northern Iran.

W. McCarthy

Crassula aquatica (L.) Schöl. (Crassulaceae)

Pigmyweed, Luibh Beag Bìodach
Status in Britain: VULNERABLE. WCA Schedule 8.
Status in Europe: Not threatened.

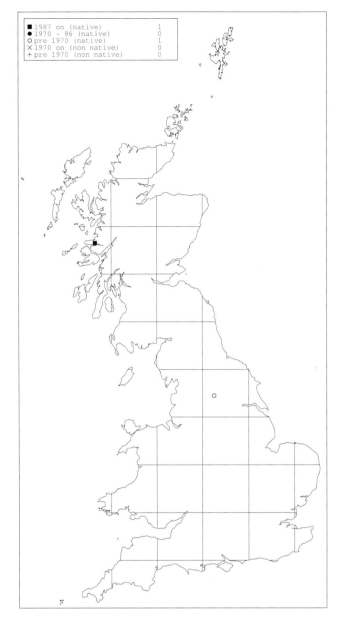

At its sole surviving British locality, *C. aquatica* grows at the side of the River Shiel, West Inverness, and in a ditch and a shallow scrape created for duck nearby. In the sheltered shallows of the river it can grow as scattered plants or in dense swards in water up to 25 cm deep, associated with frequent *Glyceria fluitans* and *Alisma plantago-aquatica*, *Callitriche* species, *Eleocharis palustris*, *Elodea canadensis*, *Hydrocotyle vulgaris*, the aquatic form of *Juncus bulbosus*, *Littorella uniflora*, *Persicaria hydropiper*, *Ranunculus flammula* and *Fontinalis antipyretica*. It also grows terrestrially as scattered plants on moist and rather bare sandy soil, with *Agrostis stolonifera*, *Galium palustre*, *Gnaphalium uliginosum*, *Juncus bulbosus*, *Ranunculus flammula*, *Viola palustris*, *Fossombronia* sp. and *Pellia epiphylla*. In Yorkshire *C. aquatica* formerly grew in shallow water at the edge of an artificial lake, with *Callitriche stagnalis*, *Limosella aquatica*, *Persicaria hydropiper* and *Rorippa palustris* on damp mud. Both sites are below 300 m.

C. aquatica is a small annual which presumably germinates in spring and summer. Plants flower and fruit in August and September. Both terrestrial and aquatic plants flower freely. The flowers are inconspicuous and are probably automatically self-pollinated.

This species was first discovered in Britain at Adel Dam, Yorkshire, in 1921 (Butcher 1921). It increased in abundance in the years following its discovery but later declined, as the soft mud on which it grew became colonised by dense beds of *Carex rostrata*, *Juncus effusus* and *Typha latifolia* (Sledge 1945). It was last seen in 1938 and had gone by 1945. The West Inverness population was discovered in 1969, when it occupied an area of about 6 square metres^2 (Sowter 1971). In 1990 it was present along a length of river of at least 200 m^2, but the population has not been surveyed sufficiently frequently to establish whether this represents a long-term increase or a natural fluctuation in numbers. The number of plants certainly varies from year to year: the plant was present in 1969 and 1971, for example, but could not be found in 1970 (Sowter *et al.* 1972).

This species is widespread in the northern hemisphere but has a disjunct distribution. The main concentrations of records are from Europe, East Asia and both western and eastern North America south to Mexico, but it is also recorded at scattered localities elsewhere. In Europe it occurs from Iceland and northern Scandinavia south to Austria.

It is difficult to account for the presence of this species in just two British sites, but their disjunct nature and their northerly location fit in with the European distribution of the species. The tiny seeds may have arrived on the feet of aquatic birds, as Druce (1922) suggests, or even on the waders of visiting salmon fishermen.

C.D. Preston, adapted from an account in Preston & Croft (1997).

Crepis foetida L. (Asteraceae)
Stinking hawk's-beard, Gwalchlys Drewllyd
Status in Britain: ENDANGERED.
Status in Europe: Not threatened.

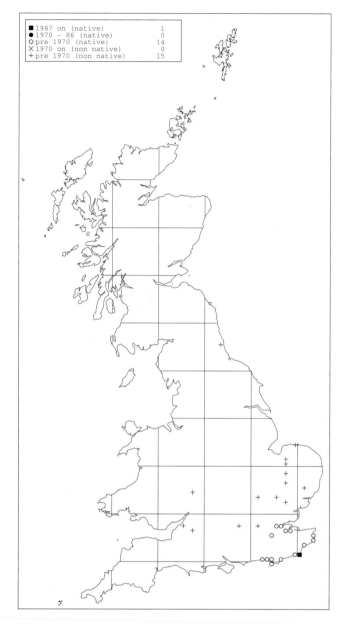

During this century, *C. foetida* has been recorded at only a few coastal sites in south-east England, typically on disturbed shingle or chalk. Most records have been from Dungeness, none involving more than a very local scattering of plants. Though it became extinct in 1980, it was re-introduced to Dungeness in 1992 by using seed from Dungeness plants that had been cultivated at Cambridge University Botanic Garden (Ferry 1995). Plants were first established in experimental plots in the area of its last recorded occurrence at Dungeness, and by 1995 it had 'escaped' onto the adjacent shingle. The shingle ridge community comprises patches of *Cytisus scoparius* in a mosaic with 'shingle heath' containing *Anthoxanthum odoratum, Festuca tenuifolia, Galium mollugo, Hypochaeris radicata, Pilosella officinarum, Silene uniflora, Teucrium scorodonia, Dicranum scoparium* and *Cladonia* species. The *C. foetida* plants, which numbered 139 in 1995 and 168 in 1996, have colonised marginal areas of this natural community where the vegetation is very sparse and the substrate a mix of pebbles and consolidated peat. Interestingly, disturbance of the shingle does not seem to have been a prerequisite for such colonisation.

Over most of its range, *C. foetida* is a biennial or an annual (Babcock 1947). At Dungeness, it routinely performs as a stressed annual if grown from seed, naturally producing diminutive plants to 15 cm tall, which lack basal rosettes and have few if any branches and only a few capitula. Plants grown in good soil *ex situ* form basal rosettes, and such plants, when transplanted onto the shingle, develop biennially and form taller, well-branched specimens with many capitula in their second year. This species is fairly easy to distinguish in the field from other similar Asteraceae by two morphological characters: nodding flower buds and the normally tightly closed, snowy white pappus.

The reasons for the extinction of *C. foetida* in England are not clear. Climatic factors may be involved, but whether the problem is one of germination, vegetative growth, flowering or seed-set remains to be answered. Viable seed is produced readily by plants in the field, and germination tests show that over 50% of newly-produced seed germinates in the laboratory, and about 30% in field experimental plots. Seed appears to remain viable for several years, so there is every likelihood that a seed-bank could persist. Dungeness has appreciably less rainfall and more sunshine than even adjacent regions of Kent, and is extremely well-drained. This near-Mediterranean set of conditions may suit the plant's needs, but it is not difficult to imagine the balance being tipped unfavourably in occasional years. Biotic factors may be significant. Rabbits seem to be partial to *C. foetida*, and the experimental plots need to be protected with wire cages to prevent transplanted biennial rosette plants from being taken by them. However, small annual plants, now escaped in some numbers onto the open shingle, seem to be ignored by rabbits. Whether the plant needs a disturbed substrate, as has often been assumed, is now less clear given the recent colonisation of undisturbed habitat around experimental plots at Dungeness.

C. foetida is widespread throughout most of Europe, occurring on open and dry, rocky or sandy sites, both coastal and inland to altitudes in excess of 1,000 m, and is more common in the southern and eastern parts of the continent (Davis 1975). It reaches its north-eastern limit in Britain and the north coasts of France, Germany and Poland. In Asia, it extends to central and southern Russia, Iran and the north-western Himalayas.

B.W. Ferry

Crepis praemorsa (L.) Tausch (Asteraceae)
Leafless hawk's-beard
Status in Britain: ENDANGERED.
Status in Europe: Not threatened.

The discovery of *C. praemorsa* in northern Westmorland in 1988 makes this species one of the most recent additions to the British flora (Halliday 1990). It occurs on the banks of a stream, on both sides of a boundary wall which crosses it and separates a hayfield from grazed pasture, the whole population occurring along a stretch of about 150 m. Most plants occur on the sloping banks of limestone drift about a metre above the stream level, in short, rather open turf. The wide range of associates on the hayfield side of the wall include *Briza media*, *Campanula rotundifolia*, *Carex ornithopoda*, *C. panicea*, *Danthonia decumbens*, *Festuca ovina*, *Gentianella amarella*, *Koeleria macrantha*, *Leontodon hispidus*, *Lotus corniculatus*, *Pimpinella saxifraga*, *Scabiosa columbaria*, *Serratula tinctoria*, *Stachys officinalis*, *Succisa pratensis* and *Thymus polytrichus*. A few plants occur at the top of the stream bank at the edge of rank vegetation in which such species as *Centaurea nigra*, *Festuca rubra* and *Sanguisorba officinalis* are prominent. Plants of *C. praemorsa* in the closely-grazed pasture are more scattered, smaller and generally with fewer leaves than those on the ungrazed banks adjoining the hayfield. It generally has fewer associates in the grazed area, but they include some of those species listed above.

C. praemorsa is a perennial and is distinct from other *Crepis* species in Britain in its leafless flowering stems, and rosette of narrowly-obovate leaves, which are shallowly-toothed in the basal part. Flowering is from late May into June. Few plants produce flowering stems, and there appears to be no seed-set, even in cultivation, suggesting that the population may be a single self-sterile clone. In cultivation, plants produce new rosettes freely alongside the old ones. It also rapidly forms a large root system, with new rosettes arising where it is damaged. This suggests that the plant can colonise effectively by vegetative means, and it seems likely that the wild population has spread along the stream banks by these means over a long period.

C. praemorsa is currently known from only one site in Britain, the whole population occurring within an SSSI. In June 1989, ten flowering plants and 150 non-flowering rosettes were seen (Halliday 1990). In August 1996, plants were difficult to find in the taller herbage at that time of year (F.J. Roberts & M.S. Porter). However, its presence was confirmed in all but one of the areas recorded in 1989, and rosettes were also found in new areas. It has been recorded in eleven discrete small colonies of 5-35 rosettes, and an additional larger one of about 70 plants, with a total population of about 200 individuals.

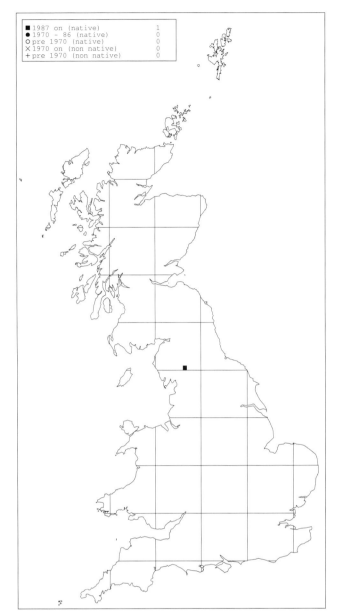

Because of its occurrence in undisturbed semi-natural grassland, it seems likely that this species is native to Britain, though some authorities (e.g. Stace 1991) have cast doubt on its native status. Searches of banks and hay meadows in other traditionally farmed areas may reveal its presence elsewhere.

It has a wide range in Europe and Asia, from eastern France, Denmark and south-east Norway, eastwards through Asia to northern China. The Westmorland site represents a considerable extension of its known range.

F.J. Roberts, M.S. Porter and M.J. Wigginton

Cynodon dactylon (L.) Pers. (Poaceae)
Bermuda grass, Gwair Bermuda
Status in Britain: VULNERABLE.
Status in Europe: Not threatened.

This species is principally a plant of sandy ground in coastal areas of southern England and Wales. Most records are from the lawns of ornamental gardens or along the sea-front of south coast resorts, where they are managed for their amenity value and are mown regularly. In such places, *C. dactylon* may be mono-dominant, or grow in a mixed sward with other robust species including *Bellis perennis, Carex arenaria, Elytrigia repens, Plantago coronopus* and *Poa annua*. However, it also occurs in semi-natural habitat in extreme south-west England. John Ray (1724) noted that it was "found by Mr Newton on the sandy shores between Pensans and Marketjeu in Cornwall plentifully". It still occurs along the shore between Penzance and Marazion, together with typical dune species including *Beta vulgaris, Festuca rubra, Honkenya peploides, Hypochaeris radicata, Plantago coronopus* and *Trifolium occidentale*. Other colonies are found on nearby roadsides with maritime ruderal species.

C. dactylon is a tough perennial grass spreading by extensive, creeping rhizomes. Flowering is in August and September, though it may not flower where severely trampled. It is not known whether there is effective recruitment by seed, but vegetative spread is likely to be the more important.

It seems certain that this species is an introduction at most of its British sites, though positive evidence of planting is hard to come by. However, the long established colonies near Penzance (and in Jersey) have a stronger claim to native status, and are mapped as such. Ryves *et al.* (1996) consider this species might be native in Britain, though Margetts & David (1981) more cautiously record it as "probably originally introduced". As a possible native species or ancient denizen, its conservation should not be disregarded, especially in its fore-dune sites. It also occurs more widely as a casual from wool shoddy and other sources.

This grass can withstand a good deal of trampling, and can survive on dry and impoverished ground, often remaining green when other grasses are parched brown. The main threat to the species may be from coastal development or local maintenance works. For instance, in 1994 part of Newton's original population was destroyed by the construction of a sewage outfall, and a long established site in Poole was destroyed by roadworks.

C. dactylon is one of the most widespread grasses in warm temperate and tropical regions. In Europe it is found throughout the west and south, and eastwards to Ukraine.

R.M. Walls

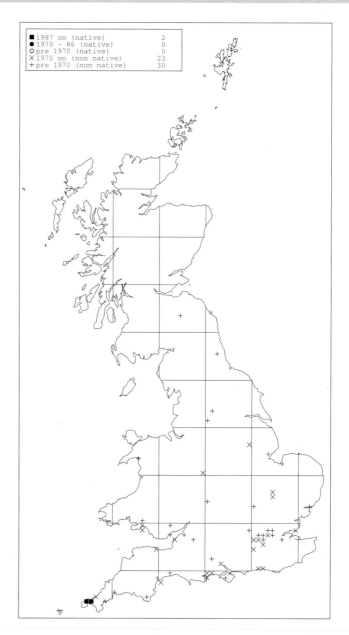

■ 1987 on (native)	2
● 1970 - 86 (native)	0
○ pre 1970 (native)	0
✕ 1970 on (non native)	23
+ pre 1970 (non native)	30

Cynoglossum germanicum Jacq. (Boraginaceae)
Green hound's-tongue, Tafod y Bytheiad Gwyrdd-Ddail
Status in Britain: VULNERABLE. WCA Schedule 8.
Status in Europe: Not threatened.

C. germanicum is native in glades in deciduous woodlands, usually in hilly regions. It is persistent in these sites, for example in the beech woods of Norbury Park in Surrey, where it has been plentiful since the seventeenth century (Merrett 1666), but the actual populations, varying from a few to several hundred plants, shift about as gaps in the canopy open and gradually close over. Similar localities occur nearby on White Hill, where this species is known to have persisted for 50 to 150 years.

In Surrey the species grows most vigorously and abundantly on moist, freely draining, shallow loamy soils over chalk. The tree canopy is dominated by beech, with ash, sycamore and field maple in the gaps and yew under the main canopy. The ground may be almost bare or with patches of *Mercurialis perennis*, but dense *M. perennis* can shade out the first year rosettes of *C. germanicum*. Other typical associates include *Arum maculatum, Brachypodium sylvaticum, Circaea lutetiana, Daphne laureola, Geum urbanum, Rubus vestitus* and *Viola reichenbachiana*. It has, however, become extinct in a few localities of this type, probably because of changes in the woodlands, such as cessation of coppicing or conversion to uniform-aged plantation. *C. germanicum* has also colonised sites that are clearly man-made. Three of its remaining localities are in oak woodlands, where it grows besides roads or tracks that contain, at least in their foundations, lime-rich materials (chalk, hard limestone or concrete) but where the natural soil is acid and has been shown experimentally to be unsuitable. Many of its old records seem to have been transient, the species acting almost as a wayside weed. It appears frequently but briefly in such sites around its main localities in Surrey. At the Oxfordshire site, *C. germanicum* grows in mixed deciduous, mainly oak-ash woodland with *Brachypodium sylvaticum, Galium aparine, Geranium robertianum, Glechoma hederacea* and *Urtica dioica*.

■	1987 on (native)	3
●	1970 - 86 (native)	1
○	pre 1970 (native)	48
✕	1970 on (non native)	2
+	pre 1970 (non native)	1

Surrey has long been the stronghold for this species in Britain, where several thousand plants were recorded at some sites in good years. The 1987 storm created much open ground, and since then the plant has been recorded at numerous locations in 14 1 km squares, mostly in woods around Mickleham. In 1997, an estimated 130,000 plants were recorded. Outside Surrey it appears to be extant at only three sites, one each in Gloucestershire, Oxfordshire and Kent. In Gloucestershire, the population is very small, with only 20-33 plants occurring between 1977 and 1987 (Taylor 1990a), and 33 in 1989 and 1993 (I. Taylor pers. comm.). The Oxfordshire site held at least 200 plants in 1985 but only about twenty in 1994. *C. germanicum* has not been seen for several years at a second site in Oxfordshire, which held several hundred plants in the 1950s and 100 in 1980 (Everett 1988). The small colony in Kent was established from deliberately sown seed (Rose 1960) and has persisted for 40 years.

C. germanicum is a biennial or sometimes a short-lived perennial, and the dark green glossy rosettes persist through the winter when the shoots of *Mercurialis perennis* have died

down. The winter leaves maintain low rates of photosynthesis in mild weather and thus benefit from the leafless state of the tree canopy. The leaves are poisonous (Mattocks & Pigott 1990) and are avoided by deer, rabbits and molluscs, but young plants are defoliated by bank voles. The large rosettes develop a thick tap-root, then flower freely in the second summer and set abundant fruit. Much of the fruit remains on the shoots for many months and is not distributed. Beside paths, however, it is carried away by larger mammals including man and dogs. In fine hair and on woollen clothing the fruits tangle and are probably lost from the whole area. On coarse hair, particularly that of roe deer, the fruits become attached but are easily detached. The great increase in the roe deer population in the last 30-50 years has resulted in the appearance of many new colonies of *C. germanicum* in central Surrey and an explosive spread after the storm of 1987.

Seeds require several weeks of low temperature before they germinate in early spring, but they then have no dormancy in most conditions beyond the first year. This is a key factor in the ecology of the species. In contrast to most species of glades, seeds of *C. germanicum* do not lie dormant in the soil. When a gap is created, fruits must be transported into it, and the probability of this occurring depends on the surrounding abundance of the species. Once the overall population becomes restricted, the chance of dispersal into gaps decreases, and eventually, as the woodland changes, the species dies out. Such small populations can be maintained only artificially.

Outside Britain, it occurs in France and Germany and in the deciduous woodland zone of the mountains of southern-central Europe. It is rare and declining in several countries.

C.D. Pigott

Cyperus fuscus L. (Cyperaceae)

Brown galingale, Ysnoden Fair Lwytgoch
Status in Britain: VULNERABLE. WCA Schedule 8.
Status in Europe: Not threatened.

C. fuscus is found on seasonally-flooded pond and ditch margins in southern England. Most of the populations grow on a peaty substrate or on sand and gravel rich in organic matter. The bulk of Britain's historic and extant colonies are associated with the Thames and Hampshire Avon (and tributaries), typically lying within the tertiary basin of these two rivers, often at the junction with the extensive southern chalk. Such sites are frequently commons with a long and continuous history of extensive grazing by livestock, which has maintained the open, poached conditions favoured by *C. fuscus*. Associates include *Agrostis stolonifera*, *Bidens cernua*, *Gnaphalium uliginosum*, *Hottonia palustris*, *Juncus bufonius*, *Limosella aquatica*, *Mentha aquatica*, *Myosotis laxa*, *Persicaria hydropiper*, *P. minor*, *Ranunculus sceleratus* and *Rorippa palustris*.

It is an annual, normally germinating in early summer, but at some sites germination has been recorded as late as August. Flowering is between July and September. Seed is presumably long-lived in the soil, since germination can be prolific following soil disturbance, even after years of apparent absence. In cultivation a relatively high temperature is required for germination (Tutin 1953), and plants may set little or no seed in a cool summer.

Always a rare plant in Britain, *C. fuscus* has only ever been recorded from twelve mainland localities, all in southern England, and one site in Jersey where it was refound in 1989. Today it is restricted to only six localities, in north Somerset, South Hampshire, Berkshire, Buckinghamshire and Surrey. Being an annual of seasonally-exposed pond margins, and often reliant on some form of disturbance, numbers are liable to fluctuate greatly from year to year. The two Hampshire colonies have consistently produced the largest number of plants in recent years, with the highest recorded population of over 36,500 plants at one site, whilst the North Somerset and Buckinghamshire colonies have consistently performed very poorly, with at best a handful of plants in the years in which it chooses to appear (Rich 1995e).

Most, perhaps all, of the extant colonies receive some form of protection, since they lie within SSSIs and/or nature reserves. In addition the plant is fully protected by law. Housing development, which destroyed its earliest known British colony in central London, does not pose a present threat to extant sites, but populations are threatened in other ways, principally from the cessation of grazing, invasion by willow and alder scrub, and the general lowering of the water table, particularly for sites in the Thames Basin. On a more optimistic note, the plant can respond vigorously to site clearance and the reinstatement

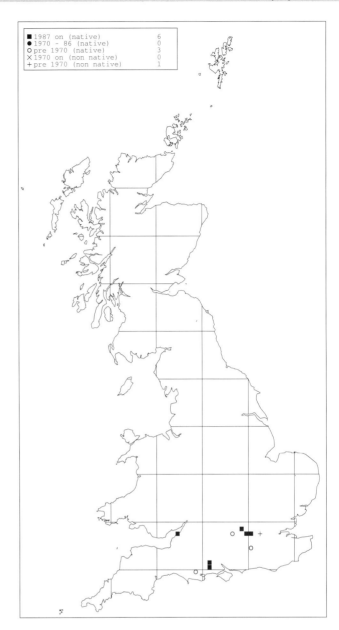

■ 1987 on (native)	6
● 1970 – 86 (native)	0
○ pre 1970 (native)	3
✕ 1970 on (non native)	0
+ pre 1970 (non native)	1

of grazing. For instance, in Surrey 250 plants appeared at one site following clearance of willows.

C. fuscus is the most widespread of all the annual species of *Cyperus*, occurring throughout most of Europe, with its stronghold in the southern parts of the continent. There is one site in Jersey. It is absent from Iceland, the Faeroes, Ireland and Fennoscandia, and is rare or threatened in the Netherlands, Denmark and Switzerland. It is also found in Asia, North Africa, Madeira and Tenerife.

A.J. Byfield

Cypripedium calceolus L. (Orchidaceae)
Lady's-slipper orchid
Status in Britain: CRITICALLY ENDANGERED. WCA
Schedule 8. EC Habitats & Species Directive,
Annexes II and IV.
Status in Europe: Vulnerable.

This most celebrated plant is one of Britain's rarest
orchids. Apart from plants propagated *ex situ*, only one
clump of native origin survives in the wild. It grows on
well-drained soil in herb-rich limestone grassland on a
fairly steep slope adjacent to woodland. *Sesleria caerulea* is
prominent in the sward (Walter 1993), and the wide range
of associated species include *Carex panicea, Epipactis
atrorubens, Galium sterneri, Helianthemum nummularium,
Leontodon hispidus, Linum catharticum, Polygala amarella,
Primula farinosa, Sanguisorba minor* and *Succisa pratensis*.
The area is grazed by rabbits.

C. calceolus is a perennial, flowering in June. It is
believed to be pollinated by small bees of the genus
Andrena (Summerhayes 1968), but the effectiveness of
natural pollination is uncertain in the remaining plant.

This species was formerly more widespread, though
local, in parts of northern England, growing on steep
slopes of open woods of oak, ash and hazel in the
limestone districts of Derbyshire, Yorkshire, Co. Durham,
Westmorland and Cumberland. It disappeared from most
of these areas during the last century and has been rare in
Craven for more than a hundred years, depredation of
native populations for garden and herbarium specimens
having been rife (Nelson 1994). A plant in semi-natural
habitat in North Lancashire could have been deliberately
planted, though it may be of native British origin.

The original clone is protected by a cage throughout
the year, though this has not protected it against the
undermining of the rhizome by small mammals. For many
years, fertilisation of the few flowers that appear has been
carried out by hand-pollination. English Nature's Species
Recovery programme, in conjunction with the Royal
Botanic Gardens, Kew, has been notably successful in
enhancing the population. At the original site in Yorkshire,
careful habitat management, together with re-establishment
of plants from *ex situ* propagation, has led to a steady
increase in the size of the colony. There are now 60 shoots
at the site as well as an eleven-year-old seedling which
first flowered in 1993 (Lindop 1996). By 1996, plants
derived from micro-propagation had been planted out at
twelve additional sites as part of the on-going recovery
programme. Other plants exist in cultivation, some
believed to be of native stock but others of known
continental origin.

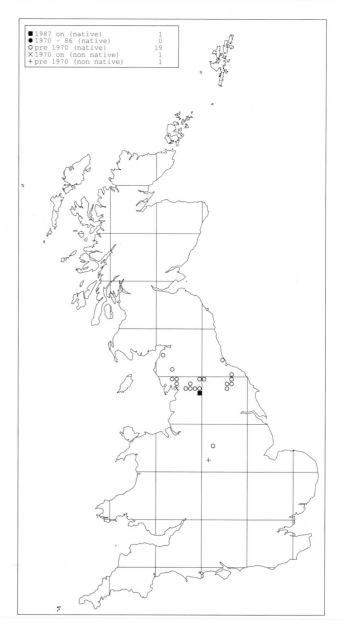

■ 1987 on (native)	1
● 1970 – 86 (native)	0
○ pre 1970 (native)	19
✕ 1970 on (non native)	1
+ pre 1970 (non native)	1

This species is widely distributed through northern,
central, eastern and south-east Europe, westwards to
Norway and the south-west Alps and eastwards to
Sakhalin Island. It is found in several types of woodland,
including those dominated by beech, larch, spruce or ash/
oak, but always on moist calcareous soils and usually on
north-facing slopes, ascending to more than 2,000 m in
altitude. It is has become rare and threatened over much
of its range, and is now legally protected in seven
European countries.

L. Farrell

Cystopteris dickieana R. Sim (Woodsiaceae)
Dickie's bladder-fern, Frith-raineach Dhicianach
Status in Britain: VULNERABLE. WCA Schedule 8.
Status in Europe: Not threatened.

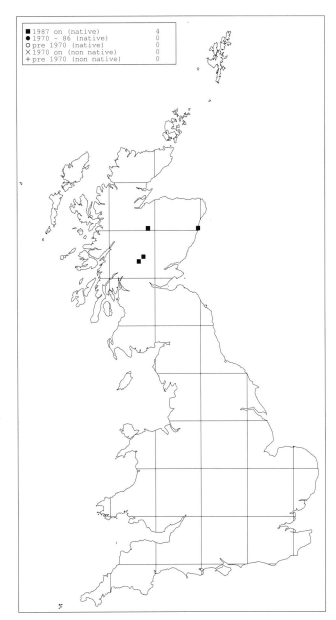

■ 1987 on (native)	4
● 1970 - 86 (native)	0
○ pre 1970 (native)	0
✕ 1970 on (non native)	0
+ pre 1970 (non native)	0

C. dickieana was, until recently, thought to be confined to sea caves and adjacent cliffs in Kincardineshire. The population in the cave that has become the *locus classicus* contains over 200 plants, and spores germinate around parent plants on a shallow matrix of sand grains and decayed filamentous algae overlying the gneiss rock. Associated species are *Asplenium marinum, Athyrium filix-femina, Dryopteris dilatata* and a few species of bryophytes. At least three other populations exist along this coast, where igneous rocks are interlain with garnet/mica-schist which weather to form a reddish, moderately calcareous soil, supporting such species as *Astragalus danicus, Geranium sanguineum, Helianthemum nummularium* and *Koeleria macrantha* (Marren 1984).

Herbarium specimens of *C. dickieana* from a few other localities have recently been detected during searches of British herbaria. However, in 1993 an extant population of considerable extent was located in central Perthshire. The site there is in a ravine on rocks associated with the Loch Tay limestones. Another site was subsequently found in Easterness, and an account of the plant in Britain is given in Tennant (1996). The populations on the cliffs in Kincardineshire are vulnerable to erosion and illegal tipping of domestic refuse, but neither of the inland sites is threatened.

C. dickieana is a pan-boreal species throughout Eurasia and in North America. Its exact status within the *C. fragilis* complex is still unclear, as are the cytogenetics of the bladder ferns generally. At least three diploid ancestors are known to have been involved in the evolution of the complex. Kincardineshire material of *C. dickieana* is known to be tetraploid; in continental Europe both tetraploid and hexaploid plants occur.

The rugose spore character has been treated as diagnostic for the species, but recent isoenzyme studies suggest that it has little genetic, and therefore taxonomic, value. However, the genetic make-up of the British populations may provide a clue to the evolution of this group as a whole and the species as we presently conceive it should certainly be conserved pending further work.

A.C. Jermy

115

Cystopteris montana (Lam.) Desv. (Woodsiaceae)
Mountain bladder-fern, Frith-raineach Beinne
Status in Britain: LOWER RISK - Near Threatened.
Status in Europe: Not threatened.

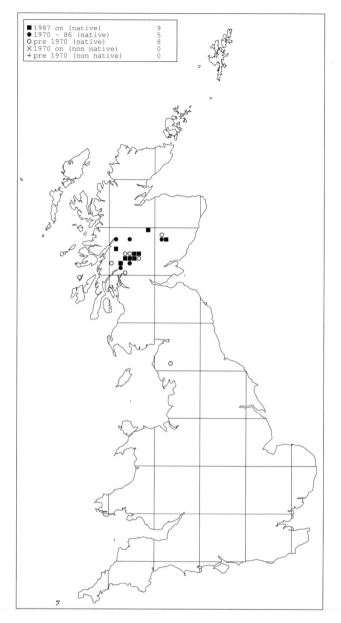

This is a fern of sheltered, humid habitats, usually on treeless, steep, north- to east-facing crags of mica-schist and limestone in sites where there is at least periodic irrigation. It is strongly calcicolous, favouring damp ledges and slopes of unstable gravel with a high calcium content and usually with a cover of bryophytes and mat-forming vascular plants. In this species-rich habitat there are numerous associates, including *Cochlearia officinalis*, *Cystopteris fragilis*, *Ranunculus acris*, *Saxifraga aizoides*, *S. hypnoides*, *S. oppositifolia*, *Sedum rosea*, *Selaginella selaginoides*, *Thalictrum alpinum* and *Ctenidium molluscum*. *C. montana* grows at altitudes from 490 m on Ben Lui to 1,070 m on Bidean nan Bian and Aonach Beag. In some places it grows on rocky banks accessible to sheep and deer, but the most luxuriant plants are on ungrazed cliff ledges.

The plant produces spores but it seems likely that the small, local populations are built up by the creeping rhizome system. Fronds seldom reach the size commonly seen in continental Europe, and the plant is best searched for in July, when the fronds are uncurled but are still green and therefore most easily seen.

C. montana is evidently extinct in the Lake District, where it was seen only once, but its status is probably little changed in Scotland. The headquarters of the plant are in the high mountains of mid-Perthshire, mainland Argyll and West Inverness, but there are eastern outliers in the Cairngorms and east of the Devil's Elbow, near Cairnwell. It is unknown north of the Great Glen. There appear to be few details available on population sizes of colonies, except for those on Ben Lawers. There, in 1993, it was found in ten sites, with the number of clumps ranging from 28 to 450. It probably survives in most of the Scottish hectads from which only pre-1970 records are available, although the plant's tendency to occur as small and isolated populations might render it prone to local extinctions.

It is an arctic-alpine circumboreal species. In Europe it extends south to the Pyrenees, the Apennines, the Carpathians and the Caucasus (Jalas & Suominen 1972). In Scandinavia it can form dominant growths on the floor of sub-montane birchwoods, where the soil is calcareous.

G.P. Rothero

Cytisus scoparius ssp. *maritimus* (Rouy) Heywood (Fabaceae)

Sarothamnus scoparius ssp. *prostratus* (C.Bailey) Tutin
Prostrate broom, Banadl Gorweddol
Status in Britain: VULNERABLE.
Status in Europe: Vulnerable. Endemic.

This segregate of *C. scoparius* grows on exposed cliffs, extending from the cliff-face into the vegetation on the cliff-top. In Cornwall, at Kynance, it is also found on rocky outcrops a little way back from the cliff-top, and at Gew Graze about 1 km inland. Typical associates are *Armeria maritima*, *Calluna vulgaris*, *Dactylis glomerata*, *Danthonia decumbens*, *Festuca rubra*, *Hypochaeris radicata*, *Jasione montana* and *Leucanthemum vulgare*.

This distinctive taxon is characterised by its procumbent habit, with densely-massed branching stems that are usually pressed close to the ground. The degree of hairiness has been used as a taxonomic character, but this is unreliable: some plants may be so hairy as to completely obscure the stem, whilst others may be no hairier than plants of the common broom (Gill & Walker 1971). It is a long-lived perennial which, on the evidence of counting growth rings, may live for more than 50 years. Flowering is normally from May to early June. Recent studies of populations in Pembrokeshire (Kay & John 1995) have shown that the production of pods is variable - some apparently healthy bushes that were examined had no set pods, whilst others growing nearby had many. It requires open ground in which to seed and, in one study area, the only available sites seemed to be under dead bushes of *Cytisus* where seedlings also received protection from grazing.

C. scoparius ssp. *maritimus* is endemic to the western Atlantic seaboard and has an extremely disjunct distribution. In Britain, records are from the Lizard peninsula and the north coast of Cornwall, Lundy, the Pembrokeshire coast and the Lleyn peninsula. Searches in many other localities have failed to detect it. There appears to be little recent information on population sizes. Elsewhere it occurs in a few locations on the Channel Islands, in Ireland, and the west coasts of continental Europe.

C. scoparius ssp. *maritimus* breeds true and maintains its character in cultivation, whilst other prostrate forms, including those from Dungeness, do not. Kay & John (1995) have shown that plants of ssp. *maritimus* from Pembrokeshire, southern Ireland and Jersey (the only ones investigated) are genetically distinct from ssp. *scoparius*. There is also a high level of genetic variability in ssp. *maritimus*, even in small populations, which may buffer it against genetic erosion. Genetic aspects are further discussed in Böcher & Larsen (1958) and Morton (1955).

M.J. Wigginton

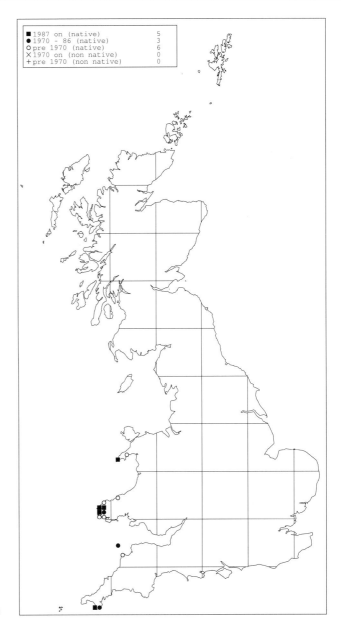

■ 1987 on (native)	5
● 1970 - 86 (native)	3
○ pre 1970 (native)	6
✕ 1970 on (non native)	0
+ pre 1970 (non native)	0

Dactylorhiza Necker ex Nevski (Orchidaceae)
Marsh orchids

The taxonomy of *Dactylorhiza* has been the subject of a great deal of research over many years, though there is still a general lack of consensus over the ranking of the taxa, and to some degree their delimitation. Difficulties arise because of the complex patterns of variation and intergradation of taxa, the relatively great variation between populations, and extensive hybridisation. Many of the studies of British species have been morphometric, but latterly, isoenzyme studies are helping to shed light on the relationships of constituent taxa. In selecting taxa for inclusion in this volume, we have followed Stace (1991). Other authorities (e.g. Allen & Woods 1993; Sell & Murrell 1996) have described additional taxa as subspecies, some of which appear to be nationally rare.

Dactylorhiza incarnata ssp. **cruenta** (Müller) Sell (Orchidaceae)

Early marsh orchid, Mogairlean Lèana
Status in Britain: ENDANGERED.
Status in Europe: Not threatened. Endemic.

The sole known British locality for this plant is in Ross-shire, where it was first recorded by Kenneth & Tennant (1984). However the plant may be overlooked elsewhere and is likely to be present in suitable habitats in other parts of Scotland. The Ross-shire plants occur in gently-sloping, lightly-grazed and more or less neutral flushed grassland at an elevation of about 300-450 m. *Molinia caerulea, Schoenus nigricans, Trichophorum cespitosum* and various *Carex* species typical of such habitats are close associates. Also associated are *Dactylorhiza incarnata* ssp. *pulchella, D. maculata, Gymnadenia conopsea, Pinguicula vulgaris* and *Tofieldia pusilla*. The Scottish habitat contrasts markedly with that of Irish and Scandinavian populations. In Ireland the plant is restricted to the vicinity of the lakes and calcareous fens of the western limestones, including the turloughs of the Burren in which it is found with *Potentilla fruticosa* and *Viola persicifolia*. In Sweden it occurs in moist richly-calcareous depressions, sometimes close to the coast.

Plants at the Scottish locality are scattered over a limited area, forming a few small populations which usually total fewer than fifty flowering plants. Flowering is in mid- to late June, sometimes extending into July. The heavy spotting on both sides of the sheathing leaves, together with the spotted bracts and ovaries and streaked upper stem, readily separate ssp. *cruenta* from other British marsh orchids. However, unspotted plants possessing the same general morphology also occur within the population, and it may possibly intergrade with ssp. *Pulchella*, which is also present (Kenneth & Tennant 1987). Similar intergradation sometimes occurs in continental European populations and this brings into question the taxonomic status of such spotted-leaved plants.

The Scottish site appears to be under no immediate threat, occurring as it does within an area of poor upland grazing. However, its habitat is easily damaged, and the proximity of a major trunk road may render it vulnerable to future roadworks. The main threat is that inherent in all small populations and especially in one occurring in such a fragile, easily damaged habitat.

D. incarnata ssp. *cruenta* occurs widely throughout central and northern Europe and Scandinavia.

M.J.Y. Foley

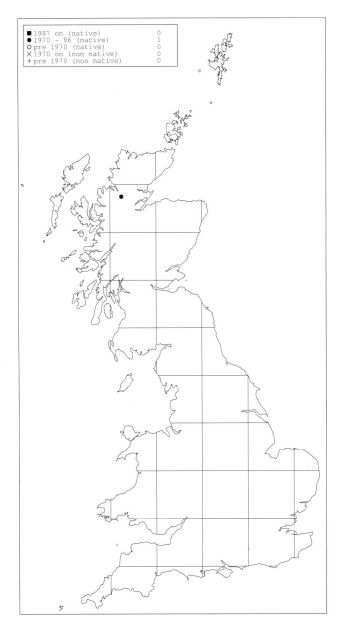

■	1987 on (native)	0
●	1970 - 86 (native)	1
○	pre 1970 (native)	0
✕	1970 on (non native)	0
+	pre 1970 (non native)	0

Dactylorhiza incarnata ssp. *ochroleuca* (Wüstnei ex Boll.) P. Hunt & Summerh. (Orchidaceae)

Early marsh orchid, Mogairlean Lèana
Status in Britain: CRITICALLY ENDANGERED.
Status in Europe: Not threatened. Perhaps Endemic.

This plant of calcareous fens shows a marked preference for moist but not permanently inundated areas where there is only moderate competition. Especially, it occupies those parts of fens that are in the process of slowly but only partially drying out and which are likely to have originated through a process of sedimentation. It forms distinct colonies, sometimes in association with other marsh orchids, in a rich community of fen plants including *Carex panicea, Filipendula ulmaria, Juncus subnodulosus, Molinia caerulea* and *Valeriana dioica.*

Flowering is in early June and reproduction is by seed, but its reproductive biology does not appear to have been studied in detail. With its pale cream-coloured flowers, it is easily confused with white-flowered variants of other subspecies of *D. incarnata*, especially ssp. *pulchella.*

It has been recorded with certainty from a few localities in eastern England. An old record from South Wales has never been confirmed and is considered dubious. Otherwise, all British records are from a limited area of East Anglia, and perhaps Surrey (Bateman & Denholm 1983). It formerly occurred in Suffolk at several fens in the Waveney valley but has been lost from these over the past 50 years or so. It was last seen there in about 1988 but, following a drought that year, has failed to reappear. The only certainly extant British population is in Chippenham Fen NNR, and even there numbers have fallen sharply. Ten years ago up to thirty flowering plants were recorded, but present numbers are substantially reduced and perhaps only a single clump survives.

The most serious threat to its survival lies in any widespread lowering of the water table, which will upset the fine balance of its moisture requirements; an adverse threat would also be posed by prolonged inundation. Encroachment of scrub and other vegetation will also place it at a competitive disadvantage. In some years plants have been bitten off or otherwise damaged, apparently by deer, and the remaining known colony is now protected by a wire cage.

D. incarnata ssp. *ochroleuca* occurs widely but locally in Europe, from the Alps northwards through Germany, Poland and Estonia to Scandinavia and north-west Russia, and possibly occurs in western Ireland (Sell & Murrell 1996). As in Britain, erroneous records proliferate owing to confusion with other white-flowered forms of *D. incarnata.* In Scandinavia it is often afforded full specific status as *Dactylorhiza ochroleuca.*

M.J.Y. Foley

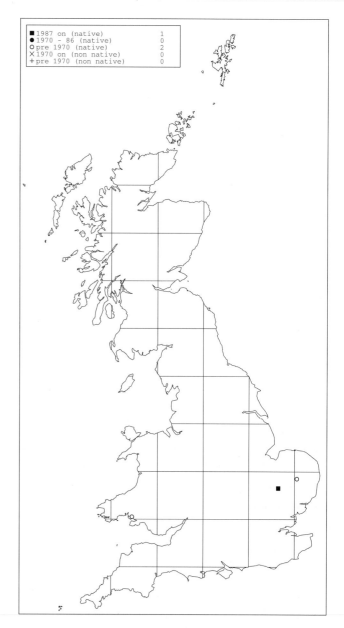

■ 1987 on (native)		1
● 1970 – 86 (native)		0
○ pre 1970 (native)		2
✕ 1970 on (non native)		0
+ pre 1970 (non native)		0

Dactylorhiza lapponica (Hartman) Soó (Orchidaceae)

Lapland marsh-orchid
Status in Britain: LOWER RISK - Near Threatened. WCA
Schedule 8.
Status in Europe: Uncertain, but thought to be endemic.

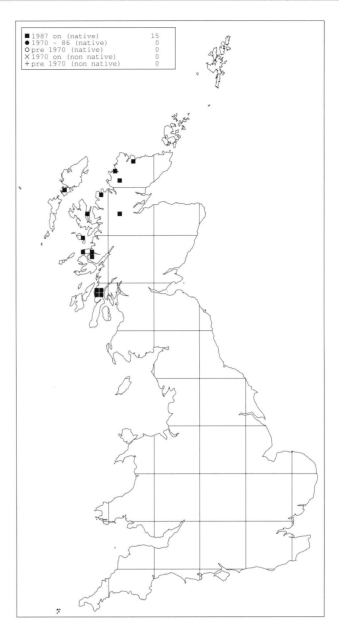

D. lapponica is found mainly in relatively base-rich hill
flushes with a pH up to 6.5, often associated with
superficially acidic and peaty soils, though it appears to
tolerate more acid conditions associated with adjacent wet
heath. Common associates in the open sedge flushes in
which it occurs often include *Molinia caerulea* and *Schoenus
nigricans* as co-dominants, together with such species as
Drosera rotundifolia, *Erica tetralix*, *Eriophorum latifolium*,
Potentilla erecta, *Selaginella selaginoides* and *Thalictrum
alpinum*.

Flowering begins in late May and can last until July.
Reproduction is by seed, but a study of Scottish
populations (Nieland 1994) has shown that only 19% of
surviving capsules contain viable seed, because of poor
pollination.

In Britain it has been found only in the Western Islands
and Highlands of Scotland, where it was first identified in
1986 (Kenneth *et al* 1988). It is found mainly at altitudes
between 150 and 310 m (for instance in Knapdale, Morvern
and Ardnamurchan), but descends to 30 m in South
Harris. There are now more than 30 known populations,
though most are very small, with fewer than 100
individuals. It is likely that further localities will be
discovered. The few populations that have been monitored
indicate relative stability, though population size may
fluctuate considerably from year to year.

The main threat to this species is from direct habitat
destruction or change (Cowie & Sydes 1995), and all
populations are vulnerable to afforestation or drainage. In
addition, most of its sites are moderately to heavily grazed
by sheep or deer, which can remove a significant
proportion of the flowers. However, grazing is absent
during the summer months at the South Harris site, which
supports one of the largest populations, and this may
favour seed production. One population, surrounded by
forestry in Knapdale, has been ungrazed for ten years, and
numbers have steadily increased. At this site, competitive
exclusion by other more vigorous vegetation may be
inevitable in the long term, but no evidence of a decline
has been detected in its core habitats of open flushes. Only
seven of the known populations are protected on reserves
or scheduled sites, though populations may yet be found
on other SSSIs in western Scotland.

It occurs sporadically throughout northern and central
Scandinavia, including Gotland, and into northern Russia.

It is also found, particularly in limestone regions, in the
Swiss, Austrian and Italian Alps. Its habitats are calcareous
open fens and wooded mires, wet meadows and
streamside gravels, ascending to 1,000 m in the northern
part of its range, and to 2,400 m in the Alps.

D. lapponica is a tetraploid marsh-orchid, in a group
that is renowned for its taxonomic difficulties. It can,
however, be identified with relative ease in the field,
though recent genetic (R. Bateman pers. comm.) and
morphometric studies (Hansson 1994) have cast doubt on
its taxonomic distinctiveness.

N.R. Cowie

Damasonium alisma Miller (Alismataceae)
Starfruit, Serffrwyth
Status in Britain: ENDANGERED. WCA Schedule 8.
Status in Europe: Not threatened. Probably endemic.

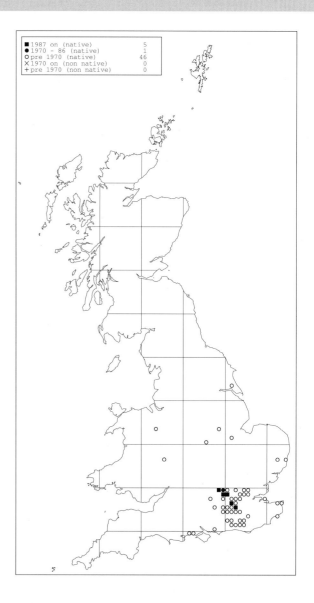

D. alisma grows in shallow, saucer-shaped ponds with fluctuating water levels, located on acid soils. It requires open vegetation with areas of bare soil, being intolerant of competition from more vigorous species. This vegetation structure is generated by disturbance such as trampling by livestock, or periodic pond clearing. It is associated with species that are tolerant of both inundation and desiccation, including *Apium inundatum, Bidens tripartita, Callitriche* spp., *Juncus articulatus, J. effusus, Lythrum portula, Myriophyllum alterniflorum, Persicaria maculosa, Ranunculus flammula* and *R. peltatus.*

In the wild, *D. alisma* generally behaves as an annual, but in cultivation, and possibly sometimes in the wild, it can live for two or three years. Its seeds germinate below water in early winter and develop into small plants with short linear leaves resembling a tuft of grass. In early summer long-petioled floating leaves are produced. If the water level falls to expose the plant above water, these leaves die and are replaced by stout, short-petioled alternatives. But if the water level remains high, the plant retains its aquatic growth-form. It flowers between June and August. The number of flowers produced by exposed plants is related to the time the mud remains moist: if it dries quickly only a single flower may be produced before the plant dies. Thus, reflecting the unpredictability of its habitat, the size and growth-form of starfruit are variable, ranging from tiny terrestrial plants with just a single flower, to large terrestrial or aquatic plants with several compound inflorescences bearing as many as 150 flowers. Plants are self-fertile and some self-pollination occurs (Vuille 1987). However, beetles and hoverflies visit the flowers and may effect cross-pollination. Fruits appear like 6-rayed stars. Only a small proportion of the seeds germinate in their first winter, the majority remaining dormant. Dormancy can be broken by a period of desiccation followed by hydration, or possibly by changes in the seed's environment following disturbance. *D. alisma* has made dramatic reappearances in ponds where it has apparently been absent for several decades, following the removal of dense stands of emergents, suggesting that dormant seeds can remain viable for very many years (Birkinshaw 1994).

Since first recorded, *D. alisma* has been found over 100 localities in 50 hectads, mainly in south-east and southern England, within an area bounded by Buckinghamshire, Hampshire, Kent and Sussex. It has never been common and, in any decade, occurred at only a few sites. However, the species continued to decline this century, until in the 1980s plants were seen in just one pond. The reasons for this decline were various: the neglect and mis-management of ponds, cessation of traditional use of ponds by livestock leading to excessive growth of emergent and submerged plants, the maintenance of constant water levels for angling, and the loss of ponds through infill. In the last few years, clearance of ponds where it was once known, including dredging out of accumulated silt and rotting vegetation, resulted in *D. alisma* appearing in six sites between 1990 and 1994 (Showler 1994; Rich, Alder *et al.* 1994). However, from a high point of several hundred

plants at these sites, it declined rapidly to only fifteen plants detected in Britain in 1994, none in 1995, and about 40 at four sites in 1996. Details of this decline, and possible reasons are discussed in Rich, Alder *et al.* (1995).

The maintenance of a healthy population of *D. alisma* in Britain is dependent on fluctuating water levels in ponds and, in the absence of livestock grazing, removal of excessive vegetation and scraping hard back to the substrate. With appropriate management, natural reappearance seems to be relatively predictable. However, *D. alisma* does not persist after restoration, and further management trials are needed to discover how to maintain populations. Currently, several ponds are being restored, and reintroduction has also been attempted. Four of its sites are in SSSIs.

Outside England, *D. alisma* occurs in southern and south-west Europe (France, Spain and Italy). The closely related *D. bourgeai* and *D. polyspermum*, which are considered by some to be subspecies of *D. alisma*, occur in North Africa, Asia Minor, southern Russia, Ukraine and southern Europe.

C.R. Birkinshaw

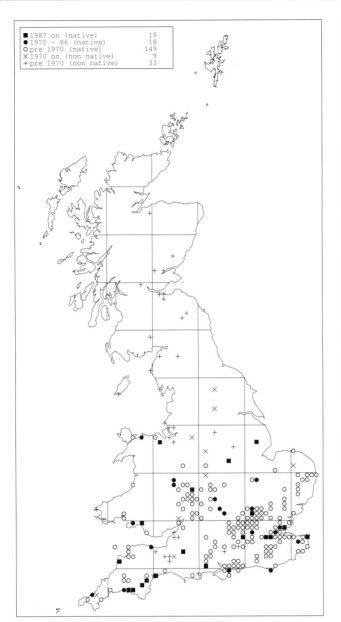

■ 1987 on (native)	19
● 1970 – 86 (native)	18
○ pre 1970 (native)	149
× 1970 on (non native)	9
+ pre 1970 (non native)	33

Dianthus armeria L. (Caryophyllaceae)

Deptford pink, Penigan y Porfeydd
Status in Britain: VULNERABLE. WCA Schedule 8.
Status in Europe: Not threatened.

This is a plant of lowland dry pastures, field borders and hedgerows. It is found mainly in short grassland where there is some open ground maintained either by grazing or some other form of disturbance and dies out when shaded by coarse grasses or scrub. It occurs on light, sandy, often rather basic soil where associated species include *Achillea millefolium, Daucus carota, Lotus corniculatus, Ranunculus repens* and *Trifolium repens*. In Nottinghamshire it grows in rough grassland with more basiphile species including *Brachypodium pinnatum, Briza media, Leontodon hispidus* and *Linum catharticum*. It formerly grew on peaty soil of pH 4.8 at Woodwalton Fen (Wells 1967), where it accompanied *Agrostis capillaris, Festuca rubra, Luzula campestris, Plantago lanceolata* and *Potentilla erecta*.

D. armeria can behave as an annual or biennial. The flowers lack scent and are seldom visited by insects and the plant is generally self-pollinated. It produces abundant seed, several hundreds per plant, with apparent high viability (70% germination *ex situ* has been shown after six months' storage at room temperature). However, there may be an inhibiting factor which delays germination for up to five months after seeds are shed (Wells 1967; Farrell pers. obs.).

This species has declined very rapidly in southern England, and the decrease seems to be continuing. Stewart *et al.* (1994) note its occurrence in only 36 hectads since 1970, but further investigation reduced this to 31 hectads encompassing 34 sites (Pearman 1997). In the 1980s it appears to have been recorded from 19 sites, and since 1990 from only 13-16 sites. Furthermore, populations are often small (1-50 plants), although there are a few sites (e.g. in Kent and Devon) with hundreds of plants, when conditions are suitable. It may be that the size of colonies can be increased by disturbing the soil to stimulate the germination of buried seed. The beneficial effects of grazing by cattle, rabbits and hares are discussed by Wells (1967). It seems surprising that there are no more precise data, and detailed investigation of past and present sites must be a priority.

D. armeria occurs widely across western and central Europe, extending eastwards to the Crimea, Anatolia and the Caucasus, and southwards to central Spain and Sicily (Jalas & Suominen 1986). It has recently been recorded for the first time in Ireland (Akeroyd & Clarke 1993). It occurs as an introduction in North America.

M.J. Wigginton and D.A. Pearman, from the account by L. Farrell in Stewart *et al.* (1994).

Dianthus gratianopolitanus Vill. (Caryophyllaceae)
Cheddar pink, Penigan Mynyddig
Status in Britain: VULNERABLE. WCA Schedule 8.
Status in Europe: Not threatened. Endemic.

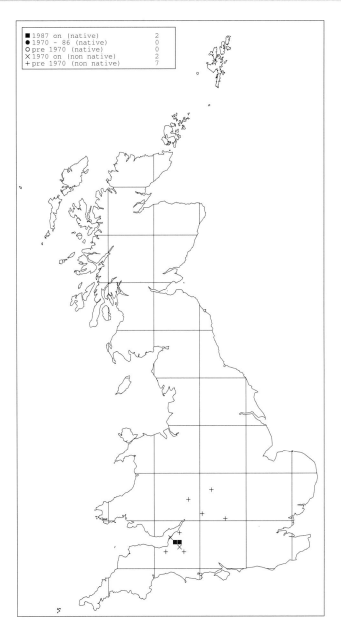

■ 1987 on (native)	2
● 1970 – 86 (native)	0
○ pre 1970 (native)	0
✕ 1970 on (non native)	2
+ pre 1970 (non native)	7

D. gratianopolitanus is native in Britain only on Carboniferous limestone in North Somerset. It has a long history in Cheddar, having been first recorded by John Ray "On Chidderoks in Somersetshire" in 1696. It is found characteristically at the edge of cliffs, rooting in thin soils or into rock crevices, and forming compact mats amongst short vegetation or trailing across bare rock outcrops. It is also to be found further away from the cliff edge in short-grazed grassland where the competition from other plants is not too great. It needs open sunny situations and does not tolerate much shade. Other plants in the species-rich limestone grassland community include *Carex flacca*, *Festuca ovina*, *Helianthemum nummularium*, *Helictotrichon pratense*, *Koeleria macrantha*, *Pilosella officinarum*, *Sanguisorba minor*, *Scabiosa columbaria*, *Sedum forsterianum*, *Thalictrum minus* and *Thymus polytrichus*. In the winter months its narrow grey-green leaves are easily overlooked in the grass turf.

It is a perennial plant flowering in June and July. The strongly scented flowers are visited by many insects, including butterflies and day-flying moths. Reproduction is by seed, which germinates in autumn.

The Cheddar Gorge population is the largest, with many thousands of clumps still present. In addition, there are several small outlying sites within about 5 km of the Gorge where it appears to be native, but it has been introduced at other sites nearby and has occurred as a casual further afield.

Its attractiveness has led to much collecting, and in the past there was a local tradition in Cheddar of gathering the flowers into posies for the tourist trade. It is now mostly to be found on high, inaccessible cliffs, the lower ledges having been stripped of their pinks many years ago. Despite special protection, picking and collecting still occur. However, the greatest threat is from the encroachment of scrub and the growth of secondary woodland that has taken place over the past 40 years. Much of the species-rich grassland has become invaded by mixed scrub and ash woodland with a dense undergrowth of ivy, shading the cliff ledges and continuing to encroach into open grassland. To arrest this succession, a large-scale programme of scrub clearance and grazing was initiated a few years ago. This has already led to the re-establishment of open species-rich swards, and in some areas populations of *D. gratianopolitanus* are increasing.

D. gratianopolitanus is a European endemic, occurring in western and central Europe, but thought to be in general decline throughout its range. It is found mainly on limestone outcrops on mountains, ascending to the sub-alpine zone.

E.J. McDonnell

Diapensia lapponica L. (Diapensiaceae)

Diapensia, Diapainsia
Status in Britain: VULNERABLE. WCA Schedule 8.
Status in Europe: Not threatened.

D. lapponica is a circumpolar arctic-subarctic plant that grows on exposed rocky ridges that are generally kept free of snow by high winds (Tiffany 1972). It was first discovered in Britain in 1951 by C.F. Tebbutt (Grant Roger 1952), growing on acidic soils in a stony fell-field on a ridge at about 800 m in Inverness, and this remains its only known site, claims of others having been disproved. About 1,200 clumps or mats are scattered over a limited area of ground on the summit ridge and on several adjacent knolls and outcrops. The most prominent associated species are *Carex pilulifera, Empetrum nigrum* ssp. *hermaphroditum, Nardus stricta* and *Racomitrium lanuginosum*; other associates include *Alchemilla alpina, Carex bigelowii, Festuca vivipara, Huperzia selago, Loiseleuria procumbens, Trichophorum cespitosum, Cetraria islandica* and *Cladonia* spp.

This species is a dwarf evergreen long-lived shrub that typically forms low mounds up to 15 cm high, or occasionally exhibits a spreading or mat-like growth up to half a metre across, sometimes dying away in the centre. The plant surface is a mass of interlocking rosettes composed of small leathery leaves. In Britain flowering is generally in late May or early June, but can vary with the season. In good flowering years, the impression may be given that there has been a population increase, but monitoring since 1980 has revealed little detectable change either in the total area covered or the number of clumps. It is uncertain whether viable seed is regularly produced, though plants have been grown *ex situ* from wild-collected seed. Young plants have not been observed, and recruitment from seed is likely to be very limited.

Studies in Newfoundland have shown that the age of plants can determined by counting growth rings in the stem (Day & Scott 1985) and there seems little doubt that many large Canadian plants are over a century old. The Scottish plants clearly pre-date their discovery by many years, though stem rings have not been counted to determine their age. In Canada, *D. lapponica* may flower twice in favourable seasons, once in late May and again in August. However, studies have shown that up to 40% of the seeds from the first flowering may be destroyed by fungal infection, whereas later flowerers are much less susceptible. Rapid germination occurred *ex situ* in seeds collected in December from the August flowers (Day & Scott 1984).

There is some sheep- and deer-grazing at the Inverness site, but it does not pose a significant threat. Small scale collecting has taken place, and there have been reports of whole plants being removed. This is clearly cause for concern because of its rarity and slow growth and our lack of knowledge of recruitment. A series of permanent

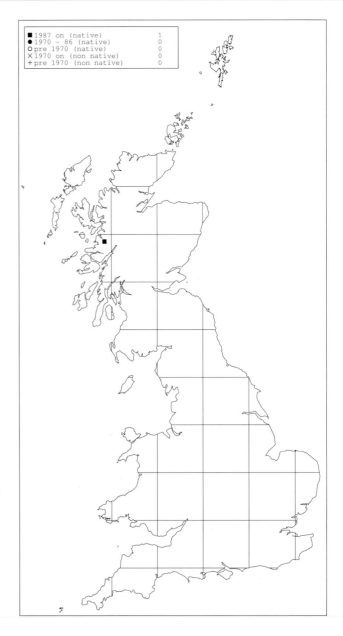

■ 1987 on (native)	1
● 1970 - 86 (native)	0
○ pre 1970 (native)	0
✕ 1970 on (non native)	0
+ pre 1970 (non native)	0

quadrats was established in 1980 in order to detect changes in the extent of the colony, and monitoring of the colony is normally carried out at two-yearly intervals. Whilst this species is afforded special protection, its site is not statutorily protected.

Two subspecies are recognised. Ssp. *lapponica* occurs in eastern North America from New Hampshire to Ellesmere Island, and in Greenland, Scotland, Scandinavia and western arctic and subarctic Russia. Ssp. *obovata* occurs in eastern arctic and subarctic Russia, in Korea, Japan, Alaska and the Yukon. The Scottish locality is the most southerly in Europe.

L. Farrell

Diphasiastrum issleri (Rouy) J.Holub
(Lycopodiaceae)

Diphasiastrum complanatum ssp. *issleri* (Rouy) Jermy,
Lycopodium issleri (Rouy) Domin
Yellow cypress clubmoss
Status in Britain: LOWER RISK - Near Threatened.
Status in Europe: Not threatened.

This taxon is found on mountainsides in rather open
communities of dwarf shrubs on shallow skeletal soils over
acid country rock. It is usually associated with shrubby
Arctostaphylos alpinus, *Calluna vulgaris*, *Vaccinium myrtillus*
and *V. vitis-idaea*, together with *Molinia caerulea* and
occasionally *Diphasiastrum alpinum*.

Although some spore abortion has been recorded,
many fertile spores are formed and there is no biological
reason why the species should not spread naturally.

Its early records in Britain (in 1866) were from lowland
heath at Bramshott, Hampshire, and later it was found in
Worcestershire and, in 1881, at Woodchester Park,
Gloucestershire. It has long since disappeared from these
sites, along with the habitat. The Woodchester plant was
described by Boswell-Syme (as var. *decipiens*) and botanists
such as Edward Marshall began recording the variety from
dwarf shrub heaths on Scottish mountains. For the next
fifty years the taxon was taken to be a form of *D. alpinum*
and was not recorded until A.G. Kenneth discovered a fine
colony on Canisp, West Sutherland. Others found it in
Strath Nethy and Glen Quoich in the Cairngorms (Jermy
1989). *D. issleri* was re-found in 1991 on Geal Charn, above
Glen Feshie, nearly a hundred years after it was first
recorded there by J.A. Wheldon and A. Wilson, and at the
810 m altitude originally recorded.

The presence of this plant in Britain has been well
documented because of its taxonomic interest, around
which there is still some nomenclatural uncertainty. It
appears to be intermediate between *D. alpinum* and *D.
complanatum* and may have arisen through hybridisation of
those species (Jermy & Camus 1991). Whatever its origin
the same taxon is found across central Europe from central
France to the Czech Republic and Poland.

The known sites in Britain, especially those above
600 m, are unlikely to be burnt and no special conservation
efforts are needed. The lowland sites have long since
changed in character or disappeared completely.

A.C. Jermy

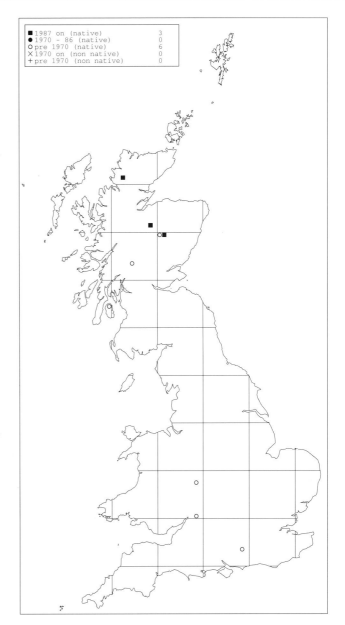

■ 1987 on (native)		3
● 1970 - 86 (native)		0
○ pre 1970 (native)		6
✕ 1970 on (non native)		0
+ pre 1970 (non native)		0

Draba aizoides L. (Brassicaceae)
Yellow whitlow-grass, Llysiau Melyn y Bystwn
Status in Britain: LOWER RISK - Near Threatened.
Status in Europe: Not threatened.

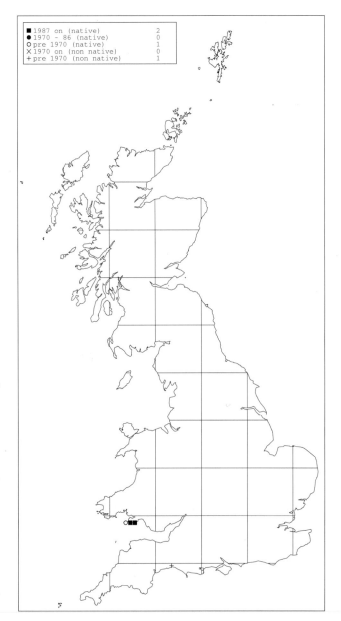

D. aizoides is a strict calcicole found only on
Carboniferous limestone on the Gower peninsula, where it
grows almost exclusively in deep crevices containing
humic calcareous soils. Young plants may be found in turf
or thin soils around outcrops, but very few reach maturity.
It occurs in small groups on the more inaccessible
limestone pinnacles and outcrops, often in association with
other scarce or rare native late-glacial relict species,
including *Helianthemum canum*, *Hippocrepis comosa*,
Potentilla neumanniana and *Veronica spicata* ssp. *hybrida*.

D. aizoides is a short-lived perennial, forming a cushion
of tightly-packed rosettes of stiff, fleshy, linear leaves.
Growth occurs mainly in winter and it flowers profusely,
usually between late February and April. The yellow
flowers are often insect-pollinated, but in their absence the
flowers can self-pollinate. Mature seeds are probably wind-
dispersed over short distances. Summer dormancy allows
mature plants to avoid drought, though seedlings are often
killed by prolonged dry weather. It is a very poor
competitor. Seedling survival near mature plants is normally
limited to a few months and established plants do not
tolerate crowding or overtopping (Kay & Harrison 1970).

First recorded from Pennard Castle in 1795, *D. aizoides*
has been regarded as a doubtful native in Britain (e.g.
Clapham *et al.* 1952), partly because of its extreme isolation
here, but genetic analysis has shown that the Gower
populations are very different from Continental ones.
There is also considerable variation within the Gower
populations. It is almost certainly a late-glacial relict that
survived the last Ice-age in Britain and which has been
isolated from Continental populations for many thousands
of years (John 1992). *D. aizoides* occurs within two main
stretches of the southern Gower cliffs. There may be a total
of just over 1,000 plants or clumps (Morgan 1989g), but
since population counts have invariably been partial in
any year and flowering is variable, an estimate of numbers
is difficult.

Because of its attractiveness in early spring, accessible
sites have been damaged by collectors, and on Pennard
Castle, for example, there are no longer plants within reach.
In recent years it has disappeared from the extreme western
limit of its range on Gower, and some colonies have
decreased in size, though the reasons for this are unclear.

In Continental Europe it has a mainly alpine
distribution, occurring between 900 and 3,400 m in the
Alps and extending eastwards from the Pyrenees to the
Carpathians, and westwards from the Alps to the Jura. A
small isolated population in the Ardennes appears to be a
pioneer one, not a relict. It is decreasing in places, mainly
owing to habitat destruction caused by road widening,
afforestation or tourist pressures.

R.F. John

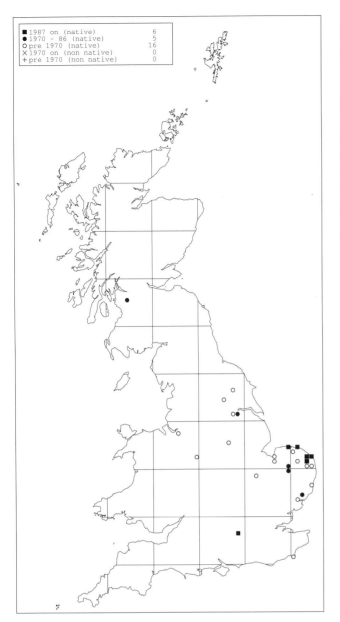

■ 1987 on (native)   6
● 1970 - 86 (native)   5
○ pre 1970 (native)   16
X 1970 on (non native) 0
+ pre 1970 (non native) 0

Dryopteris cristata (L.) A.Gray (Dryopteridaceae)

Crested buckler-fern, Marchredynen Gribog
Status in Britain: LOWER RISK - Near Threatened.
Status in Europe: Not threatened.

In Britain, this is primarily a plant of herbaceous, base-poor fens, mainly associated with acidic nuclei developed within, and often from, more base-rich fen. Such areas, with water pH about 5.0-6.5, often have a carpet of poor-fen *Sphagnum* species, including *S. fimbriatum*, *S. squarrosum* and *S. subnitens*, and a rather sparse cover of reed or other tall fen plants. Often they are readily colonised by *Betula pubescens*. In suitable localities, small 'islands' of birch within otherwise open fen can be useful pointers to *D. cristata* sites. However, in dense scrub, *D. cristata* usually either disappears or persists only in a sterile form, or becomes confined to the interface with herbaceous fen. Associates quite often include *D. carthusiana* and *Osmunda regalis* and sometimes *Pyrola rotundifolia*, along with a wide range of typical fen plants.

In a few places, *D. cristata* is not now obviously associated with *Sphagnum*, though might have been in the past.

In Broadland, all localities appear to be on fen surfaces with some vertical mobility, most often as a part-'floating' late-hydroseral phase in re-flooded peat workings (both the medieval 'broads' and nineteenth century 'turf ponds'). The capacity for vertical movement prevents much surface inundation and accounts for the occurrence of *Sphagnum* mats in close juxtaposition to base-rich (and occasionally nutrient-rich) water (Giller & Wheeler 1988). The reason for the association of *D. cristata* with areas of *Sphagnum* is not known, but *Sphagnum* surfaces may be important for establishment from spores, though germination and prothallial growth occur readily on a wide range of media.

Jermy *et al.* (1978) show that *D. cristata* has disappeared from several former sites in the north Midlands of England. The fern has also apparently been lost subsequently from other localities, for instance from

Dungeness in Kent, Bixley Heath in Suffolk (Simpson 1982) and from its Renfrewshire site (Idle 1975), even though in this last locality a seemingly suitable habitat persists. It survives in a single site in Surrey, but most of its remaining colonies are in the Norfolk Broadland and in a few nearby fenlands. In Broadland, it is most widespread in the valley of the Ant (Wheeler 1978) and, to a lesser extent, the valley of the Thurne. There are probably more than 100 separate colonies of this fern in Broadland, some with just one or two plants, others with several dozen. Many recent new records are just a product of increased investigation, but it has almost certainly expanded its range during the last 50 years as turf ponds have reached an appropriately late phase of terrestrialisation.

D. cristata may have been lost from some former sites for various reasons, including drainage, but the most likely threat to extant populations is scrub encroachment. Interestingly, populations in some comparable Dutch turf-pond sites seem less subject to birch invasion than are most in Broadland. The typical semi-floating habitat has some buffering against eutrophication, but surface flooding with enriched water is damaging.

Its European distribution is focused upon the Baltic Sea, becoming more disjunct eastwards into western Siberia and southwards to northern Italy and Romania. There are a few records from southern France and northern Spain (Jalas & Suominen 1972). The English sites are on the western edge of its European range. It occurs also in Japan and North America.

B.D. Wheeler

Echium plantagineum L. (Boraginaceae)
E. lycopsis L.
Purple viper's-bugloss, Gwiberlys Porffor
Status in Britain: ENDANGERED.
Status in Europe: Not threatened.

E. plantagineum is a conspicuous plant, growing up to a metre high, its showy nature remarked on by Druce in his description (1906) of the Jersey population: "on the slope of a hill near Beauport there were some hundreds ... the plant attracted my attention more than a mile away". It is found with other arable weeds in cultivated fields in south-west England, being most closely associated with *Chenopodium album*, *Chrysanthemum segetum*, *Fumaria muralis*, *Misopates orontium* and *Solanum nigrum*. Plants may appear amongst barley or potato crops, but it is more abundant at the field margins where it escapes herbicide treatments. In Jersey it is found on light, nutrient-rich soils that have been broken by pig- or horse-grazing.

In Cornwall and the Isles of Scilly it is typically an annual, largely autumn-germinating and flowering in the following summer and autumn, but also germinating in spring. The seedlings may be sensitive to frost, which may partly account for its restricted geographical distribution in Britain (Lock & Wilson 1996).

Most authorities consider *E. plantagineum* to be native in south-west England (Margetts & David 1981; Cuddy 1991; Stace 1991) and in the Channel Islands (Le Sueur 1984). It occurs in a few places in West Penwith, Cornwall, and perhaps very locally in the Isles of Scilly. The largest mainland population is at St Just, with over 1,000 plants appearing in suitable fields in some years. Much smaller populations (10-50 individuals) have been recorded at other sites in West Penwith. The ephemeral nature of the habitat and the long-lived seed-bank lead to very marked fluctuations in population size from year to year.

Agricultural intensification and the change from cultivation to sheep-grazing has led to a decrease in the range of *E. plantagineum* in Cornwall. In the past, native populations also existed at Grumbla (lost in the late 1960s or early 1970s) and possibly around the Mullion area. Other older records both in Cornwall and outside the county would seem to be transient casuals (Murphy 1991). The species has survived in St Just owing to the maintenance of traditional farming practices, with ripe seeds harvested with the barley, then carried into the mowhay (rickyard) and out again with the muck. Thus seed is distributed around the farm when fields are being prepared for subsequent crops (Cuddy 1988). The maintenance of an open disturbed habitat without the use of herbicides is essential if further losses are to be prevented.

E. plantagineum is a native of southern and western Europe, being extremely common in Iberia and the Mediterranean region in all kinds of marginal habitats and waste ground. It also occurs in Macaronesia, North Africa and the Caucasus. In Western Australia it can be a fearsome weed, and is known there as 'Patterson's Curse'.

P.A. Ashton

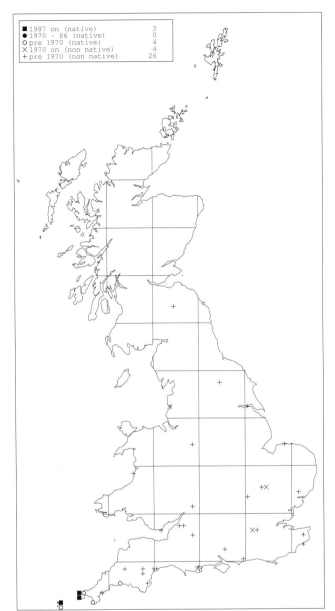

■ 1987 on (native)	3
● 1970 - 86 (native)	0
O pre 1970 (native)	4
X 1970 on (non native)	4
+ pre 1970 (non native)	26

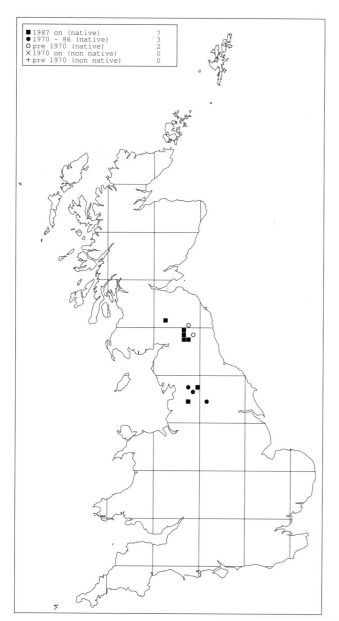

Eleocharis austriaca Hayek (Cyperaceae)

Northern spike-rush
Status in Britain: LOWER RISK - Near Threatened.
Status in Europe: Not threatened.

E. austriaca is a quite recent addition to the known British flora, and our knowledge of its distribution is still somewhat fragmentary. Almost all its localities are in the middle reaches of upland rivers, where it colonises sites in slacker water, often where there is some protection from spates, such as in shallow bays, or where side-streams join the main river. Where it occurs along the sides of the river, it is usually in deeper water (up to about 30 cm deep), where flow-rates are lower. It also occurs in ditches, pools, runnels and springs. It usually grows in a gravel substrate where there is some silt deposition, and with its stem-bases permanently submerged. A wide range of associates at some sites include aquatic and emergent species such as *Callitriche stagnalis, Carex nigra, Cirsium palustre, Equisetum palustre, Galium palustre, Glyceria fluitans, Juncus articulatus, Mentha aquatica* and *Ranunculus repens*.

It is a rhizomatous perennial, flowering from May to July, with fruit ripening between mid-June and August.

This plant was gathered in 1947 by N.Y. Sandwith in its first known site, an ox-bow pool of the Wharfe, but specimens were not identified as this species until 1960 (Walters 1963). Subsequently the plant has been found to be widely spread, if very scattered, from the Tima Water, Selkirkshire (Corner 1975), the headwaters of the Tyne (Northumberland), the Irthing (Cumberland and Northumberland), then with a considerable gap south to the Wharfe (Yorkshire) and Ribble (Lancashire). Where its distribution has been thoroughly investigated the plant has been found to be quite frequent, albeit unpredictable. It is apparently absent from many river systems that would appear to be suitable, but it may yet be found more widely.

The plant seems surprisingly capricious, with some colonies weakening and vanishing, and new ones springing up in fresh sites as conditions change. In Ribblesdale it appeared abundantly in spring-fed pools on the floor of a limestone quarry, and at Ilkley by the Wharfe on an impermanent bare patch by disused gravel workings, far downstream from its other known sites. In both sites *Eleocharis palustris* has increased at the expense of *E. austriaca*, as the conditions have become marshier. Another lowland site has recently been found by the Ribble near Clitheroe, Lancashire. Whilst some colonies probably result from rhizomes dispersed by erosion during spates, it is likely that seed-dispersal also plays an important part.

The plant seems able to tolerate the rapid changes naturally affecting the banks of upland rivers by efficiently colonising suitable sites as soon as they become available. Its rarity on lower reaches of rivers may be due in part to rapid displacement by competitors in more stable conditions. However, upland rivers seem now to be prone to more destructive spates than in the past, from the rapid run-off of storm-water from new forestry and land drainage. In conditions of extreme scour almost no plants can survive, and declines would be expected in such areas. It has not yet been possible to determine the long-term effect on populations of large-scale dredging of the Wharfe in the late 1980s.

E. austriaca is widespread in continental Europe, extending from France and Italy northwards to Norway and eastwards to Yugoslavia, the Caucasus, Ural mountains and Siberia.

F.J. Roberts

Eleocharis parvula (Roemer & Schultes) Link ex Bluff, Nees & Shauer (Cyperaceae)

Dwarf spike-rush, Sbigfrwynen Morafon
Status in Britain: VULNERABLE.
Status in Europe: Not threatened.

E. parvula is a diminutive perennial restricted to estuaries and brackish grazing marshes in large tidal rivers in southern and western Britain. There it typically forms discrete, yet often dense, patches in tidal pans and along creek margins on firm bare muddy substrates that are subject to tidal inundation. In virtually all its British localities, it grows close to the upper limits of tidal influence, avoiding the strongly saline conditions associated with many saltmarshes (Byfield 1992). This is reflected in its associates, which include *Bolboschoenus maritimus, Juncus foliosus, Limosella australis, Ranunculus sceleratus* and *Veronica anagallis-aquatica*, though it apparently never occurs with halophytes such as *Aster tripolium, Limonium vulgare* and *Salicornia* species. However, more often than not, *E. parvula* occurs in mono-dominant stands, without other associates.

In Britain, *E. parvula* has two centres of distribution: along the coast of southern England from Hampshire to Devon, and along the western coastline of North Wales. In the former area the plant is very rare, occurring in colonies of very limited extent at three localities in Hampshire and Devon. The Welsh colonies, in Caernarvonshire and Merioneth, are generally much more extensive and thriving.

The comparative health of the Welsh populations must in part be due to current habitat management. Because of its very small size, the plant is typically reliant on grazing by sheep or cattle reducing competition from coarser plant species. The Welsh colonies survive in extensive grazing marshes, where traditional grazing is still practised. The effect of grazing animals is also important as the principal mechanism by which the plant spreads: poaching breaks up the mats of *E. parvula*, which float and are redistributed at high tide when the pans are flooded. In addition, the plants produce small whitish 'bulbils', which presumably act as perennating organs and are spread by tides and on the feet of grazing livestock. It is not known how important sexual reproduction is to overall recruitment, but certainly in many localities the plant flowers and fruits very poorly. Flowering, when it occurs, is between late August and early October.

The principal threat to the survival of *E. parvula* comes from a cessation of grazing, which results in a rapid succession to brackish swamps dominated by *Bolboschoenus maritimus, Phragmites australis* or *Scirpus lacustris* ssp. *tabernaemontani*, from which *E. parvula* is quickly lost. In addition, river 'engineering' activities such as dredging and dyking, have resulted in the loss of certain British

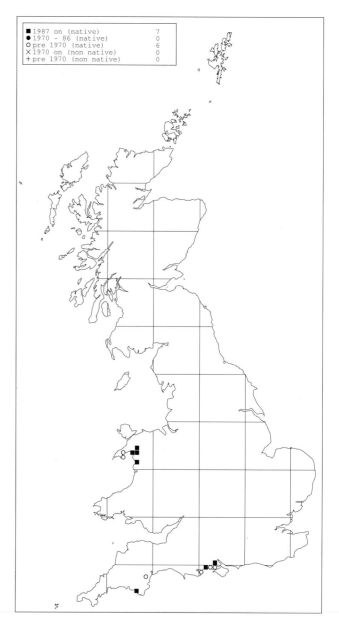

■	1987 on (native)	7
●	1970 – 86 (native)	0
○	pre 1970 (native)	6
✕	1970 on (non native)	0
+	pre 1970 (non native)	0

colonies. It is, however, an inconspicuous plant that usually grows in an uninviting habitat and may thus have escaped detection at other sites.

In Europe the plant is generally local, occurring from Fennoscandia southwards to Portugal, former Yugoslavia and south-east Russia (Walters 1980). It has been recorded from several stations in Ireland, but its current status there is uncertain. It is extinct in Germany and Switzerland. Elsewhere it occurs in North and possibly South Africa, central and eastern Asia, India and Java, and North and South America.

A.J. Byfield

Epipactis youngiana A.F.Porter & A.J. Richards
(Orchidaceae)
Young's helleborine
Status in Britain: ENDANGERED. ENDEMIC. WCA
Schedule 8.

In the 1970s, an *Epipactis* of unknown identity was
discovered in three Northumberland sites; two were in
oakwoods with an understory of *Rubus fruticosus*, and the
third on heavy metal polluted soil. These plants were
subsequently described as a new species, *E. youngiana*
(Richards & Porter 1982). It was suggested that this
partially autogamous species had evolved recently from
hybrids between the widespread outcrossing species, *E.
helleborine*, and an autogamous species such as *E. leptochila*
var. *dunensis*, which occurs nearby. The possibility that the
other parent might be *E. phyllanthes* var. *pendula*, which
occurs at one of the Northumberland sites, has since been
discounted after isoenzyme work. Two of the
Northumberland sites have now been destroyed by clear-
felling, despite strenuous efforts to save them. However, in
1995 and 1996, small numbers were recorded in two
further sites in a different part of the county. At the main
surviving site, some 50-90 stems appear annually.

In about 1987, orchids morphologically
indistinguishable from the Northumberland *E. youngiana*
were discovered on the steep, wooded slopes of two
neighbouring coal-waste bings near Glasgow. Similar sites
have since been discovered on a bing near Linlithgow, and
one near Edinburgh. Plants occur under regenerating trees,
chiefly birch, with sparse vegetation cover (Allen & Woods
1993). *E. phyllanthes* is not recorded in Scotland, but *E.
leptochila* var. *dunensis* occurs at the Scottish sites, so it is
quite likely that Scottish plants evolved separately from,
although in a similar manner to, those in Northumberland.
The largest Scottish colony of *E. youngiana* contains up to
150 plants annually; one other has 30-50, and one 10-20
plants.

In all sites where *E. helleborine* grows with *E. youngiana*,
some intermediate plants are found, and *E. helleborine*
tends to be very variable (which is typical for this species),
so some introgression from *E. youngiana* to its presumptive
parent may be occurring.

A few plants closely resembling *E. youngiana* have been
recorded in Yorkshire, together with *E. helleborine*. *E.
leptochila* does not occur in the vicinity, but there is a
station for *E. phyllanthes* not far away, so it is possible that
these few plants have a different origin from those in
Northumberland and Scotland. A colony in South Wales
occurs together with distinctive sand-dune forms of *E.
helleborine* that resemble ssp. *neerlandica*. These putative *E.
youngiana* plants need further investigation; they may have

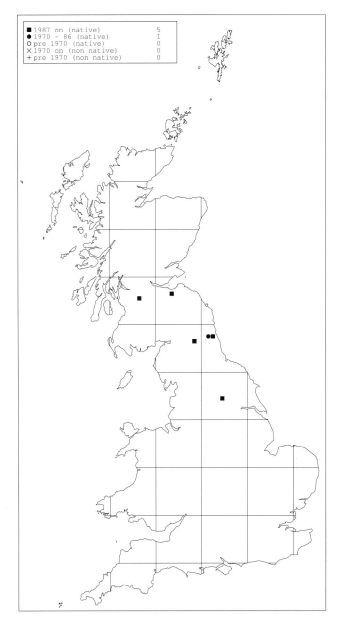

resulted from hybrids between *E. helleborine* 'neerlandica'
and a form of the *E. phyllanthes* complex such as 'E.
cambrensis'.

E. youngiana may thus represent a complex of stabilised
hybrids that are evolving towards certain, often somewhat
fugitive, man-made habitats and thus are of exceptional
interest. It is important that all colonies are conserved so
that further genetical studies, including by DNA
technology, can be undertaken.

A.J. Richards

Epipogium aphyllum Sw. (Orchidaceae)
Ghost orchid, Tegeirian y Cysgod
Status in Britain: CRITICALLY ENDANGERED. WCA
Schedule 8.
Status in Europe: Not threatened.

This is undoubtedly Britain's rarest orchid. It has not
been seen since 1986, and has appeared infrequently and
in very small numbers over the past twenty years. Its
appearances, albeit sporadic, have almost always been
under beech trees (very occasionally under oak), usually in
deep leaf-litter, with few, if any, associated species. Two
other saprophytic species, *Monotropa hypopitys* and *Neottia
nidus-avis*, have, however, occurred nearby. At the most
recent Herefordshire site, it appeared under oak, in a
wood now planted with *Picea*, *Pinus* and *Tsuga* and
managed commercially.

The environmental conditions controlling the growth of
flowering spikes are not known, though it has been
suggested that emergence might be triggered by a long
period of rain followed by warm weather. Flowers have
been recorded between April and mid-October, each
flowering spike normally bearing two. Because of the
sporadic appearance of aerial parts at historic sites after
long absences, it is presumed that the underground
rhizome can be long-lived.

The species was first recorded in Britain in 1854 in
Herefordshire. Subsequent records from that county were
in 1892, 1910 and 1982, at the last date a single flower spike
with two open flowers. There was also an unconfirmed
record from Leominster, and it was seen between 1876 and
1878 at Ringwood Chase in Shropshire. In Oxfordshire, it
was apparently regularly seen in one beechwood in the
1950s and 1960s, though the last record was in 1963. In
recent years, it has been most regularly seen in
Buckinghamshire (up to 1986), where it has occurred in
three beechwoods. Apart from an exceptional colony of 24
spikes discovered in that county by R.A. Graham in 1953,
in all sites numbers of spikes have been very few (usually
1-4 and exceptionally up to seven) in any year.

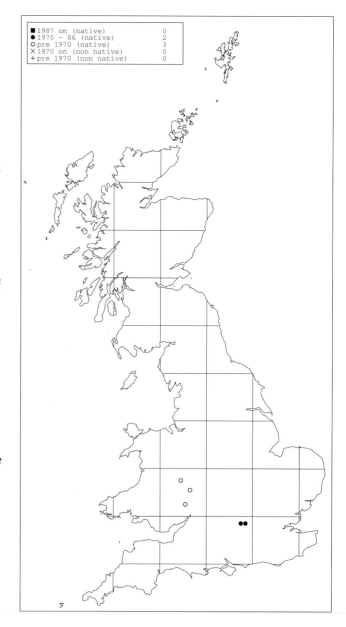

■ 1987 on (native)	0
● 1970 – 86 (native)	2
O pre 1970 (native)	3
X 1970 on (non native)	0
+ pre 1970 (non native)	0

All sites in Herefordshire, Oxfordshire and
Buckinghamshire where *E. aphyllum* has been seen recently
are SSSIs. However, since so little is known about its
ecological requirements, it is not easy to assess what
particular conservation action is appropriate. Plants are
very difficult to see against the brown background and are
certainly vulnerable to inadvertent trampling. Plants have
apparently been dug up (in 1978 and 1979) from a
Buckinghamshire wood, though the underground rhizome
may not have been destroyed. It is also attractive to slugs,
and there are several reports of it being eaten off soon
after its appearance. One site has been threatened by
increased horse-riding, timber extraction and windthrow.

However, the habitat of most of its historic and recent sites
appears to remain suitable, and there seems no reason
why it should not reappear at some sites.

On the continental mainland it occurs in both
deciduous and coniferous woods. Its distribution ranges
through northern and central Europe, extending
southwards in mountain ranges to the Pyrenees, central
Apennines, north-west Greece and the Crimea, and further
eastwards to the Himalayas. It is legally protected in
several European countries.

L. Farrell

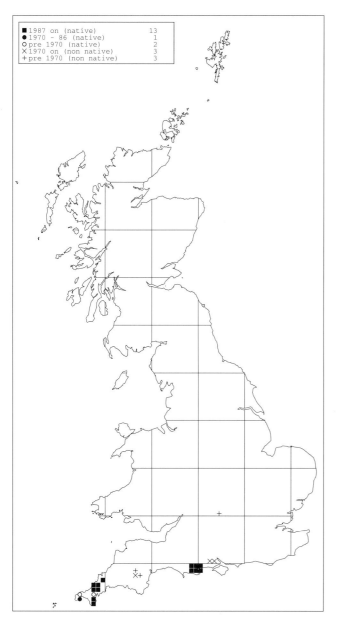

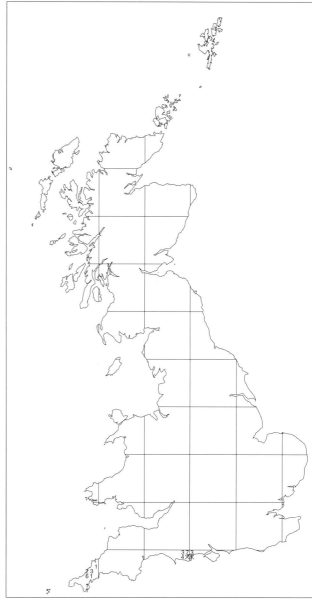

■ 1987 on (native)	13
● 1970 - 86 (native)	1
O pre 1970 (native)	2
X 1970 on (non native)	3
+ pre 1970 (non native)	3

Erica ciliaris L. (Ericaceae)

Dorset heath
Status in Britain: LOWER RISK - Near Threatened.
Status in Europe: Not threatened. Near Endemic.

E. ciliaris occurs on both acidic mineral and humic soils in a number of plant communities ranging from dry heath and acid grasslands to wet heath and peatland. In Dorset, the plant is most abundant on the wet and humid heaths where it grows in association with *Calluna vulgaris, Drosera intermedia, D. rotundifolia, Erica tetralix, Molinia caerulea* and the mosses *Sphagnum compactum* and *S. tenellum*. Rarer species in the community include *Gentiana pneumonanthe, Pinguicula lusitanica, Rhynchospora alba* and *R. fusca*. It also occurs in valley mires where, in addition to the above species, *Sphagnum papillosum* is frequent. *E. ciliaris* also occurs in drier, freely draining situations with *Agrostis curtisii, Calluna vulgaris, Erica cinerea* and either *Ulex gallii* or *U. minor*. In Cornwall it is mainly restricted to the drier *U. gallii* communities.

This species is a long-lived dwarf shrub with individual stems surviving for up to twenty years. However, the vigorous regrowth of shoots from mature rootstocks after cutting or burning suggests that the below-ground portion of the plant may be very long-lived. Seed is set annually from the third season, but seedling growth and establishment are observed only where the soil surface has been exposed and disturbed by burning, cutting or grazing of the vegetation. In undisturbed vegetation, propagation is usually vegetative, by adventitious root growth from prostrate stems.

E. ciliaris is a member of the small Lusitanian element in our flora. The majority of the British sites are in Dorset, where, however, much of its former range has been lost to conifer plantations and some to agricultural conversion. The remaining large populations are within the NNRs lying to the south of Poole Harbour. Outside of this area there are numerous small populations and isolated plants and the consensus of opinion (Chapman 1975; Haskins

1978; Chapman & Rose 1994) is that these are founder populations on the edge of an expanding range. The sites in Cornwall are thought to be relict fragments of formerly much larger populations. Colonies in other areas of Britain, most notably in the New Forest and on Dartmoor, are considered to have originated from deliberate introductions.

In recent years, many of its Dorset sites have been managed for heathland conservation, principally by grazing, and some former sites that had been planted with conifers are now being returned to heathland. In Cornwall the position is less clear, the greater fragmentation of heathland having resulted in much smaller populations of *E. ciliaris*, thus making conservation management less practicable. The influence of various management practices and environmental factors is summarised in Rose *et al.*

(1996). Hybridisation occurs where *E. tetralix* grows with *E. ciliaris*. The hybrid (*E. x watsonii*) shows a wide range of forms with a number of characters that are intermediate between its parent species, suggesting that back crossing occurs. The fertility of hybrid plants is very low, but they are more vigorous than either parent and, in places where only vegetative spread occurs, hybrids can dominate the vegetation.

E. ciliaris is at the northern limit of its range in Britain and Ireland, where it is found both in south-west England and west Galway. Its world distribution extends southwards through western France, Spain and Portugal as far as the north-west tip of Morocco.

R. Rose

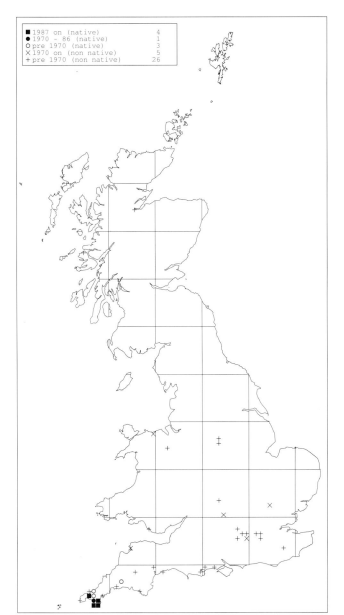

Erica vagans L. (Ericaceae)

Cornish heath, Grug Cernyw
Status in Britain: LOWER RISK - Near Threatened.
Status in Europe: Not threatened. Endemic.

In Britain, *E. vagans* occurs only on the Lizard peninsula, where it is almost restricted to four hectads. It is not merely abundant in many localities, but is co-dominant in species-rich heathlands over large areas of the plateau and in sheltered rocky coastal valleys and slopes on the ultra-basic rocks, serpentine and gabbro. It is co-dominant with *Ulex europaeus* and *U. gallii* on brown earth soils (pH 6-7) on sheltered slopes from near sea level to 100 m (Coombe & Frost 1956a), occurring with a wide spectrum of other vascular plants, including *Calluna vulgaris, Erica cinerea, Genista pilosa, Viola lactea* and basiphiles such as *Filipendula vulgaris* and *Pimpinella saxifraga*. In broad shallow valleys over much of the Lizard plateau, *E. vagans* is co-dominant with *Molinia caerulea* and *Schoenus nigricans* on silty clay and neutral to slightly acid gley soils

overlying serpentine and gabbro. This 'Tall Heath' (Coombe & Frost 1956a) is also species-rich and includes *Carex pulicaris* and *Erica tetralix*. *E. vagans* is intolerant of extreme exposure to salt-laden winds and therefore rarely occurs in the cliff-top *Calluna - Scilla verna* heaths. It survives at one site on palaeozoic shales inland in West Cornwall, but was quarried away on a basic igneous outcrop in East Cornwall. It is absent from acidic but fertile schists fringing the serpentine, and from expanses of *Ulex gallii - Agrostis curtisii* heaths on deep acidic silty loess, which overlies both serpentine and gabbro on level areas of the plateau (Coombe & Frost 1956b; Staines 1984; Rodwell 1991).

E. vagans is a rather straggly glabrous shrub up to 80 cm tall, producing its pale pink or white flowers from July to September. Seedlings and plantlets are locally frequent but often die of drought (most notably in 1949, 1955 and 1976). Bushes rarely die of old age, though about 50 annual growth rings were counted on one dead bush in an exceptionally old stand (which was burnt in 1955). Propagation by shoot-tip cuttings is easy, and it regenerates from the base after controlled winter burning.

About one-third of the Lizard heathlands were lost between 1908 and 1980 (J.J. Hopkins in Turpin 1984), and both the 'Tall Heath' and the 'Mixed Heath' vegetation have been much modified by turf-paring and repeated burning. However, large areas of the *E. vagans* heath that remains now have statutory protection, and the overall population of *E. vagans* is likely to be stable. Limited areas are grazed by cattle in the out-fields of a few farms.

The Irish population of (white-flowered) *E. vagans* in County Fermanagh is perhaps a prehistoric introduction ('archaeophyte') (Nelson & Coker 1974). *E. vagans* otherwise occurs only in western and central France and northern Spain, sometimes on calcareous soils with, for example, *Cornus sanguinea*.

D.E. Coombe

Erigeron borealis (Vierh.) Simmons (Asteraceae)

Alpine fleabane, Fuath-dheargann Ailpeach
Status in Britain: VULNERABLE. WCA Schedule 8.
Status in Europe: Not threatened.

This species was recorded in the earliest days of
botanical exploration of Ben Lawers - in 1787, by the Rev.
John Stuart (Mitchell 1992) - and remains one of the
notable species at this much visited site. It is a plant of
montane cliff ledges, on predominantly basic rocks, and
usually adjacent to grazed herb-rich grassland that
includes many of the species listed below. It occurs in two
areas: the Breadalbane district of Perthshire, and
Caenlochan SSSI in Angus. The altitude range at Ben
Lawers is 750-1,100 m, with most plants between 800 and
950 m (Payne 1981a). The most frequent associates of *E.
borealis* are *Alchemilla alpina, Cerastium alpinum, Euphrasia*
species, *Festuca vivipara, Saxifraga oppositifolia, Silene acaulis*
and *Thymus polytrichus*. Less frequent species with it on
Ben Lawers include *Alchemilla glabra, Botrychium lunaria,
Gentiana nivalis, Selaginella selaginoides, Sibbaldia procumbens,
Veronica fruticans* and *Viola riviniana*. At Glen Doll,
Caenlochan, it occurs on steep unstable rock faces where
there is little competition from other species.

E. borealis is a perennial herb, with a short creeping
stock. Flowering is in July or August, and the number of
flowering heads is very variable. Insects visit the flowers,
and seeds are usually produced. Seedlings may establish
outside the area of existing colonies. The population on the
Ben Lawers range is mostly within the NNR, and is
absolutely restricted to ungrazed ledges inaccessible to
sheep or deer. The occasional plant that establishes itself
on ground below the crags does not long survive the
intensive grazing. The population could expand
significantly in numbers and range if grazing were
removed. Collecting still occurs on a small scale, despite
protective legislation.

Since 1980, records from the Breadalbanes are mostly
within a cluster of ten 1 km squares within the Ben Lawers
NNR, with a single outlier some 5 km to the north. Within
the Caenlochan SSSI, *E. borealis* occurs in two rather widely
separated localities. An earlier record for another site on
the Meall nan Tarmachan range may indicate a decline
there, and there are early records in the Cairngorms. The
whole population on the Ben Lawers range is monitored
and between 1981 and 1991 was generally stable at about
600 plants, but the number of plants in particular colonies
may fluctuate widely from year to year. The main locality
in Caenlochan SSSI is monitored at seven separate sites. A
decline in the number of flowering heads from 274 to 59
was recorded from 1986 to 1993, with only 32 seen in 1992.
However, problems of method and access complicate the
interpretation of the figures (Geddes 1996) such that they
might not indicate a real population decline.

It is an arctic amphi-atlantic species, occurring in
eastern North America (Labrador, Newfoundland),
Greenland, Iceland, Scandinavia and arctic Russia (Benum
1958). In Scandinavia, *E. borealis* also occurs in rocky
submontane birch woodland, and at a higher altitude.

D.K. Mardon

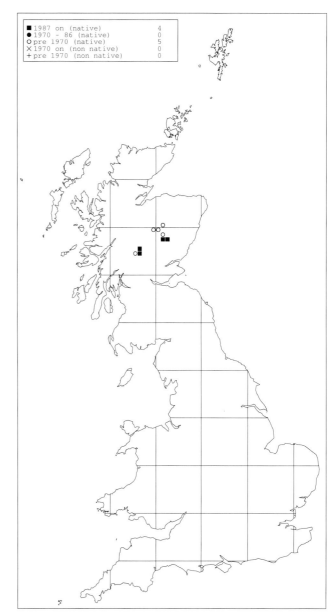

■ 1987 on (native)	4
● 1970 - 86 (native)	0
○ pre 1970 (native)	5
✕ 1970 on (non native)	0
+ pre 1970 (non native)	0

141

Eriocaulon aquaticum (Hill) Druce (Eriocaulaceae)
Pipewort, Pioban Uisge
Status in Britain: LOWER RISK - Near Threatened.
Status in Europe: Near threatened.

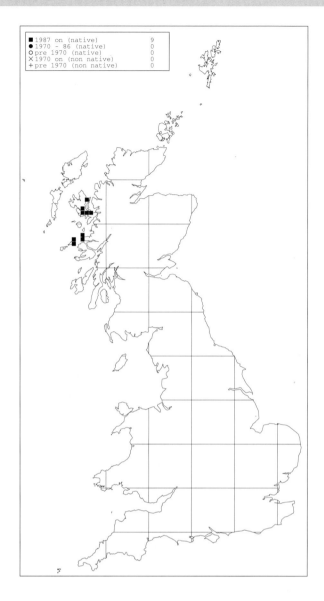

E. aquaticum grows as small clumps or extensive mats in shallow water, or terrestrially on damp and periodically flooded ground above the normal water level. It is confined to oligotrophic lakes, and is much more frequent in smaller lakes and pools than in large water bodies. It can sometimes be found in peat cuttings. *Lobelia dortmanna* is almost invariably found with it; other frequent associates include the aquatic form of *Juncus bulbosus*, *Littorella uniflora* and *Ranunculus flammula*, and scattered stems of emergents such as *Carex lasiocarpa*, *C. rostrata* or *Phragmites australis*. *E. aquaticum* never extends far above the water line (often being replaced at higher levels by *Littorella uniflora*) and descends to depths of at least 2 m. It grows over a range of substrates including peat, fine gravel, sand and silt. Although it is usually found in ditches where both the substrate and water are acidic, it can occur in slightly alkaline water over acidic substrates (Webb & Scannell 1983). It is a lowland species, ascending to 300 m in Co. Cork.

E aquaticum flowers from August onwards. Plants can flower in water at least a metre deep, as the stems elongate to carry the flowers above the surface. The flowers are structurally complex and are likely to be insect pollinated or autogamous (Cook 1988). Little is known about the pollination of plants in Britain or Ireland, nor is it known whether or not they set viable seed. The species probably spreads vegetatively within lakes in which it occurs. Large clumps or single plants of *E. aquaticum* are often found washed up on the shore, and it seems likely that the former, at least, could become established as new colonies.

The distinctive rosettes of *E. aquaticum* are unmistakable, even when they are not in flower. Although it has long been known from Ireland, Coll and Skye, the species was not discovered on the Scottish mainland until 1967 (McClintock 1968; Ferguson & Ferguson 1971): further colonies might possibly await discovery in western Scotland. Scottish populations have been surveyed in detail in recent years (e.g. Farrell 1983). The species grows in remote localities in both Britain and Ireland and its habitats do not appear to be threatened.

Within the areas in which it occurs *E. aquaticum* is frequent, but it is absent from much apparently suitable habitat both in Ireland and, more particularly, in Scotland. It seems most unlikely that any explanation based solely on current ecological factors could explain this very restricted distribution. *E. aquaticum* must be either a relict species, confined to areas from which it is unable to spread, or a recent colonist. A single grain of *Eriocaulon* pollen from Hoxnian interglacial deposits in Ireland has been tentatively referred to this species (Jessen *et al.* 1959) and pollen from a postglacial deposit in Connemara has been identified as *E. aquaticum* without qualification. There is therefore no doubt that *E. aquaticum* is a long-established member of the Irish flora (Godwin 1975; Perring 1963). Its history in Scotland is less certain. Heslop-Harrison (1953) suggested the possibility that plants might have arrived by

long-distance dispersal of seed, and Farmer & Spence (1986) refer to it as "an historically recent arrival to Scottish freshwaters". If so, it is curious that a species that could arrive by long distance transport has failed to colonise new localities in Britain, for there is no evidence that *Eriocaulon* is spreading. A comparative study of the reproductive biology of the American, Irish and Scottish populations would be instructive. The cytology of the species also requires further investigation. Löve & Löve (1958) report that European plants (2n=64) differ from American plants (2n=32) in chromosome number, and fossil and recent pollen from Iceland is larger than recent pollen from America (Perring 1963). If all American populations differ from those in Europe, the possibility of recent long-distance dispersal would be ruled out.

E. aquaticum is an amphi-atlantic species that is widespread in North America but in Europe is confined to Britain and Ireland. It is a member of the small North American element in our flora.

C.D. Preston, adapted from an account in Preston & Croft (1997).

Eriophorum gracile Koch ex Roth (Cyperaceae)
Slender cottongrass, Plu'r Gweunydd Eiddil
Status in Britain: VULNERABLE. WCA Schedule 8.
Status in Europe: Vulnerable. Endemic.

E. gracile is a plant of exacting habitat requirements. It is absent from strongly acidic bogs, favouring transitional mires, poor fens and calcareous 'brown moss' fen, typically occurring in well-illuminated conditions over liquid peats with areas of slowly moving open water. These conditions may be found within extensive bogs of varying character as well as in seepages and springhead mires. It grows as an emergent, on the sides of low tussocks of *Molinia caerulea*, and on floating *Sphagnum* rafts in the very wettest parts of bogs. Associates include *Equisetum fluviatile*, *Hypericum elodes*, *Menyanthes trifoliata*, *Phragmites australis*, *Potamogeton polygonifolius*, *Campylium stellatum*, *Drepanocladus revolvens* and *Scorpidium scorpioides*. In Surrey, *E. gracile* occurs amongst *Menyanthes trifoliata* at the margins of a pond in a heathland mire. The plant will tolerate light shade in open willow scrub. It is generally a lowland plant, though it formerly occurred at 250 m in Neroche Forest, Somerset.

It is a perennial that spreads vegetatively and grows freely from seed. The ability of the plant to colonise newly formed habitats such as Little Sea, Dorset, suggests that it may be able to recolonise former sites or new ones if conditions become suitable. Isozyme analysis of the main English and Welsh populations indicate that they are genetically distinct, though closely related, and substantially or entirely monomorphic (Kay & John 1995).

E. gracile has been recorded from some 30 sites in England and Wales since its first record in 1838, but is now restricted to about six sites in six hectads. By far the largest population occurs in Crymlyn Bog, Glamorgan, and comprises an extensive series of colonies, with a total of more than 100,000 plants (Jones 1994). In England, the New Forest formerly supported more than ten sites, but now only two are known to be extant, both with small populations of a few hundreds and a few tens of plants respectively. However, it is just possible that other colonies may be found in the extensive mire systems of that area. The Surrey site holds the largest population in England, with more than 10,000 plants. Where bogs and fens are being brought back into grazing management, it is possible that former populations may be revitalised or new ones discovered.

Many its former mire sites are still extant, but succession has changed their character, perhaps owing to the lack of grazing. Historic management that appears locally to have benefited *E. gracile* includes peat cutting and the coppicing of carr woodland. Activities that have damaged or destroyed colonies include afforestation, drainage and infilling by dumping. A recent review of English sites (Winship 1994b) highlights current threats,

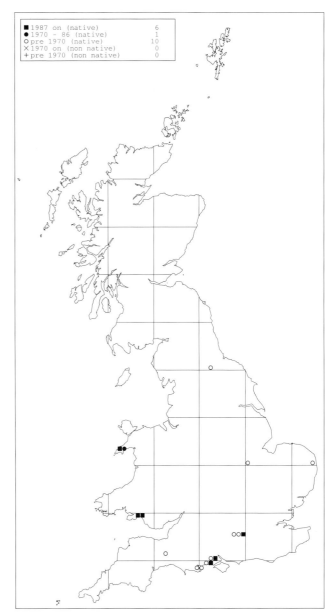

■ 1987 on (native)		6
● 1970 – 86 (native)		1
○ pre 1970 (native)		10
✕ 1970 on (non native)		0
+ pre 1970 (non native)		0

including succession to woodland, the disruption of water supply to mire sites, and perhaps agricultural run-off. Active conservation management is seeking to offset these threats. All currently known sites are designated as SSSIs, and much of Crymlyn Bog is an NNR.

E. gracile is still widespread in the Republic of Ireland, being found in Connemara, West Meath, Cork and Kerry (J. Conaghan pers. comm.). It is found in thirteen European countries from Britain and Fennoscandia southwards to Italy and Romania. It is declining and threatened in most of them and considered to be safe only in Ireland, Finland and Norway.

C. Chatters and H.R. Winship

Eryngium campestre L. (Apiaceae)

Field eryngo, Ysgallen Ganpen
Status in Britain: VULNERABLE. WCA Schedule 8.
Status in Europe: Not threatened.

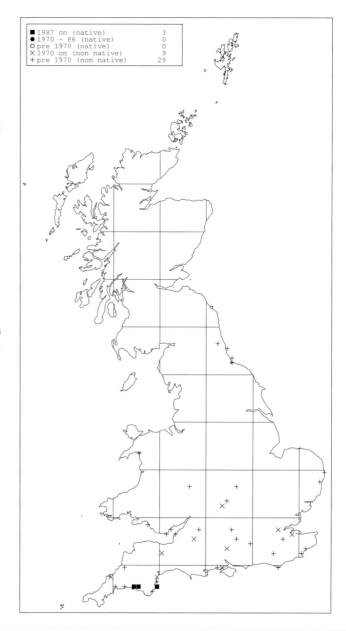

■ 1987 on (native)	3
● 1970 – 86 (native)	0
O pre 1970 (native)	0
X 1970 on (non native)	9
+ pre 1970 (non native)	29

E. campestre is a plant of neutral or calcareous grassland on light soils, often near the sea. At Scabbacombe, Devon, it grows in old pasture that has been improved by fertiliser and grazed by cattle and sheep, with undistinguished associates including *Agrostis stolonifera, Cynosurus cristatus, Cerastium fontanum* ssp. *vulgare, Ranunculus bulbosus, Lolium perenne, Plantago lanceolata* and *Veronica chamaedrys*. Near Plymouth, it occurs in grassland on Devonian limestone at two sites, both of which are used as public open space. Though the grassland is mown regularly at both sites, *E. campestre* grows mainly in areas that are less regularly mown (Proctor 1985). In addition to those listed above, a wide range of associates occurs at the Plymouth sites, including *Centaurea nigra, Festuca rubra, Galium verum, Lotus corniculatus, Pimpinella saxifraga, Sanguisorba minor* and *Trisetum flavescens*. In North Somerset, a few plants grow in rank grassland on oolitic limestone in an area planted with conifers in the 1980s. Elsewhere, *E. campestre* occurs in re-seeded grasslands and on disturbed ground.

E. campestre is a perennial, flowering from July, with a peak usually between late August and mid-September, though few flowers may appear in cool or wet summers. Some seed appears to set in hot summers, but seedlings are rare in British populations, except in those that are long-established (perhaps native) near Plymouth. Open ground is necessary for germination and seedling establishment, and light disturbance of the soil is beneficial. It is probable that the characteristic circular patches of plants in many of its sites are vegetative clones (FitzGerald 1990b). The small populations in Kent are on disturbed ground (FitzGerald 1988d, 1988e).

This species was formerly more widespread in Britain, but only nine populations are extant - in Cornwall, Devon, Somerset and Kent. Much the largest populations are in Devon (up to 10,000 at Billacombe and 2,000 at Devil's Point): other sites in Britain hold between 10 and 200 plants.

Although *E. campestre* is considered probably native at two or three of its Devon sites (at Billacombe and Devil's Point, Plymouth, and Scabbacombe near Brixham), most of the British colonies have short histories and are likely to have been introduced. For example, the Cornish populations were found only after pasture was re-seeded, presumably with contaminated grass-seed mixtures of continental origin. The population at Hinton Charterhouse, Somerset, though in semi-natural limestone grassland, was first found in 1968: re-seeding has taken place since its discovery and might also have occurred prior to it. The main threats are agricultural improvement and habitat neglect. The protection and management of the strong populations in Devon are especially important. Four sites are in SSSIs.

Conservation management would ideally include grazing (cattle or sheep) without the application of artificial fertiliser (Taylor 1990f; Scruby *et al.* 1992). Mowing can be an acceptable substitute for grazing, especially if the soil is disturbed occasionally, and the site at Billacombe is managed in this way. On very open ground the main requirement is to control rough vegetation and scrub.

E. campestre occurs throughout Europe from the Channel Islands to the Mediterranean, with its stronghold in the southern parts of its range. It extends eastwards to central Russia and Afghanistan, and southwards to North Africa.

D. Junghanns and M.J. Wigginton

Euphorbia hyberna L. (Euphorbiaceae)
Irish spurge
Status in Britain: VULNERABLE.
Status in Europe: Not threatened. Endemic.

E. hyberna is remarkably rare in Britain, considering its abundance in Ireland and in parts of western Europe. In Britain it is confined to two small areas, Nance Wood near Portreath, Cornwall, and the Lyn valley, North Devon, on the border with Somerset. It is a plant of openings in woodland, and grows best in places that receive dappled sun for at least part of the day. Nance Wood is mainly oak coppice with some sycamore, but coppicing has now been abandoned, and *E. hyberna* is in danger of being swamped by *Lonicera periclymenum* and *Rubus fruticosus*. Associated species include *Anemone nemorosa, Angelica sylvestris, Conopodium majus* and *Hyacinthoides non-scripta*. In the valley of the Lyn, the plant is locally common in ancient, formerly coppiced, oak woodland (FitzGerald 1990b).

This perennial rhizomatous species grows and flowers early, before the trees have come fully into leaf. It then remains green and leafy through the summer, before dying back to a stout tufted rootstock in autumn. Individual plants can live for at least twenty years.

E. hyberna was first found in Devon in 1840, and in Cornwall in 1883, at which time it was abundant. It has also been reported from Somerset, but has not been seen recently on the Somerset side of the county boundary with Devon (P. & I. Green pers. comm.). In 1989, more than 40 clumps were reported in Nance Wood, Cornwall. It is under no serious threat in Devon at present, but the spread of *Fallopia japonica* (which is becoming locally dominant in the Watersmeet woods) and other aggressive aliens could oust it from much of its suitable habitat. In Nance Wood, control of vigorous associates may become necessary.

E. hyberna is endemic to Europe: ssp. *hyberna* has a typical Atlantic distribution, being found in Ireland, western France, north-west Spain and Portugal; other subspecies occur in the Alpes Maritimes and on Mediterranean islands (Smith & Tutin 1968). In Ireland ssp. *hyberna* is common in the south-west, from east Cork to North Kerry, and is often a conspicuous feature of hedgebanks in these areas in spring.

No other species is so common in south-west Ireland yet so rare in south-west England. *Saxifraga spathularis* is equally common in south-west Ireland, and found in similar habitats to *E. hyberna*, but is not known in Britain. There must be some uncertainty whether *E. hyberna* is native in England, or whether it is an escape from cultivation or was brought in with cargo from Ireland to

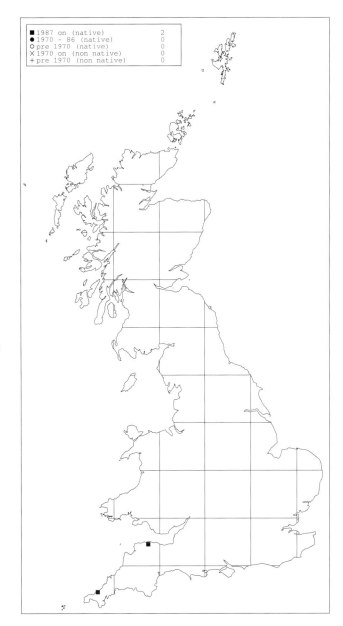

■ 1987 on (native)	2
● 1970 - 86 (native)	0
O pre 1970 (native)	0
X 1970 on (non native)	0
+ pre 1970 (non native)	0

Lynmouth and Portreath. Trade between Ireland and North Devon was well developed in the Middle Ages, and seeds could have been introduced with wool for the cloth trade. In Ireland *E. hyberna* is known as a fish poison, and there is the remote possibility that the plant could have been introduced to take fish from the Lyn. However, there are other examples of highly unbalanced distributions between south-west Ireland and south-west England (e.g. Perring 1996), and most authorities consider *E. hyberna* to be native in Britain.

M. Rix

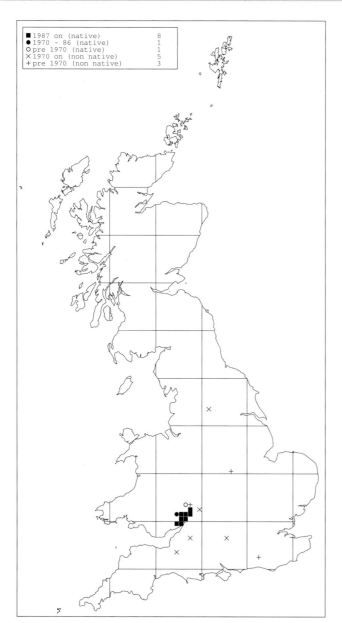

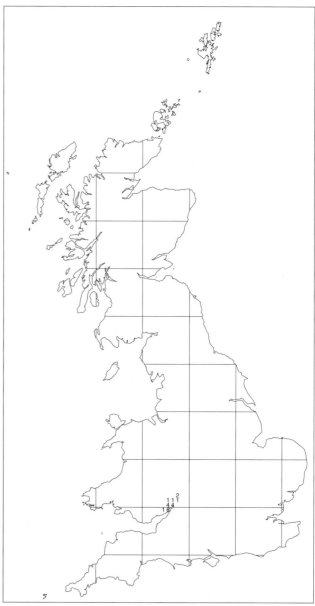

Euphorbia serrulata Thuill. (Euphorbiaceae)

Euphorbia stricta L., *nom. illegit.*
Upright spurge, Llaethlys Mynwy
Status in Britain: VULNERABLE.
Status in Europe: Not threatened.

E. serrulata is a plant of open deciduous woodland, tracks and hedge banks, growing on calcareous soils derived from Carboniferous limestones and shales, Keuper Marl, Lias clays, and occasionally alluvial gravels and clays. It grows in rides, along tracks and at the woodland edge where there is sufficient light. A wide range of associated species includes *Deschampsia cespitosa, Euphorbia amygdaloides, Geum urbanum, Heracleum sphondylium, Prunella vulgaris, Rubus fruticosus, Stachys sylvatica, Torilis japonica,* a variety of sedges and grasses, and scarce plants such as *Paris quadrifolia* and *Vicia sylvatica.*

It is a slender autumn-germinating annual or biennial, up to 70 cm high, though often shorter. Flowers appear from June onwards, and the stem leaves usually droop and fall in July. The light buoyant capsules fall when ripe and release seed to germinate near the parent or, as with other *Euphorbia* species, are frequently dispersed by ants, which carry the capsules to the nest. Seeds have also been observed to float away on woodland trickles and streams (T.G. Evans pers. obs.). *E. serrulata* requires open ground for establishment, and at least moderate light levels in relatively sheltered situations for successful development and flowering.

This species was first recorded in Britain, though erroneously as *E. platyphyllos*, by the renowned botanists the Reverend John Lightfoot and Sir Joseph Banks in June 1773 "by the Brook side, going from the Abbey (Tintern) to the Forge". It is native in Britain in east Monmouthshire and west Gloucestershire, its few occurrences outside this area

146

considered to be garden escapes, though it is well naturalised in some places. *E. serrulata* was formerly locally abundant in about 24 sites in the two counties, though it has declined markedly in recent years. Since 1987, it has been recorded at seventeen sites, but was found at only nine in 1994 (and one other in 1995), many populations comprising fewer than twenty individuals. The main colonies now appear to lie within two Gloucestershire woods.

Numbers fall as shade and competition increase and as the succession develops. However, soil disturbance will stimulate the germination of even long-buried seed, and this has been observed, for instance, where ground has been 'topped' to remove bracken, and where vegetation has been cleared from banks. Likewise, coppicing of a wood after a gap of about 40 years, during which time *E. serrulata* was not seen, resulted in a suddenly large population. But a wood where *E. serrulata* was known in the 1900s, but which was then not coppiced for 75 years, lost the species. This suggests that the limit of seed viability may lie between 40 and 75 years, and that even a long coppicing cycle would ensure its survival (C. Hurford pers. comm.).

The main causes of decline have been the replacement of old woodland with conifer plantations, the cessation of traditional woodland management, and the over-growth of colonies by vigorous and rank species. Of immediate benefit would be regular soil scarification or rotavation to allow the succession to begin again. The widening of paths and rides, and the restoration of woodland management, including coppicing and thinning, are also required.

Outside Britain, it occurs throughout western, central and southern Europe, eastwards to Turkey and the Crimea, perhaps extending to northern Iran and the Aral region of Tajikistan.

T.G. Evans

Euphrasia L. (Scrophulariaceae)
Eyebrights, Lus nan Leac

All British species of *Euphrasia* are hemi-parasitic annuals, in most cases occurring within permanent or semi-permanent grassland communities. Germination usually or always occurs in the spring, followed by attachment to the root system of a suitable host. Availability of suitable germination sites is likely to vary from year to year and early season droughts also appear to affect successful germination or host-establishment. Consequently, populations may show marked fluctuations in abundance.

The species do not appear to show any marked host specificity, but attach to a wide range of herbaceous species, though this is still to be confirmed for the rarer endemics. Populations are usually dependent on light grazing and will not survive where vegetation becomes too rank.

British species commonly hybridise within their own ploidy level and sometimes form hybrid swarms. Individual hybrid genotypes may also increase and spread through self-fertilisation, giving rise to distinctive local populations. Crosses between diploids and tetraploids are normally sterile, but occasional fertile derivatives may also produce locally distinct taxa. Small-flowered species typically self-fertilise as the flower opens (Yeo 1966) and so form hybrids less readily, although those that are formed may persist and spread. Even small-flowered species receive insect visitors, e.g. hoverflies. Speciation in the genus can be regarded as a combination of hybridisation events and subsequent adaptation. This is an ongoing process, leading to disagreements regarding taxonomic treatment, at least of the more geographically restricted taxa.

E. eurycarpa and *E. rhumica*, listed in previous editions of the British Red Data Book, are endemic to the island of Rum. Both are now regarded as taxonomically dubious and are excluded from the following accounts, pending further investigation. *E. eurycarpa* seems to be a hybrid of *E. ostenfeldii* (with which it has been confused) and may have been of no more than transient occurrence. *E. rhumica* seems likely to be an established hybrid of *E. ostenfeldii* with *E. micrantha*.

Another named taxon that may merit recognition is *E. atroviolacea* Druce & Lumb, which is endemic to Orkney, where it is currently known from three sites in thin coastal turf on shell-sand. It is threatened by extensive hybridisation with *E. confusa* and also by sand extraction at its best known site on the Links of Boardhouse.

A.J. Silverside

Euphrasia cambrica Pugsley (Scrophulariaceae)
Welsh eyebright, Coreffros Cymreig
Status in Britain: VULNERABLE. ENDEMIC.

E. cambrica is an upland species, endemic to
Snowdonia, North Wales. It occurs in short, sheep-grazed
turf dominated by *Agrostis capillaris* and *Festuca ovina* ssp.
hirtula, accompanied by such species as *Anthoxanthum
odoratum*, *Campanula rotundifolia*, *Diphasiastrum alpinum*,
Galium saxatile, *Huperzia selago*, *Thymus polytrichus*,
Vaccinium myrtillus, *Racomitrium lanuginosum* and
Rhytidiadelphus squarrosus. Sites seem to be typically well-
drained and on slopes, but *E. cambrica* can grow in
moderately flushed conditions, sometimes with *Euphrasia
rivularis* and *Selaginella selaginoides* as additional associates.
On adjacent cliffs it can be replaced by a local variant of
Euphrasia ostenfeldii. *E. cambrica* is said to take the place of
the Scottish and circumpolar *E. frigida* in Snowdonia, but
detailed habitat requirements are not the same. *E. cambrica*
was claimed by Pugsley to have occurred in the Lake
District, based on a collection from the Kirkstone Pass in
1881. This seems most likely to have been an error, based
on an apparent hybrid derivative of *E. ostenfeldii* that
occurs there.

An account of *E. cambrica* was given by Pugsley (1930),
who regarded it as "widely distributed over the mountains
of Caernarvonshire". Under natural conditions, *E. cambrica*
is a minute plant, commonly only 1-2 cm in height, with
flowers as small as any British species of the genus.
Consequently it is liable to be overlooked except by those
specifically searching for it. It has recently been confirmed
from sites on Snowdon and Cwm Idwal and there are
recent, presumably reliable, records from other sites,
including its outlier at Cader Idris. There has, however,
been a continuing history of confusion with *E. ostenfeldii*,
which resembles *E. cambrica* closely in floral characters.
Examination of herbarium material has shown that records
of *E. cambrica*, even when determined by Pugsley himself,
can be wrong. This includes all examined material of *E.
cambrica* f. *elatior*. Furthermore, at Cwm Idwal, there is a
large population of plants intermediate between the two
species and presumably of hybrid origin.

Hybrids are known only with *E. ostenfeldii* and *E.
scottica*. There is no evidence of gene-flow threatening
occurrence of the pure species. Limited critical modern
recording and the background of taxonomic confusion do
not allow a reliable assessment of the plant's current
status. However, key populations lie within NNRs and
there appears to be no reason to believe that the plant is
subject to any significant threat.

A.J. Silverside

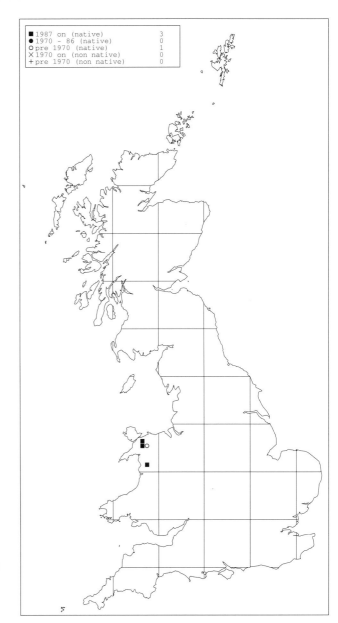

■ 1987 on (native)	3
● 1970 - 86 (native)	0
O pre 1970 (native)	1
X 1970 on (non native)	0
+ pre 1970 (non native)	0

Euphrasia campbelliae Pugsley (Scrophulariaceae)
Eyebright, Lus nan Leac
Status in Britain: LOWER RISK - Near Threatened.
ENDEMIC.

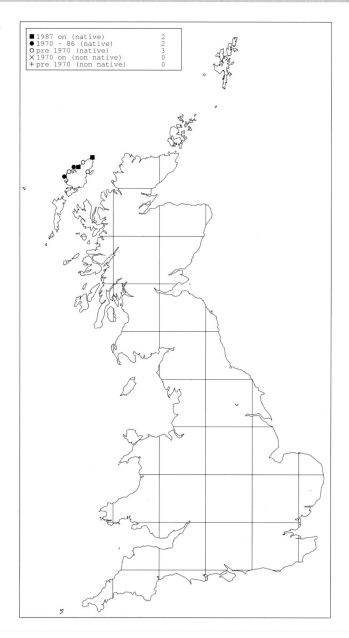

E. campbelliae is confined to Lewis in the Outer Hebrides, where it has been recorded from a number of localities, mainly along the western coast of the island. It occurs near the sea in damp, mossy, heathy turf, with *Carex panicea* and other sedges and where *Calluna vulgaris* and *Erica tetralix* are kept in check by grazing. In mosaic grass-heath vegetation it can occur on the boundaries of the heathy 'islands' where it may come into contact with *E. micrantha* in the drier heath, *E. nemorosa* in drier grass-turf or *E. scottica* in damper areas. Consequently, *E. campbelliae* can form one component of taxonomically complex populations.

E. campbelliae was first described as a species in 1940. It has generally been considered to have originated by hybridisation and it shares morphological features with local populations of *E. nemorosa*, notably the pronouncedly distal distribution of the leaf-hairs. Its origin is probably complex, also involving *E. micrantha* and quite possibly *E. marshallii* and *E. scottica*. Its very localised distribution and its occurrence as part of the diverse, much hybridised Hebridean *Euphrasia* flora may seem to cast doubt on its taxonomic validity. However, the occurrence of identical, distinctive, uniform populations in well separated, ecologically similar localities justifies its current recognition.

It is relatively small flowered, but the flowers are showy and hybrids are not rare. The more reliably recorded hybrids include those with *E. confusa*, *E. micrantha* and *E. nemorosa*, and there is some indication of hybrid swarm formation. However, the integrity of *E. campbelliae* populations does not appear to be threatened.

An assessment of the current status of the species is needed, but there are extensive areas of seemingly suitable habitat and apparently no present cause for concern.

A.J. Silverside

Euphrasia heslop-harrisonii Pugsley (Scrophulariaceae)
Eyebright, Lus nan Leac
Status in Britain: LOWER RISK - Near Threatened.
ENDEMIC.

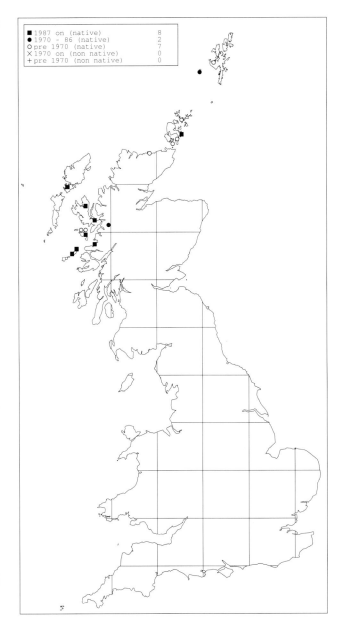

■	1987 on (native)	8
●	1970 - 86 (native)	2
○	pre 1970 (native)	7
✕	1970 on (non native)	0
+	pre 1970 (non native)	0

E. *heslop-harrisonii* is strictly coastal, occurring in
scattered mainland localities from Argyllshire northwards.
It is also known from Coll, Skye, Rona and Rum in the
Inner Hebrides, in Orkney and on Foula, Shetland. A
single, minute colony on Harris, in the Outer Hebrides, is
seemingly a recent colonist. Characteristically it occurs at
the tops of saltmarshes, often in a narrow zone
immediately above the high water mark, accompanied by
Plantago coronopus, P. maritima and other herbs such as
Leontodon autumnalis. A typical situation would be a
sheltered inlet near the head of a sea loch, where there is
no direct wave action and a somewhat sandy substratum.
It can occur in abundance on turfy islets within saltmarsh
complexes. In Orkney, it is also present on turfy banks
within reach of coastal spray.

Although first described as a species from Rum in
1945, and soon after recorded from West Ross, this
remained a poorly-known species until clarified by Yeo
(1978). It is now known to be plentiful at some sites and
further survey work is likely to increase the number of
localities. In heavily grazed turf it can remain dwarfed and
relatively featureless and so be difficult to confirm.
Identification commonly relies on the occurrence of better
developed plants in protected microhabitats. Mainland
plants tend to be more compact and broader-leaved than
those of Rum and Skye, and there is suspicion that some
of these mainland plants have been confused in the past
with E. *tetraquetra*.

It is small-flowered and usually occurs in the absence
of other *Euphrasia* species. Consequently, hybrids are rare,
though recorded with E. *arctica*, E. *confusa* and E. *nemorosa*.
In more exposed saltmarshes, E. *heslop-harrisonii* is replaced
by E. *foulaensis*, the two species hardly ever occurring
together. Another, seemingly related but as yet
undescribed, species can also occur in similar habitats in
north-west Scotland (Silverside 1991a; 1991b) but shows a
preference for areas with basic flushing.

E. *heslop-harrisonii* is dependent on the continued
existence of grazed margins of saltmarshes and is
vulnerable to changes in land use or reclamation. Such
losses have so far been of minor impact and the species is
not currently threatened.

A.J. Silverside

Euphrasia marshallii Pugsley (Scrophulariaceae)

Eyebright, Lus nan Leac
Status in Britain: LOWER RISK - Near Threatened.
ENDEMIC.

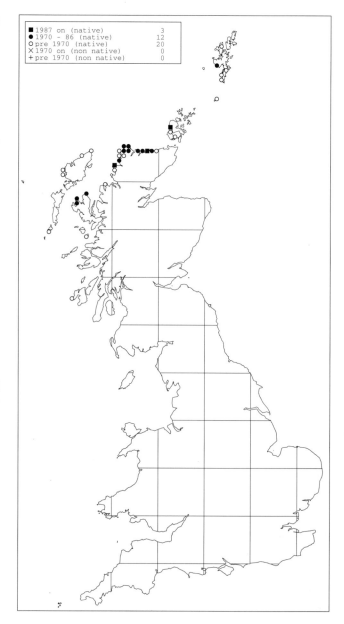

This is a strictly coastal plant of rocks and eroding sea-cliff edges, primarily of the northern and north-western Scottish mainland, with a few localities in the islands, southwards to Skye and northwards to Shetland. However, a number of past records, particularly from Orkney and Shetland, refer to other taxa (Stewart *et al.* 1994). It often occurs in moderately basic sites, with such species as *Anthyllis vulneraria* ssp. *lapponica, Carex capillaris, Coeloglossum viride, Gentianella amarella* ssp. *septentrionalis, G. campestris, Oxytropis halleri* and *Scilla verna,* but may be more typical of leached or less basic sites below maritime *Calluna - Empetrum* heath, with *Festuca rubra, Plantago coronopus* and *Thymus polytrichus. Plantago maritima* is a characteristic associate and seems likely to be a regular host. *E. marshallii* occupies equivalent habitats to *E. ostenfeldii* further north (Shetland and Faeroe Islands) and *E. tetraquetra* to the south, growing and hybridising with the latter species on Skye.

Although long known to botanists who explored the north coast, it was confused with quite unrelated taxa until described as a separate species by Pugsley in 1929. Despite continuing confusion, notably with *E. ostenfeldii,* it is a distinctive taxon, well delimited within its main range.

It appears to be regularly outcrossing and hybridises readily with other species. Where it grows close to cultivated areas, there may be substantial gene flow from *E. arctica,* at times causing considerable morphological modifications of populations. On dunes and basic cliffs, it may be entirely replaced by its hybrid with *E. nemorosa.* This hybrid is very locally abundant on Lewis, and pure *E. marshallii* requires modern confirmation. Other hybrid populations are usually of more limited extent.

Taxonomic confusion with other hairy taxa and lack of detailed monitoring make it difficult to assess any changes in distribution. However, populations appear to be relatively stable and there have been only limited losses, primarily through cultivation of cliff-tops. It requires rediscovery in some former localities, but a decline in modern records is primarily indicative of greater taxonomic insight. It is reasonable to suppose that further populations await discovery on the more remote parts of the north coast.

A.J. Silverside

Euphrasia rivularis Pugsley (Scrophulariaceae)
Eyebright, Effros Yr Wyddfa
Status in Britain: LOWER RISK - Near Threatened.
ENDEMIC.

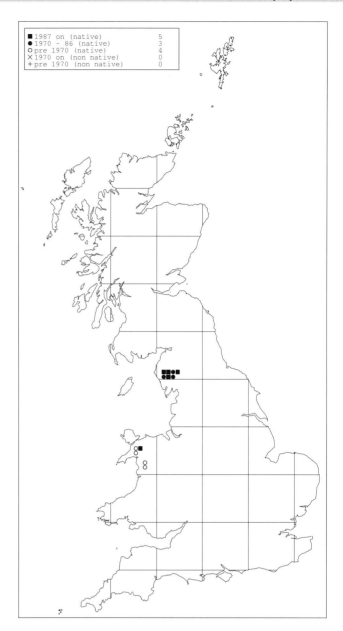

■ 1987 on (native)	5
● 1970 - 86 (native)	3
○ pre 1970 (native)	4
× 1970 on (non native)	0
+ pre 1970 (non native)	0

E. rivularis is confined to higher altitudes in Snowdonia and the Lake District. It occurs in bryophyte-rich flushes by streams, on wet, flushed slopes and on wet ledge and seepage areas on montane cliffs. In Snowdonia, it can be an attractive feature of the same basic rock-faces that support *Lloydia serotina* and other arctic-alpines. Associated species include *Arabis petraea*, *Campanula rotundifolia*, *Festuca vivipara*, *Selaginella selaginoides*, *Thalictrum alpinum* and *T. minus*. It may also be accompanied by *E. ostenfeldii*, which replaces it on the drier ledges. On flushed turfy slopes, associates include *Agrostis capillaris*, *Carex viridula* ssp. *oedocarpa*, *Danthonia decumbens*, *Epilobium brunnescens*, *Festuca ovina* ssp. *hirtula*, *Nardus stricta*, *Potentilla erecta*, *Sagina procumbens*, *Saxifraga hypnoides*, *Hylocomium splendens*, *Polytrichum formosum*, *Scleropodium purum* and *Thuidium tamariscinum*, with *Euphrasia cambrica* sometimes present on the flush margins and surrounding drier turf. In the Lake District it may be more a species of the streamside flushes, again with *C. viridula* ssp. *oedocarpa* and other sedges.

E. rivularis is closely allied to *E. officinalis sensu stricto* (including *E. rostkoviana* and *E. montana*) and may have arisen by hybridisation between one of these taxa and a tetraploid species. However, it exists as a clearly defined, independent taxon with distinctive, uniform populations. There is some variation between individual, isolated populations, as frequently occurs in alpine species.

It is a medium- to relatively large-flowered outcrossing species, but usually well isolated from other diploid species with which it might exchange genes. Hybrids are consequently rare. A single plant of *E. rivularis* x (*confusa* x *scottica*) has been noted at Cwm Idwal, where *E. rivularis* occurs as part of a complex population, and it is also known to be crossing with *E. rostkoviana* (probably ssp. *rostkoviana*) at one site in the Lake District. These hybridisation events pose no threat to the species, which is one of the most well delimited of the British taxa.

In Snowdonia, key populations lie within NNRs and appear to be subject to no obvious threat, particularly as the species is able to survive under both light and heavier grazing conditions. In the Lake District, the plant is now known in numerous localities and again seems subject to no immediate threat.

A.J. Silverside

Euphrasia rotundifolia Pugsley (Scrophulariaceae)

Eyebright, Lus nan Leac
Status in Britain: ENDANGERED. ENDEMIC.

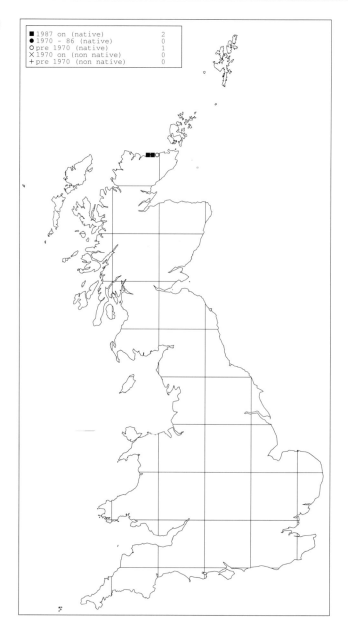

■ 1987 on (native)	2
● 1970 - 86 (native)	0
○ pre 1970 (native)	1
✕ 1970 on (non native)	0
+ pre 1970 (non native)	0

As interpreted here, *E. rotundifolia* is restricted to the northern coast of Scotland between Bettyhill and Melvich in Sutherland, with old records also known from Reay, Caithness. Currently it is certainly known at three sites, all in flushed, basic turf on sea-cliffs, with *Primula scotica* as a constant associate. Other associated species include *Agrostis capillaris, Anthyllis vulneraria, Armeria maritima, Calluna vulgaris, Carex flacca, C. pulicaris, Erica tetralix, Euphrasia foulaensis, Festuca ovina, F. rubra, Parnassia palustris, Plantago maritima* and *Scilla verna*. Populations may be in close proximity to drier cliff edges, occupied by *Euphrasia marshallii* and *Oxytropis halleri*, and *E. rotundifolia* may extend into this microhabitat. It is plausible that other colonies await discovery.

This restricted distribution is at variance with previously published accounts and reflects the uncertain taxonomic status of this species. It was described as a species by Pugsley in 1929, based on a distinctive collection from near Melvich, though he later accepted it as occurring in Shetland and elsewhere. Subsequently it has been recorded from a number of other sites in northern Scotland.

True *E. rotundifolia* appears to combine features of *E. foulaensis* and the two densely hairy species, *E. marshallii* and *E. ostenfeldii*. Examination of herbarium material and field investigation of reported populations show that records may be based on misidentifications of the two latter species or, more commonly, on recent hybrid swarms between either of them and *E. foulaensis*. As now interpreted, *E. rotundifolia* is restricted to the coast at and near its original locality. However, since dwarfed plants of *E. marshallii* in exposed sites are scarcely distinguishable from similarly dwarfed *E. rotundifolia*, the latter may yet prove to occur more extensively.

Hybrids are known with *E. marshallii* and *E. foulaensis* but do not appear to threaten the pure species. Cultivation of cliff-tops for hay, as at Melvich, has favoured *E. arctica*, resulting in gene flow into the coastal *Euphrasia* populations and a potential threat to *E. rotundifolia*, though the direct hybrid is not yet certainly recorded. *E. rotundifolia* may have a somewhat later flowering season than related taxa, which would confer some degree of reproductive isolation.

Extension of cultivation at what is believed to be the original site, near Melvich, means that very little suitable habitat now remains and the population is small and highly vulnerable, as is another restricted population near Bettyhill. However, on Strathy Point, plants are scattered over a much larger area, plentiful in some years and isolated from any effects of cultivation.

A.J. Silverside

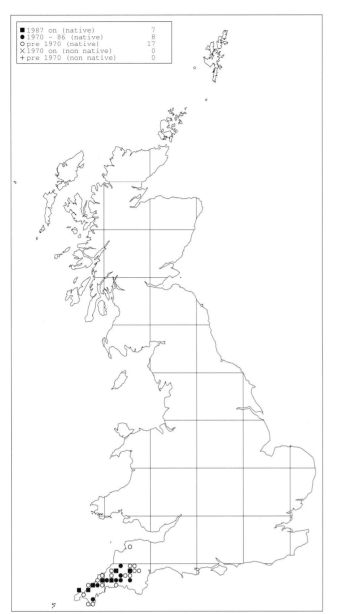

Euphrasia vigursii Davey (Scrophulariaceae)
Eyebright
Status in Britain: VULNERABLE. ENDEMIC.

E. vigursii is one of the characteristic plants of Agrostis curtisii-Ulex gallii heathland complexes in Devon and Cornwall. In coastal localities it occurs particularly in short, species-rich turf along path edges, where trampling keeps U. gallii and other robust species in check. Typical associates include Agrostis vinealis, Calluna vulgaris, Carex flacca, Erica cinerea, Festuca ovina ssp. hirtula, Scilla verna, Serratula tinctoria, Stachys officinalis, Thymus polytrichus, Viola riviniana and Hypnum jutlandicum. It apparently avoids the more open, species-poor areas dominated by Agrostis curtisii, although this species is frequently named as an associate. Inland it occurs in damp grassy heathland, including wetter areas with Agrostis canina, Carex pulicaris and other sedges, Potentilla erecta and, notably, Erica ciliaris. Current information shows Serratula tinctoria to be a regular associate at all sites.

E. vigursii is closely allied to E. officinalis sensu stricto (including E. anglica) and appears to have arisen by hybridisation between E. anglica and E. micrantha. However, it exists as distinctive, uniform populations independently of the parents and is retained at specific rank by most authorities. It has been discussed by Davey (1907) and Yeo (1956). Well grown (inland) plants are strikingly attractive. It is large-flowered and diploid, and would be expected to cross readily with E. anglica, the only other diploid taxon in south-west England. This hybrid might be expected on heathland margins but is rare or under-recorded. In at least two coastal localities in Cornwall, the hybrid with E. tetraquetra occurs in some quantity, forming a stable zone between pure E. vigursii in heathland and E. tetraquetra on adjacent cliff-tops.

Extant coastal populations may not be under immediate threat, though there are insufficient recent data to be sure, and present levels of disturbance seem to be keeping Ulex gallii in check. However, in contrast, there has been a substantial and continuing loss of inland populations through habitat dereliction; this has included taxonomically important populations discussed by Yeo (1956). Since 1987, there are records from only twelve 1 km squares (18% of the total recorded), compared with 40 since 1980 (60%) and 49 since 1970 (74%).

A comprehensive audit of *E. vigursii* in its remaining damp heathland sites is needed, and populations should be regularly monitored. Conservation management is urgently required in many sites, particularly to maintain open vegetation, such as may be provided by light grazing by ponies.

A.J. Silverside

Festuca longifolia Thuill. (Poaceae)

F. glauca var. *caesia* (Smith) Howarth
Blue fescue
Status in Britain: VULNERABLE.
Status in Europe: Vulnerable? Endemic.

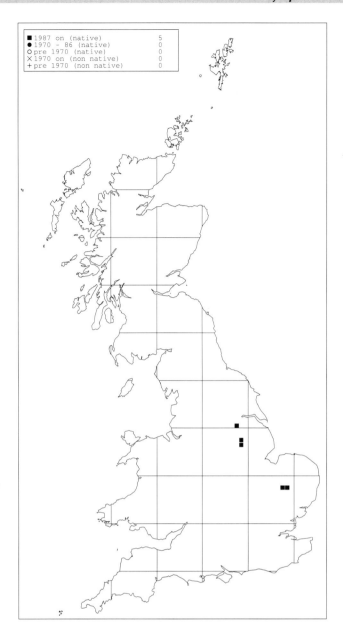

F. longifolia is one of our rarest grasses, found at only a few sites in eastern England. In Breckland, it was formerly a plant of large expanses of open heathland that found space between *Calluna vulgaris*. In its remaining open areas, its close associates are mostly small and non-aggressive, and it will not tolerate enclosure by tall grasses such as *Arrhenatherum elatius*. Its most frequent associates include *Achillea millefolium*, *Anthoxanthum odoratum*, *Carex arenaria*, *Festuca ovina*, *Galium verum*, *Koeleria macrantha*, *Rumex acetosella* and *Teesdalia nudicaulis*; others include *Agrostis capillaris*, *Allium vineale*, *Festuca rubra*, *Holcus mollis* and *Sedum acre*. The soil in Breckland is a coarse sand often with small stones over varying depths of chalk. It is loose, subject to erosion, acid and very low in mineral nutrients. On the borders of Lincolnshire and Nottinghamshire, it grows on sandy soils on roadside verges and adjacent acid heath, in some sites with *F. brevipila*. In the Channel Islands, *F. longifolia* is found in crevices and on ledges of cliff-faces and on acid rocks.

F. longifolia is a perennial, densely tufted and entirely glaucous with a blue 'bloom'. To the eye it is entirely glabrous but a lens will sometimes reveal some scabridity below the culm and minute cilia on the auricles. It is grazed by rabbits but is resistant, and in young plants the first growth remains stunted with a culm of less than 6 cm. This plant probably has a long life, as old plants expand and coalesce into clumps, especially where they have been grazed.

In Britain, it has been recorded since 1987 from seven sites in the West Suffolk Breckland, two in Lincolnshire, and one in Nottinghamshire. In Breckland, three sites have populations of 20-30 plants, a further three sites hold 100-200 and the largest site (Foxhold Heath), holds more than 2,000 plants. At its largest Lincolnshire site, near Torksey, more than 500 clumps were counted in 1981, but the present size of the population is not known. The Nottinghamshire populations, near Spalford Warren, are very small and vulnerable (D.A. Wood pers. comm. 1997). At three sites in Breckland there are restrictions of road verge space, another site is shaded by *Corylus avellana* and, at two others, conifers shade within 10 m of the colonies (they are also subject to human disturbance). With the exception of the largest site, all others in Breckland are at risk. In addition, three have been lost in the past 10 years: one from a roadside verge; one from a tumulus where the plants succumbed to rank grass growth in the absence of rabbit-grazing; one ousted by rank grass, and shade cast by an adjacent maturing pine plantation.

The Breckland scene is altering, and vast maturing conifer plantations occupying former heathland are frequently planted to the road boundary, causing a reduction of sunlight to road verges. Rabbit-grazing which formerly maintained the balance of plant competition is now considerably reduced, and many types of Breckland habitats are being overrun by coarse grasses. Populations on road verges are ever vulnerable to damaging activities. British populations, especially perhaps the outlying ones, should be carefully monitored and special conservation measures considered.

This species is endemic to Europe, where it has a very limited distribution, occurring only in England, the Channel Islands (nine sites on Guernsey and one on Sark), and northern and north-central France. Its status in continental Europe seems uncertain, perhaps because of past confusion with other glaucous taxa of *Festuca*, and it is not listed as a threatened species in Olivier *et al.* (1995).

Further information on the taxonomy and morphological characters of *F. longifolia* is given in Wilkinson & Stace (1991).

P.J.O. Trist and M.J. Wigginton

157

Filago gallica L. (Asteraceae)

Narrow-leaved cudweed, Edafeddog Culddail
Status in Britain: CRITICALLY ENDANGERED.
Status in Europe: Not threatened.

The main habitats of *F. gallica* are open, disturbed sites such as arable field margins on freely draining, sandy soils. It is a member of the dwarf therophyte community and is usually associated with such species as *Aira caryophyllea, A. praecox, Filago lutescens, F. minima, F. pyramidata, F. vulgaris, Potentilla argentea, Rumex acetosella, Spergularia rubra* and *Vulpia bromoides*. Formerly it was also recorded with *Galeopsis segetum, Gastridium ventricosum* and *Silene gallica*.

This species is probably mainly a summer annual in Britain. As seeds germinate throughout most of the year it may also behave as a winter annual, though it is prone to damping off. It is self-fertile and may form a seed-bank, at some sites occurring sporadically whenever conditions are suitable.

F. gallica has been regarded as an introduction in Britain ever since Smith & Sowerby (1812) queried its status solely because it could no longer be found in its original site, and this has been the general view ever since. However, a recent reassessment of its status has shown that it is a likely native with a long historical record, a distinct distribution and ecology, and associations with other rare species (Rich 1994a).

It was first recorded in Britain in 1696 at Castle Hedingham, Essex, and has been recorded from at least 30 sites in 19 hectads (probably introduced in four of these) in eight vice-counties, mainly in south-east England. Records from Wales and Scotland are probably errors for *Gnaphalium uliginosum*. It appears always to have been rare, and has declined because of changes in arable farming practice, and the lack of disturbance of heathland after myxomatosis. Jermyn (1974) documented its extinction at its last British site at Berechurch Common, Essex, where the last plant was seen in 1955. It has persisted on Sark in the Channel Islands since at least 1902 and still occurs there in one small quarry.

In 1994, plants originating from presumed native Essex stock were found in cultivation in a private garden. Seeds and plants from this population were re-introduced to three areas of a native site in 1994 (Rich 1995a; Rich, Gibson & Marsden 1995), and it was still persisting in one of these in 1996.

F. gallica is a variable species in Europe. It is most frequent in southern and western Europe and North Africa, with outlying localities in Britain and around the Black Sea, Cyprus and the Near East (Jordan, Israel, Lebanon) and the Canary Islands and Azores. It appears to be under threat in eastern Europe.

T.C.G. Rich

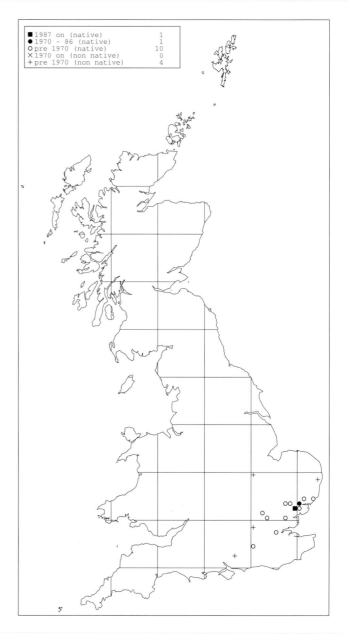

■ 1987 on (native)	1	
● 1970 – 86 (native)	1	
○ pre 1970 (native)	10	
✕ 1970 on (non native)	0	
+ pre 1970 (non native)	4	

Filago lutescens Jordan (Asteraceae)

Filago apiculata G.E. Smith ex Bab.
Red-tipped cudweed, Edafeddog Blaengoch
Status in Britain: VULNERABLE. WCA Schedule 8.
Status in Europe: Not threatened.

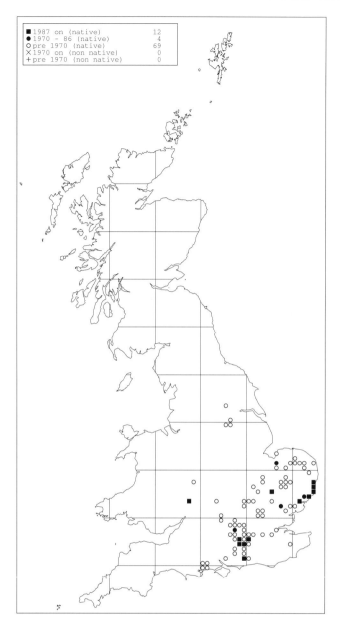

F. lutescens is a plant of light, open, usually sandy soils of low fertility, and is typically found on ground that has sporadic rather than regular disturbance. Most of the former English records were from the edges of arable fields, on fallows, trackways, heaths and commons. Many of its current sites are subject to intense grazing by rabbits, and the species grows well in the open soil of old rabbit scrapes. It is now mainly associated with species characteristic of sandy, open land rather than with arable weeds: these include *Filago minima*, *F. vulgaris*, *Geranium pusillum* and *Spergularia arvensis*.

It is mainly a summer-flowering annual, setting seed chiefly between July and October. Seed production is usually good, even in cold damp summers, although plants may be damped off in wet seasons. Plants which have flowered usually die in autumn, but occasionally survive the winter, and those plants may be able to seed the following summer. Most germination occurs between August and November, with further germination in spring. Winter seedlings overwinter as rosettes. Seed longevity is not known, although some persistence is suggested by its unexpected appearance in places after soil has been excavated or turned.

F. lutescens has declined dramatically in Britain since the 1950s. At one time it occurred in more than 200 sites in about 20 vice-counties across England, but since 1993 it has been seen at only 16 sites in 12 hectads. Rich (1994b, 1996a) has documented its occurrence in Suffolk, Cambridgeshire, Essex, Sussex, Surrey and Hampshire, and a new site was found in Gloucestershire in 1997. Its stronghold has always been Surrey, and by far the largest extant populations are near Godalming. At its remaining sites, numbers fluctuate markedly, presumably in response to the state of the habitat and the weather, but most populations are small. Between 1993 and 1997 only 4-5 sites held more than 1,000 plants, and some fewer than 20 individuals in some years. Between 1993 and 1997 the national population ranged from a few thousands to over 100,000 plants. Two sites are currently protected as SSSIs, in Surrey and Suffolk.

The main cause of its decline in Britain has been the intensification of arable farming. Whilst the centuries-old pattern of an autumn harvest, overwintering stubble and spring sowing fostered summer- and autumn-flowering annuals such as the cudweeds, modern arable cultivation leaves them no room. Densely sown, vigorous crops shade the soil, autumn cultivation and sowing destroys seedlings and overwintering rosettes and summer harvests cut down the flowering plants before seed is set. Another important factor has probably been the rabbit decline of the 1950s, which allowed much open ground to become overgrown by coarse herbage and scrub (FitzGerald 1988a). It seems to do poorly on former arable fields that have been left

uncultivated as, for example, those in long-term set-aside schemes (Rich 1994b). Further investigations are needed to determine the most appropriate means of conserving this species.

Conservation management is carried out at some sites, with the aim of maintaining open ground. At a site in Suffolk, the ground is rotavated in alternate years. Germination in response to various cultivation regimes is being studied. Active conservation management should be instigated at other sites, especially where rabbit populations have declined because of myxomatosis or viral mycorrhagia.

It is mainly a central European species, though ranging from southernmost Sweden southwards to Slovakia, northern Italy and central Spain.

C. Gibson

Filago pyramidata L (Asteraceae)
Filago spathulata auct., non C. Presl
Broad-leaved cudweed, Edafeddog Llydanddail
Status in Britain: ENDANGERED. WCA Schedule 8.
Status in Europe: Not threatened.

This is a species of arable land and other habitats with a long history of disturbance. Of the eight extant sites, three are on arable land, three in chalk quarries and two on chalk spoil adjacent to railway lines. All remaining populations are on well-drained calcareous or sandy soils, and it grows best where competition is minimal and the habitat is kept open by cultivation or drought. At some sites, the wide range of associates often includes other rare and scarce species: for example, at one arable site in Kent, *Ajuga chamaepitys, Althaea hirsuta, Anagallis arvensis* ssp. *foemina, Papaver hybridum* and *Silene noctiflora* are present. On sandy soils, *F. pyramidata* may grow with *F. minima* and *F. vulgaris*, and notable associates in quarry sites include *Cerastium pumilum* and *Iberis amara*.

F. pyramidata usually behaves as a winter annual, germinating in autumn and flowering from mid- to late summer. Seedlings can, however, germinate in spring, and in arable land *F. pyramidata* can occur in both spring- and autumn-sown crops (Wilson 1990). Abundant seed is produced, which can form a persistent seed-bank. The disappearance of *F. pyramidata* from a site may not therefore mark a permanent loss, and the return of suitable conditions could lead to its reappearance at former sites.

It has been recorded from more than 100 hectads, although from only 21 between 1930 and 1960, and eight between 1975 and 1986 (Smith 1986). Between 1993 and 1996 it was seen at only eight sites (in eight hectads), two each in Surrey, Oxfordshire and Cambridgeshire, and one in West Sussex and Kent (Rich 1995c, 1996c). Records at some former sites are thought to represent misidentifications of *Filago vulgaris* (Rich 1995b). As with *F. lutescens*, populations may vary greatly in size according to climatic and habitat conditions. Since 1993, two sites have consistently held between 10,000 and 60,000 plants, but other sites have much smaller populations, some of them in a particular year holding up to 2,000 plants, whilst others may hold just a few individuals, or none at all.

The main reason for the decline of *F. pyramidata* is likely to have been arable intensification, including autumn cultivation and the increased application of herbicides and artificial fertilisers. Infill of quarries, growth of tall herbage or scrub, and the cessation of cultivation may also have led to the loss of populations. Five of the remaining sites are SSSIs; at three of these, management appears to be suitable, one is the subject of current restoration work, whilst another is threatened by the development of a golf course.

F. pyramidata is mainly a plant of southern Europe, ranging from Iberia and France, the Mediterranean islands, Italy and Greece to North Africa, Turkey and western Asia. Its range extends northwards to the Low Countries and England, but it is rare in the northerly parts of its range (Rich 1995b).

P.J. Wilson

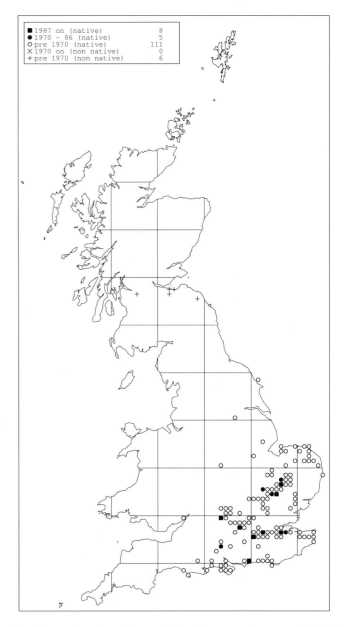

■ 1987 on (native)	8
● 1970 – 86 (native)	5
○ pre 1970 (native)	111
✕ 1970 on (non native)	0
+ pre 1970 (non native)	6

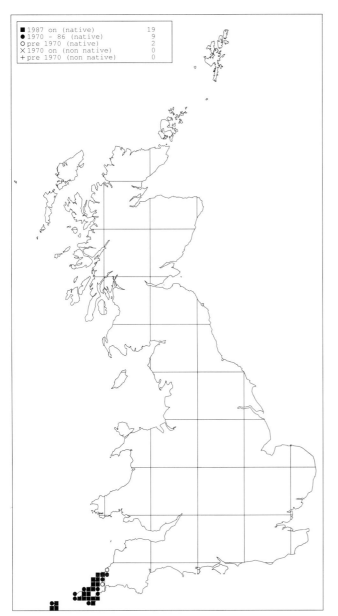

■ 1987 on (native)	19
● 1970 - 86 (native)	9
○ pre 1970 (native)	2
✕ 1970 on (non native)	0
+ pre 1970 (non native)	0

Fumaria occidentalis Pugsley (Fumariaceae)
Western ramping-fumitory
Status in Britain: LOWER RISK - Nationally Scarce.
ENDEMIC.

F. occidentalis is a locally abundant plant of field borders, stone hedges, road verges and waste places. It makes a fine show locally in Cornish lanes, frequently growing against the faces of stone and earth Cornish 'hedges' by lanesides and field borders, particularly where occasional management or disturbance keeps the community fairly open. A wide range of associated species on walls includes *Anisantha sterilis, Dactylis glomerata, Hedera helix, Galium aparine, G. mollugo, Geranium dissectum, Lonicera periclymenum, Rubus fruticosus* and *Smyrnium olusatrum*.

It is an annual, much-branched, scrambling plant, larger than the other British *Fumaria* species. The flowering season on the mainland is late May or early June, but in the Isles of Scilly it may flower as early as March or April. The distinctive whitish, later pink, flowers are borne in large racemes.

This plant was recognised as a new species by H.W. Pugsley, who described it in 1904. It remains almost entirely restricted to West Cornwall and the Isles of Scilly, with very few records in East Cornwall. In West Cornwall, most records are from the north-west Lizard peninsula and along the north coast up to about 16 km inland. It has been recorded in many localities in 30 hectads. Many new sites have been found in recent years, and it is possible that the plant is spreading as a colonist of recently disturbed ground: it has a good seed-bank and it can reappear whenever suitably open habitat becomes available. However, *F. occidentalis* can be irregular in appearance, occurring at only some of its known sites in any particular year. Some localities, such as at Lelant and Newquay, span the whole history of the plant from its first recognition 90 years ago. It was formerly plentiful in bulb fields in the Isles of Scilly, though there have been recent declines, and it is now largely confined to a few small areas on St Mary's, St Martin's and St Agnes.

In view of its persistence and the occurrence of good seed-banks, it cannot be regarded as seriously threatened. However, intensification of land management, both urban and rural, including the propensity for 'tidiness' shown by land managers, does pose a significant threat - examples being the turning of waste ground and field borders into permanent grass and the spraying of rank marginal vegetation with herbicides. As one of our endemic species, its conservation is of considerable importance.

M.J. Wigginton, from an account in FitzGerald (1990c)

Fumaria reuteri Boiss. (Fumariaceae)

Fumaria martinii Clavaud
Martin's ramping-fumitory
Status in Britain: ENDANGERED. WCA Schedule 8.
Status in Europe: Not threatened. Endemic.

F. reuteri occurs only in Cornwall and the Isle of Wight. In both areas, it grows in cultivated allotments, spreading to nearby open marginal habitats, gardens and, in Cornwall, stone hedges. Associated species (sometimes amongst vegetable crops) include the common weeds *Anagallis arvensis, Capsella bursa-pastoris, Chenopodium album, Fallopia convolvulus, Spergula arvensis* and *Stellaria media*.

It is an annual with a long flowering period (May to October), producing copious seed. Seed is long-lived, as shown, for example, by the appearance of plants following the cultivation of fields which have been in permanent pasture for many years.

This species seems always to have been rare in Britain, with historical records from Cornwall, Devon, Dorset, Somerset, Surrey, Sussex and the Isle of Wight. But conversion of cultivated land to permanent grassland and urban development has led to only two sites remaining: at Pulla Cross in Cornwall and at Lake in the Isle of Wight. Populations fluctuate greatly, reflecting cycles in the cultivation regime. It clearly does best in allotment plots that are not kept weed-free and constantly cultivated and, if allowed to do so, may become dominant. Indeed, on the Isle of Wight, one sympathetic allotment holder reported that the *F. reuteri* occurred so thickly that it kept away more pernicious weeds and acted like a good 'live' mulch to vegetable crops. In good years in the past, thousands of plants could be seen at Pulla Cross, but recently the numbers have dropped very markedly as areas have reverted to grassland and cultivated plots have been neglected. On the Isle of Wight, large populations still persist.

F. reuteri must be regarded as seriously threatened. Its survival in the two remaining sites depends largely on the tolerance of allotment holders, and a change of ownership or land use could lead to its demise. However, as a safeguard, some populations are maintained in private gardens in the Isle of Wight. The designation of the Lake allotment as an SSSI provides additional protection.

F. reuteri is endemic to western Europe, occurring in Britain, the Channel Islands, western France, Spain and Portugal. Its status is, however, unclear across most of its range. Soler (1983) assigned our plant to ssp. *martinii*, but it might not be subspecifically distinct from continental plants (Stace 1991).

M.J. Wigginton

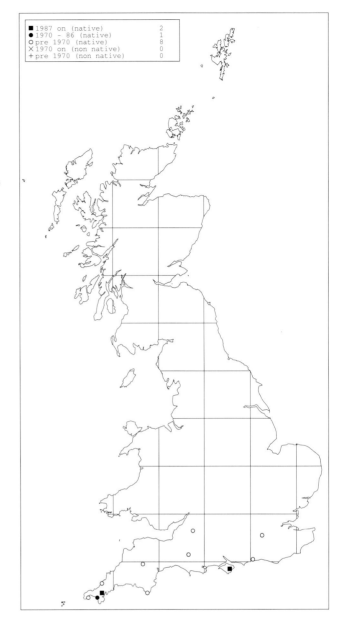

■ 1987 on (native)	2
● 1970 – 86 (native)	1
○ pre 1970 (native)	8
✕ 1970 on (non native)	0
+ pre 1970 (non native)	0

Gagea bohemica (Zauschner) J.A. & J.H. Schultes (Liliaceae)

Radnor lily, Seren y Creigiau
Status In Britain: VULNERABLE. WCA Schedule 8.
Status in Europe: Not threatened.

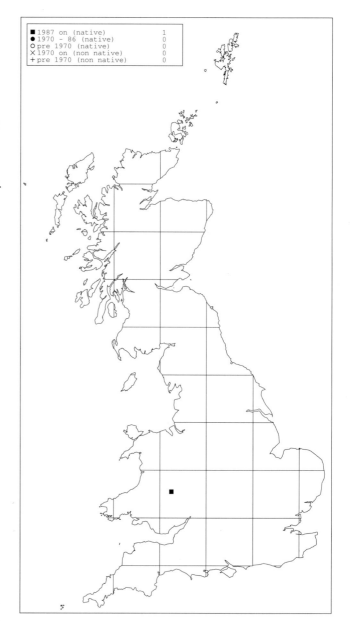

■ 1987 on (native)		1
● 1970 - 86 (native)		0
O pre 1970 (native)		0
X 1970 on (non native)		0
+ pre 1970 (non native)		0

This plant of rather cryptic habits grew undiscovered on Stanner Rocks, a much visited site in Radnorshire, until it was inadvertently collected and wrongly identified as *Lloydia serotina* in 1965. It was finally collected and correctly identified by R.G. Woods a decade later (Rix & Woods 1981). Stanner Rocks remains its only site in Britain. It grows in shallow pockets of soil in cracks and on ledges on the southern face of dolerite slopes where the continental microclimate (Slater 1990) usually ensures dry conditions for its period of aestivation from May until autumn. It is intolerant of shade and is not found in vegetation taller than about 3 cm. Associated species include *Festuca ovina, Moenchia erecta, Ornithopus perpusillus, Rumex acetosella, Sedum forsterianum, Teesdalia nudicaulis, Dicranum scoparium, Hedwigia stellata, Hypnum cupressiforme, Dermatocarpon miniatum* and several *Cladonia* species.

Its fine, almost *Festuca*-like leaves appear from the clumps of bulbs late in the year and flowers may appear as early as January. However, it is an extremely shy flowerer: from the large population of many thousands of bulbs (perhaps as many as several tens of thousands), few flowers are produced in any one year. For instance, numbers per year between 1989 and 1994 were 3, 18, 37, 21, 23 and 3, this being comparable with the variability noted since 1975 (Slater 1990). Flowering is over by April, and the leaves are dead a month later. No pollinating insects are known. Very few seed capsules have been seen and few of them escape grazing. Propagation is, therefore, almost entirely vegetative: by bulbils, sometimes 15-25 per plant, which are dispersed by soil disturbance. The population seems to be stable overall.

Scrub invasion must be, and is, controlled. The vigour of the plant does not seem to be diminished by grazing by sheep or rabbits; indeed the resulting nutrient input may be important. Rabbits not infrequently scratch out bulbils, which can lead to both destruction and dispersion. The leaves seem to be grazed casually and at random and are not sought out by herbivores. However, sheep do seem selectively to remove the flower heads, but as *G. bohemica* reproduces almost entirely by bulbils, this is probably of little biological significance, and the resulting disturbance of the ground by trampling may be a factor in dispersal.

G. bohemica has a mainly Mediterranean distribution, ranging from Syria, Israel and Turkey westwards to Portugal, though it extends northwards to Russia, and to eastern and central Europe. The Welsh locality is at the north-westerly limit of this species in Europe. There is a progressive increase in the sterility of populations towards the northern parts of its range (Uphof 1959).

F.M. Slater

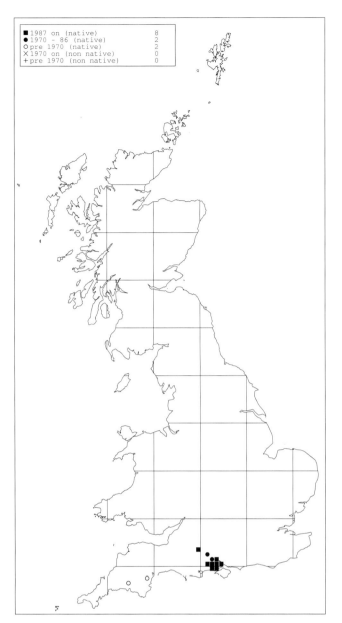

Galium constrictum Chaub. (Rubiaceae)
G. debile Desv. non Hoffsgg. & Link
Slender bedstraw
Status in Britain: LOWER RISK - Near Threatened.
Status in Europe: Not threatened.

This is a plant of open, seasonally-inundated, lowland wetland habitats, but it will tolerate light shade cast by open *Salix* scrub. Colonies are concentrated on rather basic, nutrient-rich soil on the margins of seasonal ponds, on lawns and in marl pits within the matrix of New Forest heathland. The plant can grow as an emergent, particularly when seasonal pools become inundated during the growing season. It has been recorded recently on the margin of ditches in formerly floated water meadows. Associates include *Agrostis stolonifera, Gnaphalium uliginosum, Lolium perenne, Lythrum portula* and *Ranunculus flammula*. At some sites, it occurs with nationally rare and local species including *Chamaemelum nobile, Cicendia filiformis, Illecebrum verticillatum, Mentha pulegium, Moenchia erecta, Ludwigia palustris, Pilularia globulifera* and *Radiola linoides*.

G. constrictum is a perennial herb, in which reproduction is presumed to be by seed, which is freely set. It may occur as a short (1-4 cm) plant in a permanently grazed sward, but can attain 10-15 cm where grazing is lighter, or scramble as high as 40 cm.

In Britain, almost all sites for *G. constrictum* lie within the New Forest, where there have been substantial numbers of new records in recent years, probably because of greater recording effort rather than any general increase in the numbers of sites or populations. Its habitats tend to be tightly grazed during the whole year and are subject to trampling and dunging by cattle and ponies. All the New Forest populations lie within SSSIs and seem reasonably secure despite the inherent fragility of the commoning economy, which maintains the grazing animals. Proposals by graziers to convert seasonal wetlands to permanent ponds have been successfully resisted. It has been considered extinct in the Bovey Basin, South Devon, for many years, ever since grazing ceased on the commons and scrub encroached, but the planned re-instatement of grazing may bring to light unsuspected populations. Two sites have been reported recently from Wiltshire (Gillam 1993). Records from Yorkshire are considered erroneous.

G. constrictum is widespread in Europe, including the Mediterranean islands, ranging from Iberia eastwards to Greece and Turkey and northwards to the Channel Islands and southern England. The genetic character of this species, as part of the *Galium palustre* polyploid complex, has been described in Teppner *et al.* (1976) and Klisphuis *et al.* (1986).

C. Chatters

Galium tricornutum Dandy (Rubiaceae)
Corn cleavers, Briwydden Arw
Status in Britain: CRITICALLY ENDANGERED.
Status in Europe: Not threatened.

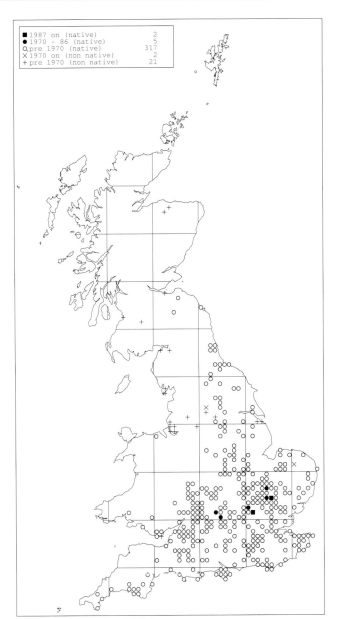

G. tricornutum is a plant of disturbed ground that, in the past, occurred mainly in arable fields in the warmer parts of southern and eastern England. There are very few recent records. At Broadbalk field, Rothamsted, it grows in an experimental plot that has never received chemical fertiliser. The Broadbalk field supports a reasonably intact traditional arable flora, including *Legousia hybrida*, *Papaver argemone*, *Ranunculus arvensis*, *Scandix pecten-veneris* and *Torilis arvensis*. At Wytham, Oxfordshire, *G. tricornutum* occurred at one time, and possibly still does, in a conservation area of arable land, together with a range of arable weeds including *Ranunculus parviflorus*, *Sherardia arvensis*, *Veronica persica* and *Vicia tenuissima*.

G. tricornutum is an annual, flowering between May and September. It is almost exclusively an autumn-germinating plant, and therefore tends to be restricted to ground cultivated in autumn and, if in a crop, to a winter cereal. It is thought to be a relatively poor competitor and in both Britain and continental Europe is typically found in rather open sites at the extreme edges of fields where the crop is weak or non-existent. Little is known about the longevity of seed in the soil but may be relatively short-lived (Grime *et al.* 1987).

G. tricornutum has declined rapidly in Britain. Recorded in 77 hectads between 1930 and 1960, it was reported from only 16 between 1960 and 1975 and 7 between 1975 and 1985 (Smith 1986), and is now known apparently in only 2 or 3. The population at Rothamsted is currently very small and threatened, with only 2, 0, 3, 0 and 4 plants seen in the years 1991 to 1995 respectively. At Wytham, Oxfordshire, its presence was first confirmed in 1983. In 1984 a minimum of 10,000 plants were estimated and in 1985 it flowered abundantly, but in 1986 no plants appeared (Everett 1988) and apparently none has been seen since. In Cambridgeshire, two plants appeared in 1996 on a roadside verge that was disturbed during roadworks. Its decline in Britain may be attributed to improved seed-screening from the end of the 19th century and progressive intensification of arable agriculture during this century, including the application of chemicals. However, its susceptibility to herbicides is not known, although the related *G. aparine* is resistant to many compounds. The species is at the edge of its range in Britain, and the probably short-lived seed-bank confers little resilience on its populations.

This species was formerly widely distributed in the warmer cereal-growing parts of Europe and the Mediterranean countries. It is still quite common in the south of Europe, but it has become very rare in northern European countries, where it is now confined to climatically favourable sites, often occurring with other rare species.

P.J. Wilson

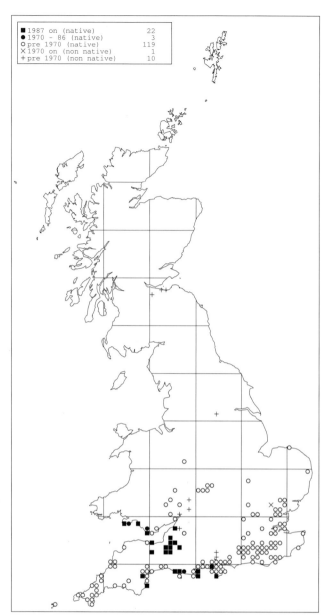

■	1987 on (native) 22
●	1970 - 86 (native) 3
○	pre 1970 (native) 119
✕	1970 on (non native) 1
+	pre 1970 (non native) 10

Gastridium ventricosum (Gouan) Schinz & Thell. (Poaceae)

Nit-grass, Llauwair
Status in Britain: LOWER RISK - Nationally Scarce.
Status in Europe: Not threatened.

G. ventricosum is a native of well-drained, open grassland on calcareous soils. These soils are invariably shallow and frequently overlie rock, their texture varying only slightly from a silty clay loam to a sandy loam (Trist 1986). Most sites are on slopes, which are often steep, and are usually south-facing, exposed to wind erosion and in close proximity to the sea or maritime influence. Some of the Somerset sites are now some distance inland (up to 26 km) but are on slopes that were once coastline when the Somerset Levels were covered by the sea. Plant associates in these habitats are *Koeleria macrantha, Pilosella officinarum, Sanguisorba minor, Thymus polytrichus,* and many other plants of well-drained calcareous soils, including local species such as *Althaea hirsuta* in Somerset, *Gentianella*

anglica and *Ophrys sphegodes* in Dorset, *Anisantha madritensis* and *Potentilla neumanniana* in the Avon Gorge and *Scilla verna* in the Gower. Bare ground is always present (Lovatt 1981). *G. ventricosum* was formerly frequently recorded as an arable colonist, with associates such as *Agrostis gigantea, Polygonum aviculare* and *Veronica persica* (Trist 1983).

G. ventricosum is an annual, germinating in autumn, and is frost-sensitive (Trist 1986). Numbers of flowering spikes vary greatly, depending on climatic conditions and competition from other species. Lovatt (1981) suggests that two more-or-less consecutive hot dry summers are required for the reappearance of *G. ventricosum* in quantity, and that the optimum conditions for seed production are a mild winter, a damp warm spring and a hot June. The longevity of the seed-bank is not known.

There seems to be little evidence of any decline in the native distribution of *G. ventricosum* in its semi-natural habitats. Indeed recent recording work, particularly in Somerset but also in Dorset and Glamorgan, has shown that the plant is much more widespread than was previously suspected. It is assumed that greater recording effort is the reason, although recent summer droughts opening up the vegetation may have contributed. Certainly *G. ventricosum* flourishes only in open situations, and grazing, especially by rabbits, is essential to its survival. It can survive for short periods amongst taller vegetation, by increasing its height as a response to deprivation of light (Trist 1986), but this is only a temporary measure; open ground is necessary for reproduction.

G. ventricosum has almost vanished from arable habitats (the source of most of its records in south-east England) because of the usual combination of cleaner seed and herbicides. But it still persists in one such locality in South Hampshire and appeared recently in enormous quantity in set-aside arable fields in Dorset, though it may have spread to these fields from a native grassland site (Preston & Pearman 1991). It has also frequently been recorded as a casual. But these arable and casual records have influenced perceptions of decline in this species. Now that its native habitat is better understood and better explored the true position of its status can be appreciated.

G. ventricosum is widespread over west and central Europe, around the Mediterranean, and eastwards to Iraq, and it is found on grassy hillsides and as an arable weed throughout this range.

I.P. Green and D.A. Pearman

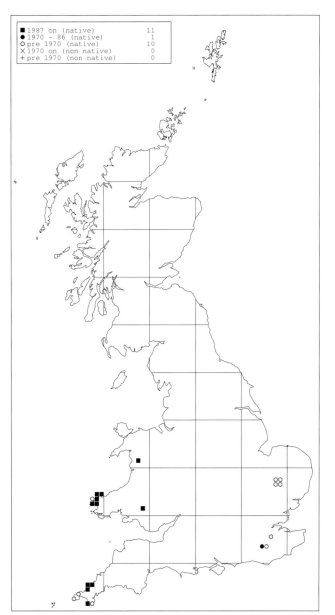

■ 1987 on (native)	11
● 1970 - 86 (native)	1
O pre 1970 (native)	10
X 1970 on (non native)	0
+ pre 1970 (non native)	0

Genista pilosa L. (Fabaceae)

Hairy greenweed, Aurfanadl Blewog
Status in Britain: LOWER RISK - Near Threatened.
Status in Europe: Not threatened. Endemic.

In Britain, *G. pilosa* occurs in a range of habitats. Most of its sites are coastal dwarf-shrub heaths and cliffs, but it also occurs inland in grasslands and on mountain rocks and crags. The substrate is usually well-drained acidic sandy or gravelly soil on heaths, but it occurs locally on rendzinas and loams overlying limestone and serpentine. On coasts, it may be major component in heath and grassland on both flat and steeply-sloping sites, together with *Armeria maritima*, *Calluna vulgaris*, *Erica cinerea*, *Festuca ovina*, *Hypochaeris radicata*, *Pedicularis sylvatica*, *Plantago maritima* and *Scilla verna*. On Cader Idris, most colonies grow on precipitous south- to west-facing crags at an altitude of 480-710 m, where they are inaccessible to sheep. Associated species include *Calluna vulgaris*, *Carex pilulifera*, *Erica cinerea*, *Festuca ovina*, *Solidago virgaurea*, *Vaccinium*

myrtillus and *Racomitrium lanuginosum*. The Breconshire colonies are in a mosaic of grassland and limestone pavement, though the suite of associated species, including *Anthoxanthum odoratum*, *Carex flacca*, *Deschampsia flexuosa*, *Helictotrichon pratense*, *Potentilla erecta*, *Thymus polytrichus*, *Vaccinium myrtillus* and *Viola riviniana*, seems to indicate that it grows on both acid and basic soils there. In southern England, it occurred on dry heaths with *Calluna vulgaris*, *Carex binervis*, *Genista anglica*, *Ulex europaeus* and *U. gallii*.

This species is a perennial scrambling or procumbent shrub, in full flower in late May and June, and pollinated by bees. It sets seed freely, with good recruitment where grazing allows.

In Cornwall, populations number many thousands on the Lizard peninsula (chiefly between Mullion and Caerthillion) and on the north coast between Godrevy Point and Cligga Head. In Pembrokeshire, there are about eleven populations holding 2,000-3,000 plants on St David's Head and Strumble Head. On Cader Idris, it occurs in about seven different locations, with a total of about 90 plants (in 1987). Numbers in particular locations on the mountain vary from about 40 to fewer than ten individuals, but because of the difficult terrain, *G. pilosa* may be under-recorded there (Morgan 1988a). In Breconshire, a total of 50-60 plants are spread between several colonies (Morgan 1988b). It formerly occurred in several counties in eastern and south-east England, but it has not been seen in its last site, the Ashdown Forest, since 1977, and is presumed everywhere extinct. Current populations appear to be stable.

A threat is posed by intense grazing by sheep at the Breconshire site preventing flowering and fruiting, but some colonies have been enclosed against grazing. Heath fires have destroyed populations in the past and are a present threat. However, *G. pilosa* appears not to be significantly threatened at most of its coastal sites. Almost all populations are in SSSIs or on National Trust properties and are thus afforded a degree of protection against land claim for agriculture, which has caused losses in the past.

G. pilosa occurs in western and central Europe from Spain, eastwards to Albania and Romania and northwards to southern Sweden. It appears to be in decline in parts of northern and central Europe.

M J Wigginton

Gentiana nivalis L. (Gentianaceae)

Alpine gentian, Lus a'Chrùbain Sneachda
Status in Britain: VULNERABLE. WCA Schedule 8.
Status in Europe: Not threatened.

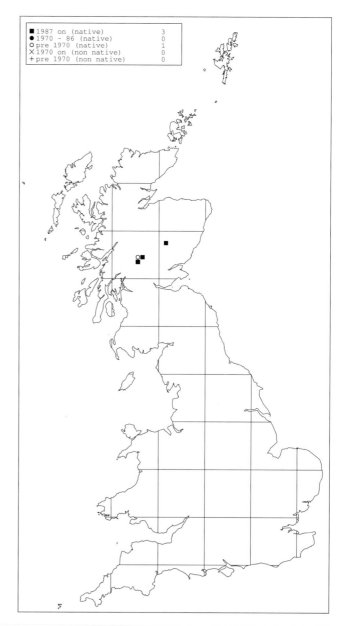

G. nivalis is a plant of rock ledges and vegetated screes on calcareous rock, restricted in Britain to two small areas in protected sites in the Scottish Highlands - Ben Lawers NNR in Perthshire and Caenlochan SSSI in Angus. At Ben Lawers the altitude range is 770-1,050 m, with the majority of plants occurring above 900 m. The principal habitat is grazed herb-rich grassland, in which conspicuous components of the sward include *Agrostis capillaris*, *Alchemilla alpina*, *Festuca vivipara*, *Minuartia sedoides*, *Silene acaulis* and *Thymus polytrichus*. Most of the Ben Lawers population is subject to intensive sheep-grazing and, at its main site, also to human trampling. Losses of 18-50% of plants by grazing during the flowering season have been recorded on Ben Lawers (Mardon 1980, 1984, 1985; Batty *et al.* 1984). However, recent experimental work has shown that this loss of plants and seed production may be more than offset by the corresponding control of plant competitors and maintenance of niches for the establishment of new plants (Miller *et al.* 1994).

G. nivalis is an annual or biennial (Raven & Walters 1956; Batty, *et al.* 1984; Morkved & Nilssen 1993). Flowering is from mid-July to the end of August, and flowers open only in warm sunshine. They are probably self-pollinated. Studies on Ben Lawers have shown that more than 75% of plants bear a single flower, but 17% bear 2-3 flowers and 5% have 4-5 flowers. Very few plants have more, though at least 32 flowers have been recorded on a single plant. Up to 1,000 seeds can be produced by a single plant, though a proportion may be shrivelled. They have a long period of dormancy resulting in a large seed-bank (Batty *et al.* 1984; Miller *et al.* 1994), which presumably buffers the population against temporary set-backs. Germination and seedling establishment probably depend on the availability of bare ground in autumn and the absence of certain competitors, particularly grasses (Miller *et al.* 1994). This in turn is affected by natural instability on cliff ledges and by grazing sheep on the more stable, vegetated scree slopes. It is possible that the numbers of plants at Ben Lawers is maintained at an artificially high density by the current level of grazing, though longer term effects of reducing seed production are not known.

Post-1980 records are in six 1 km squares at Ben Lawers NNR, all within two hectads, with older records for two other 1 km squares. At Caenlochan the plant occurs in two 1 km squares within a single hectad. Population counts at Ben Lawers usually indicate a total population of about 500-1,000 plants, though the numbers of plants in particular colonies fluctuate greatly. However, more detailed and systematic examination of small study areas, suggests that these total counts underestimate populations sizes. These studies indicate an average overall population of between 1,000 and 5,000 plants. Systematic recording in one Caenlochan site between 1982 and 1993 revealed large fluctuations in populations: numbers of plants ranged from four to more than 200 (Geddes 1994).

At Ben Lawers, the destruction of plants by human trampling has not been measured, but is thought not to be very serious at present. However, any increase in trampling could be detrimental to the population, and management seeks to avoid this. Whereas a reduction in grazing would be advantageous to the population size and range of many of the rare species at Ben Lawers, this may not be so for *G. nivalis*. Management of the NNR now includes long-term measures to control the grazing, but further research into the impacts of grazing is required.

G. nivalis is known from Iceland, Scandinavia, the mountains of central and southern Europe (Pyrenees, Alps, Apennines, Carpathians), the Balkans, Asia Minor, the Caucasus, Greenland, and the Labrador coast of North America. In other parts of the range it grows in a wider range of habitats, such as open submontane woodland and willow scrub in Scandinavia.

D.K. Mardon

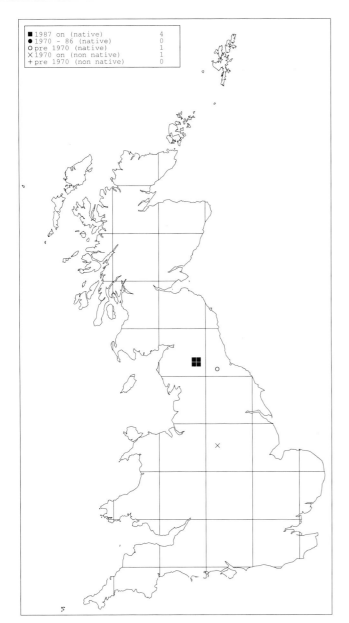

■ 1987 on (native)   4
● 1970 - 86 (native)   0
○ pre 1970 (native)   1
✕ 1970 on (non native)   1
+ pre 1970 (non native)   0

Gentiana verna L. (Gentianaceae)

Spring gentian
Status in Britain: LOWER RISK - Near Threatened. WCA Schedule 8.
Status in Europe: Not threatened.

G. verna is a plant of upland limestone grasslands in which *Festuca ovina* is normally the dominant grass and *Carex caryophyllea* and *Thymus polytrichus* frequent associates. *Sesleria caerulea* is also a regularly recorded component of these species-rich swards, which are often home to other plants of restricted distribution. On Little Fell in Cumbria, for example, *G. verna* and *Myosotis alpestris* grow together in short turf at an altitude of over 700 m. Plants also occur occasionally on hummocks in turfy marshes. The ecology of *G. verna* is discussed in Pigott (1956).

It is a perennial evergreen herb producing extensive loose mats of leaf rosettes each borne terminally on a branching stolon system. This usually lies just beneath or within the carpet of bryophytes and dense turf in which *G. verna* grows. Such a growth form makes it very difficult to estimate population sizes, as counts of rosettes considerably over-estimate the number of genetically distinct plants. These grasslands are frequently very heavily sheep-grazed with most flowers removed before seed-set, and mature capsules are rare in most populations. It is likely therefore that some 'populations' could consist of just one or a few clonal individuals.

In Britain *G. verna* is restricted to a small area of the northern Pennines on the borders of Co. Durham and Cumbria where, however, it is locally frequent. Within a range extending just 18 km north to south and 15 km east to west, it has been recorded in more than 50 1 km squares since 1970. G.G. & P.S. Graham (1993a) confirmed its continued presence at most of its known localities in Co.

Durham. The origin of plants recorded by the Tees near Darlington (Baker 1906) is unknown.

Overgrazing of some populations has undoubtedly occurred over many years, and yet some of the counts of healthy rosettes remain substantial. Nonetheless the slow attrition of these populations must be an ever present threat at such high grazing levels. The suppression of flowering is evident when impressive displays of flowers growing within enclosures are compared with the general paucity of blooms. If the level of grazing is sufficiently light, the flowers appear mostly in May (though sometimes in April at the lowest sites and not until June at the highest) but open fully only on sunny days. Seed is shed as soon as it is ripe in late June and July, and germination normally takes place in the spring (Elkington 1963). This attractive plant has been grossly over-collected in the past.

G. verna (several subspecies) ranges through the uplands of central and southern Europe, through the Balkans to the Caucasus and Iran, the mountains in central Asia (including the Tien Shan and Altai ranges) to northern Russia. It also occurs in Morocco. It is a highly polymorphic species worldwide, and includes a number of geographically well-defined and widely disjunct sub-specific taxa. British and Irish populations are now included within the widespread ssp. *verna* which also occurs throughout the uplands of central and southern Europe. Though plants in Britain and Ireland have become morphologically differentiated, such differences are not considered to be taxonomically significant, even at varietal level (Elkington 1972).

I. Taylor

Gentianella ciliata (L.) Borkh. (Gentianaceae)

Gentiana ciliata L.
Fringed gentian
Status in Britain: CRITICALLY ENDANGERED. WCA
Schedule 8.
Status in Europe: Not threatened.

In Britain, *G. ciliata* has been recorded as a presumed native at two sites, but is now extant in only one. A small population occurs near the top of a steep, west-facing, species-rich chalk grassland slope near Wendover. Species in the immediate vicinity include *Brachypodium sylvaticum, Briza media, Campanula rotundifolia, Cirsium acaule, Cynosurus cristatus, Euphrasia officinalis, Festuca ovina, Filipendula vulgaris, Galium verum, Helianthemum nummularium, Koeleria macrantha, Leontodon hispidus, Lotus corniculatus, Sanguisorba minor, Succisa pratensis, Thymus polytrichus* and *Trifolium pratense.*

It is considered to be a biennial in Britain, and requires areas of bare ground for successful germination. These conditions are provided at Wendover by sheep- and rabbit-grazing keeping the turf open. In 1982, 50 plants were noted. Numbers reached a peak in 1987 with 150 recorded, but there has been a subsequent decline and only about 10 were present in 1993, 15 in 1994 and none seen in 1996. Evidence suggests that reducing the height of the sward during the winter and the removing leaf litter might be beneficial, and it is hoped that sheep grazing, recently reintroduced by the National Trust, will provide such conditions.

The history of *G. ciliata* in Britain is of particular interest (Knipe 1988). It was first found in 1875 "in a hill near Wendover" and a specimen of that date is in the herbarium of the Natural History Museum, London. It was initially determined as *G. pneumonanthe* (Anon. 1875), but later named, correctly, as *G. ciliata* (Britten 1879). However, Druce (1926) then dismissed the record as representing *Campanula glomerata*, having failed to examine the specimen or properly investigate the record. Since the re-discovery of *G. ciliata* at perhaps the original site at Wendover by P. Phillipson in 1982, two other records have come to light. The first is represented by a herbarium specimen at Kew of a plant collected in 1910 at Limpsfield, Surrey, but the plant is considered to be an introduction there (Taylor & Rich 1997). The second is a herbarium specimen at the Natural History Museum of plants collected in chalk grassland at Pitton, Wiltshire, in 1892 (Dowlan & Ho 1995). The latter site is now a set-aside field, and intensive searches of chalk grassland nearby have been unsuccessful. That the Wendover colony remained undetected for such a long time is somewhat puzzling, especially since it is a well botanised area, but its late flowering and small population are likely contributory

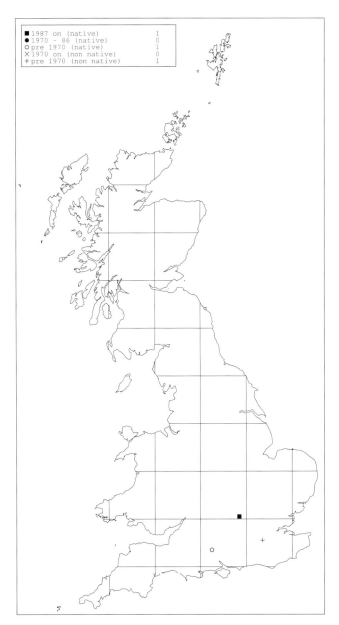

reasons. Furthermore, the colony is very restricted in area, with only a few flowers appearing in recent years, which may have always been the case.

Except in the extreme west, *G. ciliata* is widespread in continental Europe including most of the islands, extending northwards to northern France, the Netherlands and Germany (Pritchard & Tutin 1972). It occurs in meadows and wood margins, and in Limburg, the Netherlands, it occurs on steep well-drained chalk slopes.

L. Farrell and M.J. Wigginton

Gentianella uliginosa (Willd.) Boerner (Gentianacae)
Dune gentian, Crwynllys Cymreig
Status in Britain: VULNERABLE. WCA Schedule 8.
Status in Europe: Vulnerable. Endemic.

G. uliginosa is evidently restricted to open ground or short vegetation in coastal habitats. Welsh populations occur in a range of dune-slack habitats, from dry *Cladonia*-dominant slopes to occasionally submerged hollows, in association with *Epipactis palustris*, *Equisetum variegatum*, *Hydrocotyle vulgaris*, and even *Glaux maritima*. In Scotland it is known from *Schoenus nigricans*-rich machair.

This species is an annual, and not, as often stated, a biennial. It may be very short-lived and diminutive, occasionally flowering when less than 5 mm in height. It seems to be sensitive to climatic conditions and can disappear or occur in only low numbers during wet and cool summers (when plants may also be especially vulnerable to grazing by molluscs). In warm years, however, and following disturbance, *G. uliginosa* can reappear in abundance, which suggests there is often a persistent seed-bank. It normally flowers between late July and September, though occasionally as late as November. The flowers seem to be visited rarely by pollinating insects (Youngson 1986), and plants may be regularly self-pollinated.

After several false reports, *G. uliginosa* was first confirmed in Britain in 1923 from a "damp sandy pasture" in Pembrokeshire (Pugsley 1924). It was subsequently found in Gower and Carmarthenshire, and has most recently been reported from Colonsay (Rose 1998). Since 1987 it has been recorded in five dune systems in South Wales (in Pembrokeshire, Carmarthenshire and Glamorgan). In Wales, at least, *G. uliginosa* is always found associated with *G. amarella*, though they are distinct ecologically and populations are intermixed only where their habitats overlap. A range of intermediates have been reported (Ellis 1983; Pritchard 1971), but it is unclear to what extent intermediate forms occur. The presence of some *G. uliginosa* characters in the 'Bristol Channel' race of *G. amarella* has also been interpreted as indicating ancient introgression between the species. *G. uliginosa* is surprisingly local, generally in low numbers (from a few individuals to a few hundreds), and is perhaps threatened at most of its sites through undergrazing. Details of Welsh sites are given in Jones (1992). Such a small, critical, and oddly disjunct taxon could, however, also be more widespread and overlooked (Lousley 1950), as the Scottish find demonstrates. Furthermore, there has been recent confirmation of this species in England (Rich 1996d): there are herbarium specimens in the Natural History Museum, London, from Braunton Burrows, Devon (collected pre-1849, and in 1927), and, surprisingly, from Miller's Dale, Derbyshire (1898). Searches of Braunton Burrows in 1995 failed to reveal it, but it could reappear there. The inland record from Miller's Dale suggests that careful searches in damp meadows in many areas might be worthwhile.

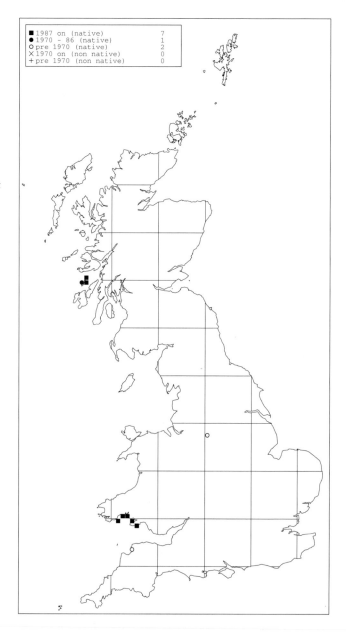

■ 1987 on (native)	7
● 1970 – 86 (native)	1
○ pre 1970 (native)	2
✕ 1970 on (non native)	0
+ pre 1970 (non native)	0

This species is a northern European endemic, most frequent in Fennoscandia and the Baltic States, and becoming increasingly rare southwards to the Czech Republic and westwards to France. Its determination and taxonomy seem to change across this range. Lousley (1950) noted that in Norway and Sweden (judging from herbarium material), it is represented by a plant of very different habit. Pritchard (1959) singled out basal branching and the distinctively pyramidal habit, but observed that the continental material he examined did not show this character. The pedicel length was intermediate between British *G. amarella* and *G. uliginosa*, and the latter species is now considered to be taxonomically doubtful in Norway and Sweden (Council of Europe 1991).

A. Jones

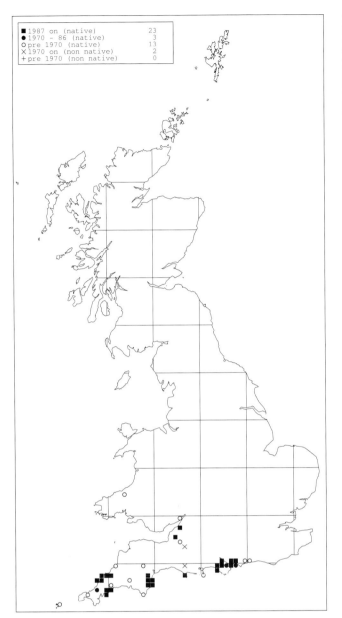

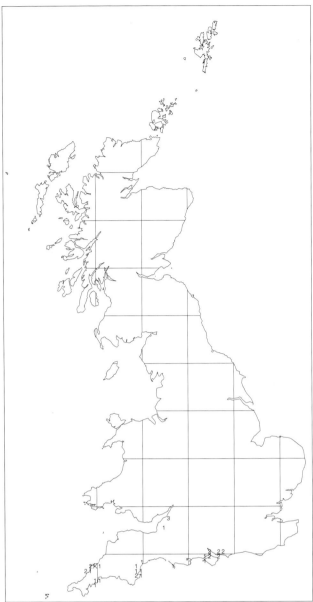

Geranium purpureum Villars (Geraniaceae)

Little robin, Llys Robert Bychan
Status in Britain: LOWER RISK Nationally Scarce.
Status in Europe: Not threatened.

G. purpureum is a plant of warm, open places, occurring in two distinct natural habitats in Britain. In the first, it grows as an upright plant in rocky places, cliffs and dry hills and hedge banks, where it has as associates *Centranthus ruber, Dactylis glomerata, Galium aparine, G. mollugo, Geranium robertianum, Hedera helix, Lonicera periclymenum* and *Rubus fruticosus*. In the second it grows as a prostrate plant in stabilised areas at the rear of shingle beaches, in a very open community with *Atriplex* species, *Geranium robertianum, Glaucium flavum, Lathyrus japonicus* and *Silene uniflora*. Both habitats are usually near the sea, and often in a rich soil (Baker 1955).

G. purpureum is an annual, usually self-pollinated, which sets copious seed. Flowering is generally from June onwards, though plants may flower (and seed) much earlier in Cornwall at least.

The plant seems to be contracting in its British range, with records since 1992 only from Cornwall, Devon, Somerset, Dorset and Hampshire. However, the Cornwall population has increased considerably within the last twenty years, possibly because it seems to respond well to modern road verge management, and perhaps also because of recent mild winters (FitzGerald 1990c). There are small populations in the Avon Gorge and Devon, but a few more records from Hampshire where it seems to be holding its own, in populations ranging from a few individuals to many hundreds of plants. The native status of the quite substantial Dorset site, which occurs along an old railway, is unclear. Here the plant is threatened by possible road development and by occasional spraying. *G. purpureum* was formerly recorded from Sussex, Gloucestershire and

Carmarthenshire. It may be somewhat overlooked, as it is difficult, if not impossible, to distinguish from *G. robertianum* in a vegetative state.

The plant is widespread in southern and western Europe, reaching its most northerly point in southern Ireland, (where plants occur on both rocks and beaches), and spreading east through the Mediterranean and North Africa to Turkey and Iran. It is naturalised in New Zealand.

Throughout its range, *G. purpureum* is a variable plant, some authorities (e.g. Baker 1955) having distinguished two subspecies, ssp. *purpureum* and ssp. *forsteri*, on the basis of vague characters. Current opinion is very much divided on the desirability of recognising these two segregates. Ssp. *forsteri* is found on stabilised shingle at several locations in Hampshire (Brewis *et al.* 1996) and formerly in Sussex, while all other British populations appear to be ssp. *purpureum*. There is also confusion between *G. purpureum* and maritime variants of *G. robertianum*, and it is likely that the former has been over-recorded in the past for that reason. In the field the combination of small flowers and yellow anthers separates *G. purpureum*, which also differs in chromosome number, but these differences are useless in herbarium specimens.

D.A. Pearman

■ 1987 on (native)   6
● 1970 - 86 (native)   0
○ pre 1970 (native)   4
✕ 1970 on (non native)   0
+ pre 1970 (non native)   0

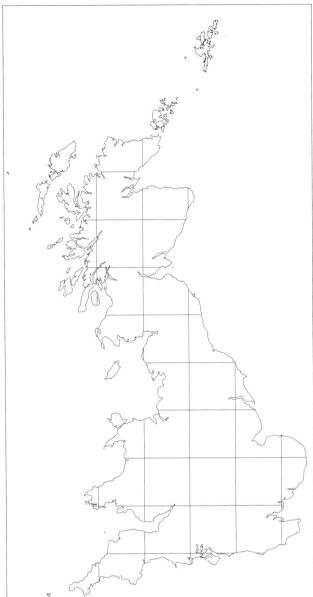

Gladiolus illyricus Koch (Iridaceae)

Wild gladiolus
Status in Britain: LOWER RISK - Near Threatened. WCA
Schedule 8.
Status in Europe: Not threatened. Endemic.

G. illyricus grows in bracken-dominated acid grass
heaths, generally on unpodsolized and stoneless brown
earths. It may occur under a canopy of *Pteridium aquilinum*
with few other associates, or in a more grassy sward in
which prominent species may include *Agrostis curtisii*,
Conopodium majus, *Galium saxatile*, *Genista anglica*, *Polygala
serpyllifolia*, *Potentilla erecta* and *Veronica officinalis*. Two
constant associates are *Hyacinthoides non-scripta* and
Anemone nemorosa, and stands of *Ulex europaeus* are
typically found in the vicinity.

It is a perennial, flowering between June and August.
Flowers are visited, and presumably pollinated, by bees
and butterflies, particularly the large skipper. Each capsule

generally contains 16-21 seeds, which usually fall within a
metre of the parent. A study of a small population in 1987
(Stokes 1987) showed that only 2% of the plants developed
flowers and set seed. If this is the general picture, then
most recruitment is likely to be by vegetative reproduction,
i.e. the formation of cormlets within the corm scales, with
wider establishment from seed a rarer event.

G. illyricus was first discovered in Britain by the
Reverend W.H. Lucas in 1856. Babington (1863) regarded it
as truly native in Britain, but Townsend (1904) suggested
that *G. illyricus* may have arrived with *Erica vagans* and
Simethis bicolor on the coast near Bournemouth with young
fir trees from Landes, in France, and subsequently spread
into the New Forest. It is now widespread, though local, in
the New Forest, with recent records from about 40 1 km
squares in six hectads. Populations differ greatly in size,
some comprising few plants, whilst others have several
thousands. It formerly occurred, as a possible native, on

the Isle of Wight, and introduced populations have occasionally been reported elsewhere.

In recent years, a considerable decline in numbers has been apparent at some sites. Factors in this decline include the encroachment of scrub and heather, attempts to eradicate bracken, and indiscriminate heath-burning (Brewis *et al.* 1996). Intensive grazing, particularly by deer and horses (but also by rabbits, cows and molluscs), is often evident. However, since large herbivores do not graze in dense bracken stands, grazing pressure tends to be much less severe in summer, thereby allowing the full development of *G. illyricus* growing under it. The bracken canopy is frequently damaged by spring frosts, leading to years of varying bracken density. In 'open' years, the increased light enables the corms of *G. illyricus* to increase in size, thus promoting flowering in subsequent seasons. Furthermore, the greater opportunities for insect pollination may lead to a higher proportion of seed being set. However, in such 'open' years, grazing increases, as also does competition from other plants. Conversely, in years of dense bracken, much reduced light levels may lead to increased mortality. Mature plants may become choked by bracken litter, and fewer seeds germinate because of the production of bracken toxins, though

grazing and competition from other plants decrease. Thus, both 'open' and 'closed' bracken canopies have advantages and disadvantages, and an alternation in bracken density may provide the most ideal conditions for populations of *G. illyricus*.

It is widespread in Europe, particularly in the south and west, extending eastwards to Bulgaria and Greece. In continental Europe, it is found in a much wider range of habitats, including heaths, scrub, open woodland and calcareous coastal cliffs.

Some authorities consider that *G. illyricus* in Britain is sufficiently different from populations in continental Europe to warrant its recognition as the separate ssp. *britannicus* (A.P. Hamilton pers. comm.). This proposal is based on genetic and floral differences. In the British populations $2n = 90$, whilst in continental Europe $2n = 60$. The only populations matching those in Britain occurred on Belle Isle in Brittany, but they were destroyed recently during the construction of a dam, though herbarium specimens from Belle Isle are held at Kew.

J. Stokes

Gnaphalium luteoalbum L. (Asteraceae)

Jersey cudweed, Edafeddog Melynwyn
Status in Britain: CRITICALLY ENDANGERED. WCA
Schedule 8.
Status in Europe: Not threatened.

G. luteoalbum is a plant of sandy fields, waste places
and sand-dunes, now found as a probable native plant
only in Norfolk, although it formerly occurred rather more
widely in the Breckland of Norfolk, Suffolk and
Cambridgeshire. In Norfolk, it occurs in a small area of
sand-dune on the north coast, where two small colonies
still exist. Both are on the margins of man-made hollows
('slacks') that were excavated in 1978/9, 1983/4 and 1992/
3 to create shallow pools for natterjack toads, the newly-
created habitat proving to be of unplanned benefit to *G.
luteoalbum* (Scampion 1993).

The two Norfolk colonies are within enclosures fenced
against cattle, but rabbits graze within them. The pools are
shallow and seasonal, often drying up in late summer and
flooding to variable levels in winter, these fluctuations
helping to maintain open conditions suitable for
germination. The substrate is predominantly sand
containing little organic content, with a pH between 6.5
and 6.8. Common species of the margins include *Agrostis
stolonifera, Carex arenaria, Centaurium erythraea, Festuca
rubra, Lotus corniculatus, Potentilla anserina, Prunella vulgaris*
and annuals such as *Erophila verna* and *Linum catharticum*.

G. luteoalbum is a short-lived herb, sometimes behaving
as an overwintering annual, though flowering in autumn
rather than spring (Watkinson & Davy 1985). In that case,
flowering is from July to September (occasionally as late as
November), seedlings normally appearing in September
and October, and the overwintering rosettes becoming
dormant in December. However, this strategy does not
appear to be constant; for instance, large recruitment of
seedlings has been noted in May. Rather, the species seems
to be opportunistic in its germination behaviour, almost
certainly correlated with the seasonal fluctuation of water
levels in the pools (Scampion 1993). The plant may also
behave as a strict biennial. Germination tests have
indicated that a high percentage of seeds are viable, and
the behaviour of wild populations suggests that a
persistent seed-bank exists. For example, prior to
excavation in 1978, no plants had been seen at the present
location since 1967.

Populations of the two colonies fluctuate annually
depending on the availability of open ground for
germination. The need for open ground was amply
demonstrated when, between 1989 and 1992, ponds dried
completely. The succession to a community dominated by
dense perennial grasses was rapid, and the loss of ground
suitable for germination was quickly reflected in the
numbers of plants recorded in one colony, declining from
more than 300 in 1989, to 47 in 1991, and only 15 in 1992.
In a different sample plot in 1991, between 47% and 75%
of the plants flowered, producing copious seed.

The status of *G. luteoalbum* in Britain has been debated
ever since its first discovery in 1820, but in view of its

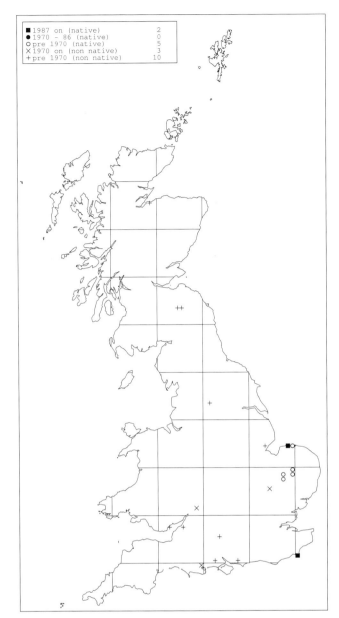

historical distribution, it seems reasonable to assume that it
is native in East Anglia. The large colony in Dorset,
discovered in 1978 on tracks, and which has spread in vast
numbers to an adjacent munitions waste-tip, is probably
not of native origin. In 1996 a population of about 100
plants appeared in the RSPB reserve at Dungeness, growing
in a hollow where shingle had been excavated and into
which silt washings had been pumped. The species had not
been known previously in Kent, and its origin is uncertain.
However, it grows in coastal habitats on the nearest parts
of the French coast, and it is possible that seed may have
been carried over the Channel by birds. Terns, at least, are
known to 'commute' between their colonies at Dungeness
and those on the coast near Calais.

Overseas, it occurs throughout Europe, north to southern
Sweden and Latvia, extending to west and central Asia.

M.J. Wigginton

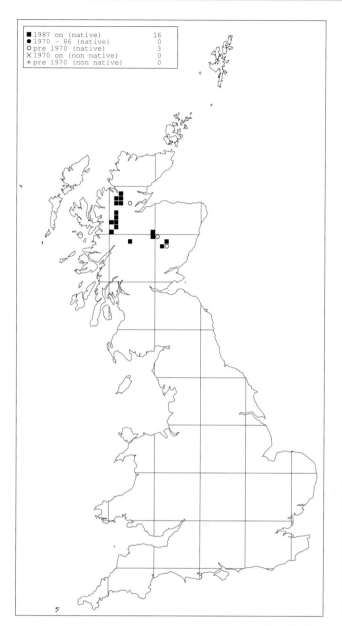

Gnaphalium norvegicum Gunnerus (Asteraceae)
Highland cudweed, Cnàmh-lus Gàidhealach
Status in Britain: LOWER RISK - Nationally Scarce.
Status in Europe: Not threatened.

G. norvegicum is a plant of gullies and stream gorges in the glens and corries of the central and western Highlands of Scotland. Colonies occur on sunlit slopes and ledges, almost invariably with an aspect between east and south and at an altitude over 700 m. Plants grow on well-drained, enriched but acidic soils with an incomplete cover of vascular plants. Frequently, its sites are dominated visually by 'yellow composites'. Common associates are *Alchemilla alpina, Anthoxanthum odoratum, Deschampsia cespitosa, Hieracium* species (usually Sect. *Alpina*), *Nardus stricta, Thymus polytrichus* and *Vaccinium myrtillus,* although, in these open and enriched areas, a great range of ferns, herbs and grasses can occur. The great majority of colonies, and certainly all the small ones, are inaccessible to sheep and red deer and their extent appears to be restricted by grazing. It is intriguing, though, that two of the largest colonies are on open grazeable hillsides.

It is a perennial herb with a short woody rootstock, and plants can live for several years. Regeneration and spread of colonies is by seed, and the presence of small rosettes in most colonies confirms that this is happening. Colonies are mainly small and isolated, and *G. norvegicum* is very much a 'relict species' in Scotland. Immediately following glaciation it was presumably much more widespread (as it is in Norway today) but there is no evidence for this in the fossil record. Climatic warming, together with pressure from grazing animals, has led to the plant surviving now in just a few mountain areas. *G. norvegicum* is present in about 30 discrete colonies on about 12 mountains, and there are old records from a further four sites. Most colonies appear to consist of fewer than 100 individuals, but three hold hundreds of plants and one, thousands. However, counting plants is problematical in that most are sterile rosettes, which are difficult to spot. Confusion in the

182

past with mountain forms of *G. sylvaticum* has meant that several old records have had to be rejected (for example, those from Perthshire and Caithness).

Sites of *G. norvegicum* are prone to snow avalanches and landslides, destabilising processes that maintain and enrich the open ground that the plant requires. Most colonies must be considered vulnerable because of their small size and unstable situation. One colony on Lochnagar consisted of 15-25 large plants and several small ones for many years, but one winter an avalanche swept most of the site away and in the following summer only six small rosettes could be found. Isolation of colonies and the distinct possibility of small colonies being extinguished by natural dynamic processes suggest that the range of the plant in Scotland is likely to be suffering steady attrition. However, the same processes may provide fresh ground for colonisation. Collecting has been heavy at only one site, Lochnagar. On that mountain the plant is now difficult to find, and the plants seen there pale into insignificance

compared with the lush, multi-stemmed specimens residing in herbaria.

Colonies are under no threat from recreational pressures, other than from the occasional adventurous hill-walker or botanist scrambling in gullies. A significant reduction in grazing levels in the hills would benefit the plant, but other changes in land use are unlikely to have any impact.

G. norvegicum occurs in the mountains of central Europe, arctic and sub-arctic Europe, Balkans, Caucasus, Greenland and north-eastern Canada. In Europe, it grows in meadows, heaths, open woods, rocky places and early-exposed grassy snow beds up to an altitude of 2,800 m (Clapham *et al.* 1962; Huxley 1967). In Norway it colonises bare stony ground beside roads and paths in the mountains, and its abundance comes as a pleasant surprise to visiting botanists from Britain.

A.G. Payne

Helianthemum apenninum (L.) Miller (Cistaceae)
White rock-rose, Cor-Rosyn Gwyn y Mynydd
Status in Britain: LOWER RISK - Near Threatened.
Status in Europe: Not threatened.

H. apenninum is a plant of steep rocky, but stable slopes over Carboniferous and Devonian limestones, where the southerly to westerly aspect and the good drainage of shallow rendzinas accentuate the warm and sunny character of an oceanic regional climate. The microclimate in which this plant occurs in Britain is probably the nearest approach on the British limestones to conditions characteristic of the Mediterranean region. Typically it occurs in short *Festuca ovina* coastal grassland that forms an open tussocky turf interspersed with patches of bare soil and rock outcrops. *Lotus corniculatus, Pilosella officinarum, Sanguisorba minor* and *Thymus polytrichus* are frequent associates, but most characteristic are annuals or pauciennials such as *Blackstonia perfoliata, Carlina vulgaris, Centaurium erythraea* and the Mediterranean-Atlantic mosses *Scorpiurium circinatum* and *Tortella nitida*. *H. apenninum* shows a markedly discontinuous distribution in Britain, a feature that it shares with a few other British rarities including *Koeleria vallesiana* and *Trinia glauca*, with which it often grows.

It is a conspicuous perennial, often growing abundantly in large masses, and flowering from April to July. Although plants flower freely and set abundant seed at all its British localities (Proctor 1956), the establishment of new individuals from seed appears to be a rare event. Seed is dispersed over very short distances, and a hot dry summer militates against the survival of seedlings. Vegetative reproduction has apparently not been observed in the field, but in cultivation cuttings root easily.

H. apenninum was first recorded in Britain in 1688 by Plukenet on "Brent Downs in Somersetshire near the Severn sea". It still grows there, at the place now known as Brean Down, at its maximum altitude in Britain of 90 m. It occurs at four localities in the Mendip Hills, Somerset, and at four near Torquay in Devon. It is abundant at both Brean Down and Berry Head, numbering many thousands, and is frequent at three of its other sites. Plants were deliberately introduced at Goblin Combe (Hope-Simpson 1987), and still survive there. The population appears to be relatively stable overall. In hot dry sites the plant does well, perhaps partly because competitors are few. In less xeric situations its survival is likely to depend on grazing keeping vigorous competitors in check. Nearly all colonies lie within protected sites, but grazing levels are not always ideal. Brean Down is grazed by both cattle and goats, but on most sites rabbits are clearly the most important or only grazers. Recreation and public pressure may become a local problem in the future but at present appear to be at tolerable levels.

The hybrid *H.* x *sulphureum* (*H. apenninum* x *H. nummularium*) is recorded from one site in the Mendips

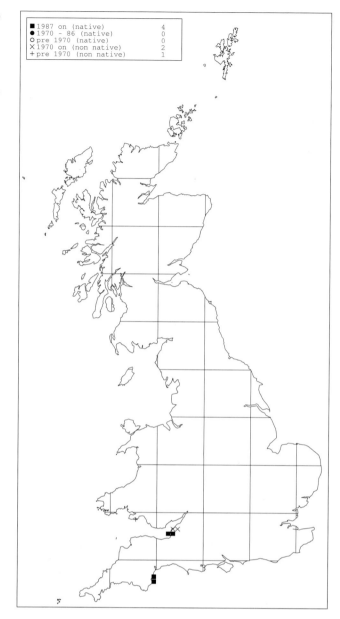

and was formerly known at Berry Head. However, the parents do not usually grow close together, *H. nummularium* generally occurring in more mesic grassland.

In continental Europe, *H. apenninum* is chiefly a montane species, occurring in the Mediterranean region from Morocco and the Iberian peninsula to the southern slopes of the Alps and eastwards to Greece and Crete. It extends in scattered localities northwards to southern England, where it reaches its northern limit.

Its distribution and ecology are further described by Proctor (1956, 1958).

R.D. Porley

Helianthemum canum ssp. *levigatum* M. Proctor (Cistaceae)
Hoary rock-rose
Status in Britain: VULNERABLE. ENDEMIC.

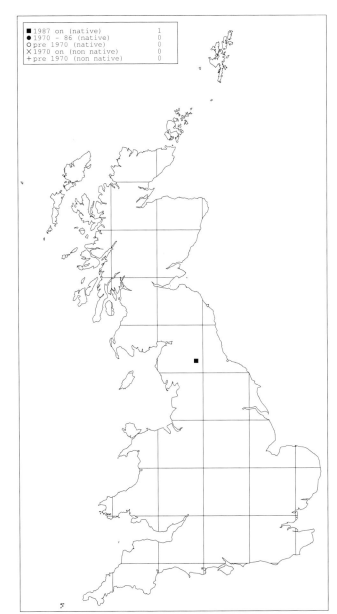

■ 1987 on (native)   1
● 1970 – 86 (native)   0
○ pre 1970 (native)   0
✕ 1970 on (non native)   0
+ pre 1970 (non native)   0

H. canum ssp. *levigatum* occurs on a few outcrops of metamorphosed granular 'sugar' limestone at about 530 m on Cronkley Fell in north-west Yorkshire. The plant colonises both the loose and open gravel of eroded limestone in more level areas, as well as fairly closed turf dominated by *Festuca ovina* or *Sesleria caerulea*. Other associates are *Carex capillaris*, *C. ericetorum*, *Galium boreale*, *G. sterneri*, *Gentiana verna*, *Helianthemum nummularium*, *Koeleria macrantha*, *Pilosella officinarum* and *Thymus polytrichus*. Plants flower profusely where they are protected within wire enclosures against grazing by sheep and rabbits. Plants do survive in the heavily-grazed vegetation beyond the wire, but as a very stunted form, and the shoots are generally bitten off long before they can produce flowers. It is abundant within its tiny range, the population consisting of thousands of plants.

This prostrate dwarf perennial shrub flowers between early June and the end of July and sets seed freely unless the inflorescences are destroyed by grazing. Vegetative reproduction by layering has been reported in *H. canum* ssp. *canum* (Griffiths & Proctor 1956), but it is not known whether this occurs in ssp. *levigatum*.

Since the locality is in the Teesdale NNR, and most plants are in protected areas within the reserve, there is no immediate threat to the survival of the subspecies other than that posed by climatic change (and the occasional rabbit which penetrates the defences). Numbers of plants within the exclosures appear to be maintained and there seems every reason that it should thrive within the exclosures. On the open fell the future is less assured unless a solution to the present overgrazing can be found. The climate of Cronkley Fell is extremely harsh and the outcrops very exposed; the development of turf on the sugar limestone is therefore a prolonged process, even in the absence of sharp hooves and grazing. Besides direct damage to the plants, rabbits are instrumental in breaking through the binding turf, initiating rapid erosion of the mineral soil beneath. Large areas of the unprotected fell show such erosion and the consequent loss of almost all plants. Rabbit populations are presently high, encouraged by recent mild winters. Maintenance of the fencing will be a long-term priority if this taxon is to be successfully conserved.

H. canum is a widespread plant, occurring from western Ireland to the Caucasus, and from the Baltic (Öland) to North Africa. Several sub-taxa occur over this large range.

M.S. Porter and F.J. Roberts

Herniaria ciliolata Meld. ssp. *ciliolata*
(Caryophyllaceae)

Fringed rupturewort, Llys y Fors Eddiog
Status in Britain: LOWER RISK - Near Threatened. Near
Endemic.
Status in Europe: Not threatened. Endemic.

In Britain, *H. ciliolata* occurs only on the Lizard
peninsula. It is found in a variety of habitats - on sea-cliffs,
in cove valleys and on heathlands - and is locally plentiful.
It is confined to short turf and other open vegetation, and
is largely restricted to slopes with a southerly aspect. It is
predominantly a coastal plant, seldom occurring more than
1 km inland. It is found in open grassland on the middle or
upper slopes of sea-cliffs in association with perennials
such as *Armeria maritima* and *Plantago coronopus*, and an
abundance of annuals including *Aira praecox, Catapodium
marinum* and, in some sites, *Trifolium incarnatum* ssp.
molinerii. On the grazed species-rich south-facing slopes of
cove valleys, *H. ciliolata* may occur with other rare species
including *Isoetes histrix, Juncus capitatus, Trifolium bocconei*
and *T. strictum*. It also occurs in maritime heath with
Calluna vulgaris, Erica cinerea and *Scilla verna*, and in
species-rich, wind-pruned *Erica vagans-Ulex europaeus* heath.
In shallow rock pans, *H. ciliolata* locally accompanies *Allium
schoenoprasum, Minuartia verna* and *Plantago maritima*. It also
occurs in some ruderal habitats, including by footpaths, on
Cornish stone-faced 'hedges', and on the gravelly margins
of airfield runways. *H. ciliolata* grows on soils derived from
a wide range of rock types including serpentine,
hornblende-schist, mica-schist, gabbro and shale, as well as
occurring on shell-sand and gravel.

H. ciliolata is a mat-forming perennial with a long tap-
root. Flowers are produced in May and June, and are said
to be pollinated by ants. Seedlings appear in autumn, but
are sparse even at sites where mature plants are abundant.
It is drought-tolerant (even of severe conditions), plants
often remaining green, though sometimes shrivelled, when
the leaves of its perennial competitors have dried, and
annuals have disappeared.

Colonies of *H. ciliolata* occur in most areas of short turf
or open heathland along the Lizard coast between
Gunwalloe and Coverack. Likewise, it is more or less
continuously distributed, up to 1 km inland, on south-
facing slopes of cove valleys, with one small population on
Goonhilly Downs 4 km from the coast. The total
population of many thousands of plants appears to be
stable overall, with little fluctuation in numbers.

Although at some sites the cessation of grazing by
domestic livestock may have caused some losses, there is
no evidence of a significant decline in the population of *H.
ciliolata* during this century. To some degree this lack of
grazing may have been compensated for by increased
tourist pressure, which has maintained areas of open
vegetation beside footpaths and on popular viewing

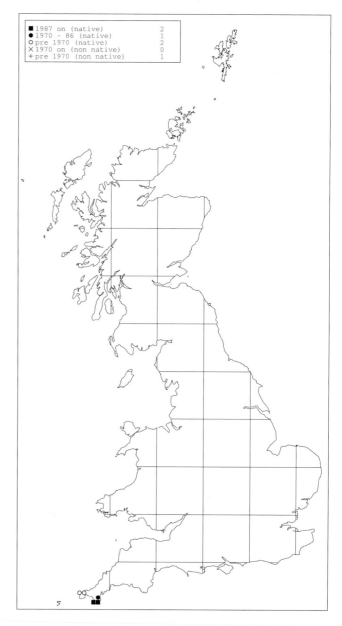

points. On cliffs, frequent summer droughts and maritime
exposure help to maintain suitably open conditions.

H. ciliolata has a hyper-oceanic, Lusitanian distribution
in Europe, being restricted to the coastal region from
northern Germany, Britain and the Channel Islands
southwards to France, Spain and Portugal. Within this
range, populations are disjunct and show differences in
their chromosome ploidy linked to morphology, a pattern
of evolutionary change characteristic of reproductively
isolated populations. Ssp. *ciliolata* is endemic to Cornwall,
Alderney and Guernsey, while ssp. *subciliata* is endemic to
Jersey.

J.J. Hopkins

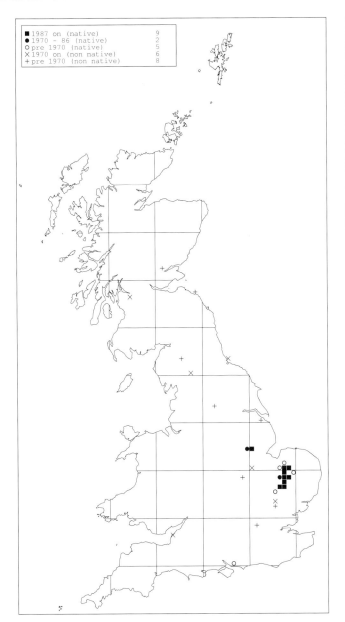

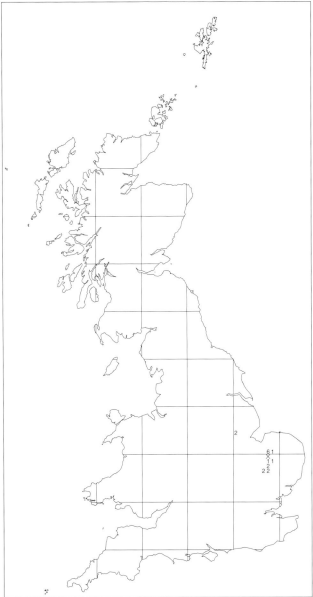

Herniaria glabra L. (Caryophyllaceae)
Smooth rupturewort, Llys y Fors
Status in Britain: LOWER RISK - Near Threatened.
Status in Europe: Not threatened.

H. glabra is a plant mainly of eastern England, where it grows in compacted sandy or gravelly soil, often with some chalk or limestone fragments. It cannot withstand close competition, so is chiefly found where disturbance by vehicles or seasonal standing water keep an open habitat for germination. In Norfolk it is generally restricted to well-used rides in pine forests, but in Suffolk and Lincolnshire it grows in disused gravel pits (especially if they hold water in winter), on golf courses, gravelled car parks, and in also in disturbed areas within short grassland. Whilst an open situation is usually favoured, healthy plants can be found in forest rides shaded by mature pines. Frequent associates include *Achillea millefolium, Daucus carota, Erodium cicutarium, Holcus lanatus, Plantago coronopus, P. lanceolata, P. major, Reseda lutea* and *Trifolium repens*.

H. glabra is a short-lived, mat-forming perennial. It is winter green, apparently unaffected by frost, and its long tap-root (up to 34 cm long) enables it to withstand drought. Mature plants flower profusely and can attain a diameter of 40-50 cm. It can flower and fruit in its first year, even as a tiny plant 2 cm across. It flowers from June to October, each flower producing a single seed. Germination is in spring or autumn, whenever conditions are suitable. Bare ground is necessary for germination, but it must be compacted to preserve moisture. Populations in a favourable site can reach several hundreds in some years, then decrease if competition is too great. Seed is viable for at least ten years, so colonies can reappear even after a long absence.

It is now confined as a native plant to the Norfolk and Suffolk Breckland, and Lincolnshire. The most recent loss was from Cambridgeshire, where sites have been destroyed and it has not been seen since 1990. In Norfolk it is maintaining a steady population in 20 localities, with new sites still being discovered. In Suffolk, one site holds populations numbered in thousands, whilst three sites have only a few plants. In Lincolnshire, the two remaining sites maintain stable, though small, populations. Sites have been lost in these counties during the last 20 years, chiefly through changes in land use, including gravel extraction and dumping, or by a lack of disturbance. *H. glabra* has appeared as a casual, persisting for several years in scattered localities north to Lanarkshire. Since it is listed in seed catalogues, these plants are presumed to have originated from gardens. It was reported in 1791 at West Sole, Weston-super-Mare, but this has been regarded as an error (Roe 1981). It was recorded nearby, in 1946, at Ellenborough Park, on the site of a 1940s army camp, where it still survives. These plants could have been introduced from the Breckland, an area long used by the army.

The main threat is a change of land use. Conservation management is difficult, since the plant requires considerable use of the site to compact the soil and to prevent more vigorous species crowding it out. In Norfolk, the rotation of forestry operations, including the use of vehicles in rides, seems ideal, and it is possible that there is effective dispersal of seed or fragments of plants on the wheels of vehicles.

It ranges from southern Europe northwards to southern Scandinavia, occurring also in North Africa and Asia as far east as India.

J.E. Gaffney

Hieracium L. (Asteraceae)
Hawkweeds

Species of the genus *Hieracium* are long-lived apomictic perennials. There are more than 250 named taxa in Britain, many of them endemic and many rare and local. The list of 80 species below comprises those species believed to be native that have been recorded from five or fewer hectads. It has not been possible to distinguish those plants

recorded in small quantity at a handful of restricted sites from those that occur in few sites but are present there in some quantity. The list includes two doubtfully native species, *H. borreri* and *H. subramosum*, which have not been seen for many years and are probably extinct. The list is somewhat provisional, but gives some focus for conservation action. In the list, the figures denote the number of hectads in which the species is known.

Hieracium species believed to be native that have been recorded from five or fewer hectads

Species	No. of hectads in which the species is known	Site(s)
Sect. Alpina		
H. backhousei	3	Angus, S. Aberdeenshire
H. calvum	2	S. Aberdeenshire, E. Inverness
H. globosiflorum	5	S. Aberdeenshire, Banffshire, E. Inverness
H. graniticola	2	S. Aberdeenshire, Banffshire
H. grovesii	2	S. Aberdeenshire, Banffshire
H. insigne ssp. insigne	1	S. Aberdeenshire
H. insigne ssp. celsum	4	S. Aberdeenshire, E. & W. Inverness
H. kennethii	2	W. Ross, W. Sutherland
H. larigense	2	S. Aberdeenshire, E. Inverness
H. macrocarpum	4	S. Aberdeenshire, E. Inverness
H. notabile	3	Perthshire, W. Inverness
H. optimum	1	Argyll
H. pensum	5	W. & E. Ross
H. pseudocurvatum	2	S. Aberdeenshire
H. pseudopetiolatum	4	S. Aberdeenshire, E. Inverness
H. subgracilentipes	3	Cumbria
H. tenuidens	1	Perthshire
Sect. Subalpina		
H. chrysolorum	2	Northumberland, Dumfriesshire
H. dissimile	3	Wigtownshire, Argyll, E. Ross
H. diversidens	4	Perthshire, W. Inverness, W. Sutherland
H. eustales	2	Perthshire
H. longilobum	1	Dumfriesshire
H. melanochloricephalum	4	S. Aberdeenshire, E. Inverness
H. westii	1	Perthshire
Sect. Cerinthoidea		
H. magniceps	1	Perthshire
Sect. Oreadea		
H. angustatum	5	Perthshire
H. cacuminum	5	Glamorgan, Breconshire, Carmarthenshire
H. cambricum	3	Glamorgan, Caernarvonshire, Denbighshire
H. cillense	1	Breconshire
H. fratum	5	Dumfriesshire, S. Aberdeenshire, E. Inverness, Argyll
H. naviense	1	Derbyshire
H. pseudoleyi	5	Caernarvonshire, Denbighshire
H. repandulare	5	Breconshire, Caernarvonshire
H. riddelsdellii	4	Breconshire, Carmarthenshire
H. stenolepiforme	1	Somerset

Hieracium species believed to be native that have been recorded from five or fewer hectads *continued*

Species	No. of hectads in which the species is known	Site(s)
Sect. Vulgata		
H. asteridiophyllum	1	Breconshire
H. caesionigrescens	3	W. Yorkshire, Dunbartonshire
H. candelabrae	1	E. Yorkshire
H. discophyllum	1	Breconshire
H. fulvocaesium	1	W. Sutherland
H. glanduliceps	2	W. Yorkshire
H. itunense	1	Cumbria
H. mucronellum	2	W. Sutherland
H. neocoracinum	1	Breconshire
H. pachyphylloides	3	Gloucestershire, Herefordshire, Breconshire
H. pauculidens	3	W. Sutherland
H. pollinarium	4	Perthshire, W. Sutherland
H. prolixum	2	W. Sutherland
H. pruinale	3	Perthshire, Angus
H. radyrense	2	Glamorgan
H. snowdoniense	2	Caernarvonshire
H. subminutidens	?2	Breconshire
H. subprasinifolium	2	Staffordshire
H. subramosum	1	Fife (extinct)
H. surrejanum	4	E. & W. Sussex, Surrey
Sect. Alpestria		
H. attenuatifolium	2	Shetland
H. australius	3	Shetland
H. breve	1	Shetland
H. carpathicum	4	Perthshire, Angus
H. difficile	1	Shetland
H. dovrense	4	Shetland
H. gratum	1	Shetland
H. hethlandiae	1	Shetland
H. mirandum	4	Shetland
H. northroense	2	Shetland
H. praethulense	3	Shetland
H. pugsleyi	4	Shetland
H. solum	2	Shetland
H. spenceanum	1	Shetland
H. zetlandicum	2	Shetland
Sect. Prenanthoidea		
H. borreri	1	Selkirkshire (extinct)
Sect. Tridentata		
H. acamptum	1	Surrey
H. cambricogothicum	4	Kent, Caernarvonshire, Moray
H. linguans	1	Breconshire
H. nidense	5	Breconshire, Merioneth
H. ornatilorum	5	Lancashire, E. & W. Yorkshire
H. subintegrifolium	1	Cumbria
Sect. Foliosa		
H. bakerianum	2	N.W. Yorkshire, Durham
H. tavense	1	Breconshire

D. McCosh

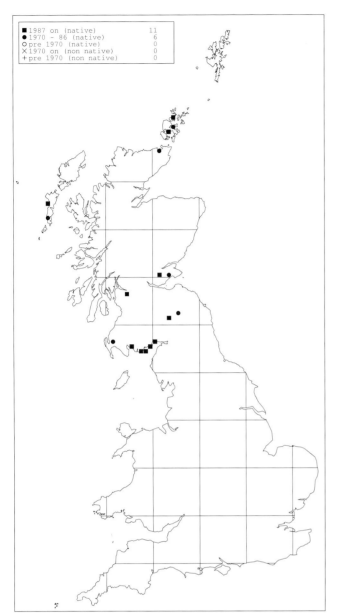

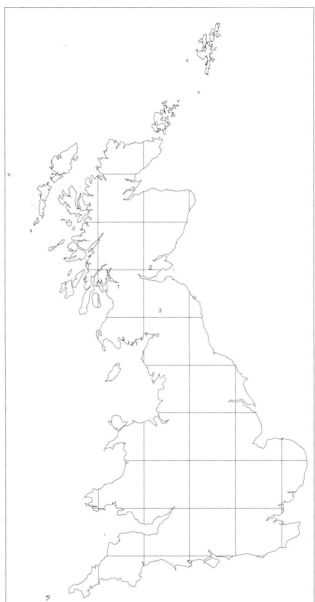

Hierochloe odorata (L.) P. Beauv. (Poaceae)

Holy-grass, Feur Moire
Status in Britain: LOWER RISK - Near Threatened.
Status in Europe: Not threatened.

H. odorata is a rhizomatous, perennial grass of northern latitudes with a curiously disjunct distribution in Scotland and Ireland. Its habitats in Scotland include base-rich fen (Blackpool Moss), raised mire (Ale Water), maritime grassland at the upper limit of tidal saltmarsh (Southwick), lakeside willow carr and sedge communities (Loch Leven), river bank (Thurso), and the base of coastal cliffs in thin, peaty saltmarsh over shingle and boulders (Ravenshall). The species, therefore, has an extremely long and unusual list of associated species, ranging from *Festuca rubra* and *Juncus maritimus* through *Deschampsia cespitosa*, *Molinia caerulea* and *Schoenus nigricans* to *Filipendula ulmaria*, *Phalaris arundinacea* and *Phragmites australis*. In south-west Scotland the sites are exclusively coastal and close to the high water mark of spring tides, but inland in the Borders

H. odorata grows up to an altitude of 260 m. The main common features of all habitats are the generally high water table and usually base-rich soils.

Flowering occurs between March and May and appears to vary considerably. In many sites this may depend on the lowering of high water tables from year to year - the plant being fully submerged for much of the winter. When not in flower, the characteristic bright green leaves are a help to finding the grass, which may be easily overlooked. The aromatic leaves contain the distinctive (almost *Briza*-like) pyramidal panicles containing many glistening spikelets, each with one terminal hermaphrodite floret and two lower male florets. Six clones sampled from Scotland were tetraploid, and although seed-set was low (around 2%), isozyme electrophoresis indicated that there is genetic variation in British populations (Ferris *et al.* 1992). Most populations show irregular male meiosis and a remarkable mixture of first and second division restitution (a cell failing to divide despite the chromosome replicating). Such irregularities produce unreduced or polyploid gametes and help to account for the low seed production.

The known sites of this grass have been well mapped and characterised, and most seem to have well-established, if small, clones which do not vary much in size from year to year. In some populations (e.g. Blackpool Moss, Ale Water), the extent of the clone(s) is very marked, whilst in other areas (e.g. at Ravenshall) the distribution appears to be patchy. Some such clones are large (at Southwick Merse plants are scattered within an area more than 160 x 30 m), while others are discrete and not obviously connected.

It is difficult to know what the main threats to *H. odorata* may be, or indeed whether they are common to all sites, but the plant appears to flourish in open areas where grazing prevents grassland from becoming too dense. Invasion of the sites by *Ulex europaeus* (in the grassland at Southwick) and by *Salix* species (at Blackpool Moss and Portmoak Marsh, Loch Leven) may be a threat. Plants in cultivation are sensitive to drying out, and both drainage of wetlands and erosion of fringing saltmarsh are likely to be major threats to specific populations. Most known sites are protected in SSSIs.

H. odorata occurs in northern and central Europe, southwards to the Alps and Moldavia, northwards to Iceland and Fennoscandia, and eastwards to northern and central Russia. It is rare in several countries and endangered in Switzerland.

A.J. Gray

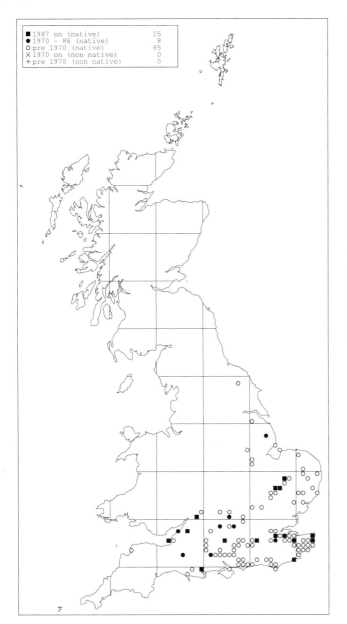

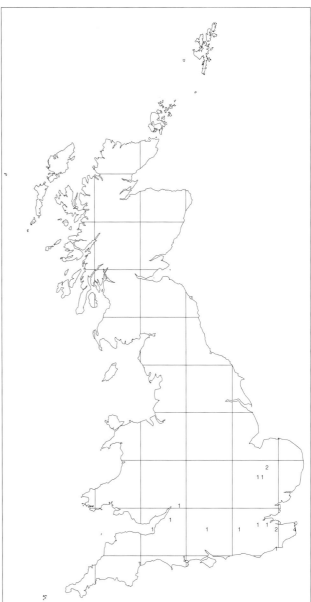

Himantoglossum hircinum (L.) Sprengel (Orchidaceae)

Lizard orchid, Tegeirian Drewllyd
Status in Britain: VULNERABLE. WCA Schedule 8.
Status in Europe: Not threatened.

H. hircinum is a plant of calcareous soils in southern England, occurring in tall, open grasslands mainly on the chalk but locally also on limestone and on dunes. In Cambridgeshire, it grows in ungrazed chalk grassland on the slopes of an ancient earthwork. Associated species in chalk grassland include *Astragalus danicus, Brachypodium pinnatum, Bromopsis erecta, Carex caryophyllea, C. flacca, Cirsium acaule, Festuca rubra, Helianthemum nummularium, Helictotrichon pratense, Leontodon hispidus, Lotus corniculatus* and *Thymus polytrichus*. On sand-dunes it may accompany *Allium vineale, Ammophila arenaria, Anacamptis pyramidalis, Elytrigia atherica, Festuca rubra, Galium mollugo, G. verum, Holcus lanatus, Pilosella officinarum, Sedum acre, Trifolium arvense* and *Vulpia bromoides*. In some of its roadside verge

sites it occurs in species-poor rank grassland with undistinguished associates.

This perennial species overwinters as a green plant. The leaves emerge in September or October and rosettes persist until May. Flowering stems sometimes reach a metre in height but are more usually about 30 cm tall. Flowering is in June and July, with a range of insects (including hoverflies, bees and moths) visiting and presumably pollinating the flowers. Plants that do not produce flowers die back in late May, and no aerial parts are visible until the following autumn. Individuals can be long-lived: some plants first observed in 1978 were still alive in 1996.

In England, *H. hircinum* was restricted to Kent until the turn of this century, after which it spread as far as Yorkshire in the north and Devon in the west. There were only a few populations extant in any one year before 1907, but by 1930 about 30 populations were recorded every

193

year. After 1934 there was a dramatic decline in the number of populations, although the extent of the geographic range was maintained. Since the 1940s there have consistently been 9-11 populations each year, although not the same ones - there have been losses and gains. In the last few years it appears that the species is becoming more common again, with seven new populations being noted since 1987. Typically, populations have been small and sometimes short-lived, with flowers appearing in perhaps only a single season. With protection, populations have become more persistent in recent years, and two have become large enough to be self-sustaining. The largest is at Sandwich, Kent: 3,000 plants in 1991 and 1,500 in 1995, the difference in numbers perhaps merely representing a natural fluctuation. The other large population (in Cambridgeshire) has been fairly stable in numbers in recent years: 200 were counted in 1995. Most of the other extant colonies are very small.

Conservation management of *H. hircinum* sites has included the control of rank vegetation and the removal of litter and moss layers to encourage the growth of seedlings. At the Cambridgeshire site rank growth has been controlled by burning quinquennially between January and March, and though there is some damage to rosettes, plants recover and flowers are produced as usual. The burning is a substitute for grazing and benefits many of the other herbs in the community. The possible benefits of raking out old vegetation and moss are also being investigated at that site. In Kent, Sussex and Somerset, *H. hircinum* grows in unmanaged dune grassland on golf courses, where populations maintain themselves without much intervention and are likely to continue to do so, provided nutrient levels are kept low. Those in Kent may be much trampled during golf championships in July, but this does not seem to have an adverse affect. Indeed, it may even be beneficial, as *H. hircinum* is dormant at that time of year and much of the damage would be to its potential competitors. This species is still targeted by collectors, and plants have been dug-up in recent years from both Kent and Cambridgeshire populations.

H. hircinum is widely distributed in Europe, extending from Iberia northwards to England and Germany and eastwards to the Balkan peninsula and the Crimea. It appears to be in decline in north-western and central Europe, and extinct in the Netherlands and the Czech Republic.

L. Farrell and P.D. Carey

Homogyne alpina (L.) Cassini (Asteraceae)

Alpine coltsfoot, Purple coltsfoot
Status in Britain: ENDANGERED. WCA Schedule 8.
Status in Europe: Not threatened.

H. alpina was first found in Scotland by George Don in 1813, at which time he described it as growing "on rocks by the sides of rivulets in the high mountains of Clova, as on a rock called Garry-barns". The plant was not seen again, or at least not recorded in botanical publications, until 1951, and Clova remains the only British site. The small colony grows there on a single undistinguished heathery ledge on a wet hillside at an altitude of about 600 m. Associated species include *Alchemilla alpina*, *Anemone nemorosa*, *Calluna vulgaris*, *Deschampsia cespitosa*, *Luzula sylvatica*, *Nardus stricta*, *Oxyria digyna*, *Saxifraga aizoides*, *Taraxacum* sect. *Spectabile*, *Thalictrum alpinum* and *Sphagnum fallax*.

It is a shortly-creeping, rhizomatous, hairy perennial with erect unbranched stems. The usual flowering time is late June, but flowers may appear between May and September. The flowers are pollinated by various insects and may be self-fertile. Seeds are adapted for wind-dispersal, but in Scotland effective reproduction seems to be only by vegetative spread (by rhizomes).

The original colony covers only a few square metres. It is impossible to determine how many individual plants make up the patch, or to what extent rhizomes connect the shoots. Flowering heads are more readily counted: there are normally between 100 and 200. This colony became increasingly threatened as its location became common knowledge, and in order to reduce the risk of damage, plants have been translocated to three additional locations near the original. It is also intended to maintain a reserve stock of plants in cultivation by propagation *ex situ*. Trampling and grazing by sheep and deer have caused damage in the past, but an exclosure has recently been erected around the original colony to prevent such damage.

H. alpina is found in the mountains of central and southern Europe, including the Pyrenees, the Alps and the Balkans. Throughout its European range, *H. alpina* occurs in a wide variety of sub-alpine and alpine habitats, including woods, streamsides, damp meadows and crags. Because the Scottish colony is a very distant geographical outlier, its status in Scotland has been much debated. Many authorities (e.g. Clapham *et al.* 1987; Stace 1991) consider it is most likely to have been introduced, though others (e.g. Raven & Walters 1956) do not discount native status. A record from South Uist is erroneous.

B.G. Hogarth

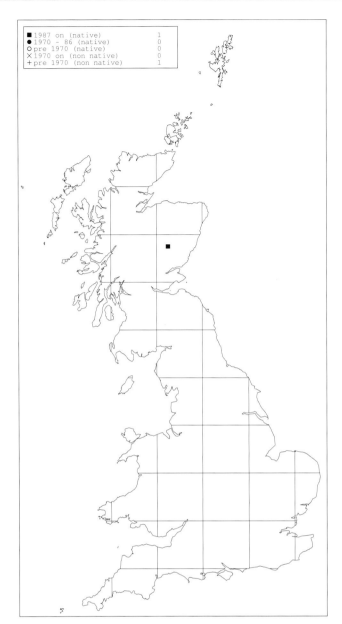

■ 1987 on (native)	1
● 1970 – 86 (native)	0
○ pre 1970 (native)	0
✕ 1970 on (non native)	0
+ pre 1970 (non native)	1

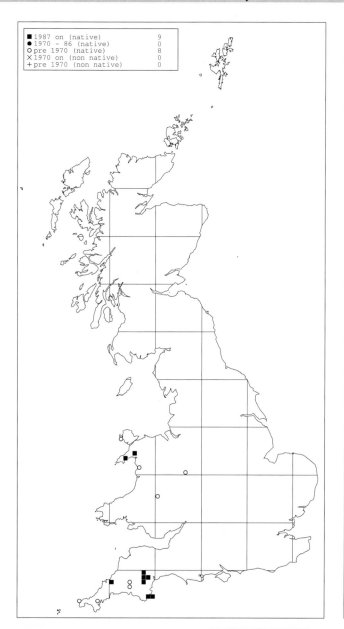

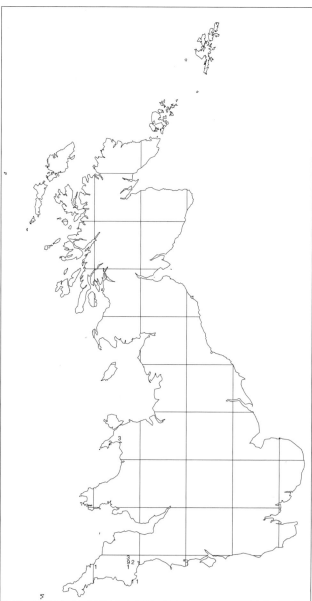

Hypericum linariifolium Vahl (Clusiaceae)

Toadflax-leaved St John's-wort, Eurinllys Culddeiliog
Status in Britain: LOWER RISK - Near Threatened.
Status in Europe: Not threatened. Near Endemic.

H. linariifolium is a plant of steep rocky slopes with a southerly aspect. It occurs on coastal cliffs and in open areas surrounded by woodland in steep inland valleys, where thin soils have developed over acid rocks. Its habitat is generally heathy in nature, and typical plants of the community include *Agrostis capillaris*, *A. curtisii*, *Aira praecox*, *Digitalis purpurea*, *Erica cinerea*, *Sedum anglicum*, *Teesdalia nudicaulis*, *Teucrium scorodonia*, *Ulex europaeus*, *U. gallii* and *Umbilicus rupestris*. Bryophytes and lichens are conspicuous elements on the rocks, which frequently have a slight seepage of water and are winter wet.

It is a perennial, though often short-lived (or even biennial), with the capacity to produce copious amounts of seed when conditions are favourable. Seeds remain viable for two years in laboratory conditions (Ivimey-Cook 1963). It needs patches of bare soil in which to germinate and is a poor competitor. This species is adapted to prolonged periods of high temperatures and drought during the summer (Muddeman 1989), and the latter is an important factor in keeping the habitat open. For example, in 1995 drought caused a significant die-back of *Erica cinerea* and *Ulex gallii*, opening up the habitat and resulting in an increase in *H. linariifolium* in 1996.

It appears to be extant only in East Cornwall, South Devon and Caernarvonshire. Its strongholds are in the Teign valley, north-east of Dartmoor (where it is currently known in thirteen 1 km squares), in the Dart valley, and on cliffs near Dartmouth. Several of the Devon sites hold populations of thousands of plants, and the total population in the Teign valley in 1996 was estimated to exceed 50,000 plants. The only extant Cornish site is in a disused railway cutting where one very small population is threatened by scrub encroachment, though a stronger

196

population occurs nearby. Recent searches have failed to find this species in the Tamar valley (where it has been known since 1849), or at Cape Cornwall. Reports from coastal rock outcrops and path sides near Prawle Point were probably misidentifications of *H. humifusum* or *H. pulchrum* (S.J. Leach pers. comm.). In Wales, small colonies occur on rocks and rocky hillsides near Tremadog and Pwllheli, but it seems to have gone from former sites in Anglesey, Merioneth and Radnorshire (McDonnell 1995a).

Past declines are most likely to have occurred because of scrub invading formerly open habitats, and uncontrolled growth of *Ulex europaeus* threatens some of its extant sites. Fire damage is also a threat. Just over half of all the currently known sites in Britain are within SSSIs or are otherwise managed for conservation. Scrub removal is ongoing at several sites, and such clearance has led to an increase in numbers of *H. linariifolium* plants. Though present in the Channel Islands and recorded from many coastal areas of Jersey in particular, it appears to have declined there in recent years and is now difficult to find in its pure state, most plants seeming to have some admixture of *H. humifusum*. Isoenzyme studies (Kay & John 1995) highlight the possibility that introgression with *H. humifusum* may also pose a threat to *H. linariifolium* on the British mainland.

It is a European endemic with an oceanic-western European distribution, occurring in Madeira, Portugal, Spain, France, the Channel Islands and Britain. It is perhaps scarce or rare in all of its surviving range.

E.J. McDonnell and J.L. Muddeman

Hypochaeris maculata L. (Asteraceae)
Spotted cat's-ear, Melynydd Brych
Status in Britain: VULNERABLE.
Status in Europe: Not threatened.

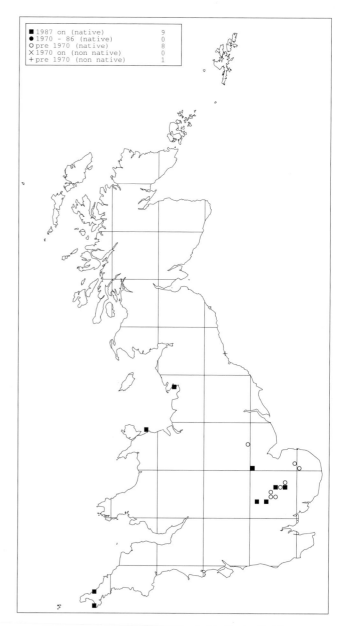

In the British Isles *H. maculata* is confined to three main habitats: grazed or ungrazed grasslands on calcareous soils in eastern England; maritime cliffs on the south-west coasts of England and North Wales; and a single site on blown calcareous sand on the north coast of Cornwall. It is restricted to soils that contain large quantities of exchangeable calcium or magnesium (pH 6.4-8.2). At five localities on the chalk and Jurassic limestone, the soil is a rendzina or brown calcimorphic loam. In contrast, at Kynance Cove, the soil is a friable, porous, red-brown calcimorphic loam, derived from serpentine. On the north coast of Cornwall, the freely draining soil is derived in part from blown calcareous sand, and at the sites in Gwynedd and Cumbria, the shallow red-brown soil is developed on dolomitised Carboniferous limestone. *H. maculatum* occurs on steep, rocky cliffs as well as on more gentle downland slopes, but its restriction to ancient earthworks and sites that have been undisturbed for many hundreds of years suggests that it is unable to recolonise more recently disturbed land. It is found on slopes of all aspects, plants on north-facing aspects producing viable seed in Bedfordshire and Cambridgeshire (T.C.E. Wells pers. obs.).

H. maculata is a long-lived, hemicryptophyte perennial with a large underground rootstock from which new rosettes of leaves are produced annually. Plants reproduce vegetatively as well as by seed. Axillary buds on the rhizome begin growth below ground, producing a small rosette of leaves near the parent plant. Up to four rosettes may arise from a single plant in this way. The proportion of plants that flower varies greatly from year to year. Good flowering is often associated with hot summers the previous year. There is a considerable variation in the number of achenes produced in each capitulum (between 40 and 270) and in total achene production per plant. Whole capitula may contain only sterile achenes, but more commonly fertile and sterile achenes are found together in the same capitula, the relative proportion probably related to the weather at the time when the florets are open and in a receptive condition for pollen. Germination and establishment under field conditions are slow and probably dependent on the soil moisture and temperature at the site where the achenes fall. In experiments to investigate these factors (Wells 1976) only 9% of seed sown produced plants, and after two and a half years, surviving plants had formed small rosettes with five to ten leaves.

During the nineteenth century, *H. maculata* was recorded from eighteen sites in the British Isles but today it is known with certainty from only nine. The largest mainland population is in Bedfordshire, where more than 5,000 plants may flower in some years in a population twice that size. About 1,000 plants occur at Kynance, and the population in Gwynedd may still number 200-300 plants. However, the other colonies (in Cornwall, Hertfordshire, Suffolk, Cambridgeshire, Northamptonshire and Cumbria) are very much smaller, three of them comprising fewer than ten plants.

The main threat to existing colonies comes from two sources: destruction of the habitat by ploughing, and inadequate management, which allows scrub invasion and the uncontrolled growth of coarse grasses. *H. maculata* is destroyed by heavy trampling by humans or by cattle, but is at least partly resistant to fire. Although it is able to survive many years of competition from coarse grasses, it ceases to flower under those conditions and eventually dies. Management to enhance and maintain extant populations should include grazing by cattle or sheep (or, where this is not possible, cutting surrounding vegetation annually), the total exclusion of herbicides and artificial fertilisers and, where populations are very small, the reintroduction of plantlets grown from seed collected from plants from that site.

In continental Europe, it has a wider sociological range, occurring in maritime heath communities, in open grasslands on a variety of soils, in woodland clearings and at the edge of scrub communities in montane regions. It is found throughout central and eastern Europe extending into Asiatic Russia as far east as the Caucasus, central Siberia and the Altai. It reaches south to Spain, eastwards to Greece, and north to the Channel Islands, Denmark and southern Fennoscandia, where it is locally abundant in dry meadows, on calcareous slopes and occasionally in heath communities. On the coastal cliffs in Jersey, *H. maculata* grows in pockets of black, acid (pH 4.4-5.1), highly organic soil, which fills the crevices and shallow depressions between the granite rocks.

Further details on *H. maculata* are given in Wells (1976).

T.C.E. Wells

Isoetes histrix Bory (Isoetaceae)
Land quillwort
Status in Britain: LOWER RISK - Near Threatened.
Status in Europe: Not threatened. Endemic.

I. histrix occurs in Britain only on the Lizard peninsula. Here it favours three habitats: the coast path, and other paths where trampling is not too severe; coastal erosion pans with fragments of serpentine and skeletal soil that are wet and preferably flooded in winter but dry in summer; and, principally, pans of bare soil on rock outcrops, mainly with a southerly aspect. On the last, conditions of winter wet and summer drought are essential, as is rough grazing. In each of these habitats there is much bare soil, but *I. histrix* will survive in short turf at up to 80% cover, together with *Allium schoenoprasum, Herniaria ciliolata, Juncus capitatus, Lotus subbiflorus, Minuartia verna, Scilla autumnalis, S. verna, Trifolium bocconei* and *T. strictum*. It is predominantly a coastal plant and, apart from four sites at Black Head, near Coverack, all are on the western, more exposed side of the Lizard. A few populations occur inland on Goonhilly Down and elsewhere. All known sites are on serpentine, with the exception of one on schist and one on the junction of schist and serpentine.

This species is a small, rosette-forming perennial, growing from an underground 'corm'. It reproduces by spores of two types, which are produced in sac-like sporangia more or less embedded in the leaf (sporophyll) bases, and ripening in April and May. Some sporangia contain megaspores (up to 480 m diameter) which, on germination, form a 'prothallus' tissue which bears the female organs (archegonia). Other leaves bear similar sporangia containing dust-like microspores (35-40 m diameter); these spores produce male gametes, which fertilise the archegonia of the germinating megaspores. As spores are formed below the soil surface, their distribution depends on disturbance of the substrate. Spores can be long-lived; they have germinated in wetted soil samples that had been kept dry for 34 years (D.A. Coombe pers. comm.). Leaves emerge with autumn rains, typically in late September and October, and wither in late May or early June.

I. histrix was not discovered in Britain until 1919, and not confirmed until 1937. Prior to this it had been known, however, in Guernsey and Alderney. A comprehensive survey on the Lizard in 1982 showed it to be present in 27 sites in 16 1 km squares (within two hectads), with a total population of 100,000 plants (Frost *et al.* 1982). Since then, an additional small site has been found in a third hectad. Almost all populations are on NNRs, SSSIs or National Trust land, and most of its recorded sites are still extant. The few losses have been caused by fire, cultivation (two sites at Caerthillian with about 200 plants were ploughed before 1982) and a lack of grazing.

The reproduction process of *I. histrix* is assisted by grazing and disturbance. Grazing keeps the turf short, and trampling (by people and animals) spreads the spores or exposes them to the wind. The bulk of the population in

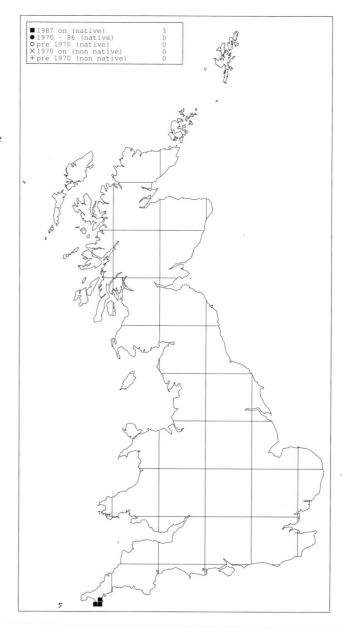

■ 1987 on (native)	3
● 1970 - 86 (native)	0
○ pre 1970 (native)	0
✕ 1970 on (non native)	0
+ pre 1970 (non native)	0

1982 (though only a small percentage of the total sites) was on areas dependent on rough grazing by livestock. Rough grazing has declined everywhere on the Lizard in the past twenty years, and it is significant that in the 1982 survey, four out of the five extinct populations were on sites where grazing had ceased. However, tourist pressure is at its heaviest from June to September, the dormant period of *I. histrix*, and the consequent trampling and scuffing of turf probably compensate to some extent for the lack of grazing.

I. histrix is principally a plant of Mediterranean coasts, where it is present in habitats that are wet in winter but dry out completely in summer. It also occurs along the Atlantic coast, reaching its northern limit in Britain.

A.J. Byfield and D.A. Pearman

Juncus capitatus Weigel (Juncaceae)
Dwarf rush, Corfrwenen
Status in Britain: LOWER RISK - Near Threatened.
Status in Europe: Not threatened.

In Britain, *J. capitatus* is now known only on the Lizard
peninsula and at a single locality in Anglesey. It is
typically a plant of open summer-droughted habitats, some
of which may be wet or shallowly flooded in winter. On
the Lizard it is more or less restricted to soils overlying
serpentine, though there is a tiny sporadic colony
associated with an outcrop of schist. Populations occur on
south-facing coastal slopes (occasionally on steep sites), in
short grassy turf, particularly on and around rock
outcrops, in serpentine rock-pans, occasionally on rock in
abandoned quarries and on rock outcrops in heathland. In
grassy turf, a rich assemblage of associates may include
many notable species including *Herniaria ciliolata*, *Isoetes
histrix*, *Minuartia verna*, *Moenchia erecta*, *Trifolium bocconei*, *T.
occidentale*, *T. ornithopodioides*, *T. strictum*, *T. subterraneum*,
and rare bryophytes including *Riccia beyrichiana* and *R.
nigrella*. *J. capitatus* is found, though less frequently, in
shallow rock pans on serpentine, in a distinctive
community together with *Allium schoenoprasum*, *Herniaria
ciliolata*, *Isoetes histrix*, *Minuartia verna*, *Plantago maritima*
and *Radiola linoides*. These rock pans are fed in winter by
seepages, which wash out finer soil particles leaving a
coarse gravelly substrate which dries out completely in
summer. A few populations also occur on ledges around
small abandoned quarries, and may mark the sites of
former undisturbed habitat for this species. At the recently
rediscovered site in Anglesey, *J. capitatus* occurs on the
margins of a seasonally flooded sand-dune hollow in very
open vegetation containing dune-heath species such as
Erica cinerea, *Lotus corniculatus* and *Salix repens* (Blackstock
& Jones 1997).

J. capitatus is a winter annual, germinating in autumn.
Capsules can be found as early as late March, but
normally mature in May, producing copious seed.
Individual populations may fluctuate from year to year,
plants being most abundant in warm wet springs,
particularly following an intense summer drought the year
before. Evidence suggests that dormant seed is likely to be
viable for long periods in the soil (Coombe 1987a).

In a typical year there are many thousands of plants of
J. capitatus on the Lizard, and more than a hundred
separate populations have been recorded. Many of
populations are small, some comprising fewer than ten
individuals, though others may number many hundreds in
good years. The Anglesey population discovered in 1995
was estimated to contain 10-20 plants. There are unreliable
records of this species from other parts of Cornwall and
the Hebrides, although it might once have occurred in the
Isles of Scilly. In its vegetative state, it is very similar to *J.
bufonius*, with which it has sometimes been confused. This
species also occurs on the Channel Islands in habitats
similar to those of the cove valleys of the Lizard.

On the Lizard, *J. capitatus* grows in areas of shallow soil
and uneven topography that have therefore escaped
agricultural reclamation, some now surviving as islands of

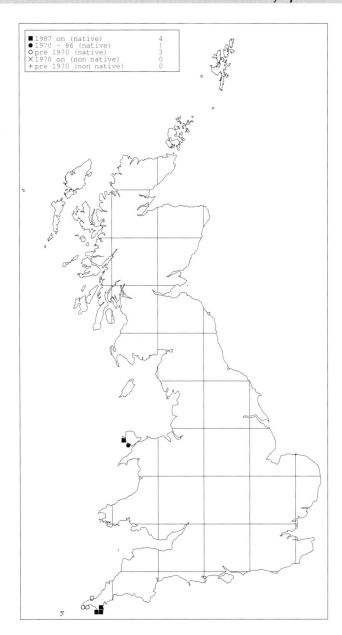

semi-natural vegetation amidst intensively managed
grassland. The continuity of grazing domestic stock is
important for the survival of many of its grassland
populations, and a cessation of grazing is likely to
encourage more vigorous plant competitors in the sward.
Losses in the past have been due to agricultural
intensification, quarrying, afforestation and building
development.

J. capitatus is found throughout most of Europe
(including the Channel Islands), though it occurs sparsely
through central Europe to southern Sweden, Finland and
western Russia. It also occurs in Africa, south-west Asia,
South America, Australia, and as a possible introduction in
Newfoundland. In view of its wide distribution in Europe,
its extreme restriction in western Britain is not easy to
explain.

J.J. Hopkins

Juncus pygmaeus L.C.M.Rich. (Juncaceae)

Juncus mutabilis Lam.
Pygmy rush
Status in Britain: ENDANGERED.
Status in Europe: Not threatened.

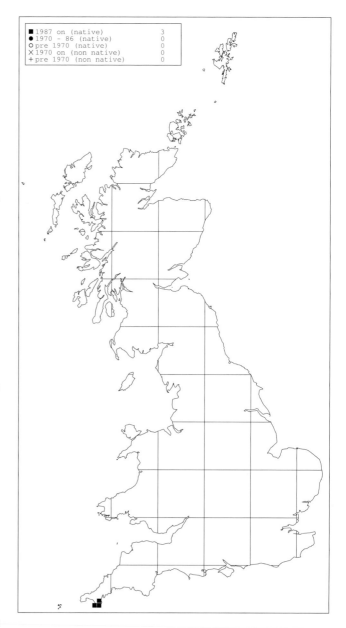

J. pygmaeus is a diminutive annual, restricted to the serpentine, gabbro and Crousa gravel heathlands of the Lizard peninsula. It is among the most threatened of the Lizard's special plants. Typically *J. pygmaeus* is a plant of seasonally-flooded rutted tracks and gateways, where the drier *Erica cinerea-Agrostis curtisii* 'short heath' grades into wetter *Erica vagans-Schoenus nigricans* 'tall heath'. On tracks it grows on shallow loess largely bare of other vegetation or in an open community of other small wetland species, including *Juncus bulbosus*. It is also more rarely recorded from the shallow margins of natural erosion pans, from cob and borrow pits, and in wet ground in serpentine quarries. In all instances, disturbance, typically rutting by vehicles on trackways (and, traditionally, poaching by cattle in gateways), is essential to the survival of this small annual. Such ephemeral habitats are of exceptional botanical importance, and rarer associates of *J. pygmaeus* include *Cicendia filiformis*, *Juncus capitatus* (infrequently), *Pilularia globulifera*, *Radiola linoides*, *Ranunculus tripartitus*, and nationally rare cryptogams such as *Cephaloziella dentata* and *Chara fragifera*.

The plant germinates in spring as water levels decline and temperature rises, maturing on the exposed mud in early summer. The rate and timing of the drop in water level are critical in determining the size and vigour of populations. In good years some of the larger Lizard colonies number thousands of individuals, whilst in poor years, for instance 1994, the plant may perform very badly, the total British population numbering only a few hundred plants.

Since 1950, *J. pygmaeus* has been recorded from about 17 1 km squares. However, since 1980 it has been reported from 13, and since 1990 from only five, only two of which lie within SSSIs. Some colonies are very small and sporadic in appearance. A main factor in its decline is the cessation of the traditional use of the ancient trackway systems, which rapidly leads to a growth of coarse perennials, including *Agrostis canina*, *Glyceria declinata*, *Juncus bulbosus* and *Scirpus fluitans*, and ultimately overgrowth by scrub species (e.g. *Salix cinerea*). However, seed may remain viable in the soil for very many years, and it is possible that *J. pygmaeus* might reappear if disused tracks are opened up. The infilling of track systems using hardcore (as witnessed recently on one protected site) is likely to lead to a permanent loss. There is an urgent need for the restoration of habitats suitable for this species.

Outside Britain, *J. pygmaeus* occurs locally in western and southern Europe northwards to Denmark, and in north-west Africa and Turkey.

A.J. Byfield

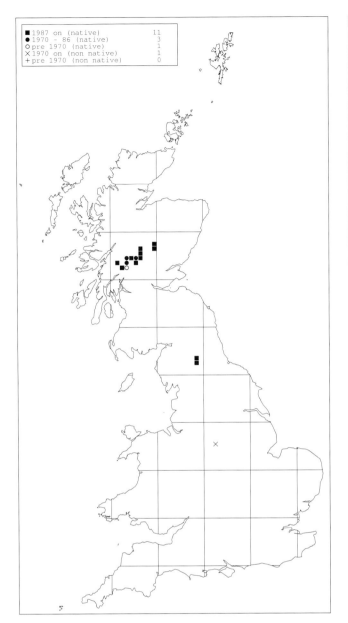

Kobresia simpliciuscula (Wahlenb.) Mackenzie (Cyperaceae)

False sedge
Status in Britain: LOWER RISK - Near Threatened.
Status in Europe: Not threatened.

This is a montane plant of calcareous stony and grassy flushes, small-sedge mires and, in Teesdale, sugar limestone grassland. In Scotland, the populations are centred on the Breadalbane mountains and on the limestones of Perthshire, the plant mainly occurring in mid- and east Perth, with outposts in Argyll. It ascends to over 1,000 m. In northern England, it occurs chiefly in Upper Teesdale, where it was first recorded in Britain in 1797. The underlying geology is invariably calcareous, the base-rock either limestone or calcareous mica-schist. Its communities are typically open and sedge-dominated, and grazed to a varying extent by sheep and/or red deer, leading to variable flowering. It grows in pure mats, or in herb-rich swards in flushed ground and mires with *Carex*

capillaris, C. dioica, C. hostiana, C. viridula ssp. *brachyrrhyncha, C. panicea, C. pulicaris, Juncus articulatus, J. triglumis, Molinia caerulea, Pinguicula vulgaris, Primula farinosa, Saxifraga aizoides, Selaginella selaginoides, Thalictrum alpinum* and *Tofieldia pusilla*. In Teesdale, it also occurs as isolated clumps with *Carex viridula* ssp. *brachyrrhyncha, Juncus articulatus* and *Minuartia verna* in open gravel flushes of sugar limestone (Pigott 1956).

K. simpliciuscula is a perennial that grows in rather dense tufts 5-20 cm in height bearing an inflorescence of 3-10 crowded spikes. It is wind-pollinated and flowers in June and July. Patches extend through vegetative growth, and individual plants are relatively long-lived. However, reproduction by seed can take place, germination occurring where the vegetation is open.

In Upper Teesdale, populations are large and extensive, comprising many thousands of plants, though a significant portion of the British population was drowned by the Cow

Green reservoir. In Perthshire, most populations are small, but in the area south of the River Tilt, there are several flush systems with hundreds of plants, and more than 10,000 at one site. There is a surprising lack of recent records from the central Breadalbanes, perhaps because it is generally assumed not to be threatened there. It is quite likely that *K. simpliciuscula* still occurs at most or all of its known sites, and that the map does not fully represent its current status.

A few populations of *K. simpliciuscula* are subject to heavy grazing, which can, however, be tolerated for many years, the plants reproducing vegetatively and producing few, if any, flowers. Its range appears not to have altered in recent years, and it is not regarded as seriously under threat. Sites are generally too high for afforestation and agricultural land claim, and most of its locations lie within SSSIs.

K. simpliciuscula also occurs in the Pyrenees, Alps, Caucasus and Altai mountains, in Scandinavia, Greenland and North America.

R.A.H. Smith

Koeleria vallesiana (Honck.) Gaudin (Poaceae)
Somerset hair-grass, Cribwellt Oddfog
Status in Britain: LOWER RISK - Near Threatened.
Status in Europe: Not threatened.

K. vallesiana has a remarkable history in Britain. It was
first discovered by Dillenius in 1726 at Uphill and Brean
Down, but his notes and specimens subsequently became
separated. It was collected again in 1773 by Rev. Lightfoot,
but its identity remained unknown, and it was not until
1904 that Druce, working on the memoirs of Dillenius,
made the connection between notes and specimen and was
able to verify the presence of the grass at Uphill. Although
not a particularly critical species, it was not recorded
further east of Crook Peak in the Mendips until some 50
years later, and was recorded at Sand Point for the first
time as recently as 1974.

It grows in thin, open, often rocky turf on generally
south-facing Carboniferous limestone slopes. Although its
sites experience mild, wet winters, the summers are harsh,
with high levels of insolation, low rainfall and parched soils.
The limestone turf is characterised by such typical associates
as *Pilosella officinarum, Sanguisorba minor* and *Thymus
polytrichus* dispersed around tussocks of *Festuca ovina*.
Particularly striking is the representation of annuals and
pauciennials, including *Blackstonia perfoliata, Carlina vulgaris,
Centaurium erythraea* and *Euphrasia nemorosa. Carex humilis*
and *Potentilla neumanniana* also occur in the community.

It is a tufted perennial grass, superficially similar to the
widespread *K. macrantha*, but its stems are swollen at the
base, with a thick covering of the fibrous remains of old
leaf shoots. Flowering is from late May to July.

The distribution of *K. vallesiana* closely matches that of
Helianthemum apenninum and *Trinia glauca*. It is currently
known at about 20 sites in 11 1 km squares on the
southern slopes of the Mendip hills, from Brean Down on
the coast inland to Shute Shelve and Fry's Hill. In recent
years, intensive searches have shown that this species has
been lost from four further sites. In at least six of its major
sites populations of *K. vallesiana* number in the thousands.
Most sites are adequately protected, though not all are
grazed by livestock. Sheep are ideal grazers, but it is clear
that rabbits are important in maintaining a thin, open turf.
It is likely that in the more steep and rocky locations the
extreme microclimate is sufficient to maintain suitable
conditions in the absence of grazing. The species is thus in
little immediate danger. It has been introduced to Goblin
Combe (Hope-Simpson 1987).

The hybrid with *K. macrantha* has been reported from
most of the *K. vallesiana* localities. It is sterile and
apparently not known outside Britain (Stace 1991).

K. vallesiana has its stronghold in southern Europe from
the Iberian peninsula to Italy, and also occurs in north-
west Africa. It reaches its northern limit in England and is
rare in Switzerland and endangered in France.

R.D. Porley

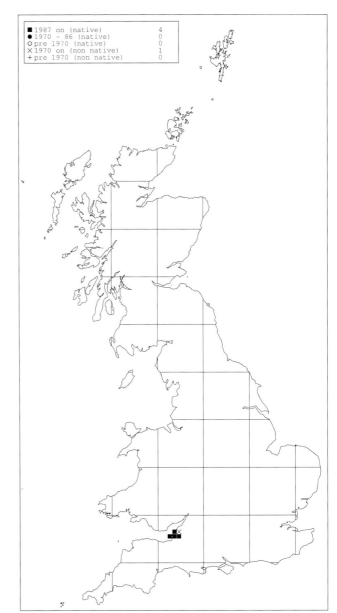

■ 1987 on (native)	4
● 1970 - 86 (native)	0
○ pre 1970 (native)	0
✕ 1970 on (non native)	1
+ pre 1970 (non native)	0

Koenigia islandica L. (Polygonaceae)

Iceland purslane, Cainigidhe
Status in Britain: LOWER RISK - Near Threatened.
Status in Europe: Not threatened.

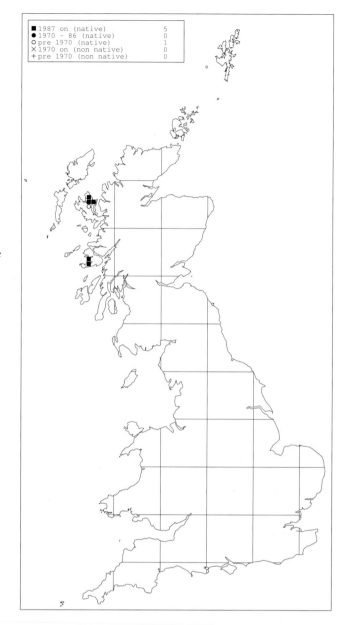

K. islandica is one of three important additions to the British montane flora that were made in the 1950s, the other two being *Artemisia norvegica* and *Diapensia lapponica*. The plant had, however, been collected in Skye in 1934, but was mislabelled as *Peplis portula* in the Kew herbarium, and its true identity not confirmed until 16 years later, in 1950. On Skye, *K. islandica* occurs in scattered localities along the Trotternish ridge, where it is chiefly confined to wet, gravelly flushes and screes of northerly and easterly aspects. However, on the highest ground near the summit of the Storr, and occasionally on Mull, it occurs on gentle south-facing slopes. In flushes the most constant associated plants are *Carex viridula* ssp. *oedocarpa*, *Deschampsia cespitosa*, *Saxifraga stellaris* and bryophytes, whilst on screes it is accompanied by such species as *Arabis petraea*, *Festuca vivipara*, *Juncus triglumis*, *Luzula spicata*, *Minuartia sedoides*, *Oxyria digyna* and *Sagina subulata*.

The plant is a small, branched annual about 1-3 cm high, with reddish stems, and superficially similar to *Montia fontana* and *Peplis portula*. The flowering period is July and August, or sometimes September, and in autumn the whole plant develops a red coloration. At some sites it can occur in an abundance reminiscent of ruderal annuals: over 250 plants per square metre have been recorded on Mull. It is particularly sensitive to high temperatures (Gauslaa 1984), which may account for its predominance on the colder north- and east-facing slopes. This species once had a wider distribution, its pollen having been detected in late-glacial deposits in the south of Scotland, Cumbria, the Isle of Man and Northumberland.

Though confined to a few hills on Skye, and the Ardmeanach peninsula on Mull, *K. islandica* is not obviously threatened at the present time. The sensitivity of *K. islandica* to higher temperatures would make it very vulnerable were any climate change to result in increased temperatures and dryness of its habitat. Indeed, rapid and locally severe declines on Mull since 1994 (in one area a reduction of 80%), documented by B. Meatyard, are believed to be linked to the recent run of dry springs. Whether this marks a real shift in weather patterns remains to be seen.

K. islandica is one of a very few annual plants with a circumpolar distribution. In the northern hemisphere, it is found throughout the Arctic and sub-Arctic regions, in Europe occurring in Iceland, the Faeroe Islands, northern Fennoscandia, Svalbard and northern Russia. It is legally protected in Sweden. It is also found in the mountains of central Asia, and in the Rocky Mountains of North America. *K. islandica* also occurs in the southern hemisphere, since *K. fuegiana*, of Tierra del Fuego, is now regarded as conspecific.

P.S. Lusby

Lactuca saligna L. (Asteraceae)
Least lettuce
Status in Britain: ENDANGERED. WCA Schedule 8.
Status in Europe: Not threatened.

L. saligna is now known in only two localities - at Rye
Harbour, Sussex, and Fobbing, Essex. At Rye it is largely
restricted to patches of disturbed sandy shingle, either
where sand has been added as a stabiliser (as along the
old railway line), or in slight hollows where shingle has
been removed for sea-wall repairs, and along the margins
of a metalled road. The vegetation is sparse, and
associates, including *Arrhenatherum elatius*, *Galeopsis
angustifolia*, *Geranium robertianum* and *Plantago lanceolata*,
are hardly competitors. At Fobbing, the bulk of the
population occurs on an old sea-wall, now some distance
from the sea, that was constructed in stepped layers from
marine alluvium. Here also the vegetation cover is patchy
with much bare soil. The most frequent associates are
Bromus hordeaceus, *Crepis capillaris*, *Leontodon autumnalis* and
Lolium perenne, but *Bupleurum tenuissimum* and *Torilis
nodosa* are locally common. Other smaller populations
occur on irregular mounds of alluvium, bucket-dredged
from Fobbing Creek, and one colony has established itself
recently on the sea-wall of nearby Canvey Island.

It is an overwintering or spring-germinating annual.
The flowers, produced from July to September, close before
midday and are self-pollinated. Seed is believed to be
viable for only about one year (Prince & Hare 1981) and
both overwintering and spring-germinating plants are
vulnerable to hard winters. Under ideal conditions, plants
can become basket-like, producing numerous flowering
stems up to a metre high and bearing upwards of 1,000
achenes; however, most plants are smaller. A poor
competitor under normal conditions, its ability to survive
the hottest summers in dry exposed situations gives it a
competitive edge and probably explains the upturn in
numbers in recent years. At Rye, the bulk of the
population consists almost entirely of tiny entire-leaved
plants, but the few surviving the winter become taller and
more bushy. At Fobbing the plants appear to be tillered by
grazing, and have several flowering stems ascending
laterally from a central rosette. In damp weather the
capitula fail to open, entrapping the developing seeds, and
biting insects also attack the flowering heads making them
sticky with droplets of latex, again preventing many of
them from releasing their achenes.

L. saligna has been recorded as a presumed native from
Middlesex, Kent, Essex, Cambridgeshire, Norfolk, Suffolk
and Sussex. Its strongholds were the low crumbling
London Clay cliffs, sea-walls of the Thames estuary, the
river embankments of the fen country in eastern England,
and maritime shingle on the south coast. However, since
1950 the refurbishment of sea-wall defences has resulted in
a catastrophic decline. At Rye the population is scattered
over a large area, and the tiny plants are easily overlooked.
Permanent quadrats installed there in 1991 indicated a
total population that year of 50,000 plants, but following a
dramatic decline, only about 1,000 plants remained in 1995.
Inundation of its habitats by sea water, which became

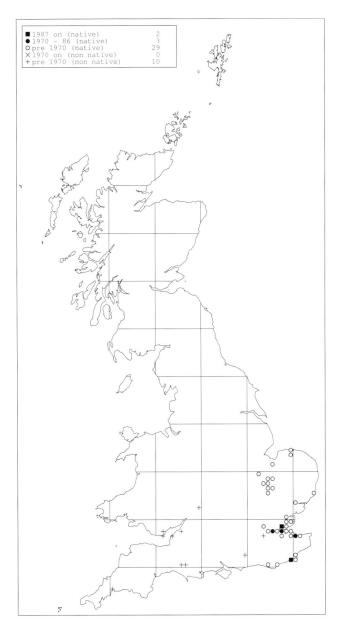

impounded behind the sea-wall during onshore storms,
appears to have been the main cause of this decline. In
contrast, the population at Fobbing has increased from
about 6,800 plants in 1978 (Hare 1986) to 14,500 in 1994
(Otley College 1990; Adams 1994). About half the
population occurs on a 600-m segment of sea-wall that is
regularly grazed, and forms the more or less stable core
area and upwind seed source. Other smaller sub-
populations fluctuate in numbers depending on the level
of grazing. Though currently known only from Rye and
Fobbing, it is possible that small populations remain
undetected along the many km of remote sea and river-
walls of south-east England.

At Rye, disturbance of the sandy shingle above the
level of saltwater flooding is advantageous, and fencing
against hares and rabbits is beneficial. At Fobbing,
trampling and grazing by cattle are essential to its survival
there, and on areas of sea-wall where grazing is excluded

L. saligna is quickly ousted by *Arrhenatherum elatius* and other robust competitors.

L. saligna is widely distributed in Europe, most frequent in the south, but ranging northwards to the Netherlands, Germany, Denmark and northern Russia. It extends eastwards to the Altai and Himalayas and southwards to North Africa, from the Canaries to Ethiopia. It has been introduced to North America.

K.J. Adams and A.D.R. Hare

Lavatera cretica L. (Malvaceae)

Cretan mallow, Môr-Hocysen Fychan
Status in Britain: VULNERABLE.
Status in Europe: Not threatened.

In Britain, *L. cretica* is regularly recorded only from the Isles of Scilly. It occurs mainly in open habitats including bulb fields, roadsides, old quarries and disturbed ground. Among its common associates in the Isles of Scilly are *Allium triquetrum*, *Chrysanthemum segetum*, *Erodium moschatum*, *Gladiolus byzantinus*, *Montia perfoliata*, *Oxalis pes-caprea* and *Silene gallica*. It grows with *Fumaria occidentalis* in at least one site.

In Scilly, this annual species most commonly germinates in autumn and winter, with flowering beginning in April and the plants dying off in late June or early July (Lousley 1971). However, germination may take place at other times, and the occasional flowering plant can be seen at almost any time of year. In the shelter of walls or hedges the plants can grow to over a metre tall, but in cultivated fields the plants usually get sprayed with herbicide or ploughed up before they have a chance to grow tall and set seed.

Since its discovery, this species has been found on only three of the inhabited islands in the Isles of Scilly: Tresco, St Mary's and St Agnes (with one record on St Martin's in 1989 but not since), and has not spread to the other islands. Its range has not changed markedly over the past hundred years, except for a slight spread on St Mary's, though the plant has undergone severe fluctuations in abundance (Lousley 1971). At present, populations in the Isles of Scilly appear to be relatively stable. It occurs only sporadically on the Cornish mainland, with recent records from the Helford river (Margetts & David 1981) and between Penzance and Marazion (Margetts & Spurgin 1991). The population in Tenby is thought to have originated from introduced plants, and it has persisted there since about 1941 (Jones 1992). Some authorities (e.g. Stace 1991) regard *L. cretica* as native in the Isles of Scilly and perhaps in West Cornwall, whilst others (e.g. Webb 1985) consider it to be an introduction in Britain.

At one time in the Isles of Scilly, the plant seemed likely to be lost from many of its cultivated habitats. It suffered from the widespread use of herbicides intended to eliminate arable weeds from the cultivated crop. Lately plants have recolonised from remnant populations in hedges and field margins, and the species is now common again in some of its old localities. However, *L. cretica* remains vulnerable, as it does not seem to have increased its range, despite apparently suitable habitat, and in many situations it is mistaken for the common *Malva sylvestris*, a rather more showy plant considered a pest by farmers.

L. cretica is predominantly a plant of southern Europe, extending eastwards from the Azores across the Mediterranean region to Turkey and the Near East. It also extends northwards along the west and north coasts of France to the Channel Islands.

R.E. Parslow

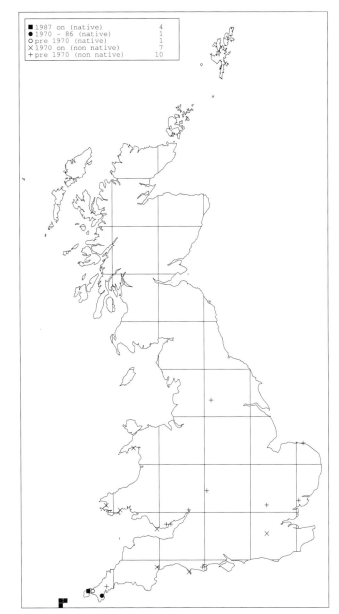

■ 1987 on (native)	4
● 1970 – 86 (native)	1
○ pre 1970 (native)	1
✕ 1970 on (non native)	7
+ pre 1970 (non native)	10

Leersia oryzoides (L.) Sw. (Poaceae)
Cut-grass
Status In Britain: ENDANGERED. WCA Schedule 8.
Status in Europe: Near threatened?

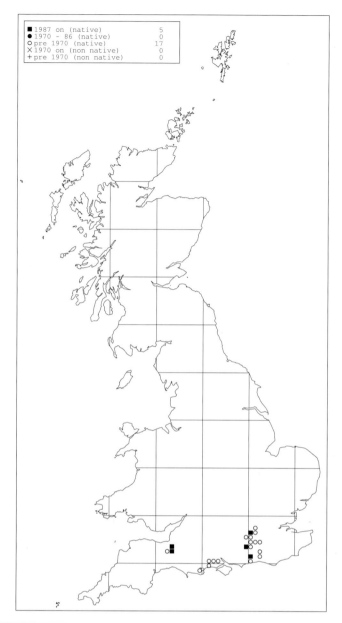

L. oryzoides occurs on the margins of drainage ditches, dykes, canals and ponds, and formerly also in marshy fields. It grows on nutrient-rich mud of acid to neutral pH, often close to the water's edge (or occasionally as an emergent), and seems to do best in a rather open community, sometimes in cattle-poached ground. It may occur amongst emergent perennials such as *Alisma plantago-aquatica, Glyceria fluitans, G. maxima, Phalaris arundinacea* and *Sparganium erectum*, or lower growing perennials or mud-annuals including *Agrostis stolonifera, Bidens cernua, Deschampsia cespitosa, Galium palustre, Lycopus europaeus, Lythrum salicaria, Mentha* x *verticillata, Persicaria hydropiper* and *Stachys palustris*. Though the habitats in Britain are quite varied, features in common include the presence of nitrogen-rich mud, the proximity of stagnant or slow-flowing water, seasonal inundation, and regular disturbance, which provides areas of bare mud and maintains an open vegetation structure (Birkinshaw 1990b).

This rhizomatous perennial forms loose tufts and patches. It is distinctive when seen at close quarters, and FitzGerald (1988c) notes that "not only the startling cutting power of the finely-spined leaf margins, but the characteristic yellow-green colouring, and the jaunty angle made by the single leaf which tops each stem, become easily recognisable when surveying the plant", but it can be readily passed by if not looked for specially. In Britain it has a long dormant period, and starts to shoot only in late spring. In most years the panicles are not exserted from the sheaths, and are then cleistogamous. Only in years with high temperatures throughout the spring and summer will panicles open fully for cross-pollination, and fruits do not mature until well into September in our climate. Birkinshaw (1990b) has shown that warm summer temperatures may be more important than warm spring temperatures in promoting this development.

Since 1985, *L. oryzoides* has been recorded at only four sites in Britain. Two are in West Sussex (Shillinglee Lake, where it grows along 120 m of shore-line, and Amberley Wild Brooks), one in Surrey (the Basingstoke Canal at Sheerwater); and one in Somerset, by the Taunton-Bridgwater Canal. The only sizeable population in Britain is at Amberley Wild Brooks, where it occurs in many ditches over an area of 300 hectares. The Sheerwater colony, already very tiny, was destroyed during canal re-construction, but in anticipation of this, plants were successfully transplanted to three other sites by the Basingstoke Canal. In West Sussex, *L. oryzoides* is no longer known in Arundel Park (where its pond was filled in), and has not been seen since 1982 at Waltham Brooks. In Somerset, it was discovered in 1959 near North Newton in several localities along the Taunton-Bridgwater canal, on both sides of a hectad boundary. It was still present at most, if not all, of these in 1981, but by 1988 only one clump remained. It has not been seen there since 1990 and may now be extinct. Recent searches confirm its extinction

in Dorset, and attempts in 1988 and 1990 to re-establish the plant in the New Forest seem to have been unsuccessful.

Drainage of wetland habitats, the cessation of traditional methods of water-course management, and the pollution of waterways have contributed to the decline of *L. oryzoides*. Regular dredging of water courses provides areas of mud for colonisation, and this has been carried out at Amberley Wild Brooks for many years. The maintenance of high water quality is also essential. At Amberley, monitoring of *L. oryzoides* over the past 15 years has shown that the plant grows best on banks of managed, clean-water ditches that are trampled and grazed by cattle.

L. oryzoides has been recorded from southern Finland to Spain and eastwards to temperate Asia and North America. It has markedly decreased throughout the whole of Europe.

M. Briggs and P.A. Harmes

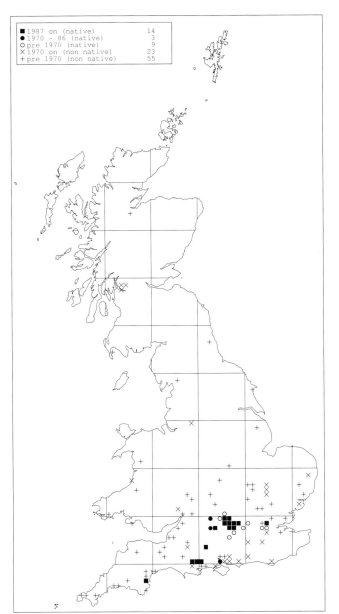

Leucojum aestivum L. ssp. *aestivum* (Liliaceae)

Summer snowflake, Eiriaidd
Status in Britain: LOWER RISK - Near Threatened.
Status in Europe: Not threatened.

Most populations of *L. aestivum* ssp. *aestivum* occur in *Salix* carr or *Alnus glutinosa* woodland along muddy river banks and ditches or on islets in rivers, often in areas that are flooded in winter. In most sites, associates are generally a few other spring-flowering species such as *Arum maculatum* and *Ranunculus ficaria*, with *Galium aparine* and *Urtica dioica* usually present, but at some sites it occurs in a more diverse community (FitzGerald 1990d). Other wetland associates may include *Caltha palustris*, *Carex riparia*, *Filipendula ulmaria*, *Glechoma hederacea*, *Oenanthe crocata*, *Phalaris arundinacea*, *Phragmites australis*, *Ribes sylvestris*, *Rumex sanguineus* and *Symphytum officinale* (Farrell 1979). A few populations occur in other habitats, including damp woodland rides and under hedges.

L. aestivum ssp. *aestivum* is a perennial, with leaves appearing above ground in February or March. The flowering period is generally from late March into April, but it may be earlier in a mild year. The large seeds are thought to be dispersed by water.

Most of its sites are by the Thames and its tributaries, where it is found in more than 40 1 km squares in two main areas: between Reading and Windsor, and between Goring and Abingdon. A survey in the 1970s showed that over 100,000 clumps, 77% of the total British population, were contained within six sites on the Thames from Reading to Marlow, and by Loddon near Twyford. Most of its sites hold at least several hundred clumps, though a small number may hold as few as a single clump (Farrell 1979). Outside its main area, native populations occur along the Stour valley in Dorset (several thousand plants), at a few sites in Wiltshire (FitzGerald 1993), at Littlehempston in Devon (more than 4,000 plants), and

211

perhaps a site in Monmouthshire. In the past there has been uncertainty about the status of *L. aestivum* in Britain, mainly because the morphological differences between the native ssp. *aestivum* and the introduced ssp. *pulchellum* were not fully understood. The latter has smaller flowers and a narrower spathe and flowers later, and the edges of the scape are not denticulate. Subspecies *pulchellum* is widely naturalised, mainly in southern Britain but extending northwards to East Inverness.

Many sites for ssp. *aestivum* are vulnerable to development, with possible threats stemming from the construction of artificial riverbanks to control erosion, the straightening of river courses and the development of

marinas and boat-moorings. Trampling of river bank vegetation could pose an additional threat, and the digging up of bulbs would be particularly damaging to the smaller colonies.

L. aestivum is widely distributed in marshes and wet meadows in Europe, from Ireland eastwards to the Netherlands and the Czech Republic, and southwards to Sardinia, Greece and the Crimea. It is legally protected in Austria, Belgium, France, Germany, Netherlands and Switzerland. A decline is reported in Russia and eastern Europe, mainly because of drainage and river engineering.

L. Farrell and M.J. Wigginton

Limonium bellidifolium (Gouan) Dumort.
(Plumbaginaceae)
Matted sea-lavender
Status in Britain: LOWER RISK - Near Threatened.
Status in Europe: Not threatened.

L. bellidifolium occurs in an intermediate zone between saltmarsh and sand-dune, growing on firm sandy/silty substrates. It grows best where the base is distinctly shingly, and in such sites frequently occurs in association with *Frankenia laevis*, the two species sometimes forming an almost complete sward. *Suaeda vera* is a potential competitor there, but in that habitat it seldom occurs as more than scattered plants. On sandy substrates *L. bellidifolium* often occurs with *L. binervosum* ssp. *anglicum*, but wind-blown sand is an ever present threat, and once above the reach of the highest tides, *L. bellidifolium* fails to thrive. It is able to colonise muddy ground, but after a few years is usually ousted by the coarser growth of *Suaeda maritima* and *Salicornia* species, which not only shade it out but also trap and build up the mud, making the site unsuitable.

It is a perennial, deciduous herb, usually profusely branched, and with a matted, wiry appearance. The leaves wither early, and by the time the flowers appear in July, the plant is usually leafless. Reproduction is by seed and by vegetative spread.

This species was formerly found in Cambridgeshire, where it was last seen in the nineteenth century, and Lincolnshire, where it was last seen, at Gibraltar Point, in 1967. Almost all remaining sites are in West Norfolk, from Holme-next-the-Sea to Blakeney, though it just extends into East Norfolk. It has been recorded in 20 1 km squares since 1980. At some sites, the disposition and extent of the populations of *L. bellidifolium* do not remain static, but some colonies may disappear, whilst others appear and spread. Such changes occur in response to the shifting of sand, particularly during northerly gales. Sand deposited at the edge of the saltmarsh, trapped by *Suaeda vera*, may smother colonies of *L. bellidifolium*, becoming colonised instead by sand-dune plants such as *Elytrigia atherica*. This has occurred at several sites on the Norfolk coast, but sand removal elsewhere often compensates for this and creates new sites for *L. bellidifolium*.

The main human threat comes from trampling in the more popular areas, the tendency being to choose the easiest walking, which is on the firm drier areas at the edge of the saltmarsh. The loss of the plant at Hunstanton is likely to have been from this cause. Vehicles, though few, follow the same firm track. Currently, trampling and vehicles are only minor problems on the Norfolk coast, and as almost all existing sites are within an NNR or on

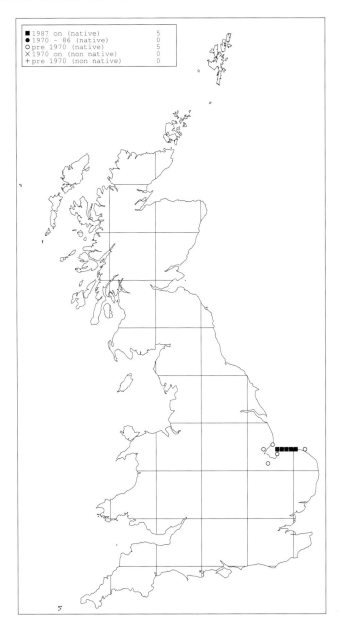

■ 1987 on (native)	5
● 1970 – 86 (native)	0
○ pre 1970 (native)	5
✕ 1970 on (non native)	0
+ pre 1970 (non native)	0

National Trust property, further damaging tourist developments are unlikely. However, increased storminess and sea-level rise, both possible consequences of global warming, could pose a serious threat.

L. bellidifolium occurs along the coasts of western and southern Europe, from England southwards to Spain and eastwards to Romania, Greece and the Crimea. In southern Russia it also occurs inland on saline soils.

G. Beckett

Limonium binervosum agg. (Plumbaginaceae)
Rock sea-lavender, Lafant y Morgeigiau

The rock sea-lavenders have long been regarded as a difficult and somewhat 'critical' group. Until recently the *L. binervosum* aggregate was considered to comprise four species in Britain:- *L. binervosum sensu stricto* - acknowledged as being "very variable ... with many named varieties" (Clapham *et al.* 1987) - and three rare endemics, *L. paradoxum, L. recurvum* and *L. transwallianum*. However, a recent thorough revision of the complex (Ingrouille & Stace 1986) has redefined *L. binervosum sensu stricto* and delimited a further five species (*L. britannicum, L. dodartiforme, L. loganicum, L. parvum* and *L. procerum*) and numerous infraspecific taxa.

Of the nine species now recognised, six qualify for inclusion in the Red Data Book, while the remaining three (*L. binervosum sensu stricto, L. britannicum* and *L. procerum*) also qualify in that they each have one or more nationally rare subspecies. Almost all the taxa included here are British endemics, only *L. binervosum sensu stricto* being known to occur elsewhere in Europe (north-west France).

A detailed taxonomic account of the complex is given by Ingrouille & Stace (1986) and is usefully summarised by Stace (1991). Readers are also referred to the *L. binervosum* agg. species account and maps given in Stewart *et al.* (1994).

S.J. Leach

Limonium binervosum (G.E. Smith) Salmon (Plumbaginaceae)

Rock sea-lavender, Lafant y Morgeigiau
Status in Britain: LOWER RISK - Nationally Scarce. Two subspecies (*cantianum*, *mutatum*) are considered to be VULNERABLE, and three subspecies (*binervosum*, *anglicum*, *saxonicum*) are currently considered to be LOWER RISK. All subspecies except ssp. *binervosum* are ENDEMIC. Status in Europe: Uncertain. Endemic.

L. *binervosum sensu stricto* occurs on the coasts of eastern England between East Sussex and Lincolnshire and (as ssp. *mutatum*) at one outlying site in south-west England (Stewart *et al.* 1994). Ingrouille & Stace (1986) recognised five subspecies in Britain (a sixth, ssp. *sarniensis*, is restricted to the Channel Islands), all of which appear to qualify for inclusion in the Red Data Book. Nevertheless, subspecies are extremely difficult to distinguish, and some taxa are almost certainly under-recorded (particularly ssp. *saxonicum* and ssp. *anglicum* in East Anglia).

The subspecies have reasonably well defined geographical limits and favoured habitats: ssp. *binervosum* on chalk cliffs and saltmarshes in East Sussex and south-east Kent; ssp. *cantianum* in similar habitats in north-east Kent; ssp. *saxonicum*, typically on shingle-sand-dune-saltmarsh transitions in north-east Essex (and possibly elsewhere in East Anglia (P. Sell pers. comm.); ssp. *anglicum* on sand-dune-saltmarsh transitions in Norfolk and Lincolnshire; and ssp. *mutatum*, restricted to coastal rocks at Lannacombe, South Devon.

L. *binervosum sensu stricto* is characteristically found in patchy vegetation close to saltmarsh-sand-dune interfaces and in tidally inundated 'slacks', often where blown sand and inwashed silt overlie shingle. In north-east Essex ssp. *saxonicum* occurs at Hamford Water and Colne Point with species such as *Catapodium marinum*, *Parapholis incurva* and *P. strigosa*. In Norfolk ssp. *anglicum* is associated with *Armeria maritima*, *Atriplex portulacoides*, *Puccinellia maritima*, *Suaeda vera* and (locally) *Frankenia laevis* and *Limonium bellidifolium*.

On many saltmarsh sites, grazing, especially by rabbits, is important in keeping the vegetation open; reduction of grazing can lead to dominance by such species as *Atriplex portulacoides*, *Elytrigia atherica* and *Suaeda vera*. At other sites L. *binervosum sensu stricto* populations may be periodically subject to 'roll-over' of adjoining sand-dunes. Whilst this may destroy some plants, it also creates new habitat for colonisation, although in the long term progressive sea-level rise may pose a significant threat.

C. Gibson and S.J. Leach

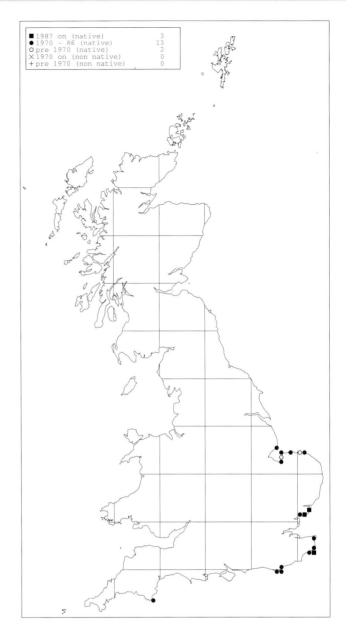

■ 1987 on (native)	3
● 1970 – 86 (native)	13
○ pre 1970 (native)	2
✕ 1970 on (non native)	0
+ pre 1970 (non native)	0

Limonium britannicum Ingrouille
(Plumbaginaceae)

Rock sea-lavender, Lafant Prydeinig

Status in Britain: LOWER RISK - Nationally Scarce.
ENDEMIC. Four subspecies (*britannicum, coombense, transcanalis* and *celticum*), all of which currently considered to be LOWER RISK - Near Threatened or Nationally Scarce.

L. britannicum is one of the more widespread members of the *L. binervosum* aggregate, occurring in scattered localities in western Britain between South Devon and Lancashire (Stewart *et al.* 1994). Four subspecies have been distinguished, all of which currently qualify for inclusion in the Red Data Book. However, the subspecies (and their numerous named varieties) are extremely difficult to separate in the field, and are undoubtedly under-recorded.

The subspecies are readily defined on the basis of their currently known geographical limits: ssp. *britannicum* on sea-cliffs on the north coast of Cornwall; ssp. *coombense* in similar habitats in South Devon (Bolt Head-Prawle Point) and the south coast of East Cornwall; ssp. *transcanalis* on saltmarshes, shingle and sea-cliffs in North Devon and Pembrokeshire (perhaps elsewhere in South Wales); and ssp. *celticum* on saltmarshes and sea-cliffs in Caernarvonshire, Anglesey, Cheshire and Lancashire.

It is anticipated that further investigations will increase the number of known sites for the *L. britannicum* subspecies. There are numerous populations of *L. binervosum* agg. in western Britain still requiring critical examination, and it is likely that many of these would be referable to this species or *L. procerum* ssp. *procerum*.

S.J. Leach

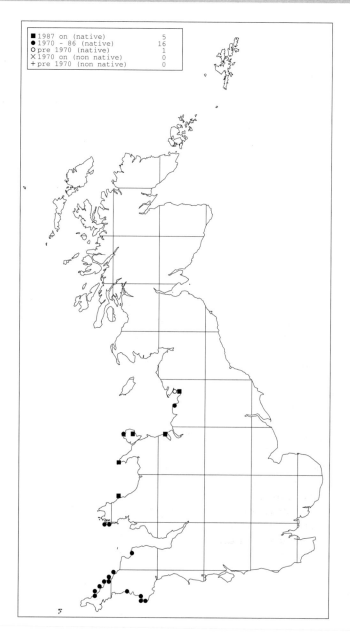

■ 1987 on (native)	5
● 1970 - 86 (native)	16
○ pre 1970 (native)	1
✕ 1970 on (non native)	0
+ pre 1970 (non native)	0

Limonium dodartiforme Ingrouille
(Plumbaginaceae)
Rock sea-lavender
Status in Britain: VULNERABLE. ENDEMIC.

This is one of the more easily distinguished members of the *L. binervosum* aggregate, with a basal rosette of relatively large leaves and somewhat lax flowering spikes on tall, often well-branched stems (Stace 1991). *L. dodartiforme* is restricted to Dorset where, apart from *L. recurvum* on the Isle of Portland, it appears to be that county's sole representative of the *L. binervosum* aggregate (Stewart *et al.* 1994). However, some *L. binervosum* agg. populations in Dorset have yet to be critically assessed (the apparent absence there of *L. procerum* is particularly surprising, especially as it occurs just over the border at Beer Head, Devon).

L. dodartiforme occurs in two quite different habitats. There are numerous colonies (thousands of plants) on chalk sea-cliffs between White Nothe and Arishmell; it favours cliff-faces with large expanses of bare chalk, and in such situations has few associated species (e.g. *Brassica oleracea*, *Crithmum maritimum* and *Spergularia rupicola*). It also occurs on stabilised shingle at Chesil Beach, where recent surveys confirm the existence of six colonies (of up to 20 plants) between Small Mouth and Langton Herring. Shingle associates include *Armeria maritima*, *Cerastium diffusum*, *Cochlearia danica*, *Festuca rubra*, *Geranium robertianum*, *Plantago coronopus*, *Silene uniflora* and the lichen *Cladonia furcata*.

All currently known populations lie within SSSIs and are not considered to be under significant threat. It is possible that examination of other, as yet undetermined, *L. binervosum* agg. in Dorset would reveal further populations of *L. dodartiforme*.

S.J. Leach and D.A. Pearman

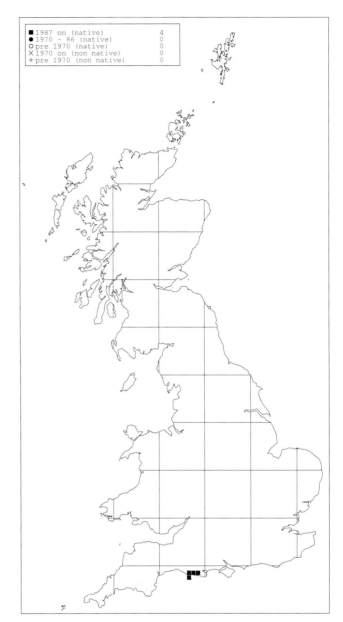

■ 1987 on (native)	4
● 1970 – 86 (native)	0
○ pre 1970 (native)	0
✕ 1970 on (non native)	0
+ pre 1970 (non native)	0

Limonium loganicum Ingrouille
(Plumbaginaceae)
Rock sea-lavender
Status in Britain: VULNERABLE. ENDEMIC.

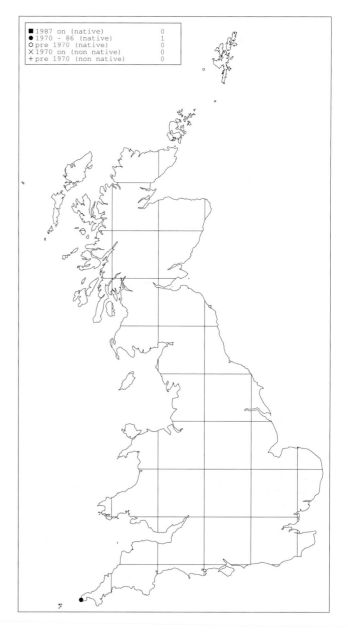

L. loganicum is restricted to a short stretch of coast to the south of Land's End, Cornwall: at Logan Rock (Treen Cliff), Porthgwarra, Gwennap Head and Carn Boel (Nanjizal). It occurs there on granitic rock outcrops, cliffs and coastal scree where its associates include *Armeria maritima*, *Crithmum maritimum*, *Inula crithmoides*, *Plantago coronopus* and *Spergularia rupicola*.

This is a medium-sized, well-branched species, distinguished from its congeners by the relatively lax flowering spikes and rough stems (Ingrouille & Stace 1986) - cf. *L. britannicum* and *L. procerum*, which are the only other segregates known to occur in Cornwall (Stewart *et al.* 1994). *L. loganicum* appears to be the sole representative of the *L. binervosum* aggregate in West Penwith (Margetts & Spurgin 1991), and further localities for it may yet be found. Recent records of '*L. binervosum*' from Pellitras Point and Porth Chapel (St Levan) are likely to be referable to this species.

L. loganicum is possibly the most endangered of the British rock sea-lavenders (M.J. Ingrouille pers. comm.). All known populations lie within SSSIs, but the area is a popular climbing resort and many colonies are very small and vulnerable to trampling.

S.J. Leach and R.J. Murphy

Limonium paradoxum Pugsley
(Plumbaginaceae)
Rock sea-lavender, Lafant Tyddewi
Status in Britain: VULNERABLE. ENDEMIC.

This is a rather short, well-branched species, similar in many respects to *Limonium britannicum*, but distinguished by its 'knotted' spikes, clustered spikelets and relatively long outer bracts (M.J. Ingrouille pers. comm.). *L. paradoxum* is a long-recognised segregate of the *L. binervosum* group (Pugsley 1931), confined in Britain to a single locality: along about 1 km of coast between St David's Head and the east side of Porthmelgan Bay, Pembrokeshire. In the past it was also recorded from sea-cliffs near Malin Head, East Donegal (Ireland), but these populations are now considered to be a separate taxon (Ingrouille & Stace 1986).

At St David's Head, *L. paradoxum* occurs on basic igneous rock outcrops and cliffs, occurring in more or less accessible sites at the cliff-edge, and in inaccessible ones on cliff-ledges. Associated species include *Agrostis stolonifera, Anthyllis vulneraria, Armeria maritima, Atriplex prostrata, Crithmum maritimum, Cochlearia officinalis, Festuca rubra, Limonium procerum, Plantago coronopus, Silene uniflora* and *Spergularia rupicola*.

In a detailed survey, Morgan (1989e) recorded about 11 populations of *L. paradoxum*, with an estimated total of 400-600 plants, all within a single 1 km square. Though there has been no recent count, a previous survey, in 1979, showed a similar population size, suggesting that the population is probably stable. The whole population is within an SSSI.

S.J. Leach

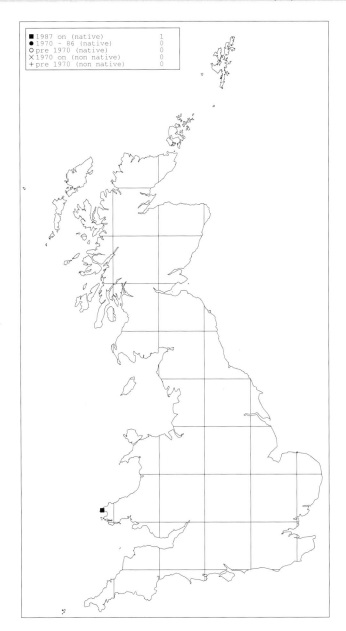

■ 1987 on (native)	1
● 1970 - 86 (native)	0
○ pre 1970 (native)	0
✕ 1970 on (non native)	0
+ pre 1970 (non native)	0

Limonium parvum Ingrouille (Plumbaginaceae)
Rock sea-lavender, Corlafant Penfro
Status in Britain: VULNERABLE. ENDEMIC.

One of the more easily distinguished members of the *Limonium binervosum* aggregate, *L. parvum* has a compact growth form, with tight 'cushions' of basal leaves and dense flower spikes on very short, usually unbranched stems less than 7 cm high. The species is described from a single population at Saddle Point, Pembrokeshire (Ingrouille & Stace 1986). It occurs on Carboniferous limestone in rock crevices and on heavily rabbit-grazed steep rocky cliff-slopes with a southerly or south-westerly aspect. Associates include *Armeria maritima*, *Cochlearia danica*, *Crithmum maritimum*, *Festuca rubra*, *Limonium procerum* ssp. *cambrense*, *Plantago coronopus* and *Spergularia rupicola*.

In 1993, three further populations of *L. parvum* were found between Flimston and Stackpole Head. The species may in fact be widely scattered on limestone sea-cliffs between Linney Head and Stackpole Quay. All known populations lie within SSSIs, two of which are in NNRs. Old records of *Limonium transwallianum* from this area are probably referable to *L. parvum* or to *L. procerum* ssp. *cambrense*.

S.J. Leach

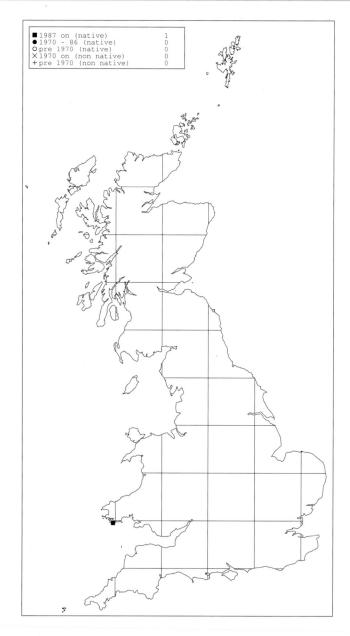

■ 1987 on (native)	1
● 1970 – 86 (native)	0
○ pre 1970 (native)	0
✕ 1970 on (non native)	0
+ pre 1970 (non native)	0

Limonium procerum (Salmon) Ingrouille (Plumbaginaceae)
Rock sea-lavender

Status in Britain: Lower Risk - Nationally Scarce.
ENDEMIC. Three subspecies, of which two (ssp. *devoniense* and ssp. *cambrense*) are VULNERABLE. (Ssp. *procerum* is widely distributed and assumed to be LOWER RISK - Nationally Scarce).

L. *procerum* is, together with *L. britannicum*, the most widespread member of the *L. binervosum* aggregate in western Britain. Ingrouille & Stace (1986) assigned most populations within this species to ssp. *procerum*, a variable taxon occurring in a range of habitats including sea-cliffs, dock walls, shingle banks and saltmarshes. Two sea-cliff populations have, however, been recognised as separate subspecies: ssp. *devoniense*, recorded from near Torquay, Devon; and ssp. *cambrense*, known only from the Carboniferous limestone between Stackpole Head and Linney Head, Pembrokeshire (where it sometimes occurs with *Limonium parvum* and *L. britannicum* ssp. *transcanalis*).

Separation of *L. procerum* subspecies in the field is problematic, although ssp. *devoniense* is said to have wavy stems, and ssp. *cambrense* has narrower leaves than the other subspecies (Stace 1991). It is suspected that botanists have so far paid little attention to these taxa, and our knowledge of their distribution and extent is therefore likely to be incomplete.

S.J. Leach

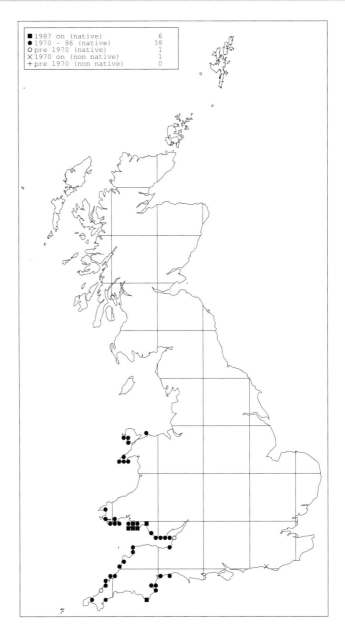

■ 1987 on (native)	6
● 1970 - 86 (native)	38
O pre 1970 (native)	1
X 1970 on (non native)	1
+ pre 1970 (non native)	0

Limonium recurvum Salmon (Plumbaginaceae)
Rock sea-lavender
Status in Britain: VULNERABLE. (ssp. *portlandicum*, ssp. *recurvum*); LOWER RISK - Near Threatened (ssp. *humile*). Four subspecies are recognised, all endemic to Britain and Ireland; ssp. *recurvum* is endemic to Britain.

Apart from outlying populations in Cumberland and Wigtownshire recently assigned to this species (ssp. *humile*), *L. recurvum* is restricted in Britain to the Isle of Portland, Dorset (Ingrouille & Stace 1986; Stace 1991), where it was discovered in 1832 and long ago recognised as a separate taxon (Salmon 1903). Two subspecies are recognised there: ssp. *recurvum*, known only from Portland Bill; and ssp. *portlandicum*, recorded elsewhere on Portland, and also from North Kerry, Ireland (Ingrouille & Stace 1986). A fourth subspecies, ssp. *pseudotranswallianum*, is recorded only from Co. Clare.

L. recurvum is a distinctive and elegant species, with relatively short, rough and often well-branched stems bearing dense arching or recurved flowering spikes. On the Isle of Portland it occurs on crumbling oolitic limestone cliff-slopes, quarry cliffs, ledges and stabilised quarry spoil. Associates include *Armeria maritima*, *Beta vulgaris* ssp. *maritima*, *Catapodium marinum*, *Crithmum maritimum*, *Festuca rubra*, *Inula crithmoides* and *Plantago coronopus*.

L. recurvum has been thoroughly surveyed on the Isle of Portland, with nine sites holding a total of more than 1,000 plants; all known sites lie within SSSIs and are not threatened, as long as protection from further quarrying can be guaranteed. In the past, collecting may have been responsible for the loss of some populations.

S.J. Leach

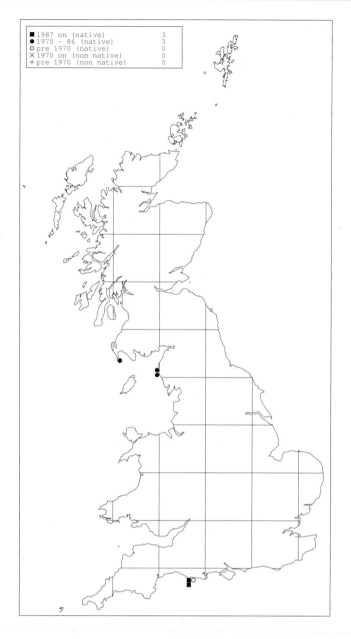

■ 1987 on (native)	3
● 1970 - 86 (native)	3
○ pre 1970 (native)	0
✕ 1970 on (non native)	0
+ pre 1970 (non native)	0

Limonium transwallianum (Pugsley) Pugsley (Plumbaginaceae)
Rock sea-lavender, Lafant Penfro
Status in Britain: VULNERABLE. ENDEMIC.

This is another long-recognised member of the *Limonium binervosum* aggregate, although considerable confusion has arisen in the past between *L. transwallianum* and other segregates. It is now certainly recorded only from steep, south-facing Carboniferous limestone cliff-slopes at Giltar Point, Pembrokeshire. Associated species include *Armeria maritima, Crithmum maritimum, Festuca rubra, Inula crithmoides, Limonium procerum* ssp. *procerum, Plantago coronopus* and *P. maritima*. Alleged *L. transwallianum* from Linney Head to Stackpole Quay is probably referable to other taxa (*L. britannicum* ssp. *transcanalis, L. parvum* and *L. procerum* ssp. *cambrense*), whilst a rock sea-lavender in Co. Clare, Ireland, long suspected of being *L. transwallianum*, is now regarded as a segregate of *L. recurvum* (ssp. *pseudotranswallianum*).

All currently known populations of *L. transwallianum* lie within SSSIs and are not considered to be under significant threat. Close examination of other populations of *L. binervosum* agg. in Pembrokeshire may yet reveal further *L. transwallianum*; it is certainly a difficult taxon to separate in the field, although the narrow petals are said to be diagnostic (Stace 1991).

S.J. Leach

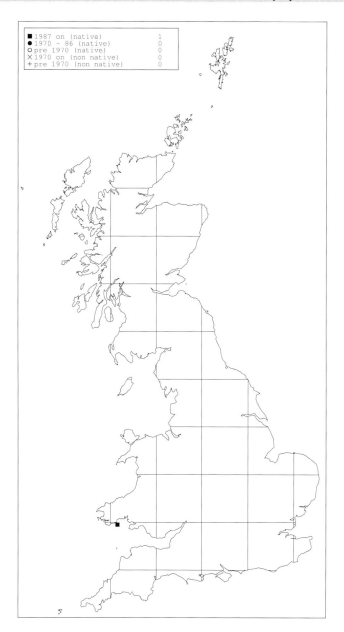

■ 1987 on (native)	1
● 1970 – 86 (native)	0
○ pre 1970 (native)	0
✕ 1970 on (non native)	0
+ pre 1970 (non native)	0

Limosella australis R.Br. (Scrophulariaceae)

Welsh mudwort, Lleidlys Cymreig
Status in Britain: VULNERABLE. WCA Schedule 8.
Status in Europe: Vulnerable. Occurs only in Britain.

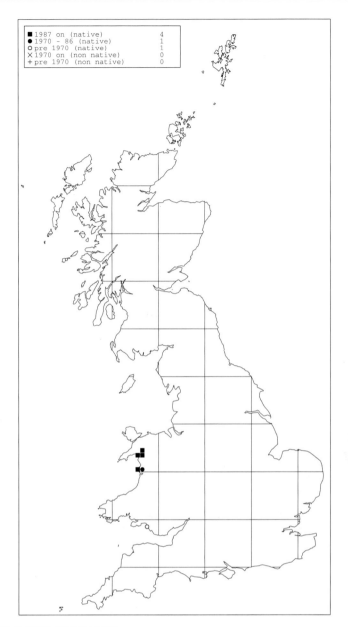

The major population centres for *L. australis* are now at the mouths of the Glaslyn and Dysynni rivers (Caernarvonshire and Merioneth), where it is dominant along several 100 m stretches of river margin, and widespread in the Glaslyn over mudflats and saltmarsh pools. On the lower saltmarsh, *L. australis* is scattered and loosely spreading, occurring with sparse *Juncus gerardii*, *Suaeda maritima*, *Triglochin maritimum* and other halophytes. By contrast, in upper saltmarsh pools, and conspicuously between sheltered *Schoenoplectus tabernaemontani* stands, it forms a dense smothering 'turf'. In Britain, the species usually occurs in only a narrow range of salinity conditions, between the dominance of *Juncus maritimus* downstream, for instance, and *Littorella uniflora* or *Callitriche* species upstream.

The rosette is 1-8 cm across, with roughly cylindrical awl-shaped pale green leaves, shallow fibrous roots and rapidly spreading runners. Its flowers are small (2-3 mm), solitary, and scented when fully open, but are cleistogamous, self-pollinated in the bud underwater. However, most plants, and particularly those submerged in more than 50 cm of water, are vegetative and seem to be at least short-lived perennials.

The first record in Britain for the species was from Glamorgan in 1897 (when it was thought to be a form of *L. aquatica*), and plants were not found in the Glaslyn until 1921, and in the Dysynni since 1955. The new 1990 record for the Dwyryd adds to the current picture of mobility. It is sometimes abundant, but is apparently sensitive to habitat change, which can cause large annual fluctuations in population size. There are also smaller populations on the neighbouring Dwyryd estuary, and records of other, perhaps temporary, colonies on saltmarsh at Morfa Harlech. *L. australis* used to be found near Port Talbot - in Kenfig Pool, Crymlyn Bog and nearby Morfa Pools - but these populations had probably died out by the 1940s.

L. australis was described originally from Australian material in 1803, and occurs widely throughout the southern hemisphere, north Pacific and eastern North America, but it is unknown from elsewhere in Europe. In Australia, South Africa and Chile it has been recorded far inland, and at up to 3,000 m altitude in New Zealand. The late notice of this quite conspicuous species in Britain, with its fluctuation, mobility and wide overseas range, indicates a relatively recent arrival. The association with ports and man-made habitats suggests boat traffic, perhaps as a ballast-alien (Jones 1991).

A. Jones

Liparis loeselii (L.) L.C.M. Rich (Orchidaceae)

Fen orchid, Gefell-Lys y Figen
Status in Britain: ENDANGERED. WCA Schedule 8. EC
Habitats & Species Directive, Annexes II and IV
Status in Europe: Vulnerable.

In Britain *L. loeselii* is restricted to fens in the Broadland district of East Anglia and dune-slacks on the South Wales coast. In Broadland, *L. loeselii* is apparently confined to vegetation in which *Phragmites australis* or *Cladium mariscus* is usually dominant in a species-rich community of fairly tall fen plants (Wheeler & Shaw 1987). This community occurs on infertile soils in areas with a relatively high and stable groundwater regime and develops as a transient stage in the hydroseral colonisation of shallow re-flooded peat cuttings (Giller & Wheeler 1986). All the remaining fenland populations are located in sites that have in the past been cut for peat (Doarks 1993). In South Wales it occurs mainly in successionally young dune-slacks characterised by *Salix repens* and a rich flora including *Agrostis stolonifera, Carex flacca, Carex viridula* ssp. *viridula, Hydrocotyle vulgaris, Juncus articulatus, Mentha aquatica, Ranunculus flammula, Samolus valerandi, Aneura pinguis, Calliergonella cuspidata, Campylium stellatum, Pellia endiviifolia* and *Petalophyllum ralfsii.*

This perennial species has a basal pseudobulb that is replaced each year. Additional daughter bulbs and shoots are frequently produced by vegetative propagation. Flowering shoots typically range from 3-18 cm in height and bear between one and 10 flowers on a stem, though up to 17 have been recorded. Non-flowering shoots are common and often comprise the majority of the population in dune-slacks. Flowering occurs mostly between late June and early July, though it may extend to mid-September. Autogamy has been observed, and this may be aided by impaction by rain drops (Catling 1980). Individuals represented by a single shoot can live for up to eight years, although genets comprised of multiple shoots probably survive for much longer. In dune populations, demographic studies suggest that recruitment from seedlings is most prevalent in young slacks where the plant can become established within the first 15 years of plant succession on bared sand. The dune-slack populations of *L. loeselii* are represented by a distinct taxon (var. *ovata*), which is distinguished from the fenland taxon (var. *loeselii*) by its generally shorter stature and broader ovate-elliptical blunt leaves.

L. loeselii has suffered a severe decline in the British Isles. Although once known from at least 30 sites in Norfolk, Suffolk, Kent, Lincolnshire and Cambridgeshire, the fenland variety is now confined to three Broadland stations with a total known population of fewer than 250 plants (Norfolk Wildlife Trust 1996). Whilst habitat destruction has resulted in the loss of some populations, the cessation of peat cutting is probably the most important contributory factor that has led to the decline of *L. loeselii* at intact Broadland sites (Wheeler 1993). The cessation of summer mowing management may also have been significant, as this helps prevent scrub encroachment and retards the ongoing process of terrestrialisation.

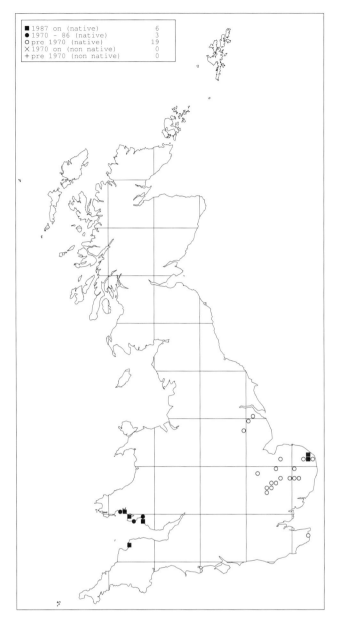

■ 1987 on (native)	6
● 1970 - 86 (native)	3
○ pre 1970 (native)	19
✕ 1970 on (non native)	0
+ pre 1970 (non native)	0

By far the largest populations occur on the South Wales coast. During the 20th century it has been recorded from eight Welsh dune systems (Jones *et al.* 1995), but is now confined to four sites. Although the Welsh population is known to exceed 10,000 plants, long-term decline is evident. Undergrazing is undoubtedly an important factor, resulting in the loss of suitable supporting vegetation types through succession. The stabilisation of many dune systems to a point where few new slacks suitable for colonisation are being formed is also of key importance and is a function not only of undergrazing but also of sediment starvation and past management policy, which favoured stabilisation. The var. *ovata* once occurred at Braunton Burrows, North Devon (Willis 1967), but has not been seen since 1987 and may be extinct there.

L. loeselii has a circumboreal distribution, occurring throughout Europe and Asia to North America. In Europe the species occurs from Britain and southern Fennoscandia southwards, extending from south-west France eastwards to the Balkans and southern Russia. In Europe, *L. loeselii* is threatened throughout its range and is legally protected in many countries.

P.S. Jones and B.D. Wheeler

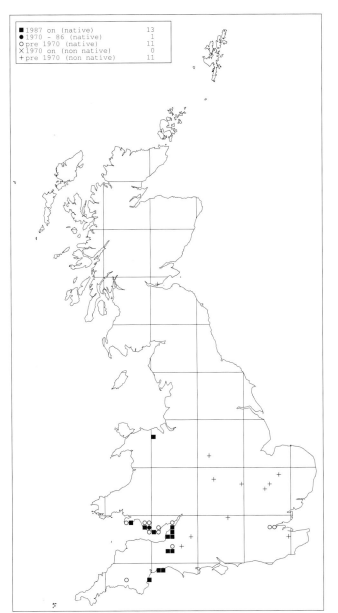

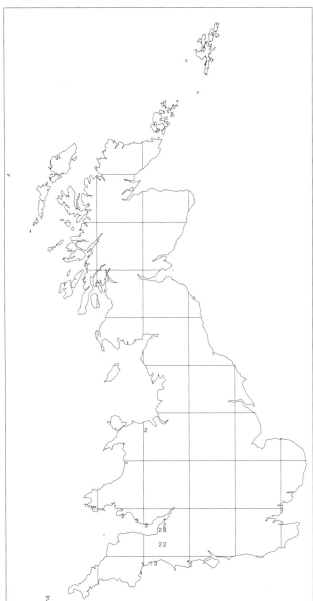

Lithospermum purpureocaeruleum L.
(Boraginaceae)
Buglossoides purpureocaeruleum (L.) I. M. Johnston
Purple gromwell, Maenhad Gwyrddlas
Status in Britain: LOWER RISK - Near Threatened.
Status in Europe: Not threatened.

This plant of limestone and chalk occurs in two distinct habitats, although it is principally a plant of woodland edges and rides, and of lanesides and banks in partial shade. It maintains vigorous growth where there is some protection afforded by other vegetation, as, for instance, in open bushy areas or under a light coppice canopy, but it becomes weak in deep shade. Common associates include *Arum maculatum, Brachypodium sylvaticum, Geum urbanum, Hedera helix, Ligustrum vulgare, Melica uniflora, Mercurialis perennis* and *Rubia peregrina*. In Glamorgan it occurs mainly along the coast on slumped limestone cliffs, slopes and crags. In these places, it usually occurs in lightly-grazed or ungrazed naturally-dwarfed scrubby vegetation dominated

by *Cornus sanguinea, Crataegus monogyna, Ligustrum vulgare, Prunus spinosa* or *Ulex europaeus*, which provide protection without too much overshading. Devon populations are in similar habitats on a chalk undercliff.

L. purpureocaeruleum is a rhizomatous perennial, flowering mainly in May, but extending into June, and readily producing seed. Decumbent non-flowering stems may root where they touch the ground, and provide another means of vegetative reproduction.

An area of wooded limestone ridges, bounded by Cheddar, Weston-super-Mare and Bristol, was historically considered to be the most important area for this species, encompassing the largest number of sites.
L. purpureocaeruleum still occurs in about 20 of them, though in many the populations are now small and vulnerable. Elsewhere, the plant occurs at four sites in Devon, about eight in Glamorgan, and one in Denbighshire. Most of the Glamorgan sites have strong

populations of many thousands of plants, as do the Devon coastal sites. In Denbighshire, the tiny colony, covering only a few square metres, may represent a number of independent plants or a single one with inter-connected creeping stems (Evans & Ellis 1994b). It formerly occurred as a native plant in Kent, and has been recorded as a casual in a few places.

L. purpureocaeruleum has declined in recent years in Somerset, and colonies remain much threatened locally by changes in woodland management. In the past, regular coppicing of *Corylus avellana*, *Fraxinus excelsior* and *Tilia cordata* provided the necessary light for strong growth and flowering. Many woods are now overgrown and neglected and have an impoverished ground flora. Plants of *L. purpureocaeruleum* may survive in a weak or depauperate vegetative state for many years under a fairly dense canopy, but will eventually succumb. Revitalisation of woodland populations should be a high conservation priority. The strong coastal colonies do not appear significantly threatened.

It is mainly a plant of southern and central Europe, common in the Balkans and Iberia, extending northwards to Britain and Poland (rare in both countries) and eastwards to the Caucasus and the Lake Baikal area.

M.J. Wigginton

Lloydia serotina Reichb. (Liliaceae)

Snowdon Lily, Lili'r Wyddfa
Status in Britain: VULNERABLE. WCA Schedule 8.
Status in Europe: Not threatened.

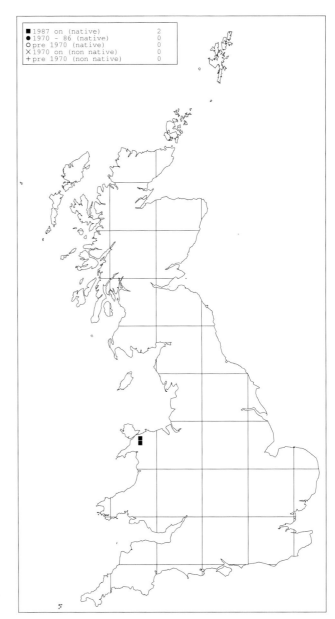

In Britain, *L. serotina* is confined to a few cliff-faces in Snowdonia, North Wales. Most sites face north or north-east, and are generally shaded and damp. It grows out of vertical and horizontal cracks and on tiny ledges, often sheltered by overhangs. The most frequent associates include *Campanula rotundifolia, Carex viridula* ssp. *oedocarpa, Festuca ovina, Selaginella selaginoides, Thalictrum alpinum, Breutelia chrysocoma* and *Racomitrium lanuginosum*. Unlike many other arctic-alpine species in Britain, *L. serotina* does not require base-rich rocks and may grow with few other species on a relatively acid substrate. In North Wales it is found on rock types ranging from basaltic to rhyolitic in composition, on substrates of pH 5.4-6.2 (B. Jones in prep.). In the Alps a wider range of pH has been observed, ranging from 4.9 to 6.7.

Flowers are usually solitary, rarely two, the two basal filiform flexuous leaves overtopping the flowering stem. In the first two years only one leaf is produced, followed by two leaves, and after four to six years the plant may begin to flower. However the leaves are often the only visible parts, as generally only 10-20% of plants flower in any one year, and less than 2% produce seed capsules. Seeds readily germinate in suitable conditions, but first year mortality is probably high. Most reproduction appears to be vegetative, with rhizome-like structures producing new plants close to the parent. This process may have produced extensive clones,which, although comprising hundreds of plants, may have low genetic diversity. Flowering and seed production are rather lower than those of populations in the North American Rocky Mountains, where 20-50% of plants flower and 5-10% produce seed capsules.

L. serotina has always been rare in Britain. It was first recorded by, and subsequently named after, Edward Llwyd in the late 17th century and was of great interest to 18th and 19th century botanists. Many of the localities mentioned in old botanical accounts still hold populations of *L. serotina*, but as very few plants can now be reached without ropework, it must be presumed that the more accessible ones were all removed by these early plant hunters, who sometimes went to extreme lengths to find and collect specimens for their herbaria. At the present time some sites still hold hundreds of plants, but others have very small numbers. Despite the early and continuing interest in this plant, very few studies have been carried out, and little is known in detail of its ecology and biology in Britain.

Sheep-grazing may have restricted *L. serotina* to its present habitats, whilst collecting has reduced it on the cliff-faces where it has survived. Population studies are being carried out to assess trends and to determine whether its genetic status and reproductive strategies will ensure its survival in the face of possible climatic change. The latter seems likely to constitute the main threat to the species, as also to many other arctic-alpines. Increasing numbers of feral goats, and recreational activities, are more immediately manageable.

This species has a highly disjunct distribution (Woodhead 1951), being found at high altitudes and latitudes in North America, the European Alps, Carpathians, Caucasus, Urals, Himalayas, Arctic Russia, Japan and China; it is however absent from Scandinavia. In countries in the centre of its range it can be found growing in abundance in high alpine tundra 'meadows' above the treeline. These gently sloping, species-rich, high insolation habitats are in contrast to those in which *L. serotina* grows in North Wales, where it has the characteristics of being a glacial relict, isolated from its nearest neighbours by nearly 1,000 km.

B. Jones

Lobelia urens L. (Campanulaceae)

Heath lobelia, Bodoglys Chwerw
Status in Britain: VULNERABLE.
Status in Europe: Not threatened.

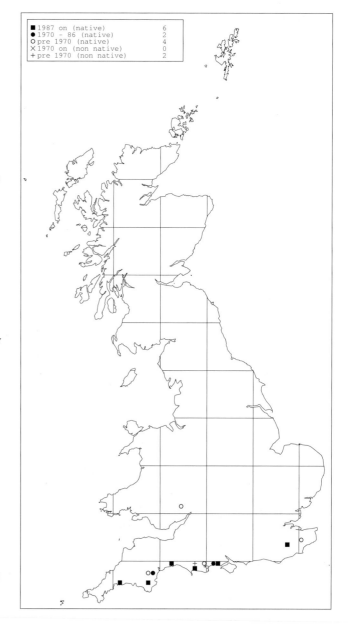

L. urens grows in open communities of rough pastures and grassy heaths, often on the margin of woodland on infertile acid soils. Extant populations are in low-lying terrain, often valley bottoms. The soils of such areas are often seasonally waterlogged, and although the surface horizons at some sites are predominantly freely draining sandy loams, they overlie more clayey horizons (Findlay *et al.* 1984). Associated species include *Juncus articulatus, J. conglomeratus, Lotus pedunculatus, Mentha aquatica, Molinia caerulea, Potentilla erecta, Pulicaria dysenterica* and *Salix cinerea*.

This species is a rhizomatous perennial herb. Overwintering rosettes of lanceolate, slightly serrate leaves each produce a single flowering spike 10-100 cm high the following spring. The spike typically has 40-80 purple flowers between June and October. The flowers are entomophilous but can self-pollinate, and seed is copiously produced. After seed ripening the flowering spikes die back and new rosettes emerge from the rhizome. In damp weather dehiscence and dispersal may be delayed.

L. urens is confined to southern coastal counties of England - Cornwall, Devon, Dorset, Hampshire, Sussex and Kent, with a single record from Herefordshire. The number of localities has declined steadily from 19 to six. Andrew's Wood, Devon, has historically held the largest population of *L. urens* in Britain and currently supports about 3,500 plants. Hurst Heath, Dorset, has sustained a similar number over the last 20 years through active management of the habitat, including rotavating. The population at Flimwell, Sussex, increased dramatically from fewer than 200 to over 2,000 plants, following woodland clearance in 1990 for a bird park. The remaining three extant populations in Hampshire, East Devon and Cornwall are small, each with fewer than 100 plants. Populations are monitored at most sites. Erratic fluctuations in the size of colonies increase their vulnerability.

Five colonies in southern England have been lost this century through afforestation and cultivation. The main threat to today's extant populations is lack of appropriate management. *L. urens* requires open ground for its establishment from seed. Ideally this would be achieved by disturbance, either using horses and cattle (many sites were originally used as rough grazings) or by coppice rotation. However, adult plants are very susceptible to such disturbance and therefore a balance between recruitment and maintaining established plants must be reached. Adults are shade-tolerant and have a life span of more than six years. The seeds of *L. urens* are small and

compact and are thus likely to be persistent in the soil (Thompson *et al.* 1993). Hence occasional disturbance would allow the establishment of recruits from the seed-bank and a subsequent enhancement of the population.

L. urens has a Lusitanian distribution, extending from Morocco, Madeira and the Azores along the Atlantic coast through Portugal, Spain and France to Belgium (Brightmore 1968; Daniels 1990; Tutin 1976). It reaches its northern limit in Britain.

Genetic aspects are discussed in Daniels *et al.* (1996).

J. Dinsdale

Lonicera xylosteum L. (Caprifoliaceae)

Fly honeysuckle, Gwyddfid Syth
Status in Britain: ENDANGERED.
Status in Europe: Not threatened.

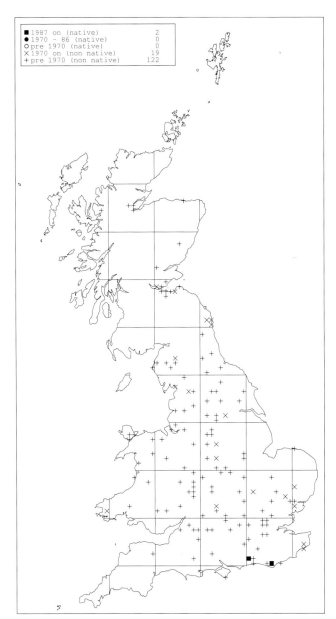

L. xylosteum is accepted as a possible native along a few kilometres of the South Downs chalk scarp near Amberley and Storrington in West Sussex (FitzGerald 1988c; Hall 1980), where it was first recorded in 1801 by William Borrer. All the West Sussex sites are on the chalk in ancient woodlands and hedgerows, or in old scrub within species-rich chalk turf. In several places the plants are close to old trackways, which often retain a relict woodland flora. These woodlands contain much *Fraxinus excelsior*, often as large coppiced stools, with many shrubs including *Corylus avellana*, *Crataegus monogyna*, *Euonymus europaeus*, *Viburnum lantana* and *V. opulus*. The ground flora typically includes *Mercurialis perennis* and other ancient woodland indicators, such as *Adoxa moschatellina*, *Anemone nemorosa*, *Campanula trachelium*, *Hyacinthoides non-scripta*, *Lamiastrum galeobdolon*, *Primula vulgaris* and *Sanicula europaea*. Of particular interest is the presence, at one site, of native *Tilia platyphyllos* in the form of groups of trunks growing from ancient coppice stools close to *L. xylosteum* bushes. C.D.Pigott (*in litt.*) notes that these two species are frequently associated in chalk woodlands in France.

L. xylosteum forms a loose, non-climbing, deciduous bush, normally 1-2 m high, but sometimes exceeding 3 m. The flowers appear in May, and berries are produced in abundance during the summer. However, there is little evidence of regeneration from seed, and spread seems to be mainly by layering. The young stems are upright, but mature branches tend to sprawl over, rooting at the nodes and giving rise to clusters of bushes, which can form extended colonies.

The West Sussex population consists of about 120 plants, almost all of them mature. Other sites of interest in south-east England where *L. xylosteum* might also be native include a small woodland near Wilmington, East Sussex, which is similar in character to the West Sussex locations. In 1901, it was recorded scattered throughout that wood, but it is now reduced to one large plant. There is also a group of sites near Godalming, where this species has been known for more than 150 years, some of the bushes here being notably large. Elsewhere in Britain, plants are considered to be either direct introductions or bird-sown from cultivated stock: indeed, the species has been in cultivation since at least the late 17th century. Although most sites have statutory protection, local damage has occurred during the past 10 years from excessive poaching by cattle and from woodland clearance.

L. xylosteum occurs throughout much of Europe, extending from southern Scandinavia to Spain and Sicily, and from France through central and eastern Europe to western Siberia.

F. Abraham

231

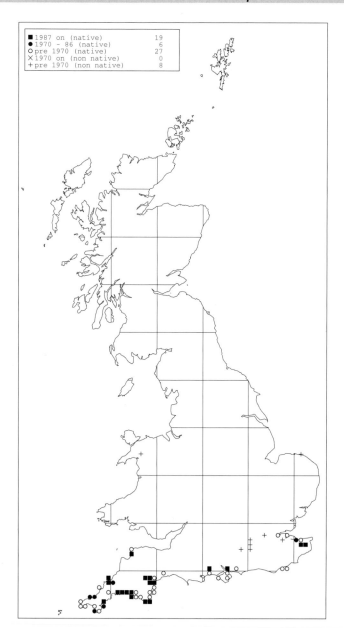

Lotus angustissimus L. (Fabaceae)

Slender bird's-foot trefoil
Status in Britain: LOWER RISK - Nationally Scarce.
Status in Europe: Not threatened.

L. *angustissimus* typically occurs in grassland amongst cliff-top scrub, around rocky outcrops and on grassy banks by the sea. It is a plant mainly of short open-textured and rather 'scruffy' swards over thin drought-prone soils on sheltered slopes of a southerly or south-westerly aspect. At many of its localities cliff-slopes are now covered by dense scrub, and then it is usually confined to narrow pathside banks and verges, and old trackways. Common associates include *Agrostis capillaris, Crepis capillaris, Dactylis glomerata, Hypochaeris radicata, Ornithopus perpusillus, Plantago coronopus, Rumex acetosella, Sedum anglicum* and *Vulpia bromoides*, with patches of *Ulex europaeus, Rubus fruticosus* and *Pteridium aquilinum* rarely far away. It is often found with its nationally scarce congener *L. subbiflorus*; indeed, the two species may be "so interwoven . . . that they

cannot be separated without tearing them to pieces" (Marquand 1901). Johns (1982) observed on Alderney that "if *L. subbiflorus* isn't around you haven't much chance of finding *L. angustissimus*", and this is a useful rule of thumb when searching for it in mainland Britain too.

L. *angustissimus* is an annual, usually germinating in autumn (August - October) and spring (April - May) and flowering between June and October. Observations suggest that autumn-germinating plants flower in June and July, whereas those germinating in spring do not flower until August. Exceptionally, in a wet summer seed produced in June can germinate almost immediately, producing plants which, weather permitting, flower and set seed in early winter. Patches of bare soil are essential for seed germination, and scrub cutting, trampling, fire and especially drought may all be of benefit for this reason (Leach *et al.* 1994). Whilst summer droughts help to keep the sward open, they may also cause spring-germinating plants to shrivel and die before they have a chance to flower. It is notoriously erratic in its appearance: one year there may be thousands of plants, whilst in the next it can be almost impossible to find. The reasons for such dramatic fluctuations are usually obscure, although timing and severity of droughts may be important, the largest numbers of plants often occurring in the two summers following a bad drought year. As with *L. subbiflorus*, it has a persistent bank of buried seed, enabling it to appear intermittently whenever conditions are suitable.

Most *L. angustissimus* populations are small, and recent surveys indicate that at some sites, particularly in West Cornwall, it may have been lost altogether. Undoubtedly it has declined as a result of scrub encroachment following the demise of traditional management practices such as grazing and burning, and a few populations may have gone owing to agricultural improvement of cliff-top grasslands. Its erratic appearance makes it difficult to assess current status, but it has been seen in 50-60 localities in Britain since 1980, mainly in Devon and East Cornwall but with a few outliers in Hampshire and Kent. The main populations are confined to three areas: Pentire Point (200-300 plants in 1994), between Polruan and Penlee Point (more than 600 plants in 1994), and between Prawle Point and Start Point (more than 6,500 plants in 13 localities in 1993/94) (Leach 1995). This last, a relatively short stretch of coastline, is currently the plant's main stronghold in Britain. There are several inland sites in South Devon, in the Teign valley and near Bishopsteignton and Dawlish. Its few extant localities in Hampshire and Kent are mostly inland, associated with sand and gravel workings. There are old records from Sussex and Caernarvonshire, unsubstantiated records from the Isle of Wight (Bevis *et al.* 1978), and an erroneous report from Dorset (Good 1984).

Some of its largest populations are well protected, occurring within SSSIs and on coastlines owned by the National Trust. However, even on these sites there is a risk that *L. angustissimus* will continue to suffer through lack of management. Much will depend on prompt action being taken to prevent sites becoming rank and overgrown and, in particular, to limit the spread of scrub.

L. angustissimus is an oceanic-southern species. It is widely distributed in southern Europe, extending throughout the Mediterranean region and northwards to Britain and all the larger Channel Islands, and eastwards to the Ukraine. It is found on the Azores, in northern Africa and in parts of western Asia. Generally associated with dry lowland grasslands, in parts of southern Europe it is also found in damp grassland on mountains.

S.J. Leach

Ludwigia palustris (L.) Elliot (Onagraceae)

Hampshire purslane
Status in Britain: LOWER RISK - Near Threatened.
Status in Europe: Not threatened.

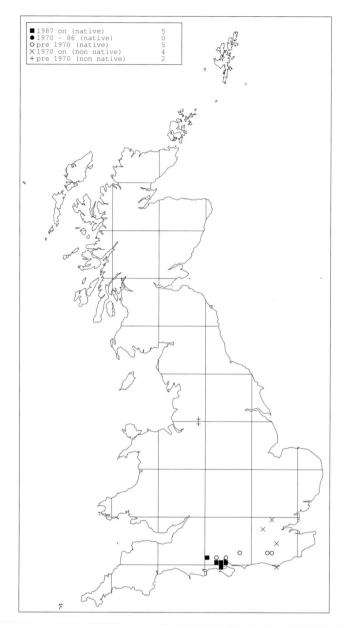

L. palustris occurs in a wide variety of wetland habitats, particularly where underlain by relatively base-rich Oligocene clays. Most sites are seasonally inundated pools and marl pits, where it may form free-floating rafts, or be submerged in permanently deep water, or grow on grazed and trampled pond margins (Salisbury 1970). Other colonies are found within relatively base-enriched runnels within valley mires, in ponds and poached areas in pasture or woodland glades and, rarely, in backwaters in riparian woodland. It benefits from well-illuminated conditions but will tolerate some shade. With a few exceptions, most sites are open to grazing and trampling by livestock, which maintain suitably open ground for germination and growth. The seasonal pools tend to be heavily grazed and are usually sited within New Forest lawns adjacent to settlements. The exclusion of livestock can lead to the extinction of colonies of *L. palustris* by the rank overgrowth of more vigorous species. A wide range of associated wetland species includes several that are local and rare, including *Apium inundatum, Galium constrictum, Littorella uniflora, Mentha pulegium, Pilularia globulifera* and *Pulicaria vulgaris*.

It is normally perennial, flowering and setting seed from June to August. The rather inconspicuous axillary flowers are probably self-pollinated, and plants produce copious seed. The plant is normally decumbent, roots freely from nodes and readily grows from fragments. It may occur as an annual pioneer on exposed muds. The growth of plants from seeds and fragments helps to explain the sporadic appearance of small populations in the vicinity of substantial permanent populations (Salisbury 1972).

Native sites have long been considered restricted to the New Forest, mostly within and between the catchments of the Beaulieu and Lymington rivers, where it is extant in 20 1 km squares. Its populations have been remarkably stable in recent decades, and very few sites have been lost. However, the historical record since 1843, when it was first described in the New Forest, seems to indicate a significant expansion. This seems to have occurred at the same time as a population shift from the central parts of the Forest to the southern parishes, the reason for which is unclear. All of the New Forest populations (and the Dorset one) lie within SSSIs. Populations which formerly occurred in Sussex, and on the Wealden heaths in Hampshire (where it was first recorded by Goodyer in 1667) have sometimes been regarded as native, and are shown as such on the map. These colonies appear to have died out in the nineteenth century, probably as a result of drainage and the infilling of wetlands. In recent years, plants recorded as *L. palustris* have appeared at a number of sites further afield and have generally been regarded as introductions (e.g. Briggs & Maurice 1993); as a plant attractive to aquarists, further casual occurrences are likely. However, following the re-determination of plants from Putney

Heath, Surrey, as the hybrid *L. x muellertii* (Clement 1997), it seems entirely possible that other outlying records are of this taxon. The newly discovered Dorset colony (Cox 1997) should also be reassessed. The colony in Epping Forest, discovered in 1976, disappeared in the 1980s. *L. palustris* was known in Jersey from 1830s, but was lost by 1920.

L. palustris is widespread, though local, in western, central and southern Europe, occurring from Britain, France and Spain to Poland and Russia, with its stronghold in southern Europe and the Balkans. It also occurs in Turkey. It has declined, particularly in northern Europe, and is threatened in some countries. It also occurs in North Africa and West Asia and in somewhat different forms in North, Central and South America. The Central American species, *Ludwigia uruguayensis*, is naturalised in Spain, France and Belgium.

C. Chatters and M.J. Wigginton

Luzula arcuata Sw. (Juncaceae)
Curved wood-rush, Learman Crom
Status in Britain: LOWER RISK - Near Threatened.
Status in Europe: Not threatened.

This plant is found in one of the most uncompromising habitats on the highest of the Scottish hills. It grows in open, stony fell-field and rock debris on windswept ridges and plateaux where protective winter snow is stripped away by the wind, leaving a scoured and harsh niche in which few vascular plants can survive. It appears to be calcifuge, growing on coarse, leached alpine soils derived from granite, quartzite, granulite and acidic schists. The most frequent associates are *Carex bigelowii* and *Juncus trifidus* with, less frequently, *Gnaphalium supinum* and *Luzula spicata*. The moss *Racomitrium lanuginosum* is also a common associate, along with some lichen species. In higher and more open corries *L. arcuata* can extend into more snow-loving vegetation with *Salix herbacea* and a variety of mosses and liverworts. The altitudinal range is from 760 m on Slioch up to 1,290 m on Cairn Toul, making this one of the most exclusively montane of all British plants.

L. arcuata is a tufted perennial species with short stolons and a creeping rhizome system. It flowers in June and July.

There is nothing to suggest that this species is under any significant threat, and it probably still occurs in most of the hectads for which only pre-1987 records are available, in which case it would be classified as nationally scarce. However, because there are records from only 15 hectads since 1970, it is placed in the Near Threatened category for the time being. In its Cairngorm stronghold it is still locally frequent. The map indicates the need for information on its current status.

This is an arctic species with a circumpolar distribution. It reaches its southern European limit in Scotland.

G.P. Rothero, in Stewart *et al.* (1994)

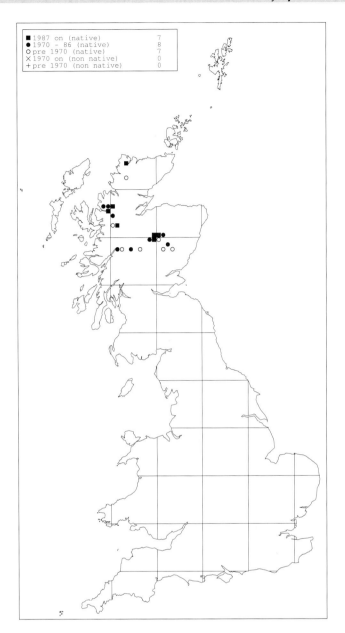

■	1987 on (native)	7
●	1970 - 86 (native)	8
○	pre 1970 (native)	7
✕	1970 on (non native)	0
+	pre 1970 (non native)	0

Luzula pallidula Kirschner (Juncaceae)

Luzula pallescens auct., non Sw.
Fen wood-rush
Status in Britain: VULNERABLE.
Status in Europe: Not threatened.

This species was first detected in Britain at
Woodwalton Fen, Cambridgeshire, in 1907. At the time of
its discovery, some doubt was expressed as to its status,
but there seems no reason to suppose it is not a true
native. It was subsequently located at Holme Fen 5 km to
the north, and these remain its only known sites. In both,
it has always been very localised. It is a plant of fens,
fenny banks, open peat, grassy rides in damp woodland,
and woodland glades. It appears to favour open,
preferably disturbed sites, and at Holme Fen has a history
of being abundant in years following, for example, gross
disturbance of rides, or peat-digging. At such times, more
than 1,000 plants have been recorded, though the current
population at Holme probably numbers about 100-300
plants. At Woodwalton, it is currently in very low
numbers.

L. pallidula is a perennial, flowering in May and June
and setting copious seed, which is long-lived in the soil.
Kirschner (1995) concludes that the plant is likely to be
generally autogamous. Some observers have remarked on
the difficulty of distinguishing putative *L. pallidula* from *L.
multiflora* in some populations at Holme, which may
suggest the possibility of local hybridisation or,
alternatively, that the differentiating characters were not
appreciated. Kirschner & Rich (1993) have pointed out that
many of the characters used to identify species in *Luzula*
Sect. *Luzula* in British Floras are unreliable, and that pale
forms of *L. multiflora* are not rare.

At the present time there is no management specifically
for *L. pallidula*, but some disturbance by rotavating or
shallow ploughing is proposed at Holme Fen. Both of its
sites are NNRs.

L. pallidula is widespread in central and northern
Europe, extending eastwards through Asia to Japan, and
also occurs in eastern North America. Tetraploid
populations in Ireland probably represent an alloploid
derivative of *L. campestris* x *L. pallidula* parentage
(Kirschner 1995).

M.J. Wigginton

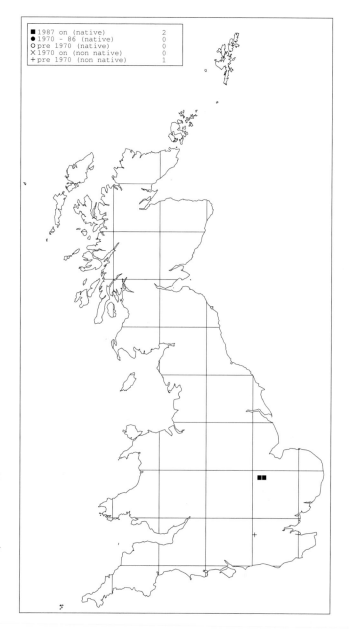

■ 1987 on (native)	2
● 1970 - 86 (native)	0
O pre 1970 (native)	0
X 1970 on (non native)	0
+ pre 1970 (non native)	1

Lychnis alpina L. (Caryophyllaceae)
Alpine catchfly, Coirean Ailpeach
Status in Britain: VULNERABLE. WCA Schedule 8.
Status in Europe: Not threatened.

Like several others of our choice mountain plants, *L. alpina* was first found by G. Don, who discovered it on Little Kilrannoch in 1795. For a while thereafter, the population suffered severely from collecting, but survived against all odds. In Britain it is currently known only from that site and one in Cumbria. They are very different in character.

The Cumbrian population is by far the smaller, and only about 50-60 plants remain from a former colony at least twice that size (Taylor 1987a). There it clings precariously to the steep faces of a high crag of acidic Skiddaw Slate at 600-700 m in the northern Lake District, where it is not accessible to sheep or deer, and not overtly threatened. Associates include *Alchemilla alpina, Calluna vulgaris, Cryptogramma crispa, Deschampsia flexuosa, Festuca ovina, F. vivipara, Solidago virgaurea, Vaccinium myrtillus* and, notably, the 'copper moss' *Grimmia atrata* (Ratcliffe 1960). On Meikle Kilrannoch, Angus, it occurs on a broad summit plateau at 870 m, growing on skeletal soil on serpentine debris, in an open fell-field community of alpine species including *Armeria maritima, Carex bigelowii, Cerastium fontanum* ssp. *scoticum, Cochlearia micacea, Minuartia sedoides* and *Racomitrium lanuginosum*. This population currently totals about 65,000 plants. Soils in both sites are highly metalliferous, though markedly different in composition (Johnston & Proctor 1980), but leaves of *L. alpina* were found not to accumulate metal (Proctor & Johnston 1977). A tiny population on Rum was thought to be introduced and no longer persists, and the 1870 record on Coniston Old Man, Cumbria, has never been confirmed.

L. alpina is perennial, with a simple flowering stem arising from a crowded rosette of sub-spathulate to linear leaves. In flower it is unmistakable, but vegetative rosettes can be remarkably similar to those of *Armeria maritima* (Wright & Lusby 1994). The plant perennates as a tap-root, but the leafy rosettes die down completely in winter. If, however, leaves remain green under snow, then they will die with the onset of spring, and new ones appear. Flowers are protandrous and perhaps usually outcrossing, bumble-bees and flies appearing to be the main pollinating agents (Raven & Walters 1956). Self-pollination has been suspected in Kilrannoch plants, but to what extent it occurs is not clear, and reproduction is exclusively by seed (Proctor & Johnston 1977). The Lake District plant is large, lanky and pale-flowered (mostly pink, rarely white), and said to be similar to a form from Labrador. In contrast, Angus plants are smaller, with deep carmine-red flowers, similar to those in Norway.

This species has a discontinuous distribution in Europe. In the north, it occurs in Iceland and throughout

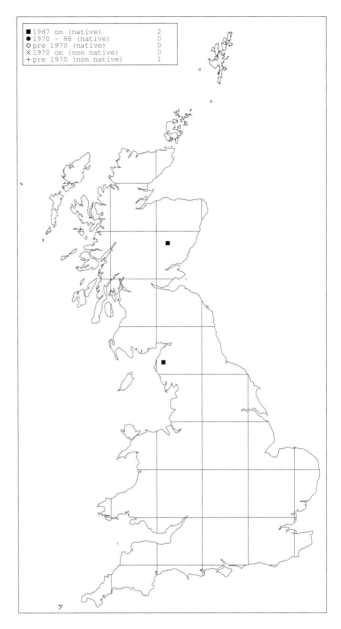

■ 1987 on (native)	2
● 1970 - 86 (native)	0
○ pre 1970 (native)	0
✕ 1970 on (non native)	0
+ pre 1970 (non native)	1

Fennoscandia, especially on serpentine, extending eastwards to the Altai mountains. In the southern part of its range it occurs in montane regions of the Pyrenees, Alps and Caucasus. Elsewhere it is known on the coasts of Greenland and eastern Canada. In Sweden it has been used as an indicator for locating metal-rich ore for the mining industry.

Further details are given in Böcher (1963), Brooks *et al.* (1979) and Nagy & Proctor (1996). The population genetics of *L. alpina* in Europe and North America is discussed in Haraldsen & Wesenberg (1993).

M.J. Wigginton

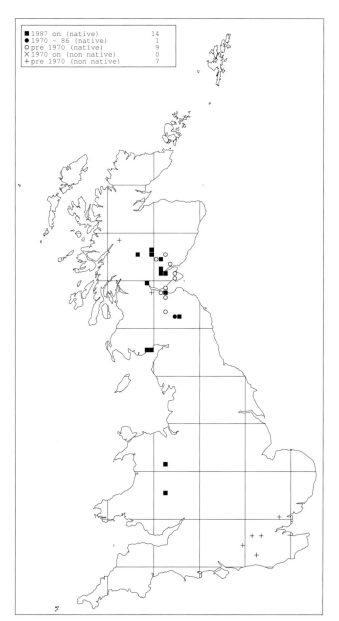

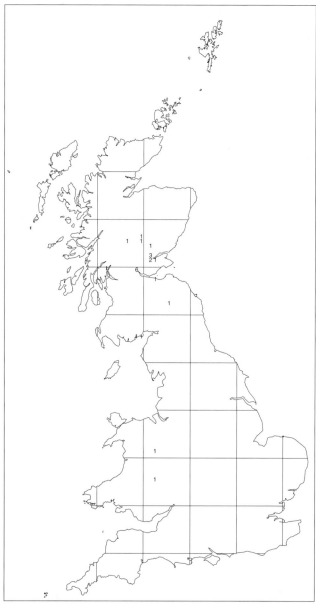

Lychnis viscaria L. (Caryophyllaceae)

Sticky catchfly, Coirean Leantalach, Lluglys Gludiog
Status in Britain: VULNERABLE.
Status in Europe: Not threatened.

In Britain, *L. viscaria* is mainly found in open sites on dry south- to west-facing basic and intermediate igneous rocks, although it occasionally occurs on sedimentary and metamorphic rocks, as in Kirkcudbrightshire and Perthshire respectively. An interesting feature of the associated flora of *L. viscaria* is the proximity of species that do not normally grow together because of their preferences for either acid or basic soils. Various combinations of the following plants may occur depending on the locality: *Calluna vulgaris, Cytisus scoparius, Erica cinerea, Helianthemum nummularium, Helictotrichon pratense, Origanum vulgare, Scabiosa columbaria, Teucrium scorodonia, Thymus polytrichus* and *Ulex europaeus*. The combination of the inherent chemistry of the soil, the pH range within which most nutrients are in an available form, and small-scale variation in the soil probably account for the juxtaposition of calcicole and calcifuge plants. Other contributory factors include the relaxed competition in these open sites and the differential rooting depth of plants, calcifuges such as *Calluna vulgaris* rooting nearer the more leached surface (Jarvis 1974).

This species is a tufted, winter-green perennial with upright flowering stems that are viscid beneath the nodes. Panicles of flowers are produced mainly from late May to July, with butterflies and bumble-bees the chief pollinators. *L. viscaria* can withstand drought well and recovers after severe wilting but is very sensitive to shading by surrounding vegetation. Although *L. viscaria* is perennial, it does not have an effective vegetative propagation mechanism and relies on copious seed production and seedling recruitment for continued survival. Where populations are accessible to grazing animals, flowering can be severely reduced with consequent loss of seed production.

There are at least 19 extant sites for *L. viscaria*, ranging from fewer than five clumps in at least two sites to several hundred in the western Ochil Hills. The plant has recently disappeared from two sites in Selkirkshire and Fife through scrub encroachment and is threatened to varying degrees in at least a further four sites (Wright & Lusby 1993; 1994). Fire has severely reduced the population at its oldest known British locality in Holyrood Park. In Wales, an apparently stable population of several hundred plants occurs on south-facing ledges and in crevices in the dolerite cliffs of Stanner Rocks (Morgan 1989c; Woods 1993). At its other Welsh site, at Breidden, quarrying has much reduced the extent of its habitat and only about 70 plants remained in 1986 (Morgan 1989a), with perhaps fewer now. 'Recovery' programmes, including habitat management and translocations, have been carried out at some threatened sites in Scotland and Wales, including the long-known site on Arthur's Seat, Edinburgh. *L. viscaria* is not uncommon in cultivation.

L. viscaria extends north to beyond the Arctic circle to 70°N, south to the Alps and east to the Novosibirsk region of central Russia (Wilson *et al.* 1995).

P.S. Lusby

Lythrum hyssopifolia L. (Lythraceae)

Grass-poly, Gwyarllys Isopddail
Status in Britain: VULNERABLE.
Status in Europe: Not threatened.

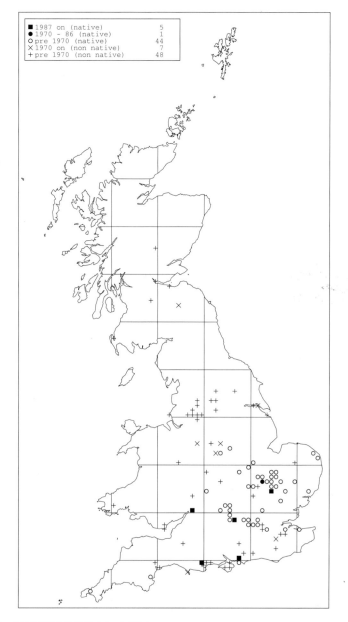

This is a plant of disturbed ground that is flooded during the winter months. It is usually found in hollows, ruts and other low-lying areas in arable fields, in sites where autumn-sown crops have been killed by flooding or where the ground was too wet for crops to be sown in spring. In these areas, *L. hyssopifolia* is often accompanied by other spring-germinating annuals, including *Juncus bufonius, Persicaria maculosa, Plantago major* ssp. *intermedia, Polygonum aviculare,* and ephemeral bryophytes such as *Physcomitrella patens* and *Riccia cavernosa* (Preston & Whitehouse 1986). However, the largest British population of *L. hyssopifolia* does not occur on cultivated land, but on winter-flooded ground at Slimbridge that is disturbed in summer by waterfowl. In addition to its occurrence in winter-flooded sites, this species is sometimes found as a casual in a range of other sites to which seeds have been introduced as a contaminant of imported grain or in other ways.

L. hyssopifolia is an annual which germinates after the water recedes in spring. The flowers are slightly protogynous, but if cross-pollination fails they are automatically self-pollinated and almost always set seed. Like many annuals, plants may differ greatly in size and fecundity. Populations may also vary greatly in abundance from year to year. In hollows in arable fields near Cambridge *L. hyssopifolia* and its associated species may fail to appear after dry winters, but are present in abundance in favourable years (Preston 1989). If the flood water disappears but the hollows remain damp the plants may grow luxuriantly; if the hollows dry out rapidly they flower and fruit as small plants. The seed may remain viable in the soil for many years.

L. hyssopifolia has been recorded in seasonally flooded habitats from scattered sites in southern England. There are particular concentrations of records in the London, Oxford and Cambridge areas, perhaps because these were well recorded areas in the 17th and 18th century when the species may have been more frequent than it is today. It is impossible to obtain a clear picture of its former abundance, as it is an inconspicuous plant which may appear and reappear at a site at irregular intervals and is also able to colonise new sites. The species was certainly rare by the time that botanical recording increased in intensity in the middle of the 19th century. It is currently known in five British sites (Callaghan 1996). The large population of an estimated several hundred-thousand plants at Slimbridge, Gloucestershire, is continuing to spread, and is colonising newly excavated hollows made for wildfowl. Substantial populations of up to 10,000 plants occur in hollows in arable fields south of Cambridge, and smaller populations are known in Dorset, Oxfordshire [but v.c. 22] and Sussex. It also occurs in Jersey.

The ecological requirements of *L. hyssopifolia* are specialised, and it relies for its continued occurrence on the flooding and disturbance of its sites. If the arable sites are not cultivated they rapidly become overgrown by perennial species and, eventually, by *Salix* scrub. The absence of disturbance is currently a threat to the small Oxfordshire population (although the long viability of the seeds may allow the plant to withstand a period of unfavourable management).

L. hyssopifolia is widespread in central and southern Europe, North Africa and West Asia. The northern limit of its native distribution is unclear: in Britain it may be a native species or a long-established introduction. It has been introduced in many areas outside its native range, including South Africa, North and South America and Australia (Meusel *et al.* 1978).

C.D. Preston

Maianthemum bifolium (L.) F.W. Schmidt (Liliaceae)

May lily, Lili Fai
Status in Britain: VULNERABLE.
Status in Europe: Not threatened.

In Britain, *M. bifolium* is a component of the ground layer of oak-birch woods, growing on well-drained acidic soils, being commonly found with species such as *Ceratocapnos claviculata*, *Dryopteris dilatata*, *Luzula sylvatica*, *Pteridium aquilinum* and *Rubus fruticosus*. Its shade-tolerant nature allows the species to maintain populations under *Larix kaempferi* and *Pinus sylvestris* following the development of conifer plantations on some of its English sites.

It is a hemicryptophyte, spreading by slender rhizomes. Tens of shoots may represent one individual. Flowering is in May and June with the leaves persisting until autumn. The main energy input is into vegetative reproduction, and the frequency of flowering is very low. For instance, in the Norfolk population only 8% of the shoots bore flowers in 1994. The seed-set is even lower, no seed being recorded from Norfolk in 1993 or 1994 (Scampion 1994), this low figure being typical of the other British populations.

There is also some evidence that *M. bifolium* was once more widespread than at present. Parkinson (1640) wrote "It groweth in . . . many places of the Realme". However, since Victorian times *M. bifolium* has been considered a rare plant. The evidence supporting the native status of *M. bifolium* in Britain was reviewed by Jackson (1913). Based on ecological conditions, longevity of records and the distance from other sites with human influence he concluded that *M. bifolium* is native at three of its current English sites: Cockrah Wood in North Yorkshire, Fulsby Wood in Lincolnshire and Hunstanworth in Co. Durham. Each of these sites currently supports at least one healthy population. Over the last 20 years the Fulsby population has remained relatively static, typically numbering several hundred leaves each year. In the same period of time three populations at Cockrah Wood, covering areas from 1-25 square metres, have maintained their size, although there has been both movement and fragmentation of some of the clones. However the original native population documented at the site since 1857 was lost in the 1980s. Other supposedly native populations were known, the most celebrated being on Hampstead Heath, this population being lost prior to the Second World War (Fitter 1945).

Records of the plant at other sites are likely to be introductions, the most notable extant ones being at Allerthorpe Common and Swanton Novers Great Wood (Swann 1971), though Ratcliffe (1977) considers it native in the latter site. The Allerthorpe population, known since 1981, was probably introduced accidentally with conifers and is on the verge of extinction through grazing by slugs and shading by rank *Deschampsia flexuosa*. A decline in the population at Swanton Novers has been offset by creating

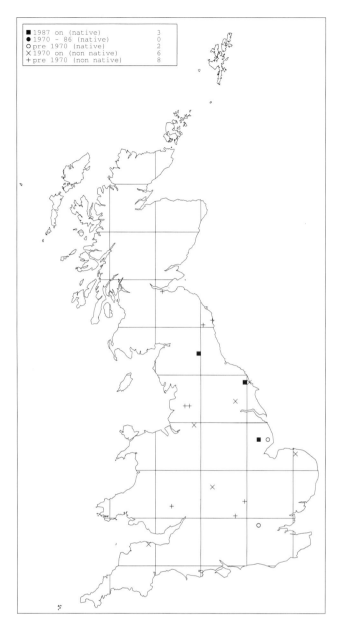

■ 1987 on (native)	3
● 1970 - 86 (native)	0
○ pre 1970 (native)	2
✕ 1970 on (non native)	6
+ pre 1970 (non native)	8

dappled shade (by the selective removal of *Ilex aquifolium* and other shrubs and trees), thereby allowing sufficient light for the vegetative spread of *M. bifolium* while still suppressing potentially out-competing ground vegetation. Such management may be appropriate at other sites.

M. bifolium is a continental species growing at the western limit of its range in eastern Britain. Its distribution covers northern Europe through Siberia to the Far East, being one of the commonest plants in the Eurasiatic boreal coniferous woods (Hultén 1962).

Further information on *Maianthemum* is found in Ietswaart & Schoorl (1985), Kawano *et al.* (1986) and Raatikainen (1990).

P.A. Ashton

Matthiola sinuata (L.) R.Br. (Brassicaceae)

Sea stock, Murwyll Tewbannog Arfor
Status in Britain: VULNERABLE.
Status in Europe: Not threatened.

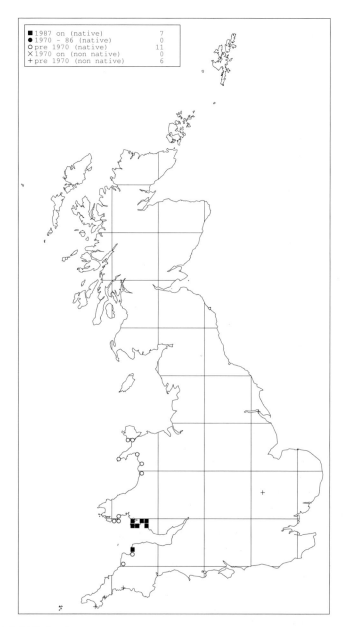

M. sinuata is a conspicuous plant of sand-dunes and sea-cliffs. Most colonies are on young, fairly mobile dunes, often near the drift-line or around low-lying, relatively damp blow-outs, where its local populations fluctuate rapidly in numbers and are often short-lived. *Ammophila arenaria, Calystegia soldanella, Elytrigia juncea, Eryngium maritimum, Festuca rubra, Lotus corniculatus, Ononis repens* and *Senecio jacobaea* are its typical associates. Cliff populations have apparently always been small but perhaps more stable in numbers.

It is biennial in cultivation, but in the field rosettes often take several years to reach the flowering stage and, although normally monocarpic, sometimes perennate after flowering. At the rosette stage, and also in plants that are starting to flower, adventitious rosettes are produced freely from the lower stem or fleshy roots if the leafy shoots are severed by grazing or wind-throw. Flowering is from June to August. The large, night-scented, pale purple flowers, borne in a spreading, branched inflorescence up to about 70 cm high, are adapted for and visited mainly by Lepidoptera, especially noctuid moths. British plants are self-compatible, with high pollen and seed fertility. After flowering, the dry inflorescence, which retains some viable seeds within undehisced siliquae for several months, may be dispersed both by wind-tumbling across dunes or cliffs and by floating on water. Long-distance dispersal with flotsam to new drift-line sites, although infrequent, seems likely to have been a major factor in its history and distribution.

M. sinuata was formerly known from several coastal areas in western Britain from the Isles of Scilly to Anglesey, but there are recent records only from Devon and Glamorgan. It was rediscovered in the latter county in 1964, after having been unrecorded since 1848. In both Devon and Glamorgan, scattered populations occur across groups of extensive dune-systems, most lying within protected sites. In south-west Glamorgan, estimates in 1994 showed that the main and outlying populations exceeded 2,000 plants (predominantly non-flowering rosettes), with 50-150 plants in each of five or six other sites along about 45 km of the coast. In North Devon, it occurs along a 7 km stretch of coast, also in local abundance. However, the history of the species in Britain and Ireland since the 17th century shows that all but one (the North Devon population) of the local populations that have been discovered have subsequently become extinct, in several cases after a period when it appeared to be well-established. The causative factors are uncertain, although

disease, predation and fluctuation (random or weather-driven) of population size below critical levels could all have been involved.

It is a plant of Mediterranean and Atlantic coasts, widespread in the Mediterranean and at its northern limit in Britain and Ireland. It is still locally frequent on the western coasts of France, but perhaps decreasing. It extends beyond Europe to North Africa, Turkey and Israel.

Q.O.N. Kay

Melampyrum arvense L. (Scrophulariaceae)

Field cow-wheat, Gliniogai'r Maes
Status in Britain: ENDANGERED. WCA Schedule 8.
Status in Europe: Not threatened.

M. arvense is typically a species of grasslands in which prominent species are *Arrhenatherum elatius, Brachypodium pinnatum* or *Bromopsis erecta*. Soils are usually calcareous, and sites in both Britain and mainland Europe are often on field-edges or road verges where there has been recent disturbance and where the grassland has not become too rank. These grasslands can often be relatively species-rich: associates include *Festuca rubra, Heracleum sphondylium, Rubus fruticosus* and *Tussilago farfara* and, on the more chalky soils, also *Clematis vitalba, Galium mollugo, Helianthemum nummularium, Hippocrepis comosa* and *Origanum vulg*are. One of the British sites is now in a garden sited in the corner of a former arable field. This species was formerly found in arable fields, particularly on the Isle of Wight, where it was said to be "so abundant as to render the bread discoloured and unwholesome, the seed being ground up with the wheat" (Townsend 1904), and locally called 'poverty weed'.

This species is an annual hemi-parasitic herb with a wide variety of host plants ranging from *Salix alba* to cultivated cereals (Oesau 1975). Flowering occurs between June and September, depending on climatic conditions, and bees are the usual pollinators. Its seeds are similar in size and weight to those of cereals and have poor dispersal characteristics. Up to half the annual seed production may germinate in the following year, while seed may remain dormant in the soil for up to two years (Matthies 1991) or longer (L.Farrell pers. comm.).

M. arvense was formerly found in a number of sites in southern and eastern England but was never common except very locally in Norfolk, Essex and the Isle of Wight. It is now known from only four sites, in the Isle of Wight, Wiltshire and Bedfordshire, but with flourishing populations at all sites. In 1996 between 1,000 and 3,000 plants were recorded growing on a chalk cliff-face in the Isle of Wight. It has been recorded from ten other sites since 1940, although at most of these it has not persisted for long (Wilson 1993).

As an annual species it is intolerant of both strong competition and heavy grazing. At one site the open conditions are preserved by natural erosion but at the other three only by human intervention, so that natural succession is a potential threat if conservation management is discontinued. Populations in Bedfordshire and the Isle of Wight were much reduced when the surrounding vegetation was allowed to become rank, but numbers have recovered following conservation management. More

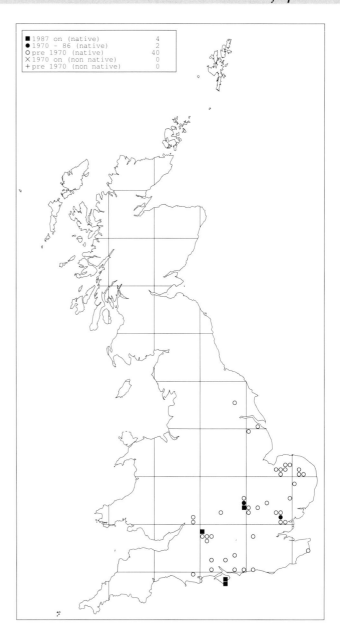

■ 1987 on (native)	4
● 1970 – 86 (native)	2
○ pre 1970 (native)	40
✕ 1970 on (non native)	0
+ pre 1970 (non native)	0

efficient seed cleaning, and agricultural intensification, including the greater use of herbicides, have led to the disappearance of *M. arvense* from arable fields (Wilson 1993).

In Europe, its range extends eastwards to the Urals, as far north as southern Sweden and southwards to the northern Mediterranean (Matthies 1991). It has declined in recent years over the whole of western Europe, although it is still widespread in eastern Europe.

P.J. Wilson

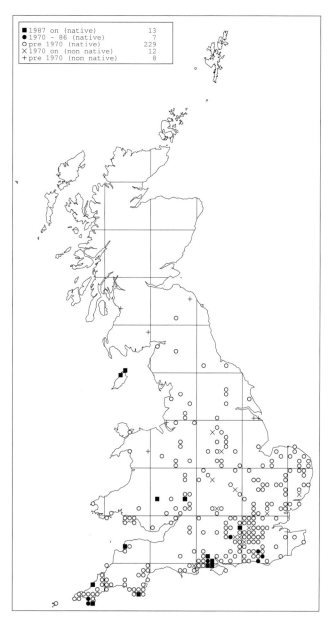

■ 1987 on (native)	13
● 1970 - 86 (native)	7
O pre 1970 (native)	229
X 1970 on (non native)	12
+ pre 1970 (non native)	8

Mentha pulegium L. (Lamiaceae)

Pennyroyal, Brymlys
Status in Britain: VULNERABLE. WCA Schedule 8.
Status in Europe: Not threatened.

This is a short-lived perennial herb of seasonally inundated grassland, usually appearing within and around ephemeral pools and runnels. The grasslands supporting *M. pulegium* are very short turf overlying clay and silt and are subjected to intense all-year-round grazing, trampling, dunging or disturbance by livestock or vehicles, causing poaching and ruts. This habitat is found within traditionally managed lowland village greens, settlement-edge lawns adjacent to open heath, and the verges of unmetalled trackways. Relict populations appear to persist where there is rutting and poaching in the absence of hard grazing. The ephemeral pools are typified by broken ground with grasses such as *Agrostis stolonifera* and the herbs *Chamaemelum nobile*, *Gnaphalium uliginosum*, *Lythrum portula* and *Ranunculus flammula*. The habitats supporting

M. pulegium also contain a wide range of rare and scarce plants including *Cicendia filiformis*, *Galium constrictum*, *Pilularia globulifera* and *Pulicaria vulgaris*.

A mature plant of *M. pulegium* consists of a central group of rooted stems giving rise to a mass of weakly rooting arching non-flowering stems. When subject to trampling, rutting or 'mulching' by dung, the stems readily root. As each fragment of *M. pulegium* is short-lived, this process needs to be continuous if the plant is to persist. If conditions are suitable, the plant may become the dominant species in the turf. *M. pulegium* flowers freely and sets seed but seedlings have rarely, if ever, been seen in the wild.

As traditional management of village greens has become very scarce in the lowlands, this species has declined. It is only in the New Forest, where a pastoral economy persists, that the plant is found in the abundance historically reported from commons throughout the lowlands. *M. pulegium* is just one of a suite of species associated with village greens to have declined (Bratton 1991; Hare 1990). Even within land owned or managed by conservation agencies, it is important to achieve a change in perception to permit the survival of these superficially untidy habitats (Byfield 1991). Threats come from ornamenting greens with flowering trees, excavating seasonally wet areas to 'save' village ponds, and filling tracks with hardcore. This declining species has been recorded as a native species in only 15 hectads since 1980, and is perhaps still extant in 12 hectads. Seed of *M. pulegium* is available commercially, and there are several records of it

turning up in recently reseeded grassland, presumably as a contaminant in seed mixtures (Kay, G.M., 1996). The alien stock is apparently more robust and erect than our native plant, though it is possible that some alien populations are shown on the map as 'native' records.

M. pulegium is widespread in Europe, north to Ireland and Poland, and also present in Macaronesia, North Africa and the Near East. It has declined elsewhere in Europe.

C. Chatters, in Stewart *et al.* (1994).

Mibora minima (L.) Desv. (Poaceae)
Early sand-grass, Eiddil-welltyn Cynnar
Status in Britain: LOWER RISK - Near Threatened.
Status in Europe: Not threatened.

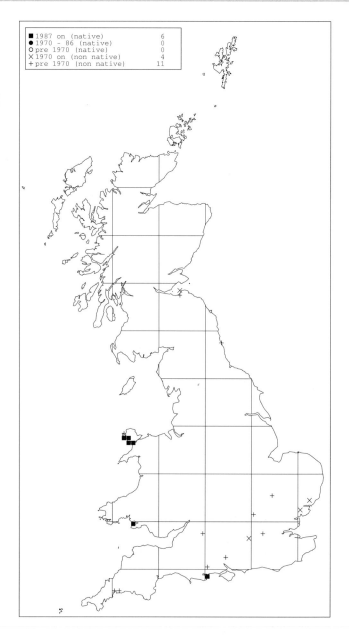

In Britain this species is confined, as a native, to coastal dune systems in Wales. It grows on nutrient-poor sandy soils that are freely draining but damp in winter, and is typically found in small isolated patches of open ground set within a matrix of unsuitable habitat. It occurs in small erosions and rabbit scrapes, stabilising blow-outs or patches of bare sand, and on pathsides. It tolerates a wide range of soil pH levels. Of a wide range of associated herbs and bryophytes, *Cerastium diffusum*, *Erophila verna*, *Ononis repens*, *Thymus polytrichus* and *Tortula ruralis* ssp. *ruraliformis* are the most constant. In the Channel Islands it is locally frequent and occurs there on thin, gravelly soils on cliff-slopes.

M. minima is a diminutive annual. Flowering begins in December during mild winters and ends by mid-March, each plant producing a few synchronous, often reddish inflorescences. Despite its small size (often less than 2 cm) it is wind-pollinated and habitually out-breeding: levels of genetic variation are high in British populations, although lower than on mainland Europe (John 1992). Some seeds are shed upon maturity in March to May; others remain attached to the parent plant, which dies immediately after seed production. Seed dispersal is often very limited. Germination normally occurs in late summer or autumn, and temporary chilling is necessary to initiate flowering (Pemadasa & Lovell 1974a, 1974b).

M. minima occurs as a native species at several localities on Anglesey and at Whiteford on the Gower peninsula. Its small size and early flowering period make it easily overlooked, and it may occur on other dune systems or even cliff sites. Other dune records are from Studland Heath, from Southport, and from Dirleton, East Lothian, but it is assumed not to be native at any of these sites, and recent investigations have revealed that it was deliberately introduced at Direlton (A.J. Silverside pers. comm.). As a naturalised weed, probably introduced from the Netherlands, it occurred in plant nurseries in Dorset and Suffolk (perhaps also elsewhere), and rarely grows as a casual.

Since they require areas of suitable open sand, which are often temporary, populations on mature dune heath are vulnerable to over-fixation and so depend upon the creation of suitable habitat; small-scale erosion and/or the presence of rabbits is beneficial.

M. minima is largely confined to western Europe and the Iberian peninsula, where it also occurs on old fields and waste ground inland. It is scarce in Belgium and Germany, and most Dutch populations are naturalised introductions, perhaps from southern European stock. Isolated populations also occur around the Mediterranean and in Russia.

R.F. John

Minuartia rubella (Wahlenb.) Hiern
(Caryophyllaceae)

Mountain sandwort, Gaineamh-lus Artach
Status in Britain: LOWER RISK - Near Threatened.
Status in Europe: Not threatened.

M. rubella is a plant of relatively dry, base-rich, gravelly soil and rock faces at high altitude. Most sites are strongly calcareous, on sloping ground, and subject to frost-heave. The resulting instability leads to incomplete plant cover and so provides the niches in which *M. rubella* may establish and maintain itself as a pioneer. Although plants may persist in closed vegetation, they do not thrive in a well-established sward with taller competitors. Associated species include *Cerastium fontanum, Minuartia sedoides, Poa alpina, Saxifraga aizoides, S. oppositifolia, Thalictrum alpinum* and *Racomitrium lanuginosum*.

M. rubella is a short-lived perennial forming discrete tufts frequently 2-3 cm wide, rarely up to 6-7 cm. The homogamous flowers are freely produced, but open only on warm or sunny days and many have imperfectly formed petals. On steep sites plants often hang by a conspicuous stock up to 3 cm long, and the strong tap-root enables plants to grow in unstable habitats. Monitoring on Ben Lawers in 1981, 1987 and 1992 showed that, whilst the total population remained roughly constant, large fluctuations in numbers occurred at some colonies. This reflects the dynamic nature of the habitat in which sporadic natural catastrophes occur, which do, however, provide new niches for recolonisation.

The largest and best documented populations are in Ben Lawers NNR, where several colonies are known on three separate mountains at altitudes between 900 and 1,190 m. Since 1980, records from Ben Lawers have been in five 1 km squares, with older records in two other 1 km squares (Payne 1981b), and with 1,000 or more plants recorded in regular monitoring (Payne 1981b; Hogarth 1987; Mardon 1992). In the Ben Alder range there are two separate sites, with records since 1990 in two 1 km squares. At one site a population size of about 500 plants was estimated in 1994, but at the other only five plants were noted in 1977 and only one in 1995. There are two outlying sites away from its Breadalbane centre of distribution: at Cairnwell in the eastern Highlands, three colonies of 200, 100 and 30 plants were recorded in 1993, and at Ben Hope in Sutherland a small population of some 30 plants was seen in 1995.

At Ben Lawers many plants are destroyed by trampling, by both people and sheep, although the sheep may not often eat the plants. Plants are crushed underfoot, or the stock broken, or dislodged by acceleration of the natural erosion of the unstable habitat. The main site is much visited by botanists, who have some adverse impact,

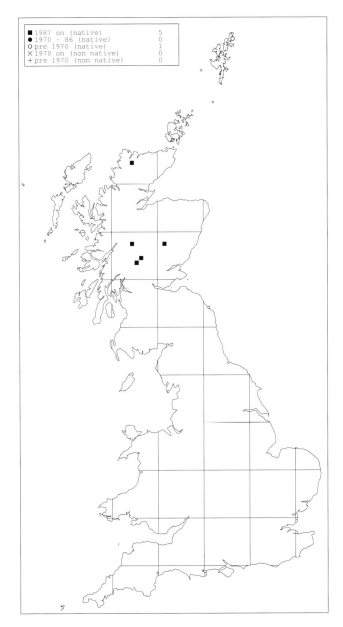

■ 1987 on (native)	5
● 1970 - 86 (native)	0
○ pre 1970 (native)	1
✕ 1970 on (non native)	0
+ pre 1970 (non native)	0

but it is bypassed by very large numbers of hill walkers. Although the population is apparently not depressed by current levels of disturbance, an increase in visitor pressure could be detrimental, as could a marked increase of grazing animals. The effects of visitors on populations at the other sites is not known but are probably of less concern.

It has an arctic circumpolar distribution in North America, Greenland, Iceland, Scandinavia, Arctic Russia and northern Asia.

D.K. Mardon

Minuartia stricta (Sw.) Hiern (Caryophyllaceae)
Alsine stricta (Swarz) Wahlenb.
Teesdale sandwort
Status in Britain: ENDANGERED.
Status in Europe: Not threatened.

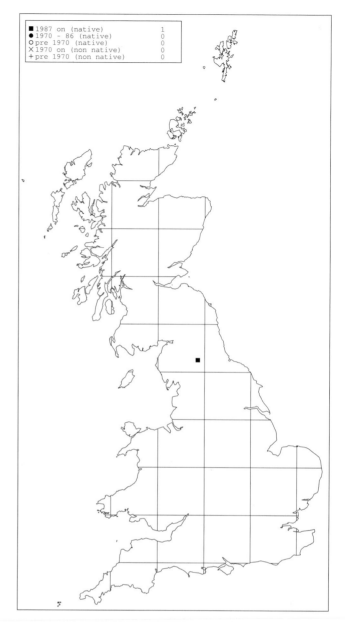

One of our rarest, but least conspicuous, plants, *M. stricta* is famously confined to Widdybank Fell in Co. Durham. James Backhouse Jr. generously attributes its discovery to G.S. Gibson of Saffron Waldon, who was accompanying him on a survey of that area in 1844 (Graham 1988). A specimen of the puzzling plant was sent to J.D. Hooker, who identified it as *Alsine stricta*. It occurs sparingly in the open gravelly and stony flushes, and eroding margins of three sikes, that are associated with outcrops of sugar limestone at 490-510 m. These wet, soft areas of oozy mud and rubbly limestone are only partially vegetated and are characterised by scattered large hummocks of the mosses *Catoscopium nigritum* and *Hymenostylium recurvirostrum*. *M. stricta* is usually confined to a narrow zone around the base (less commonly the sides and tops) of the moss hummocks, and accompanying species may include sparse occurrences of *Carex capillaris*, *C. viridula* ssp. *brachyrrhyncha*, *Juncus alpinoarticulatus*, *J. triglumis*, *Kobresia simpliciuscula*, *Minuartia verna*, *Primula farinosa*, *Saxifraga aizoides*, *Thalictrum alpinum* and *Tofieldia pusilla* (Clapham 1978; F.J. Roberts pers. comm.). It cannot withstand much competition.

M. stricta is sometimes perennial, though it may often behave as an annual. Its frequency fluctuates from year to year, presumably depending on the success of germination and survival of seedlings. Seeds can remain viable for at least five years in the soil (Pigott 1956).

The populations on Widdybank Fell appear in recent years to have been relatively stable in numbers, though individual colonies have fluctuated quite markedly, even doubling in size over a three year period but then returning to former numbers. Monitoring in 1985, 1988, 1992 and 1994 has revealed totals of 280, 311, 374 and 282 plants (S. Headley pers. comm.). Because of the very fragile habitat, a census is not carried out every year.

The main potential threat to *M. stricta* is likely to be from visitors trampling the fragile community of the gravel sikes. This could rapidly lead to severe damage, and the current restriction of access to the site is essential. There is no evidence that sheep-grazing at its present levels is adversely affecting the population.

M. stricta occurs throughout arctic and sub-arctic Europe, primarily in Norway and the Urals south to 60°N. It also occurs in Sweden, Iceland, Svalbard, Greenland and arctic North America. Though formerly known in the mountains of south-west Germany and in the French and Swiss Jura, it is probably now extinct in those areas.

M.J. Wigginton

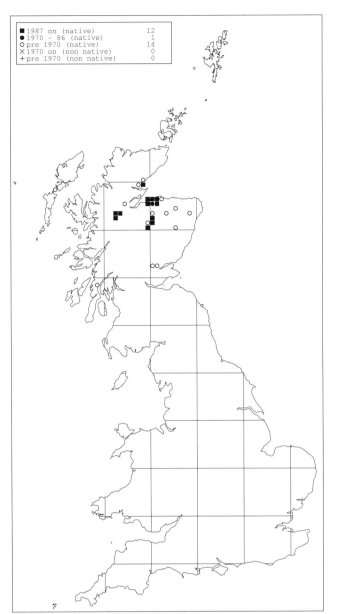

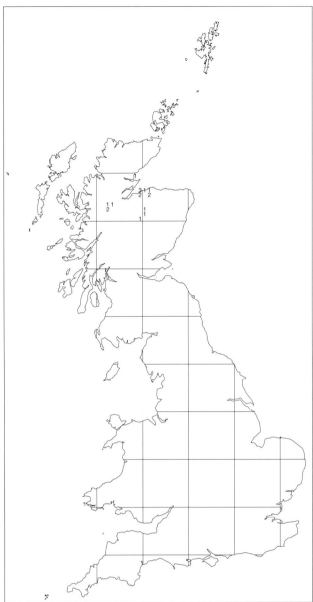

Moneses uniflora (L.) Gray (Pyrolaceae)

One-flowered wintergreen, Glas-luibh Chùbhraidh
Status in Britain: VULNERABLE.
Status in Europe: Not threatened.

In Britain, *M. uniflora* is the rarest herb of pinewoods. It is predominantly a plant of the north-eastern Scottish pinewoods, although nearly all known populations occur in old established plantations rather than in the remaining fragments of native pine forest. Within the forest, *M. uniflora* is restricted to the most humid areas of the forest floor, rosettes being found within the moist bryophyte and litter layers. The field layer is usually a mosaic of dwarf shrubs, which include *Calluna vulgaris, Empetrum nigrum, Vaccinium myrtillus* and *V. vitis-idaea,* one or other of which may be locally dominant. Besides these, the main constants are *Deschampsia flexuosa* and a range of common bryophytes, chiefly *Dicranum scoparium, Hylocomium splendens, Pleurozium schreberi, Rhytidiadelphus loreus, R. triquetrus* and the less common *Ptilium cristacastrensis.* Among the associated herbs

are *Galium saxatile, Luzula multiflora, Potentilla erecta* and *Trientalis europaea,* and *Linnaea borealis* accompanies it at a few localities. One colony of *M. uniflora* occurs in a sparsely wooded mire where the vegetation is dominated by *Calluna vulgaris, Erica tetralix* and *Molinia caerulea.* However, in that site *M. uniflora* does not appear to be very healthy as it displays leaf chlorosis (not seen at any other of its sites), and it is probably at the limit of its wetness tolerance.

This plant is a clonal wintergreen herb, flowering between June and August. It spreads within the forest mainly by a network of fine, branching roots, but the production of dust-like seed provides opportunities for long distance dispersal. It is highly adapted to the nutrient-poor shaded conditions of the forest floor: its wintergreen habit allows it to photosynthesise in all seasons and in low light conditions. Unlike closely related members of the Pyrolaceae such as *Orthilia secunda* and *Pyrola* species, which possess rhizomes, *M. uniflora* lacks such storage organs and relies on a regular and continuous supply of nutrients via its mycorrhizal associates.

M. uniflora has undoubtedly suffered a significant decline since its discovery in 1792. It is extant in three vice-counties but has formerly been recorded from nine, with doubtful records from a further three. The centre of its distribution is in Moray, and it reaches furthest west in Strathfarrar and Glen Affric. Some 80-90% of the British population of many thousands of rosettes occurs in one site. Most of the other populations are small, ranging from fewer than 50 rosettes to a few hundreds.

Historically, collecting has had a major impact on its distribution and abundance, but during this century, changes in land use and land management have become the major causes for concern. The restriction and adaptation to the conditions of the coniferous forest floor microhabitat renders *M. uniflora* extremely vulnerable to large-scale disturbance. The plant is threatened by any activity that causes widespread disruption of the forest ground-layer, resulting in increased dryness and higher light levels. Further, because of its heavy dependence on mycorrhizae for nutrition, any factor that causes a decline in the health of its fungal associates will also lead to a decline in the health of *M. uniflora*. The establishment of forest plantations has evidently provided areas for colonisation in the past, but modern harvesting by large-scale clear-felling could destroy many of the small extant populations. The continued survival of *M. uniflora* is highly dependent upon effective communication between conservationists, foresters and forest managers, not least because of the difficulty in locating the plant. *Ex situ* conservation is not an option at present because *M. uniflora* is extremely difficult to propagate in cultivation.

M. uniflora has a circumboreal distribution with its main centre in north-central Europe and Scandinavia. On the American continent it extends south to near Mexico City and in Asia it reaches the Lashai Hills of India. Little variation occurs in this monospecific genus, but one apparently poorly-differentiated variety, var. *reticulata* (Nutt.) Blake, has been recognised on the American Pacific coast.

P.S. Lusby

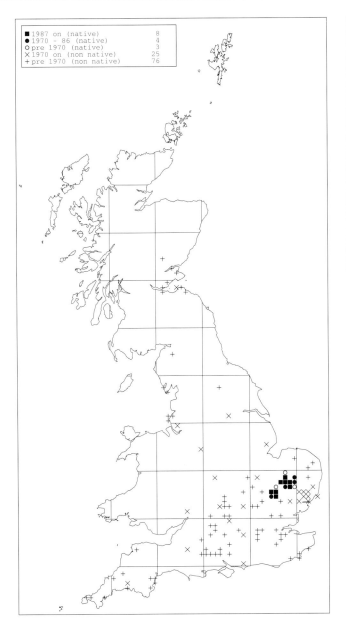

■ 1987 on (native)   8
● 1970 – 86 (native)   4
○ pre 1970 (native)   3
✕ 1970 on (non native)   25
+ pre 1970 (non native) 76

Muscari neglectum Guss. ex Ten. (Liliaceae)

M. atlanticum Boiss. & Reuter
Grape-hyacinth, Clychau Du-las
Status in Britain: VULNERABLE.
Status in Europe: Not threatened.

M. neglectum grows on a wide variety of soils, ranging from acidic, nutrient-poor sands to fertile, calcareous loams. Though mostly of open habitats, it sometimes occurs in quite dense shade. In West Suffolk and Cambridgeshire (its presumed native range), habitats include grassy roadside verges, hedgebanks, and margins and headlands of cultivated fields, and in the former county there are colonies on wind-blown sand beneath *Pinus scotica*, and in an old sand-pit. Associates are generally ubiquitous plants of grassland and edges, including *Anthriscus sylvestris, Ballota nigra, Chaerophyllum temulentum, Dactylis glomerata, Festuca rubra, Helictotrichon pratense, Holcus lanatus, Pastinaca sativa* and *Stellaria media*, but at a few sites it occurs with the rare Breckland species

Phleum phleoides, Silene otites, Veronica praecox or *V. verna*. In contrast with populations in West Suffolk, those in East Suffolk and Essex are more or less confined to grassy places near gardens, cottages or former habitation and are assumed to be relatively late introductions. At Chadlington, Oxfordshire, a large population has been known for many years in allotments, and smaller numbers in rank grassland on oolitic limestone on field verges nearby. In 1996, a small population of about 80 plants remained on a single field verge (C. Lambrick *in litt.*).

The leaves appear in autumn and may reach 30 cm by the following June. Flowering is from early April to mid-May, and by July seed is set and the leaves wither. Plants reproduce freely from seed and from offset bulbils, and it can become a pest in cultivated fields and gardens. It competes successfully in grass swards up to about 10 cm tall, but becomes scarce in rank grassland in competition with such species as *Arrhenatherum elatius* and *Elytrigia repens* (Watt 1971).

M. neglectum was first recorded in Britain in 1776, at Pakenham and Hengrave in the West Suffolk Breck. Henslow discovered it near Hinton, Cambridgeshire, in 1828, and the earliest Oxfordshire records date from 1835. Druce (1927) recorded it as "very abundant over a considerable portion of a large upland pasture near Ditchley Park [Oxfordshire] where it has all the appearance of being native", but local opinion is now that Oxfordshire populations are unlikely to be native. Since 1800, there has been a decline in the number of native sites and a contraction of its range. In 1995, 14 native sites, including two new ones, were recorded in West Suffolk (mainly around Lakenheath, Tuddenham and Culford), but in south-east Cambridgeshire (Gog Magog Hills, Fulbourn and Babraham) nine believed-native sites have been reduced to only three since 1980. Extant native colonies vary in size from just a few plants to 100-200, the largest being in West Suffolk. Introduced populations are widespread, but their range seems to have declined in recent years.

The decline of *M. neglectum* is attributed to the neglect of grassland habitats, agricultural intensification, and habitat destruction from roadworks and housing development. At the remaining sites, conservation management is best directed towards maintaining a short grassy sward, together with occasional disturbance to provide open soil for germination.

M. neglectum is widespread throughout southern Europe to northern France and the Rhenish vineyards, south-central Russia and North Africa. It has been found in rocky places up to 2,200 m on Mt. Olympus in Greece.

G. Crompton, A.J. Dunn and Y. Leonard

Myosotis alpestris F.W. Schmidt (Boraginaceae)
Alpine forget-me-not, Lus Midhe Ailpeach
Status in Britain: LOWER RISK - Near Threatened.
Status in Europe: Not threatened.

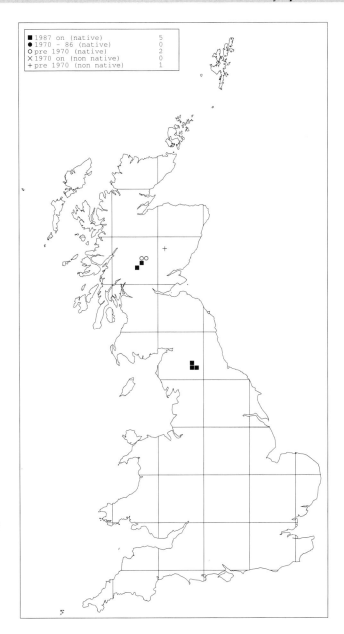

M. alpestris is a calcicolous montane plant, occurring at
an altitude of 700-1,180 m. It is found in two distinct types
of community, in two discrete areas of highland Britain. In
the northern Pennines, it is predominantly a plant of
heavily-grazed limestone grassland, occurring on base-rich
mull soils, sometimes in areas flushed with water
percolating down from limestone cliffs above. The plant
grows best in turf that is not completely closed. Other
typical plants of the community include *Carex caryophyllea*,
Campanula rotundifolia, *Festuca ovina*, *Galium sterneri*, *Luzula
campestris*, *Minuartia verna*, *Thymus polytrichus*, *Racomitrium
lanuginosum* and *Tortella tortuosa*. It is known in three
separate hills in the northern Pennines, one of which is
estimated to hold a very large, probably stable, population
of up to 100,000 plants (Taylor 1987a), but some colonies
on the other hills seem to be declining.

By contrast, in the Scottish Highlands it is principally
a plant of mica-schist ledges and slopes below cliffs,
frequently in open communities on substrates largely
composed of mica flakes and rock fragments, and often
in places inaccessible to sheep and deer. *Alchemilla alpina*,
Festuca ovina and *Thymus polytrichus* normally occur in
the community, together with such alpine species as *Draba
norvegica*, *Minuartia sedoides*, *Persicaria vivipara*, *Saxifraga
oppositifolia*, *Sibbaldia procumbens* and *Silene acaulis*. It is
known from about nine sites in the Breadalbane
mountains, with populations ranging from several
thousand plants at some sites to a few hundreds or just a
few tens of plants at others. It may be extant in hectads for
which there are only pre-1987 records, and further
information is needed. *M. alpestris* is considered to have
been introduced to one locality in Angus.

M. alpestris is a perennial, reproducing exclusively by
seed. Germination is in spring with flowering mainly in
June and July. Self-pollination takes place under
greenhouse conditions, but it is not known whether this
is usual in wild populations. It seems likely, however, that
the flies and butterflies that visit the flowers carry out at
least some cross-pollination. Seed is freely produced and
is shed in August and September. Plants grown from seed
first flower in their second year, but the longevity of
individual plants is not known, neither is the length of
time that seed remains viable in the soil (Elkington 1964).

The English and Scottish populations from rock ledges
are morphologically distinct, those in the Pennines being
smaller in all parts, and this distinction is apparently
maintained in cultivation. However, plants from grazed
slopes in Scotland are reported as being much smaller than
those growing on ledges, though it is not known whether

these differences are genotypic. It may be that grazing
pressure has led to the development of dwarf ecotypes
(Elkington 1964). *M. alpestris* may be vulnerable to
excessive grazing, but the plant is not seriously threatened
in Britain as a whole.

It is common in many montane regions of Europe from
southern Spain to the Alps, Tatras and Carpathians, the
British sites being distant outliers of its main range. The
plant is very variable throughout Europe, with many local
forms, and there are numerous very closely-related species
in Europe, Asia, and North America.

M.J. Wigginton

Najas marina L. (Najadaceae)

Holly-leaved naiad, Hollyweed
Status in Britain: VULNERABLE. WCA Schedule 8.
Status in Europe: Vulnerable.

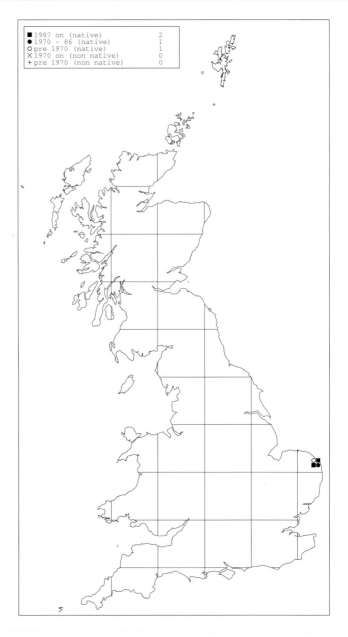

This aquatic species occurs in a few Norfolk Broads, where it is confined to clear, mesotrophic and relatively unpolluted water. It is usually found in sheltered waters with minimal boat traffic, since turbulence or strong wave action renders its brittle stems vulnerable to breaking. The late development of the 'leaf canopy' means that it cannot grow well alongside more aggressive waterweeds that mature much earlier, as is often the case in more nutrient-rich habitats. *N. marina* tends to grow best at water depths of between 0.5 and 1.5 m, generally rooting into fine organic sediments. It may form pure stands or grow with such species as *Ceratophyllum demersum, Myriophyllum verticillatum, Potamogeton pectinatus*, or charophytes including the rarities *Chara intermedia* and *Nitellopsis obtusa*. In some places it can form mono-dominant stands in which filamentous algae may be its only associates. Areas of bare substrate are readily colonised where conditions are suitable.

N. marina is a dioecious annual or short-lived perennial, a most striking and distinctive plant when mature. Male plants have, however, not been detected in Britain. It is fully developed in mid- to late summer and flowers in July and August. Fruit is common, and it sets good seed in the absence of male plants, suggesting the female plants may be apomictic.

This plant was formerly much more widespread in Broadland and has occurred in most of the Broads. Outlying sites in the last 40 years include Alderfen Broad, Barton Broad and Hoveton Great Broad, but it occurs in none of these today (Jackson 1981). It is now largely restricted to the Thurne valley and is particularly abundant in Hickling and Martham Broads. In Heigham Sound it is very scarce and has been found only a few times recently, despite much searching. Blackfleet Broad supported good populations in 1985 but encroaching reedswamp and increased sedimentation are making the site less suitable. It is likely to be surviving in a number of the ponds scattered amongst the extensive marginal reedswamp of the Upper Thurne. In Upton Broad, an isolated site in the Bure valley, *N. marina* is locally dominant, and it has now reappeared further upstream at Pound End and Cockshoot Broad, where it is increasing (Kennison 1993).

In recent years, nutrient loading has been reduced by phosphate stripping at key sewage treatment works, and some Broads have had their nutrient-rich sediments removed by suction-dredging. Experimental lake management through biomanipulation of fish populations has recently brought about the clear water that *N. marina* requires, and the strengthening colony at Cockshoot Broad bears testimony to the success of this work. *N. marina* has

seen a small recovery in the 1990s, but as poor water quality can lead to a very rapid decline, this plant remains vulnerable.

N. marina is a cosmopolitan species, occurring in both temperate and tropical regions in Europe, Asia, Africa, Australia and North and Central America. It is a variable plant, and many sub-taxa have been recognised in the Old World (Triest 1988). The British plant is ssp. *intermedia*. Though *N. marina* (also including ssp. *marina* and ssp. *armata*) is widespread in lakes across Europe, it is rare or endangered in many regions, particularly in northern Europe.

G.C.B. Kennison

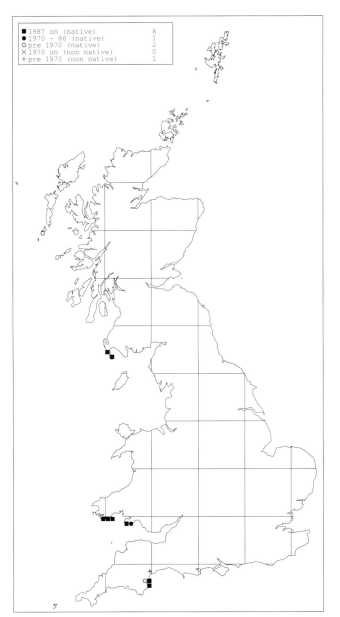

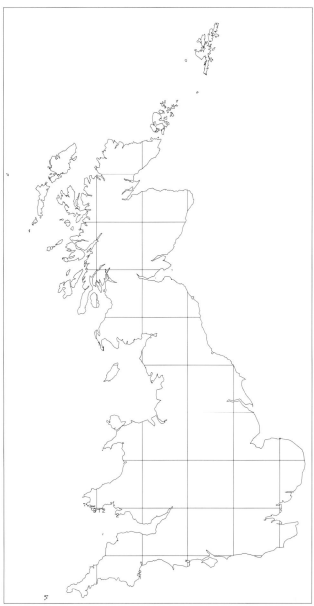

Ononis reclinata L. (Fabaceae)

Small rest-harrow, Tagaradr Bach
Status in Britain: VULNERABLE. WCA Schedule 8.
Status in Europe: Not threatened.

O. reclinata is a short-lived annual which, in Britain, has a very narrow ecological niche. It grows only on thin, dry, stony soils, usually with a low organic content, on sheltered south-facing coastal cliff-slopes of eroding limestone or (in Scotland) of thin-bedded calcareous greywacke (Lang 1977). Vegetation cover is usually sparse, consisting of species such as *Anthyllis vulneraria, Armeria maritima, Dactylis glomerata, Festuca rubra* ssp. *juncea* and *Sedum acre*.

The flowering period is very variable (depending upon the weather) and flowers may appear at any time between April and October. In Wales, and perhaps elsewhere, more than one cohort may appear in some years, resulting in at least two flowering periods. The extensive roots aid drought tolerance up to flowering, but once seed-set has begun plants desiccate rapidly from ground level in dry conditions, sometimes before the upper flowers are fertilised. Seeds may germinate at any favourable time and seedlings may overwinter (Davis & Evans 1980), being tolerant of light frost, though winter conditions will kill older plants that have not fruited. *O. reclinata* is habitually inbreeding, the flowers remaining at least partially closed in wet or cold weather but still achieving good seed-set if conditions improve. In sunny weather the flowers open fully but are rarely visited by insects. The calyx at maturity dries and opens explosively, shooting the seeds up to a metre away. However, in wet summers seeds frequently rot in the pods before they dry out sufficiently to open, resulting in almost total seed failure. Seedlings can establish only on bare ground, and self-thinning is very rapid (John 1992). It is intolerant of competition at all stages.

This species occurs in South Devon, the Gower peninsula, Pembrokeshire and Wigtownshire. Genetic studies on single populations from each area have shown that whilst they are all closely related, each population tested is both monomorphic and genetically unique (John 1992). Population size fluctuates widely: one increased from about 70 to more than 1,000, fell to zero for two years and recovered to over 40 within a decade (Davis & Evans 1980). Similar patterns occur elsewhere, with frequent gaps in the records. The reasons for such fluctuations are unknown, and there appears to be little link with the current or previous years' weather patterns.

Although several new populations have been found recently and others may remain undetected, this species is still very vulnerable both to storm damage on exposed sites and to overgrowth of suitable habitat. Both factors have caused the recent loss of populations in the Channel Islands. Whilst seeds can persist in the soil for some years, successful germination apparently requires areas of bare soil among short turf maintained by rabbit- or sheep-grazing, or at ungrazed sites where open conditions are maintained by strong winds, drought and erosion. The plant is unpalatable to rabbits. In Pembrokeshire the green calyx surrounding developing pods is sometimes grazed, apparently by snails, though the leaves are untouched.

In Europe, *O. reclinata* is largely a Mediterranean species, growing in dry, sandy and rocky areas, though it is uncommon across much of its range. It is missing from the northern Iberian coast, but occurs along the west coast of France as far north as Brittany, and in Alderney. Three varieties and one possible subspecies were described by Sirjaev (1932), who assigned British plants to var. *linnaei*. It also occurs eastwards to Iran, Syria and Israel, southwards to Arabia and Ethiopia, and in North Africa and the Canary Islands.

R.F. John

Ophioglossum lusitanicum L. (Ophioglossaceae)

Least adder's-tongue fern
Status in Britain: VULNERABLE. WCA Schedule 8.
Status in Europe: Not threatened.

In Britain, *O. lusitanicum* occurs only on St Agnes in the Isles of Scilly. It was discovered there in 1950, since which time many botanists have searched unsuccessfully for it on the other islands and on the British mainland. The plant is confined to heathland in the southern part of St Agnes, growing in discrete patches in open acid grassland. It usually occurs in thin turf on peaty soil overlying granite, but also occasionally in patches of bare peat or amongst *Polytrichum* species. Associated species include *Aira praecox, Anagallis arvensis, Armeria maritima, Danthonia decumbens, Festuca rubra, Plantago coronopus, P. lanceolata, Radiola linoides* and *Sedum anglicum*.

In a wet warm autumn, fronds may appear as early as August, but they normally appear in October. They grow throughout the winter, turning yellow and disintegrating in spring. The plant is easily overlooked since the fronds are extremely small, the exposed tips usually less than 1.5 cm long. It can be mistaken for *O. azoricum*, which shares many sites, but the fronds of *O. lusitanicum* are often still green to some extent when *O. azoricum* starts to emerge in spring. As a rule of thumb, if the fully grown plant is the same size as or smaller than the average rabbit dropping, there is the strong possibility that it is *O. lusitanicum*! A useful diagnostic feature is that *O. lusitanicum* often has fewer than six sporangial slits on the fertile frond.

None of its colonies covers more than a few square metres, and the number of fronds in a colony varies from only one or two, to several hundreds. Most plants comprise 1-3 sterile fronds arising from very short erect underground stems with long spreading roots that may bear more such stems, so that dense patches emerge from the moss or turf around the margins of large granite slabs. The total population is estimated to be fewer than 2,000 plants but in some years only about half that number are found (estimates assume that plants have on average two sterile fronds). There is usually a total of only a few hundred fertile fronds in a particular season. Grazing and other damage may soon reduce numbers, and it is not unusual to find colonies at some sites with only one or two fertile fronds, and often none at all. It is not known whether the population has increased recently or is stable, as most of the colonies have been found only in the last ten years. Furthermore, the number of fronds produced each year at individual sites can be extremely variable. In both *O. lusitanicum* and *O. azoricum*, the rootstock seems to remain healthy for several seasons underground, without producing any fronds above ground. Coombe (1992) has observed that in cultivation it is extremely prolific, fertile throughout the year, and can recover from prolonged (18-month) desiccation.

Because the turf is very thin, colonies are vulnerable to erosion and trampling, especially since several colonies

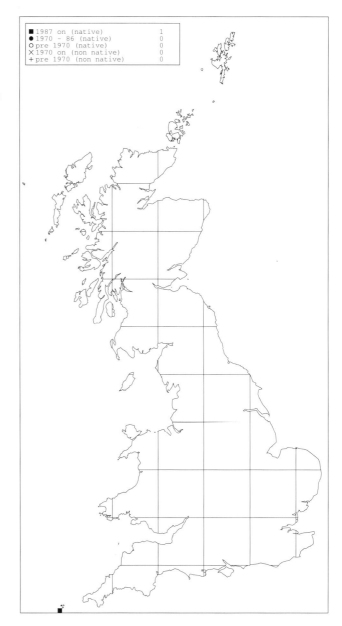

grow in the short grassland favoured by holidaymakers. Fronds are brittle and easily broken off, and at least one colony has been severely damaged by trampling. Light trampling on level ground is, however, less damaging than on slopes. With all the colonies restricted to such a small area they could also be vulnerable to uncontrolled burning, especially when fires penetrate the peat. Rabbit grazing helps to maintain the open grassy sward which favours *O. lusitanicum*.

O. lusitanicum occurs in the Channel Islands, around the western and southern coasts of Europe from France and Iberia eastwards to Greece, on most of the Mediterranean islands, in Turkey, and in the Canary Islands and the Azores.

R.E. Parslow

Ophrys fuciflora (Crantz) Moench & Reichenb. (Orchidaceae)

Late spider orchid
Status in Britain: VULNERABLE. Schedule 8.
Status in Europe: Not threatened.

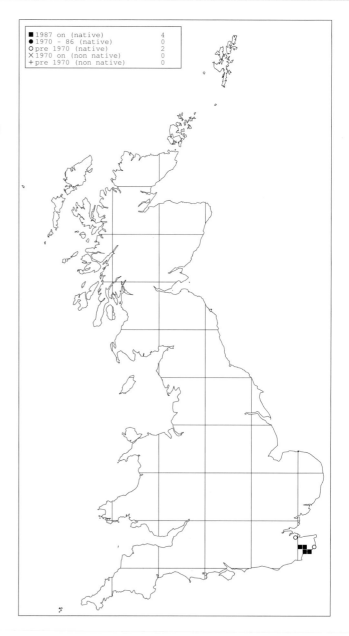

O. fuciflora grows in herb-rich grassland on freely draining soils overlying the chalk of the North Downs. The swards are dominated mainly by *Bromopsis erecta* and *Festuca ovina*. Associated species include *Achillea millefolium*, *Centaurea nigra*, *Gymnadenia conopsea*, *Linum catharticum*, *Plantago media* and *Polygala vulgaris*. Though *O. fuciflora* does not require short turf and grows satisfactorily in swards up to about 15 cm high, in some former sites it appears to have been unable to compete successfully with rank *Brachypodium pinnatum* and deep leaf litter, and may now be extinct. Some populations of *O. fuciflora* occur in chalk grassland that has developed on previously disturbed areas. In these, it tends to grow in discrete colonies, and does not appear able to colonise adjacent areas despite there being no obvious difference in land management, microtopography, habitat or soil chemistry (Ingram & Ingram 1981).

Plants flower from the end of June to mid-July (Lang 1989), and after flowering the above-ground parts of the plant die back. Small basal rosettes are occasionally produced in September and may persist through the winter (D.A. Stone pers. obs.). Studies by Russell (1993) and Stone (unpublished) suggest that this species may persist below ground in a vegetative state for two or three years, perhaps reproducing vegetatively, and that individual plants may live for at least nine years and possibly as long as 30 years. It appears that, unlike *O. apifera*, flowers of *O. fuciflora* are not self-pollinated.

In Britain this orchid has been found only in East Kent. The few records from outside this area are considered to be erroneous, and are probably all of *Ophrys apifera*. *O. fuciflora* is restricted to the North Downs chalk escarpment between Folkestone and Wye and a few nearby localities on the plateau chalk, reaching its northernmost limit there. Many of its populations have a long recorded history: for instance, one on the Folkestone escarpment has been known for over 160 years. *O. fuciflora* has been recorded from 20 localities in East Kent but now regularly occurs at only five, though it appears sporadically in a few others. In 1994 the total population was approximately 400-500 plants, about half of which occurred on the Wye Downs. At Wye, because a very small proportion of flowers set seed naturally, hand-pollination has been carried out to increase seed-set (Duffield 1979).

All known persistent localities of this species lie within SSSIs, in habitats that are currently sympathetically managed. Historically, the major threats to it and its habitat were changes in agricultural practice, particularly the improvement of grassland or its conversion to arable use, and the cessation of grazing. The major threat at the present time is likely to be the growth of rank grass, particularly *Brachypodium pinnatum*, which becomes dominant when grazing lessens or ceases. Some inflorescences of *O. fuciflora* are grazed by stock animals, and occasionally by rabbits.

O. fuciflora has its stronghold in southern and central Europe but is widely distributed, occurring from France and Spain eastwards to Romania, Greece and Turkey. It is endangered in Hungary and the Czech Republic and is vulnerable in Switzerland.

D.A. Stone and P.E. Taylor

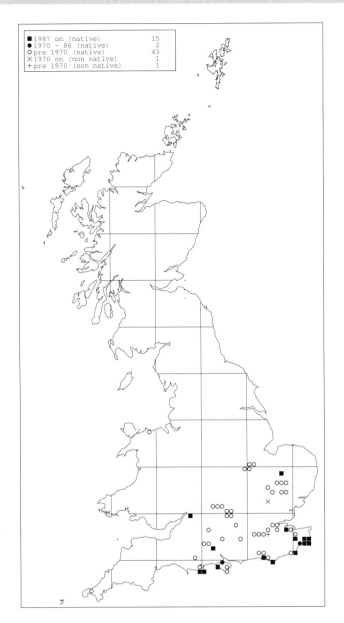

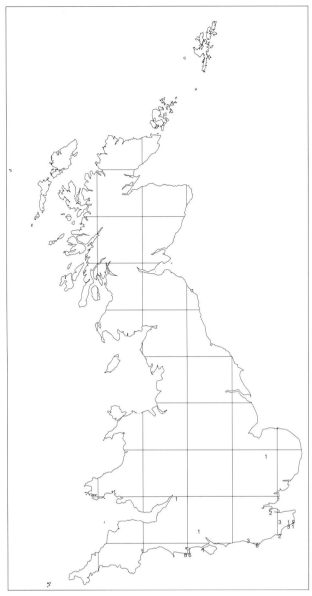

Ophrys sphegodes Miller (Orchidaceae)

Early spider orchid, Tegeirian-corryn Cynnar
Status in Britain: LOWER RISK - Near Threatened. WCA
Schedule 8.
Status in Europe: Not threatened.

O. sphegodes is a plant of ancient, species-rich grassland on chalk and Purbeck limestone. Its preferred habitat is short turf, where it grows with a wide range of associates, including *Asperula cynanchica, Bromopsis erecta, Festuca ovina, Gentianella amarella, Hippocrepis comosa, Koeleria macrantha, Lotus corniculatus, Origanum vulgare, Polygala vulgaris, Sanguisorba minor, Thesium humifusum* and *Thymus polytrichus*. It can survive in taller swards, even in *Brachypodium pinnatum*-dominated grassland, and can colonise disturbed ground, being found, for example, in limestone quarries, in turf developing on old spoil heaps, and by tracks.

This species is perennial, although the above-ground parts are analogous to a winter annual. Rosettes emerge above ground in autumn and are fully grown by October. They overwinter and produce flower spikes early in the year (generally April in Dorset, and May in Sussex and Kent), and senesce soon after, particularly if no seed is produced. Some pollination is assumed to be by Hymenoptera, but neither Summerhayes (1951) nor Lang (1989) name pollinating insects, and self-pollination is likely to be much more common. A long-term demographic study has shown that over 70% of plants flower in the first year of emergence, and this proportion increases with age until the fourth year, when almost all the non-dormant plants flower (Hutchings 1987b). Individual plants show irregular patterns of emergence, and it is likely that no more than 50% of a population is above ground during the flowering period. Few individual plants flower for more than three successive years (though they may rarely live for ten), and those that have not appeared for three years may be assumed to be dead (Hutchings 1987a).

This species has always been local and, with the exception of one site in North Wales, confined to southern England. Formerly, it was found in Cornwall, Northamptonshire, Bedfordshire, Cambridgeshire, Essex, Oxfordshire and Denbighshire, but it has declined sharply in the past fifty years and no longer occurs in these counties. It is now very largely restricted to Dorset, East Sussex and Kent, with most populations at or near the coast, but there are isolated populations in Gloucestershire, Wiltshire and West Suffolk. The largest are in Dorset, where most are on coastal grasslands between St Aldhelm's Head and Durlston. Numbers vary, but more than 15,000 plants may flower there in the best years, representing perhaps two-thirds of the British population. Other significant populations of a few thousand plants occupy similar habitats on the South and North Downs. By contrast, only a single plant appears at the West Suffolk site.

Grazing management seems to be the key factor for its survival. A long-term study of a population in Sussex showed that winter grazing by cattle caused a rapid loss of plants, with deaths exceeding recruitment every year.

However, sheep-grazing throughout the year, except for the three months while plants flowered and set seed, reversed this trend, and the population more than recovered its original status (Waite & Hutchings 1991). In Dorset, the re-introduction of grazing has helped at some sites (D. Pearman pers. comm.). Hutchings (1987a) found that no measures of population performance or recruitment showed convincing correlation with any climatic variable. Most populations are in SSSIs, and thus have a degree of protection against ploughing and other adverse agricultural practices. It is possible that local trampling and picking might be minor threats.

O. sphegodes has its stronghold in southern and central Europe. It ranges from Iberia eastwards to Bulgaria and Greece, including most of the Mediterranean islands, thence to southern Russia, Turkey and northern Iran. It reaches its northern limit in England, Belgium and the Netherlands and is threatened in several countries.

A. Horsfall and M.J. Wigginton

Orchis militaris L. (Orchidaceae)
Military orchid
Status in Britain: VULNERABLE. WCA Schedule 8. Var.
tenuifrons is ENDEMIC.
Status in Europe: Not threatened.

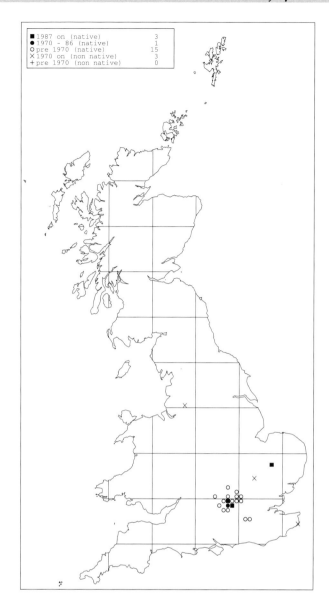

In Britain, *O. militaris* is a plant of grassland, scrub and woodland glades on chalk, occurring regularly at one site in Buckinghamshire and one in Suffolk. When first discovered in the former county, it grew in open chalk downland, but woodland has since developed on the site (planted beech, and naturally occurring ash, field maple, hawthorn, yew), and the orchid has become confined to a glade. Associated herbs are woodland and wood-edge species, including *Brachypodium sylvaticum, Chamerion angustifolium, Festuca rubra, Fragaria vesca, Hedera helix, Mercurialis perennis, Rubus fruticosus, Viola riviniana* and robust mosses including *Rhytidiadelphus triquetrus.* Its Suffolk site is an old chalk pit, now surrounded and partly shaded by a plantation of mainly sycamore and birch with an understorey of bramble and privet. A patchy ground cover of herbaceous species, including *Daphne mezereum, Festuca rubra, Inula conyza, Mycelis muralis, Poa pratensis, Ranunculus repens* and *Torilis japonica,* is interspersed with patches of abundant hypnoid mosses and some bare ground (Farrell 1985).

New shoots may first appear above the ground in early January as small crocus-like buds, but growth is slow until May, when the flower stalk rapidly elongates. The first shoots to emerge are usually the ones that flower, later ones remaining vegetative. Flowering is generally between the end of May and early June. Rather few insect species visit the flowers in Britain (Farrell 1985), but ants feed on the nectar. The main pollinators are likely to be hoverflies and bumblebees, the latter having been observed removing pollinia (Farrell pers. obs.). Seedpods ripen and turn purplish-green in August, though a relatively small percentage of plants set seed in Britain: 2-28% reported by Summerhayes (1951), and 3-10% observed in Suffolk. Because of poor seed-set, vegetative reproduction is likely to be important in British populations. Like many orchids, development from germination to flowering is likely to take several years, and individual plants can live for at least 15 years.

The Suffolk site holds by far the largest population, with up to 2,000 flowering spikes. In Buckinghamshire the colony is much smaller, currently comprising about 50 plants. Two other native sites are known, both in scrubby chalk grassland in the Oxfordshire Chilterns, but in those sites the appearance of plants has always been sporadic, and no more than six have appeared in any year. It formerly occurred in Surrey, Hertfordshire, Middlesex and Berkshire, but records from Kent are almost certainly erroneous (Philp 1982). Sell & Murrell (1996) assign the Suffolk population to var. *militaris,* widespread in Europe, and the Chilterns plants to var. *tenuifrons,* which they regard as endemic to Britain and therefore the more important taxon in conservation terms.

Habitat management in recent years has resulted in significant increases in both of the main populations. Because plants in dense shade may not flower or produce any aerial parts, conservation management has been directed towards keeping the canopy open in woodland glades by felling trees and lopping branches. Recent clearing in Buckinghamshire has resulted in the emergence of *O. militaris* in re-opened areas. Also important is the regular control of shrubs and removal of moss (Farrell 1991). At the Suffolk site it seemed that small mammals were nibbling off many shoots, taking advantage of the protective cover of blanketing moss and making runs under it. Removal of the moss reduced the damage.

O. militaris is widespread and locally abundant in Europe, ranging from south-east Sweden and Russia through all the central European countries to northern Spain, central Italy, former Yugoslavia, Romania, Bulgaria and European Turkey, and eastwards to the Altai Mountains and Lake Baikal. Habitats in continental Europe are more varied and include sand-dunes, roadside verges, wet meadows, river shingle and abandoned cultivated ground, mainly on calcareous substrates.

L. Farrell

Orchis simia Lam. (Orchidaceae)

Monkey orchid
Status in Britain: VULNERABLE. WCA Schedule 8.
Status in Europe: Not threatened.

O. simia is a plant of south-facing chalky banks in southern England. It occurs mainly in open, grazed chalk grassland, although it can tolerate some degree of shade from scrub and sometimes grows at the woodland edge. The normally herb-rich turf supports a wide range of calciphilous species, amongst which are *Bromopsis erecta*, *Carex flacca*, *Cirsium acaule*, *Festuca ovina*, *Knautia arvensis*, *Koeleria macrantha*, *Leontodon hispidus*, *Linum catharticum*, *Origanum vulgare*, *Sanguisorba minor* and *Thymus pulegioides*.

This species is a long-lived perennial. The leaves emerge early in the year, the stem elongates in April, and the pale pink flowers open from late May into June. Inflorescences are irregular in appearance, and the flowers open more haphazardly, often from the top downwards, the opposite way to most orchid species. Seed-pods ripen in late July. Some natural pollination occurs in Britain. In the Netherlands, Willems (1982) observed that it took seven years to produce a flowering plant from seed. A similar time scale has been observed in Kent populations. Individuals can live for many years, and Willems & Bik (1991) record a plant that has flowered for 19 consecutive years. Willems (1982) found that, in the Netherlands, the population development is strongly influenced by variable weather conditions, and that the long-term survival of populations depends on the longevity of one or a few individuals in the isolated populations. The colour and stature of the Kent and Oxfordshire plants are variable, the more robust Kentish plants most resembling those in continental Europe, though there is no evidence to suggest that the variation is anything more than phenotypic.

The first record for *O. simia* in Kent was from near Faversham in 1777. It was subsequently recorded from West Sussex, Surrey, and the Thames valley, where it was relatively abundant until the 1840s but then declined dramatically. It was thought to be extinct in Britain until rediscovered near Faversham in 1955 (Wilks 1960). Assisted by hand pollination, the population reached 200 flowering plants by the mid-1960s (Wilks 1966). Seed from this population was introduced to several other sites in Kent and a population became established at one of them. The first flowering plants appeared seven years after the seed was sown. Both populations in Kent are now stable and increasing, with over 200 plants at the native site and more than 100 plants at the other, although only a small proportion flower. The only other extant site is in scrubby chalk grassland in Oxfordshire. The colony there is a remnant of a much larger one that was ploughed up in the early 1950s, only a few plants having escaped the plough at the top of a steep slope. Fencing of the site against rabbits has encouraged a population increase to over 100 plants, and it appears to be spreading down the slope following scrub clearance. A few plants appeared in 1974 in dune grassland in south-east Yorkshire, increased in subsequent years to a maximum of 25 plants with nine

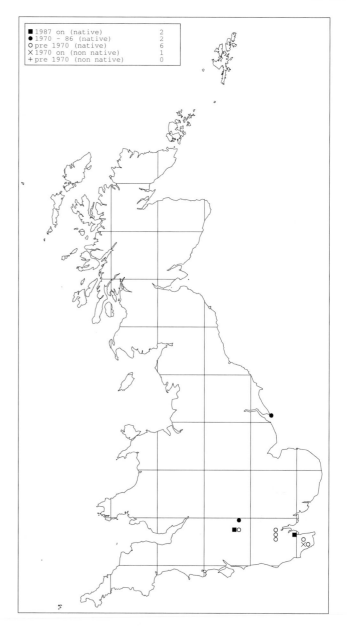

■ 1987 on (native)	2
● 1970 - 86 (native)	2
○ pre 1970 (native)	6
✕ 1970 on (non native)	1
+ pre 1970 (non native)	0

flowering (Crackles 1990), but the colony persisted only until 1983.

Short turf and open ground for seedling establishment provides ideal habitat for *O. simia*. Grazing or mowing maintains those conditions, but at some sites, flowering plants are protected by wire cages to prevent rabbits nibbling off the flowering heads. The main threats to *Orchis simia* are from a deterioration of the habitat through lack of grazing.

O. simia is widespread across southern and western Europe, north to Britain, and in the Mediterranean region to North Africa. It is abundant in France, where it frequently hybridises with *Orchis militaris*.

J.M. Church and L. Farrell

Ornithopus pinnatus (Miller) Druce (Fabaceae)
Orange bird's-foot
Status in Britain: LOWER RISK - Near Threatened.
Status in Europe: Not threatened. Near Endemic.

In Britain, *O. pinnatus* is found only in the Isles of Scilly. It has been recorded from all the larger islands, except Samson, though it now appears to be extinct on St Mary's. It is a plant of dry heaths, grassland and disturbed sandy areas, growing on granite carns and dry, heathy slopes, and also on disturbed sandy soil as, for instance, in neglected bulb fields and gardens. On the slopes of the granite carns it typically grows with *Erica cinerea, Lotus corniculatus, Ornithopus perpusillus, Plantago coronopus* and *Sedum anglicum*. In some places it is found in association with the scarce *Lotus subbiflorus, Ophioglossum azoricum* and, at one site, *O. lusitanicum*. On dune heath and other sandy sites it may grow amongst *Aira caryophyllea, Calluna vulgaris, Carex arenaria, Erodium cicutarium, Lotus corniculatus, Viola riviniana* and colonising dune plants. In some localities it can form extensive patches, trailing across the other vegetation in long skeins over 30 cm in length.

O. pinnatus is a decumbent annual, germinating in spring and flowering generally in April and May. Although the flowers are small, their brilliant colour can be quite eye-catching. In wet seasons, plants may persist until October, but in dry years plants can be difficult to find and disappear early. Such variable longevity makes any population change difficult to assess.

Although it can be found at its traditional sites every year, it can also appear sporadically in suitable habitats elsewhere. It does not appear to be under any major threat at present, though isolated populations may be vulnerable to uncontrolled burning of its heathland habitat. Other populations can apparently be lost, especially where fields have been abandoned and have reverted to grass and tall vegetation. However, it is possible that buried seed remains viable for some time, since plants have been known to reappear after cultivation.

On one island where *O. pinnatus* is quite widespread it colonised the stony bare ground exposed after an accidental heath fire and was very successful there for a number of years, although it declined to a low level as dwarf shrubs returned. Natural cycles of this kind are possible only where the plants are relatively common, emphasising the importance of maintaining all extant colonies, including those in arable fields.

In Europe this species is found in the Channel Islands, on the Atlantic coast of France, in the Mediterranean region from Spain eastwards to Greece, and in north-west Africa and Macaronesia. Its stronghold is in southern Europe.

R.E. Parslow

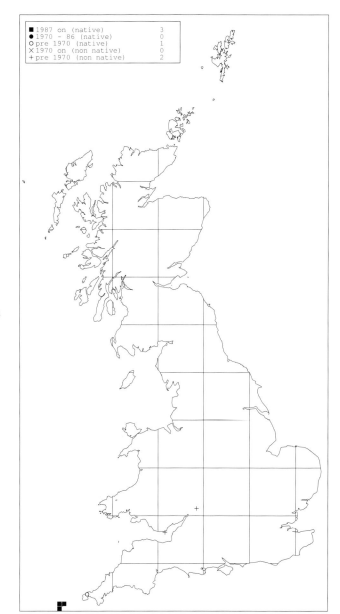

■ 1987 on (native)	3
● 1970 – 86 (native)	0
O pre 1970 (native)	1
X 1970 on (non native)	0
+ pre 1970 (non native)	2

Orobanche artemisiae-campestris Vaucher ex Gaudin (Orobanchaceae)

Orobanche loricata Rchb. pat., *O. picridis* F.W. Schultz ex Mert. & Koch [synonyms *sensu* Stace 1991]
Oxtongue broomrape
Status in Britain: ENDANGERED. WCA Schedule 8.
Status in Europe: Not threatened.

O. artemisiae-campestris now appears to be confined to unstable coastal chalk cliffs in southern England. In cliff-top grassland it occurs in species-rich turf with *Filipendula vulgaris*, *Picris hieracioides* and *Sanguisorba minor*, and typical associates in its cliff-face habitats include *Anthyllis vulneraria*, *Galium verum*, *Onobrychis viciifolia* and *Ononis repens*.

This species is normally an annual and is in flower in late June and July. On exposed maritime cliffs the flowers quickly wither and plants appear to remain in peak flowering condition for only a few days. It parasitises species of Asteraceae, especially *Picris hieracioides*. Other apparent hosts include *Leontodon hispidus* and *Pilosella officinarum*, but in most cases verification of the host has not been possible. On occasions it has been seen at a considerable distance from its traditionally accepted hosts, and so other host species may be involved.

This species is currently extant in Kent and the Isle of Wight, but has not been confirmed recently in West Sussex. It was formerly thought to have had a wider range, having been recorded from Breconshire and Worcestershire eastwards to Oxfordshire, Buckinghamshire and Suffolk. However, because of its similarity to the variable *O. minor*, most or all records away from the south coast are considered likely to be erroneous (e.g. Stace 1991) and are not mapped. Its present-day sites currently hold very few plants, usually fewer than 20. The colony at Freshwater, Isle of Wight, was formerly much larger, the plant having been described there as "plentiful" in 1883 (Bevis *et al.* 1978). Apparently few plants survive today, though erosion has made access and inspection of the colony very difficult. In 1986 about twenty spikes appeared in the chalk grassland on the cliff-top at Tennyson Down, but the population declined rapidly and no plants have been seen there in recent years. In East Kent there are two surviving populations in similar habitats to that on the Isle of Wight. One at Dover held up to 20 plants in 1994, scattered over a distance of 100 m on more or less unstable cliffs and ledges. Further east at what is sometimes considered to be the classic British site, numbers are lower and plants can be atypical, perhaps because of hybridisation with *O. minor*. There are considerable stretches of unstable, inaccessible cliffs elsewhere on the south coast where occasional records have been made, and the plant may still be locally present in small numbers.

The main threat to its survival lies in the unstable nature of its cliff-face habitat. Its relative inaccessibility provides some protection, although at one locality there is some recreational pressure. In the past, collecting was a

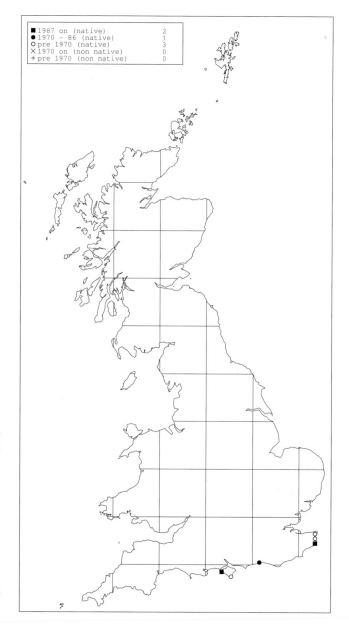

serious threat, as evidenced by the number of specimens in British herbaria. In East Kent, construction work for the Channel Tunnel has been cause for concern, but so far it appears to have had little effect on the population.

O. artemisiae-campestris is a member of a critical group of plants allied to *O. minor*. It is widely distributed throughout central and southern Europe, though there are different views on the taxonomic affinities of this and related species, especially *O. picridis* and *O. fuliginosa*. Kreutz (1995), following Gilli (1966), treats the British taxon as a separate species (*O. picridis*). He regards *O. loricata* as synonymous with *O. artemisiae-campestris*, a central and southern European taxon. This taxonomic uncertainty obscures the distributions of the species.

M.J.Y. Foley

Orobanche caryophyllacea Sm. (Orobanchaceae)
Bedstraw broomrape, Clove-scented broomrape
Status in Britain: VULNERABLE. WCA Schedule 8.
Status in Europe: Not threatened.

In Britain, *O. caryophyllacea* is restricted to East Kent; the few records from outside the county representing errors in identification or mis-labelling of herbarium specimens. The population at Sandwich Bay is on an established dune system, whilst its other sites in Kent are hedgebanks or scrub on chalk downs and undercliffs.

This species mainly parasitises *Galium mollugo* and *G. verum*. It is perennial and, once established, may be long-lived. It flowers in June and July, with flowers at the start of anthesis being noticeably clove-scented. Most flowers produce seed, indicating good pollination. It appears to be an outbreeding species, as isolated single inflorescences or multiple inflorescences from the same clone do not set seed, suggesting a high degree of self-incompatibility. A range of insects, including bumble-bees, wasps and stiletto flies, have been recorded visiting the flowers, and all are likely pollinators. The plant is also able to reproduce vegetatively.

O. caryophyllacea is most numerous at Sandwich Bay, where it can regularly be seen in considerable numbers, with more than 1,000 inflorescences. It has been recorded at six other sites since 1980, all within the Dover-Folkestone area, though one is now lost to development. Most of these populations are small (from one to 20 spikes), although in the latter part of the 1980s, over 60 inflorescences were recorded at two of these sites. However, the chalk cliffs are a fragile and difficult habitat to explore, and it is possible that it may be more frequent in this area than the few records suggest.

Currently, the population at Sandwich Bay seems reasonably secure, but populations on the South Downs may be threatened by habitat destruction as a result of road widening or other development associated with the Channel Tunnel link.

O. caryophyllacea has a widespread distribution in central Europe, extending eastwards to Poland, Latvia, the Caucasus and Iran, southwards to North Africa and northwards to Sweden and perhaps Norway. It is difficult to explain its limited British distribution, since its occurrence in the northern Alps and Scandinavia suggests that its range is not limited by a cooler climate.

M. Jones

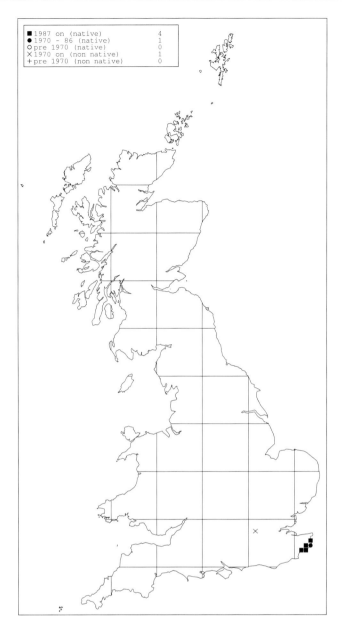

■ 1987 on (native)	4
● 1970 – 86 (native)	1
○ pre 1970 (native)	0
✕ 1970 on (non native)	1
+ pre 1970 (non native)	0

Orobanche purpurea Jacq. (Orobanchaceae)
Purple broomrape, Gorfanc Glasgoch
Status in Britain: VULNERABLE.
Status in Europe: Not threatened.

O. purpurea is found mainly on dry, unmanaged and undisturbed grasslands, on soils that are at least slightly basic. The more natural habitats include cliff-top grassland, and many records are from roadside verges and banks that have escaped disturbance. However, it is also a plant of disturbed habitats occurring, for instance, in rough grassland on a reclaimed industrial site, on the level top of a river embankment, and by a car-park. Its grassland habitats are varied and include both tall rank swards dominated by *Arrhenatherum elatius,* and shorter swards in which *Agrostis capillaris, Dactylis glomerata, Elytrigia repens, Festuca rubra* or *Holcus lanatus* may be prominent. Among a wide range of associates, usually undistinguished, are *Centaurea nigra, Cruciata laevipes, Daucus carota, Heracleum sphondylium, Hypochaeris radicata, Hypericum perforatum, Lotus corniculatus, Plantago lanceolata* and *Rhinanthus minor. Orobanche minor* occurs with it at some sites.

O. purpurea is parasitic on *Achillea millefolium.* It is an annual or short-lived perennial, with a flowering period extending from the end of May until August. Seed production is prolific, though seed may be poorly dispersed. Like other *Orobanche* species, it is thought that seed can remain viable for long periods in the soil, and this assumption is supported by the re-appearance of plants at former sites after long absences. The flowers may normally be self-pollinated.

Despite the abundance of its host plant, *O. purpurea* is very localised. Between 1990 and 1994 it was recorded from 17 localities in Dorset, Hampshire, the Isle of Wight, Suffolk, Norfolk, Kent, Pembrokeshire and Cumbria. It has been lost from its natural sites in Lincolnshire owing to land-fill, though it was successfully translocated to nearby sites, where it continues to thrive. Most extant populations are small, ranging from a single spike to a few tens of individuals. However, more than a hundred flower spikes have occurred in one population in the Isle of Wight (Everett 1988), and newly discovered populations at Maryport docks are large and increasing, with up to 360 spikes recorded in 1993. It has also reappeared in counties where it was thought to be extinct, as, for instance, at a previously known site in Dorset after an absence of 40 years, and at Manorbier in Pembrokeshire in 1991, also after a gap of 40 years. The appearance of this plant may be erratic at many of its sites, in some years appearing in very small numbers or not at all.

Inappropriate grassland management and land development are potential threats to some populations.

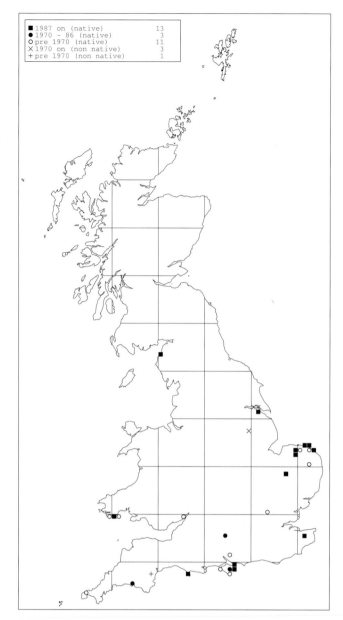

■ 1987 on (native)		13
● 1970 – 86 (native)		3
○ pre 1970 (native)		11
× 1970 on (non native)		3
+ pre 1970 (non native)		1

The rapid erosion of coastal cliffs threatens the largest Isle of Wight population and some colonies in Norfolk.

O. purpurea occurs throughout Europe (to central Russia) and is not uncommon in the Mediterranean region. However, it appears to have declined sharply in the northern parts of its range and is threatened in several countries, including Germany, Poland and the Low Countries. It is still locally frequent in some of the Channel Islands.

A. Pickering

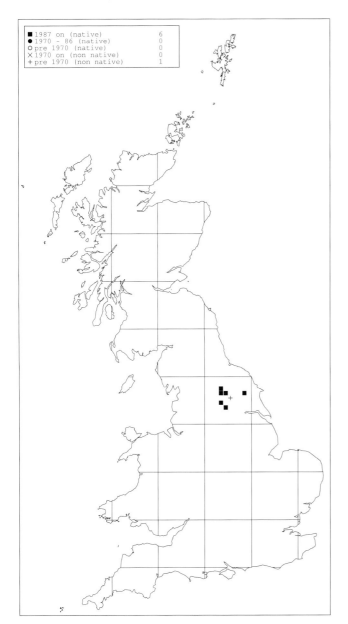

Orobanche reticulata Wallr. (Orobanchaceae)
Thistle broomrape
Status in Britain: LOWER RISK - Near Threatened. WCA
Schedule 8. The British taxon is perhaps ENDEMIC.
Status in Europe: Not threatened.

In Britain, *O. reticulata* is restricted to Yorkshire, where
it was first identified as a British plant in 1909, though
herbarium specimens pre-date this by at least 50 years. The
great majority of its localities are on or close to the
Magnesian limestone formation, with an outlier further
east on the chalk. It is a root parasite of thistles, almost
always (99% of plants) on *Cirsium arvense*, but very
occasionally on *C. eriophorum, C. heterophyllum, C. palustre,
C. vulgare* or *Carduus nutans*. Typical habitats are rough
pastures, road verges, semi-natural grassland and,
especially, the banks of rivers and flood plains with
associated light scrub, though not where ground is
permanently waterlogged. Its most frequent associates are
Arrhenatherum elatius, Cruciata laevipes, Dactylis glomerata,

Festuca rubra, Glechoma hederacea, Ranunculus repens and
sparse *Rubus fruticosus* and *Urtica dioica*. In moister
habitats, it may occur with *Filipendula ulmaria, Impatiens
glandulifera* and *Petasites hybridus* (Abbott 1996).

This species is possibly perennial in certain situations
but otherwise behaves as an annual or biennial, appearing
above ground in July. Seed setting is usually good, and the
small seed is readily dispersed. Water-borne seed dispersal
appears to be very efficient, as evinced by its tendency to
occur at or close to river margins, although existing plants
along with their host may additionally be displaced and
relocated by flood water. Nevertheless, even in areas where
it is frequent, vast stands of thistle remain unparasitised,
the controlling ecological factors remaining largely
unknown. It tends to be absent from the more dense
vegetation and usually parasitises smaller, relatively
immature host plants. Dead spikes can persist well into the
winter and sometimes remain until the following flowering
season. Its appearance in newly-cleared sites (even

woodland) strongly suggests that seed remains viable for many years (Abbott 1996).

Most of its locations are in the valleys of the Ure and Wharfe, where it seems to have increased in recent years; however there is at least one strong colony on a roadside verge away from river influences. A recent survey established that, of 70 separate recorded populations, 54 were extant and more than 85% of these were within the influence of river flood-water (Foley 1993). Numbers vary considerably from one year to the next, but it has the ability to persist even in small colonies, some of which are still extant after a period of 80 years or more. Most populations are small (1-50 plants), but a few regularly exceed 100, though the largest held 787 spikes in 1995, following soil disturbance. It seems to do particularly well in hot summers. Its occurrence close to Roman roads and settlements has led to suggestions (e.g. Pugsley 1926) that it might have been an early introduction, but more recent studies support its claim to native status.

The main threat to its survival results from the deliberate destruction of its hosts, many of which are serious agricultural weeds. Other damaging activities, especially ploughing and spraying, road construction and gravel extraction, all pose additional threats. It will not thrive in shade, or in tall grass, and clearance of dense stands of tall *Urtica dioica* would benefit some colonies (Abbott 1995). Opening up the turf and disturbance of the ground have resulted in increased populations at some sites. Rabbit activity may likewise be a positive asset.

O. reticulata is recorded from much of Europe and especially the alpine region. It apparently extends eastwards into Russia and western Asia and is also recorded from North Africa. Towards the northern limits of its range it exhibits a more disjunct distribution pattern, as shown by its occurrence in Britain, Scandinavia, the Baltic states, the Netherlands and northern Germany. Outside Britain, several variants and subspecies have been recorded. The exact status of the British plant, which was formerly referred to var. *procera*, is not fully understood and the taxonomy of the European members of this group is currently under investigation. Kreutz (1995) regards the British plant as a separate species - *O. pallidiflora* Wimmer & Grabowski (*O. reticulata* ssp. *pallidiflora* (Wimmer & Grabowski) Hayek).

M.J.Y. Foley

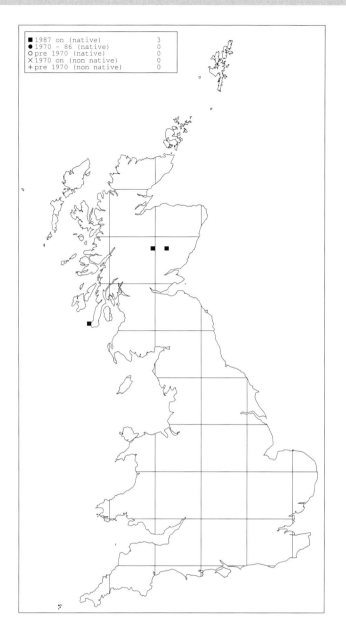

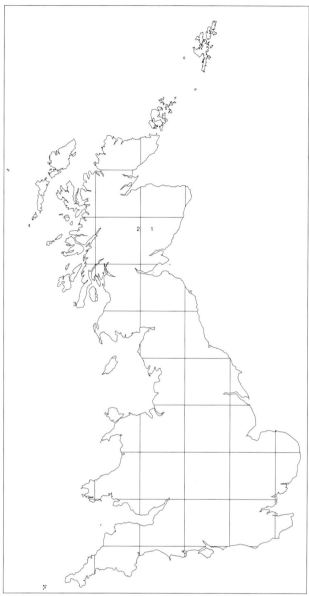

Oxytropis campestris (L.) DC. (Fabaceae)
Yellow oxytropis, Ogsatropas Buidhe
Status in Britain: VULNERABLE.
Status in Europe: Not threatened.

O. campestris appears to be a strict calcicole in Britain. In Coire Fee, Glen Clova, it grows on calcareous hornblende schist, at Loch Loch on limestone, and at Dun Ban on limestone and calcareous schists; all these rocks belong to the Dalradian series. Its montane associates include *Alchemilla alpina, Carex capillaris, Dryas octopetala, Galium boreale, Persicaria vivipara, Polystichum lonchitis, Saxifraga aizoides, S. oppositifolia, Sedum rosea, Silene acaulis* and *Veronica fruticans.* Accompanying lowland plants are *Campanula rotundifolia, Carex caryophyllea, Festuca ovina, Helictotrichon pratense, Thymus polytrichus* and *Viola riviniana.* In its coastal site in Kintyre, it grows on limestone cliffs from 25 to 180 m above sea level, growing best on south- and west-facing slopes (Cunningham & Kenneth 1979). Inland sites are at altitudes between 500 and 650 m.

O. campestris is a tufted perennial, flowering between May and late July. Recruitment is by seed, which sets freely.

This species was long thought to be confined to two mountain cliffs at moderate elevations in the Scottish Grampians, but a third colony on sea-cliffs near the Mull of Kintyre, originally described as *Oxytropis halleri*, was later referred to *O. campestris.* The other localities are in East Perth and Angus. Each station has populations of probably thousands of plants, but growing over a fairly restricted area and found mainly in open rock face communities of cliffs with a south-south-east to west aspect. Some plants grow on adjoining rocky slopes, but all habitats are dry and well-drained.

In its inland localities, *O. campestris* seeds into earthy screes and closed grasslands below the cliffs, but is then vulnerable to grazing and appears to have limited powers of spread under present conditions. A true relict, it has been much collected in the two mountain stations, but has maintained its populations fairly well. In Kintyre, the numerous wild goats seem not to graze the plants to any extent (Cunningham & Kenneth 1979).

O. campestris is a widespread Palaearctic arctic-alpine. It occurs in northern Europe down to near sea-level but is restricted to montane regions of central and southern Europe. In North America, it is found in the Rockies, in Maine and in Labrador, but it is absent from Greenland.

In Scandinavia, *O. campestris* occurs in a wider range of habitats, including coastal sand-dunes and machair, grassland, road verges, rock faces and debris, and fell-fields, growing on soils that vary from calcareous to only slightly base-rich. It also shows wider variation in flower colour, from cream to pink and purple.

D.A. Ratcliffe

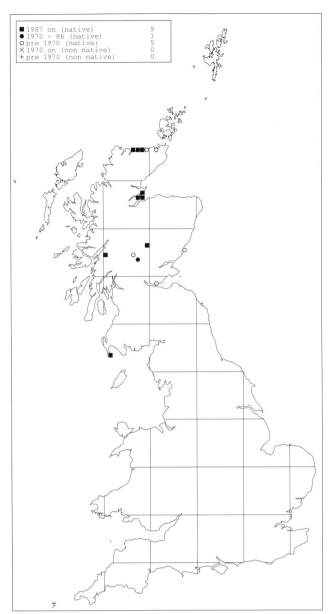

Oxytropis halleri Bunge ex Koch (Fabaceae)
Purple oxytropis, Ogsatropas Corcarach
Status in Britain: LOWER RISK - Near Threatened.
Status in Europe: Not threatened. Endemic.

Most populations of *O. halleri* occur close to sea level on the north and east coasts of Scotland. There it grows on dry cliffs formed from basic Old Red Sandstones and conglomerates and on calcareous sand-dunes. Associated species on cliff ledges include *Geranium sanguineum*, *Helianthemum nummularium*, *Sedum rosea* and *Thymus polytrichus*, whilst on cliff-tops it grows in short turf with *Carex flacca*, *Lotus corniculatus*, *Plantago coronopus*, *P. maritima* and occasionally *Scilla verna*. On dunes, associates include *Anthyllis vulneraria*, *Astragalus danicus*, *Galium verum*, *Koeleria macrantha*, *Linum catharticum*, *Lotus corniculatus*, *Thalictrum minus* and, in a few places, *Primula scotica*. Inland, *O. halleri* grows on Dalradian limestones and schists at altitudes of 600-760 m, often in rich swards with such choice species as *Astragalus alpinus*, *Botrychium*

lunaria, *Cerastium alpinum*, *Draba incana*, *Persicaria vivipara*, *Saxifraga oppositifolia* and *Silene acaulis*. Whilst most sites are rather basic, some are not. For instance, it grows on an acidic dyke on Ben Chonzie, and some of its coastal bank habitats are not obviously basic.

This perennial herb flowers between May and July, often a month later in montane sites than on the coast. Small plants frequently occur in the vicinity of larger ones, so reproduction from seed can be inferred. Flowering is inhibited by grazing. Plants in heavily grazed swards can be diminutive and quite different in appearance from the luxuriant profusely-flowering clumps seen on inaccessible cliffs.

O. halleri was first reported in Britain by Lightfoot in 1777 growing on Carboniferous limestone cliffs near North Queensferry, Fife and on Ben Sgulaird, Argyll. The former site was destroyed by the building of the railway cutting to the Forth Bridge, but the plant is still on Ben Sgulaird. It has a relict distribution in Britain, and is now confined to the north coast of Sutherland, the east coast of Ross-shire, a single site on the Mull of Galloway and three mountains in Perthshire and Argyll. There are eighteen extant localities, some of which support more than one population. Populations vary in size from a few plants at most east coast sites to some large colonies of many thousands extending along hundreds of metres of cliff on the north Sutherland coast. In the absence of detailed long-term monitoring, population trends cannot be discerned. There may be other as yet undiscovered colonies, particularly along the more remote stretches of the north Scottish coast.

On coastal cliff ledges *O. halleri* is generally protected by its inaccessibility. Elsewhere, encroachment by gorse and scrub and, conversely, the use of fire to control scrub, may both present a threat. One site is in danger of being overshadowed by a commercial forestry plantation. Dune and cliff-top habitats are susceptible to overgrazing by rabbits and domestic livestock. At one locality, the supplementary feeding of livestock has led to excessive nutrient enrichment near a colony of *O. halleri*, which is now threatened by the vigorous overgrowth of rank grass. In montane habitats grazing could be restricting its distribution.

Elsewhere in Europe *O. halleri* is strictly montane, being found in the mountains of central Europe, including the Pyrenees, Alps, and Carpathians. There is one known locality in eastern Albania.

R. Scott

Petrorhagia nanteuilii (Burnat) P. Ball & Heywood (Caryophyllaceae)

Kohlrauschia nanteuilii (Burnat) P. Ball & Heywood, *K. prolifera* auct., non (L.) Kunth

Childing pink, Penigan Ffrwythlon

Status in Britain: ENDANGERED. WCA Schedule 8.
Status in Europe: Not threatened.

P. nanteuilii has been recorded in a range of coastal habitats, including shingle and sand-dunes, sandy banks and pastures, verges of tracks and roads, and on waste ground. The plant is now mostly confined to more or less stabilised shingle in thinly-vegetated areas. It occurs where there has been temporary disturbance, but not on unstable shingle. A wide variety of associated species includes *Arrhenatherum elatius*, *Beta vulgaris*, *Crepis vesicaria*, *Daucus carota*, *Glaucium flavum*, *Myosotis ramosissima*, *Sedum acre*, *Senecio jacobaea*, *Trifolium pratense*, bryophytes and lichens.

It is an annual with stems to 50 cm high bearing a compact ovoid head of flowers that normally open one at a time. Flowers appear mostly in late June and July, though some may remain until September, and are visited by butterflies and other insects. The seeds when fully developed are relatively heavy and fall not far from the parent plant. Such poor dispersal may significantly limit the plant's ability to extend its range. Morphological, cytological and geographical data strongly suggest that *P. nanteuilii* is an allopolyploid taxon derived from *P. proligera* and (the non-native) *P. velutina* (Akeroyd 1975; Akeroyd & Beckett 1995).

The history of *P. nanteuilii* is complicated by the fact that until recently it had been widely confused with the closely allied *P. prolifera* (Ball & Heywood 1962; 1964). *P. nanteuilii* has been recorded in a number of places between Pagham and Selsey on the Sussex coast, and it was formerly abundant around Hayling Island, South Hampshire. It was once thought to be native in the Isle of Wight, Middlesex, Berkshire, Buckinghamshire and Norfolk (being casual at several other widespread locations), but these records are of *P. prolifera*. It is considered to have been native at Hythe in Kent, where it occurred until 1960 in shingle communities similar to those in which the plant is found today. It was last seen in Hampshire in 1968 (Brewis *et al.* 1996) and now occurs only at Pagham, West Sussex. A self-maintaining introduced colony in Glamorgan has been known since about 1930.

At Pagham it is restricted to two shingle spits and sandy ground on either side of the harbour mouth (Carver 1990). In recent years there have been some 15 discrete populations, but they are very limited in extent and fluctuate considerably in size year by year and may shift their position. In good years, tens of thousands of plants may occur, but in cold wet years, colonies may be small (even just a few individuals), with some disappearing altogether.

The survival of *P. nanteuilii* at Pagham seems precarious, with individual populations being threatened by natural erosion and shingle movement. Massive

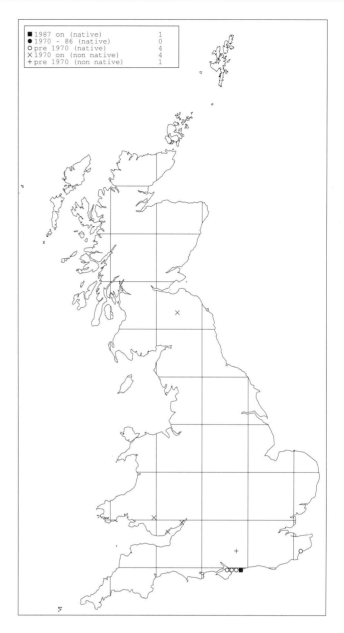

■ 1987 on (native)	1
● 1970 – 86 (native)	0
○ pre 1970 (native)	4
✕ 1970 on (non native)	4
+ pre 1970 (non native)	1

movement of shingle bars has occurred in the past and may do so again. Remedial or preventative coastal defence works may be detrimental, causing physical damage and excessive consolidation of shingle. The largest populations on the southern spit are currently protected by being in a bird breeding area to which human access is limited during the summer months. Encroachment by sea, coastal defence works and other building development were the main causes of loss from former sites.

In the Channel Islands, it is still common in Jersey. Elsewhere in Europe, it is most frequent in Iberia, where it is common both inland and on the coast. In the Mediterranean region, it is scattered along the coasts of France and Italy, and in Corsica, Sardinia and the Balearic Islands. It also occurs in Morocco, Madeira and the Canary Islands, and perhaps in western Asia.

M.J. Wigginton

Petrorhagia prolifera (L.) P. Ball & Heywood (Caryophyllaceae)

Kohlrauschia prolifera auct.

Proliferous pink

Status in Britain: CRITICALLY ENDANGERED.

Status in Europe: Not threatened.

This species has been recorded from scattered sites throughout southern England. Its status in Britain has been a matter for debate, and it has not always been regarded as a native plant. Whilst it has been undoubtedly of casual occurrence in some places, it is now generally regarded as a relict native in Norfolk and, less convincingly, in Bedfordshire (Akeroyd & Beckett 1995). In Norfolk, its history is well documented from the 1840s, all records being on the light soils of West Norfolk, and most referring to parishes within 5 km of its present known site.

P. prolifera can grow to 50 cm but is usually only 10-20 cm tall. Its narrowly linear, grass-like leaves are difficult to spot when the plant is not in flower, especially when it is growing amongst grass. It is an annual, flowering in June to August, the flowers opening singly and not very widely. The seeds have a distinctive reticulate pattern, which differentiates them from those of *P. nanteuilii*, which are tuberculate.

In Norfolk a number of records were made in the 1950s, but the plant is currently known at only one site, where it was discovered in 1985 (Beckett 1992). The colony holds it own in the very open vegetation on droughted ground at the edge of a concrete roadway leading into a former military camp. It also spreads into the adjacent ranker grassland on roadside banks, and within the adjacent SSSI. In 1994, about 75 plants were counted: many flowering stems, which were cut during the trimming of road verges, flowered again in early autumn. Another locality nearby is now dense grass at the edge of afforestation and *P. prolifera* is no longer found there. However, it is likely that its seed is long-lived, and that the plant may yet re-appear there if the habitat is opened up. In Bedfordshire, *P. prolifera* grows on light sandy soils by a sand pit. The original site was lost in about 1980 through further sand winning, and it was feared that it had become extinct as a result. But twelve plants were seen in 1995 close to the site of the original colony.

P. prolifera is widespread throughout central and southern Europe, extending north to southern Sweden, and also occurs in North Africa. It is naturalised in Chile, New Zealand and Australia, where it is sometimes a prolific weed on dry, sandy soils.

G. Beckett

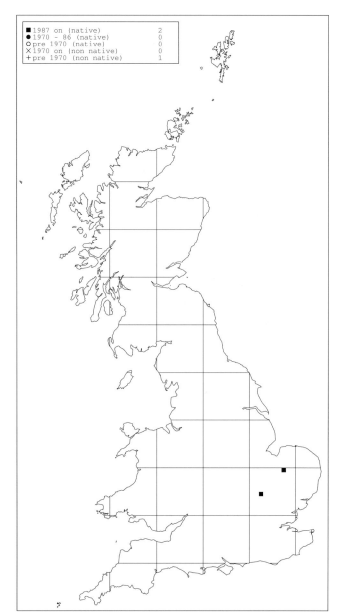

■ 1987 on (native)	2
● 1970 - 86 (native)	0
O pre 1970 (native)	0
✕ 1970 on (non native)	0
+ pre 1970 (non native)	1

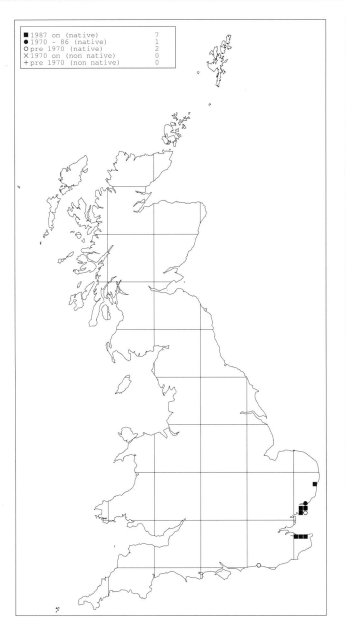

Peucedanum officinale L. (Apiaceae)

Sea hog's-fennel
Status in Britain: LOWER RISK - Near Threatened.
Status in Europe: Not threatened.

In Britain, *P. officinale* is exclusively a plant of coastal grassland, from the highest saltmarshes up to, exceptionally, 1.5 km from saline water. It forms conspicuous mounds of ferny foliage amongst coarse grasses such as *Arrhenatherum elatius* and *Elytrigia atherica*, but generally dies back to the deep rootstock in winter. In places, it occurs abundantly along the edge of *Prunus spinosa*, *Rosa canina* and *Rubus fruticosus* scrub, where it may benefit from reduced grazing, trampling and mowing. The main habitats are inherently species-poor, although it frequently grows alongside local species such as *Genista tinctoria*, *Ononis spinosa* and *Trifolium squamosum*. Soils are London Clays, boulder clay and recent alluvium (Thornton 1990).

P. officinale is a robust perennial, flowering from July to September. High summer temperatures appear to be required for seed to ripen fully, conditions not occurring every year. Even when seed does ripen, germination is rather patchy, being best where the sward is open and contains bare ground.

It is mainly restricted to two coastal districts in Britain - the north Kent coast from Faversham Creek to Reculver, a distance of some 20 km, and the Walton Backwaters in north-east Essex. Most plants are found on sea-walls and former estuarine land claimed for grazing, apart from one Kent site (Tankerton Cliffs), where it dominates 2 hectares of unstable cliff-slopes. Because shoots arise from spreading rhizomes, population sizes are difficult to judge. However, the Kent populations are considered to be stable at about 10,000 plants, and Essex is estimated to hold 60% of the total British population, the proportion having risen from only 30% in the 1970s. Numerically minor localities

275

are Lee-over-Sands (Essex), first reported around 1978, and Southwold, Suffolk, where it was discovered in 1990. The plant was also formerly present in two other Essex estuarine systems (Stour, and Holland Brook) and in West Sussex, where it was recorded in 1666 (Ray 1677) but not subsequently.

Colonies may be damaged by 'engineered' improvements to sea-walls, although disturbance is not always a long-term problem, since one of the largest Essex populations is on a former explosives testing site. New coastal defence policies of managed retreat and setting back sea-walls, whilst possibly extending the area of saltmarsh, could adversely affect colonies on the front-line defence. Other threats include cliff stabilisation works and also the mowing of sea-banks, since the plants that grow after mowing may flower but do not set seed. *P. officinale* is the sole larval food plant for two nationally rare moths, a micromoth *Agonopteryx putridella* and the vulnerable Fisher's estuarine moth *Gortyna borellii*. The latter is found only in the main Essex population, where it was identified in 1970. The mature larvae of *G. borellii* feed in the rootstock, and a small but significant threat to the Essex population arises from occasional illegal depredations by unscrupulous moth collectors digging up the plant.

Some doubt has been cast upon the native status of *P. officinale*, particularly in view of its disjunct distribution with its main sites being near Roman ports. Certainly it has been of economic importance - its roots yield a stimulant resin (Tutin 1980) - and it was considered good fodder for pigs. However Britain may equally be considered a natural outlier of its predominantly continental distribution. In southern Brittany, it is found in similar upper estuarine sites, although further south in Iberia, and eastwards to western Siberia, it occupies a wider range of grassland habitats, up to 1,800 m in altitude.

Further details are given in Randall & Thornton (1996).

C. Gibson

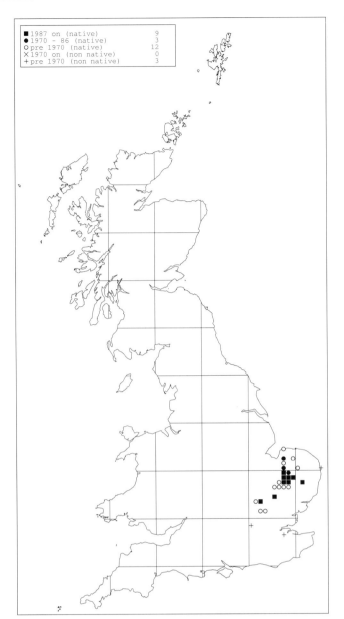

Phleum phleoides (L.) Karst. (Poaceae)

Purple-stemmed cat's tail
Status in Britain: LOWER RISK - Near Threatened.
Status in Europe: Not threatened.

At most of its East Anglian sites, *P. phleoides* grows on very freely draining light sandy 'Breck' soils where the chalk is relatively close to the surface. It occurs in grazed, open grassland in a wide range of habitats, including grass heaths, roadside verges, tracksides, edges of pits, along forest rides, near rabbit burrows and on other disturbed ground. On downland in Hertfordshire it grows on glacial sand overlying chalk. Typical associates include *Achillea millefolium, Festuca ovina, Galium verum, Helictotrichon pratense, Koeleria macrantha, Lotus corniculatus, Luzula campestris, Plantago lanceolata, Ranunculus bulbosus, Sagina nodosa* and *Scabiosa columbaria*.

This perennial grass flowers chiefly in June, and usually (at least in Breckland) a week or so before *Phleum bertolonii*, with which it sometimes grows. *P. phleoides* is remarkably tolerant of drought. After a long dry spell, the flowering panicles may fail to open properly, but the leaves remain firm and green when all other grasses are becoming brown and shrivelled. Open ground is necessary for the establishment of seedlings.

P. phleoides was first recognised in Britain near Newmarket sometime before 1775. It is a Breckland speciality, now almost entirely confined to Norfolk and Suffolk, but still occurring at one locality in Cambridgeshire, and away from Breckland at one locality in Hertfordshire. Although the colonies outside Breckland are small and vulnerable, within Breckland it is thriving in open habitats. A recent survey (Leonard 1993) showed it to be present at some 40 sites in Norfolk and Suffolk within seven hectads.

Rabbits appear to graze this species preferentially, perhaps because it tends to stay green longer in summer than do other grasses, and where grazing is heavy it may fail to flower for many years. When grazing was heavy before myxomatosis, *P. phleoides* was erroneously thought to be much rarer. Although new populations have been found recently, all are considered likely to have been overlooked in the past, rather than representing new colonisation. The maintenance of short, open grassland by grazing is a principal requirement, and in those that have been improved by the addition of fertiliser, it is quickly crowded out by more vigorous species.

It is widespread across Europe, except in the south, ranging from Scandinavia into central Asia in Turkestan. It also occurs in North Africa.

G. Beckett

Phyllodoce caerulea (L.) Bab. (Ericaceae)

Blue heath, Fraoch a' Mhèinnearaich
Status in Britain: VULNERABLE. WCA Schedule 8.
Status in Europe: Not threatened.

This calcifuge, dwarf shrub of montane heaths occurs on acidic freely draining soils on very steep ground between 670 and 840 m, almost always with a northerly or easterly aspect. All localities bear late snow-beds, which normally clear in June. It occurs in communities typical of moderately late snow-beds, dominated by dwarf shrubs, in which *Vaccinium myrtillus* is the most frequent and generally the most abundant species, though *Empetrum nigrum* may be locally dominant. More rarely it occurs in open areas dominated by low grasses and herbs such as *Sibbaldia procumbens*, and is occasionally found at the edge of grasslands dominated by *Nardus stricta*.

P. caerulea is a low bushy evergreen shrub up to 20 cm tall, normally flowering in July, though some flowers may be found as late as October. Flowering is, however, irregular, and may be sparse or absent in some plants in some years. Both self- and cross-pollination occur. Capsules in all stages of maturity remain at the end of the season and seed production is generally low and variable. Seed capsules have been significantly grazed by ptarmigan in most years on the Sow of Atholl, but Ben Alder populations appear to have remained ungrazed. Plants can spread vegetatively by layering of basal shoots or spread of rhizomes, though monitoring over the past decade has not detected any new plants created in this way. It is likely that late snow cover protects the buds from frost damage in winter. It may be readily confused with *Empetrum nigrum* when not flowering.

It is known from six locations, in three adjacent hectads. On the Sow of Atholl, plants are scattered over about 1 km of hillside. Although this represents the largest assemblage of plants in one area, the population is probably too fragmented for cross-pollination to occur, since plants flower at slightly different dates apparently depending on their emergence from the winter snow cover. In the remote Ben Alder range, five smaller populations occur in a small area (McBeath 1967). Some years ago, *P. caerulea* was reported from other sites in the vicinity, but these have never been confirmed. However it is possible that it occurs elsewhere, since it inhabits difficult terrain buried by snow for more than half the year. Its discovery in 1810 and its subsequent history are described in Nelson (1977).

The total known British population numbers fewer than 300 plants, though some of these may represent several individuals growing together. About half are on the Sow of Atholl. Detailed monitoring since 1985 has revealed just six new plants. Quite a number of other small plants occur that were established sometime before 1985, and which are developing very slowly indeed. The overall picture from monitoring in the past decade is of the survival of long-lived individuals. Losses are mainly because of local erosion, which occurs when the soil becomes saturated as the snow melts. Negligible

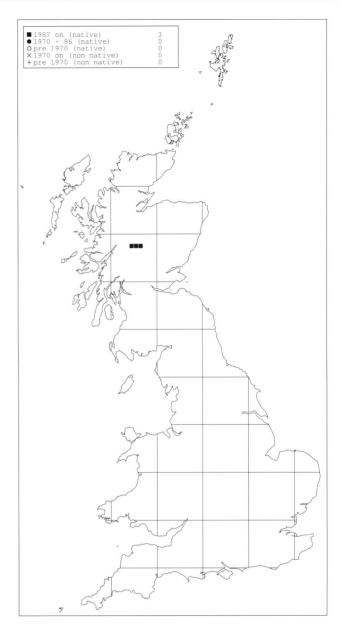

recruitment of new plants at the present time means that this species is undergoing an overall decline. Grazing is slight, despite considerable increases in red deer in the Grampians and the presence of sheep on the Sow of Atholl. Plants are, however, often damaged in the autumn, and this appears to be caused by deer moving through the habitat in the stalking season (Sydes & MacKintosh 1990).

All known colonies lie within two upland SSSIs, but the protection afforded by this designation is limited. Three substantial landslips on the Sow of Atholl have caused considerable damage over the past 20 years, possibly related to mild winters with rapid thaws. Aerial photographs reveal no evidence of slips there in the previous 30 years. The plant could be severely affected by acid deposition, for snow can concentrate pollutants, which are released in bursts of very acid water as melting starts, and this species occurs on soils that have low buffering capabilities. Extreme acidic pollution has been recorded in snow near British *P. caerulea* sites (Davies, Abrahams *et al.* 1984). Spring muirburn is not at present a serious problem as all known plants are normally under snow until well after the legal muirburn season. However, extensive fires have occurred close to the Sow of Atholl. Autumn muirburn could pose a threat but does not currently take place on the two estates with *P. caerulea* populations. It has been collected by botanists and gardeners ever since it was first discovered, but the close monitoring of the past decade has provided no evidence of the removal of complete plants.

In Europe its centre of abundance is in Scandinavia. It also occurs in Greenland, Iceland and eastwards through Russia, and is very rare, perhaps extinct, in the Pyrenees. In other parts of the world it occurs with other members of the genus and is found in North America and through northern Asia to Japan.

Further details are given in Coker & Coker (1973).

C. Sydes and M.J. Wigginton

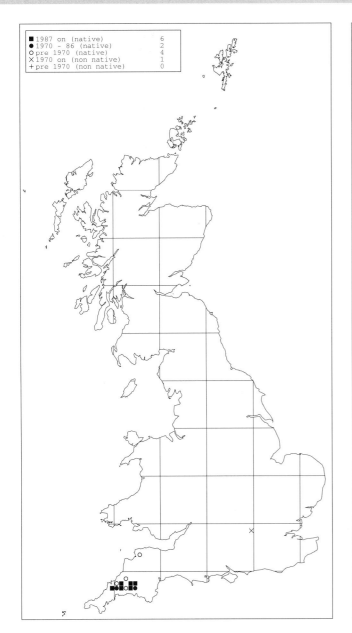

Physospermum cornubiense (L.) DC. (Apiaceae)
Bladderseed
Status in Britain: VULNERABLE.
Status in Europe: Not threatened.

P. cornubiense occurs in a range of habitats, including open woodland, *Ulex europaeus* scrub on heaths, rough grassy slopes (often in stream valleys), scrubby *Molinia* heath and roadside banks where there is sheltering scrub or woodland. Nearly all of the early records were from woodland, the plant being most abundant where the canopy is open, or in clearings. It still occurs in this habitat, but not to the same extent as formerly. The largest populations currently occur in *Molinia caerulea* heath and on heathy or scrub-covered slopes. In those habitats, associated species include *Agrostis curtisii*, *Calluna vulgaris*, *Chamerion angustifolium*, *Erica cinerea*, *Galium saxatile*, *Potentilla erecta*, *Pteridium aquilinum*, *Rubus fruticosus*, *Teucrium scorodonia*, *Ulex europaeus* and sometimes *U. gallii*. Good populations occur in open birch and gorse scrub on some heaths.

P. cornubiense is a perennial producing leaves in May and flowering usually in July and August, though in some years the first flowers can appear as early as the end of June. Fruiting is correspondingly variable in timing, and the fruits may fall between early September and late October. It does well in open habitats but fails to flower in dense shade. Plants heavily overgrown with other vegetation may not even produce leaves, but the rootstock can be remarkably persistent and seed can remain viable in the soil for years. Thus, after burning or clearance, plants can appear in their hundreds.

During its maximum range, it was known in East Cornwall, South and North Devon and Buckinghamshire (Margetts & David 1981; Ivimey-Cook 1984). The old record for West Cornwall might have been in error, and it has not been seen in North Devon since 1858. It was last seen in South Devon in about 1977. Since the late 1970s, the number of sites in Cornwall has declined from nearly 50 to about 20, most of them in the south-east of the county. Populations have also declined, from thousands to hundreds or merely tens of plants in many of its locations. For instance, populations in Silver Valley near Callington, which used to total more a thousand plants, dropped to a few hundreds by 1989, with a recent survey (Friesner 1994) revealing a further drop. A few plants are still be found in woods by the Fowey and in some by the Tamar. However, some heathland sites (for instance, on Pinsla Downs and near Cadsonbury) still hold populations of 1,000 plants or more. The single colony in Buckinghamshire appears to be thriving, though it is restricted to a small area of about 30 x 50 m. It seems unlikely to be native there, however. Druce (1926) considered it possible that the colony had originated from bird-sown seed from a nearby population planted in 1810 at Bulstrode Park.

Changes in woodland management, afforestation, scrub clearance, burning and the loss of grazing may all have contributed to the decline of *P. cornubiense*. However, conservation management has bolstered some populations as, for instance, in the Luckett Reserve by the Tamar, an oak woodland that is managed principally to maintain habitat for the nationally rare heath fritillary butterfly. Clearings are made in the wood on a five-year cycle, and this management is also benefiting *P. cornubiense*.

P. cornubiense is a European-Temperate species of southern Europe, extending northwards to southern England and eastwards to Hungary and south-central Russia.

R.J. Murphy

Phyteuma spicatum L. (Campanulaceae)
Spiked rampion, Cyrnogyn Pigfain
Status in Britain: VULNERABLE. WCA Schedule 8.
Status in Europe: Not threatened. Endemic.

P. spicatum is a plant of damp, fertile, acid soils, its typical habitat being by streamsides in coppiced woodland. However, it has disappeared from many such sites, since they are no longer coppiced and have become overgrown. It is now mainly found on damp, summer-shaded roadside verges, with some populations in woodland rides and on stream banks in unworked woodland. It is usually found in good quality habitat, with many species characteristic of ancient woodland, including *Anemone nemorosa*, *Hyacinthoides non-scripta*, *Lathyrus montanus*, *Lysimachia nemorum*, *Oxalis acetosella*, *Potentilla sterilis*, *Pteridium aquilinum* and, in damper spots, *Carex laevigata, C. remota* and *Dryopteris carthusiana*.

This species is a long-lived perennial with a large tap-root, bearing an erect stem that can exceed a metre in height, although it is usually 50-80 cm tall. Flowering is normally in late May and June, depending on the warmth of the season, and seed is set in July and August. At the present time, seedlings appear to be rare, and recruitment negligible at some sites. Seed can remain viable in the soil for at least five years, and probably much longer.

In Britain, established populations of *P. spicatum* have always been confined to a small area of East Sussex, where, however, it was much more abundant in the past. During the nineteenth century it was reported as growing in hedgerows "scattered for miles" (Branwell 1872), but it is now confined to a few localities in two small areas of the county, near Heathfield and Hailsham. The largest extant population occurs on a steep roadside verge near Hadlow Down (one of three such verges where *P. spicatum* occurs), where 134 plants were counted in 1994, having decreased from 198 in 1980. The smaller populations also appear to be declining, and perhaps only 300-400 plants now survive in Britain.

P. spicatum declines and largely disappears in deep shade, but can reappear when the canopy is opened up. A lack of coppicing and woodland management is generally cited as the main reason for its decline in Sussex, and this assumption is supported by observations in similar woodland in northern France. An area in the Forêt d'Othe, near Sens, showed an enormous increase in *P. spicatum* after coppicing and its rarity and restriction to edges in adjacent old coppice apartments (C.D. Pigott pers. comm.). In Sussex it has disappeared from many of its former sites where the canopy has closed over, and most populations are now in woodland rides and on verges where there is sufficient light. It is possible, however, that small colonies are surviving in unworked woodland and could be revived by opening up the canopy. Roadside verge and woodland ride populations are also threatened by inappropriate management: for example, flowering plants have been mown or cut before they have had the chance to set seed, even on roadside verges designated as local nature reserves.

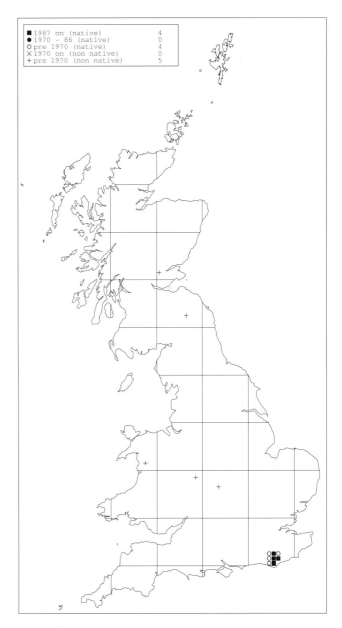

■ 1987 on (native)	4
● 1970 – 86 (native)	0
○ pre 1970 (native)	4
✕ 1970 on (non native)	0
+ pre 1970 (non native)	5

P. spicatum has long been known as a medicinal plant. Indeed, the first reference to it is in Gerard's 'Herball' (1597), in which, under the name of *Rapuntium maius*, it is described as a garden plant. Arnold (1887) considered it might have originally been an escape from Warbleton or Michelham Priories, though there is no firm evidence for this. Most authorities consider *P. spicatum* to be a native species in Britain.

P. spicatum is endemic in Europe, extending from Britain, southern Norway and the Baltic states southwards to northern Spain, Italy and Romania. It occurs in forest-edge habitats, meadows and mountain pastures up to 2,000 m in altitude.

B.R. Wheeler & D.A. Pearman

Pilosella flagellaris (Willd.) P.D.Sell & C.West ssp. *bicapitata* P.D.Sell & C.West (Asteraceae)

Shetland mouse-ear hawkweed
Status in Britain: VULNERABLE. ENDEMIC.

This plant of rocky slopes, pastures and outcrops is endemic to Shetland, and was first described in 1962. It is very localised, being known from only three sites (Scott 1968), despite its ability to grow in a variety of habitats. It grows amongst limestone rocks in rocky pastures in many discontinuous occurrences along 3 km of the White Ness peninsula, particularly on the west side, where it is associated with such species as *Anthyllis vulneraria*, *Polygala vulgaris* and *Primula vulgaris* (W. Scott pers. comm.). It is, however, rarely able to flower there in the heavily grazed sward, except in one or two rocky places inaccessible to sheep. At West Burrafirth, it occurs on a large heathy rocky knoll of granulitic gneiss with *Calluna vulgaris* and *Solidago virgaurea*. Flowering was formerly more frequent there but is now very limited because of increased grazing. The other known site, at Ronas Voe, is on feldspathic rock adjoining granite on steep rocky sea-banks, where it grows with rare *Hieracium* species (Scott & Palmer 1987). There are at least 120 plants at Ronas Voe, in three colonies, and flowering is more likely at this site because of the absence of grazing.

Where allowed to flower, *P. flagellaris* ssp. *bicapitata* will produce flowers and achenes abundantly. This is the only *Pilosella* known on Shetland: *P. officinarum*, common on Orkney, appears to be entirely absent.

M.J. Wigginton

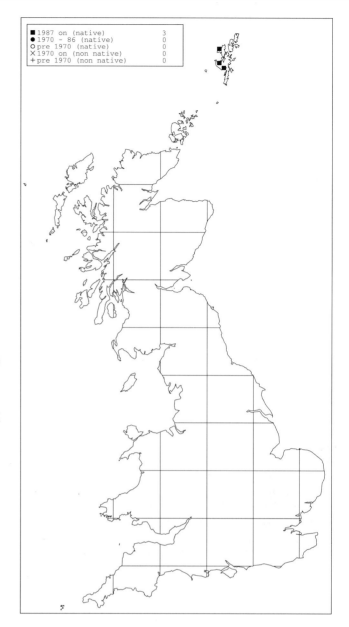

■ 1987 on (native)	3
● 1970 - 86 (native)	0
○ pre 1970 (native)	0
✕ 1970 on (non native)	0
+ pre 1970 (non native)	0

Pilosella peleteriana (Mérat) F. Schultz & Schultz Bip. (Asteraceae)
Shaggy mouse-ear hawkweed
Status in Britain: VULNERABLE. (all subspecies)
Status in Europe: Not threatened. Endemic.

This perennial hawkweed is known in few locations in Britain, with recent records from Dorset, the Isle of Wight and Montgomeryshire. It is, however, generally the commonest *Pilosella* in the Channel Islands. Three sub-taxa are distinguished, currently at subspecies level, though varietal status might be more appropriate (Stace 1991).

P. peleteriana ssp. *peleteriana* occurs only in Dorset and the Isle of Wight, though was formerly also in Kent. It is a plant of chalk soils, with one known locality on limestone on Portland. It occurs on south-facing, steep, and thus well-drained slopes with undistinguished calcicolous associates including *Carex flacca, Helianthemum nummularium, Leontodon hispidus, Plantago media, Sanguisorba minor* and *Thymus polytrichus.* It often grows on the edges of paths or terracettes, where drainage is at its sharpest. In the Channel Islands it is far more widespread and grows on soils overlying granite, particularly on coastal cliffs.

P. peleteriana ssp. *subpeleteriana* is known only from Craig Breidden in Montgomeryshire, growing there on shallow, drought-prone basic soils overlying dolerite. Populations occur mainly on the dry rock shelves at the south-western end of the hill, with smaller colonies on the northern part. It also grows freely on quarry waste, spreading by rooting stolons. In cultivation, it is attractive to slugs, which in spring keep the plants well in check and might destroy them entirely if not controlled (R. Woods pers. comm.). It is possible that in natural habitats such predation could be a factor controlling its spread to moister habitats from the dry outcrops where slugs are scarce.

P. peleteriana ssp. *tenuiscapa*, characterised by its narrower leaves and smaller flowering heads, seems more elusive and has been much neglected by botanists in recent years. It has been recorded, and may still occur, on the borders of Derbyshire and Staffordshire (in both counties) and perhaps in Yorkshire, but is likely to have gone from Devon (L. Spalton pers. comm.). Pugsley (1948) considered this plant much more common than ssp. *peleteriana* in Jersey, though Le Sueur (1984) has not differentiated them.

Like *P. officinarum*, the single head of yellow flowers is borne on a stem arising from a rosette of lanceolate leaves, but the rare species is more robust, with larger flower heads and short thick stolons often ending in a rosette of large crowded leaves, which themselves are often notably shaggy with long hairs. It is an altogether more striking plant. It spreads by stolons from a thick, long-lasting rootstock. In garden conditions the stolons (of ssp. *peleteriana*) can elongate by up to 10 cm a year

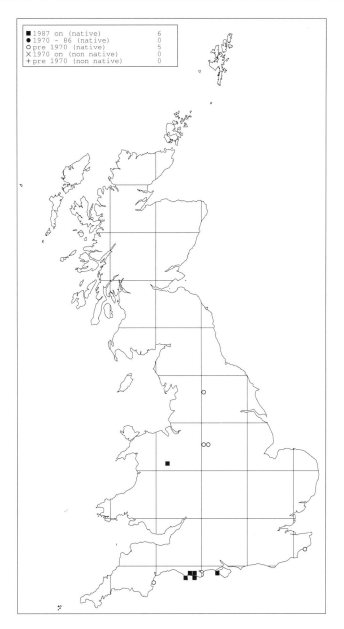

■ 1987 on (native)	6
● 1970 - 86 (native)	0
O pre 1970 (native)	5
X 1970 on (non native)	0
+ pre 1970 (non native)	0

(D. Pearman pers. obs.), but they probably spread less rapidly in the wild. Its main threat on the south coast is from the spread of *Brachypodium pinnatum*, which now covers vast areas of ungrazed turf, restricting *P. peleteriana* to areas grazed by rabbits or kept short by human activities, such as trampling on path edges. The main populations on Craig Breidden seem reasonably secure, provided quarrying does not encroach.

P. peleteriana occurs on the western and southern coasts of Europe. Its ecology relative to that of *P. officinarum* is not clearly understood, but the two species are known to hybridise.

D.A. Pearman and M.J. Wigginton

Poa flexuosa Sm. (Poaceae)

Wavy meadow-grass, Tràthach Casta
Status in Britain: VULNERABLE.
Status in Europe: Not threatened. Endemic.

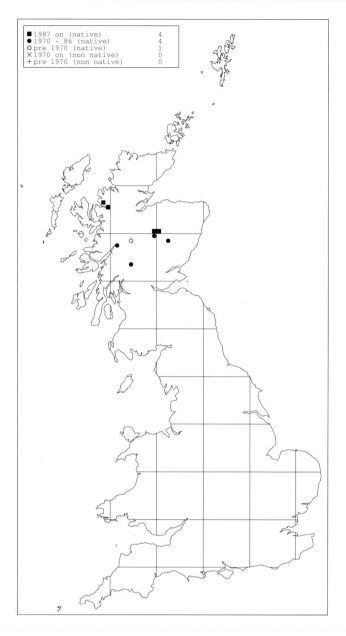

■ 1987 on (native)	4
● 1970 – 86 (native)	4
○ pre 1970 (native)	1
✕ 1970 on (non native)	0
+ pre 1970 (non native)	0

P. flexuosa is a montane grass occurring on acidic rocks from 760 to 1,100 m. Clumps grow between stones on plateaux, on open scree and on stone ledges, mostly in isolation without immediate associates. However, plants growing in the vicinity may include *Alchemilla alpina* and *Festuca vivipara*, and less commonly *Huperzia selago*, *Juncus trifidus*, *Luzula spicata*, *Saxifraga stellaris* and *Silene acaulis*.

It is a tufted perennial, lacking rhizomes, flowering in July and August. Little seems to be known of the longevity of plants or of its reproductive biology, including the factors controlling seed-set and recruitment.

During the past 20 years, *P. flexuosa* has been recorded in about 12 sites in eight hectads, in Aberdeenshire, Moray, Ross-shire and Inverness, and may still be extant in all sites for which there is only a pre-1987 record. Records from Perthshire are now considered to be doubtful (H.A. McAllistair pers. comm.). Much the largest known population, of 200-300 plants, occurs on scree slopes on Ben Nevis. Elsewhere, populations are small: for example, colonies of 36 and eight plants on Liathach in 1995, 17 on Beinn Alligin the same year, and two in Coire an Lochan, Cairngorm, in 1991. The rare hybrid *P. x jemtlandica* (*P. flexuosa* x *P. alpina*) grows at some of the *P. flexuosa* sites.

Small populations are threatened by human activities, including climbing (in the more popular localities), whilst in Ross-shire damage has been caused by deer trampling the tussocks (Perring & Farrell 1983).

P. flexuosa is endemic to montane regions of northern and western Europe, occurring only in Iceland, Scotland, Norway and Sweden. Plants from Greenland and the Alps, hitherto considered to be forms of *P. flexuosa*, are now treated as distinct species.

P.J.O. Trist

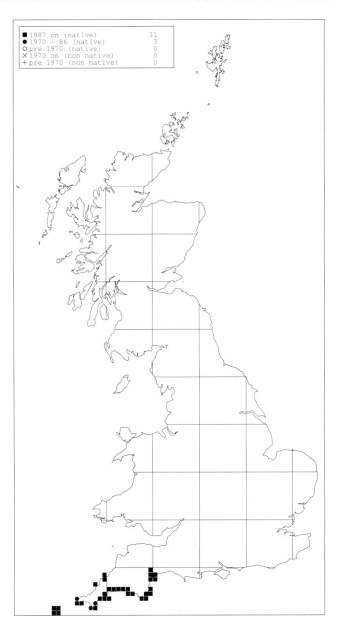

Poa infirma Kunth (Poaceae)

Early meadow-grass
Status in Britain: LOWER RISK - Nationally Scarce.
Status in Europe: Not threatened.

P. infirma is predominantly a coastal plant, occurring on cliff paths, tracksides and sandy marginal ground and also in short turf, amenity lawns, dune grassland and semi-fixed dunes. A wide range of associates includes *Bellis perennis*, *Cerastium diffusum*, *Montia fontana*, *Plantago coronopus*, *P. lanceolata*, *Poa annua*, *Trifolium ornithopodioides* and *T. scabrum*.

P. infirma is a short-lived annual, appearing early in the year, usually between late February and early April. However, flowering and senescence can occur much earlier; for instance, Takagi-Arigho (1994) reported flowering plants in January and others that had fruited, senesced and almost disappeared by early March. Conversely, plants on blown sand or dunes may flower

later than normal, and furthermore, colonies may be sporadic in appearance and variable in the date of flowering in successive years. For these reasons, it is likely that the great extension recently of its known range is, at least in part, a product of increased recording, rather than a real colonisation eastwards. When seen at its best, *P. infirma* has a characteristic yellowish-green colour, and the small plants are flattened against the ground in a star-shaped pattern, often showing up as distinct lighter patches in the turf. However, these and other qualitative characters should not be relied on for field identification, as they can all be shown by the polymorphic *Poa annua*. The only reliable character to differentiate the two species is the size of the anthers: 0.2-0.5 mm in *P. infirma*, 0.6-0.8 (-1.3) mm in *P. annua* (Sell & Murrell 1996).

At one time *P. infirma* was thought to be nationally rare on the mainland, confined to a few locations in West Cornwall, though it has always been widespread in the Isles of Scilly, where it occurs on all inhabited islands and

287

perhaps some of the uninhabited ones. Since the early 1980s, the grass has spread rapidly in the Isles of Scilly (Parslow 1994), and recent surveys have shown it to be present in many places along the Channel coasts of Cornwall and Devon. Doubt has been cast on the few records that have been made in Dorset and Hampshire (Brewis *et al.* 1996). Its distribution on the south coast appears to be circumscribed by the isotherm of the February mean daily minimum temperature of 2.5°C (Takagi-Arigho 1994). It remains a rare plant on the north coast of Cornwall and, despite searches, has yet to be found in North Devon or Lundy. It is currently known in over 30 hectads.

P. infirma occurs in western and southern Europe, ranging from Britain and the Channel Islands south and west to Spain, Greece and Turkey. It also occurs in North Africa and south-west Asia.

M.J. Wigginton

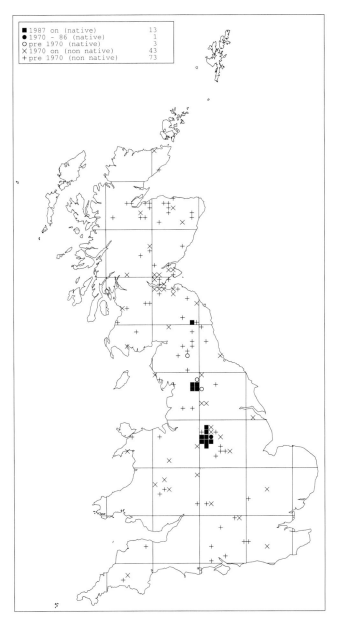

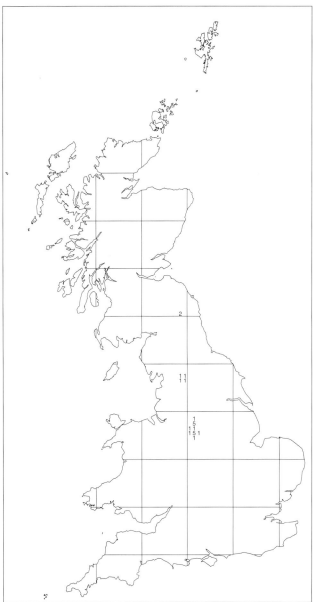

Polemonium caeruleum L. (Polemoniaceae)
Jacob's-ladder, Ysgol Jacob
Status in Britain: LOWER RISK - Near Threatened.
Status in Europe: Not threatened.

P. caeruleum is found mainly on cool, moist, north-facing slopes of Carboniferous limestone. It grows on thin soils, composed almost entirely of black crumbly organic particles, which overlie blocky limestone scree or crags. It is a component of a distinctive tall-herb community characterised by *Arrhenatherum elatius, Filipendula ulmaria, Heracleum sphondylium, Mercurialis perennis, Urtica dioica,* and a carpet of bryophytes including *Brachythecium rutabulum, Lophocolea bidentata* and *Plagiomnium undulatum.* Seedlings of *P. caeruleum* seem to be especially vulnerable to wilting, particularly the seedlings. Ideal localities therefore tend to be on steep, north-facing slopes at the base of cliffs. An open canopy of *Fraxinus excelsior* may also help to maintain a humid environment. In Northumberland it grows on clay soils along river banks,

and many ephemeral occurrences of *P. caeruleum* outside its normally accepted native range are in this habitat.

This species is a long-lived polycarpic perennial which reproduces by seed. It flowers in June and July (exceptionally into September) and is pollinated by bumble-bees, although it is also self-fertile. Seed is released from the dry capsules from early autumn to late winter. Germination occurs mainly in spring. Mature plants die down during the winter but the dead flower stems remain erect, eventually releasing the seeds.

P. caeruleum is currently found as a native in only three areas in Britain, and populations seem stable overall. The largest populations are in the White Peak of Derbyshire and Staffordshire, with at least 40,000 plants estimated in the 17 localities (Taylor 1990b), and a second smaller cluster of sites on the Craven limestones of Yorkshire. It also appears to be native in two riverside locations in the Cheviot Hills of Northumberland. Our understanding of its native distribution in Britain is confused by the widespread occurrence of garden escapes and introductions, although the floral and foliar characters differ in garden and naturalised forms (Pigott 1958).

This species is highly palatable to grazing mammals, but the large populations of snails, so characteristic of its main habitat, appear not to affect it significantly. Conservation management presents something of a dilemma. Many sites are grazed, and flowering in such situations is seriously suppressed, although basal leaves may continue to be found in abundance. On the other hand, ash woodland could develop in the absence of grazing. At most sites, cessation of grazing would be desirable, but on those sites prone to successional development to woodland, very light and occasional grazing might be needed to maintain the desired vegetation structure.

P. caeruleum is part of a species complex which in total has a near circumboreal distribution. *P. caeruleum* itself is found throughout the uplands of central and northern Europe with strongholds in the Alps, Finland, Norway and Russia, extending eastwards to Lake Baikal. Other closely related species range across eastern Siberia, China and North America. *P. caeruleum* is also widely introduced, a consequence of its popularity amongst gardeners, and is well-established throughout Eurasia and in eastern North America. On the continental mainland, it is found in a much wider range of habitats than in Britain, including alpine meadows in Switzerland, birch woodland in Scandinavia, *Picea abies* forest in Slovakia and even lowland fens in northern Germany.

I. Taylor

Polycarpon tetraphyllum (L.) L. (Caryophyllaceae)
Four-leaved allseed, Gorhadog
Status in Britain: LOWER RISK - Near Threatened nationally,
but VULNERABLE in mainland Britain.
Status in Europe: Not threatened.

P. tetraphyllum occurs in both semi-natural and artificial
habitats, the former including steep therophyte-rich south-
facing banks with high levels of insolation on the Lizard
peninsula, West Cornwall, in South Devon, and on
compacted shingle on Chesil Beach, Dorset. Man-made
habitats include bulb-fields, gardens and the base of
roadside walls. In all its situations it favours open
conditions with low levels of competition, subject to
summer droughting and with a low incidence of frost. Its
associates include a high proportion of annuals and
biennials such as *Anagallis arvensis*, *Aphanes inexpectata*,
Arenaria serpyllifolia, *Bromus hordeaceus*, *Cerastium
glomeratum*, *Erodium moschatum*, *Trifolium ornithopodioides*,
T. scabrum, *T. subterraneum* and *Verbascum virgatum*.

In Britain, this plant is normally a summer annual.
Flowering can extend from April until as late as November
or December, and, unlike many annuals, it is apparently
highly tolerant of drought. However, it may also behave as
a winter annual, germinating in late summer or early
autumn, the germination time varying according to the
prevailing weather conditions.

In mainland Britain, *P. tetraphyllum* is very rare and is
currently known from only about six regular sites, in West
and East Cornwall, South Devon and Dorset.
Notwithstanding its habit of reappearing after long
absences, this species does seem to have undergone a
severe decline. Even in the 1920s it was becoming very
scarce on the south coast. Many of its extant localities are
long-standing ones (for instance, its Chesil Beach locality
dates from 1774, whilst on the Lizard it was first found in
1872), yet populations are often very limited in extent,
numbering only a few hundred plants annually, and have
never spread. This scarcity on the mainland contrasts
markedly with its status in the Isles of Scilly, where the
plant is generally widespread in ruderal habitats and in
bulb fields. The plant was not discovered in the Isles of
Scilly until 1928, despite relatively high levels of botanical
survey, and during the period 1928-1940 the plant
expanded its range markedly (Lousley 1971).

In the Isles of Scilly, *P. tetraphyllum* seems reasonably
secure, and only on the mainland can the plant be
regarded as seriously threatened. A number of populations
have been lost to urban expansion and through excessive
overgrowth by coarse grasses and scrub brought about by
the cessation of grazing and lack of disturbance.

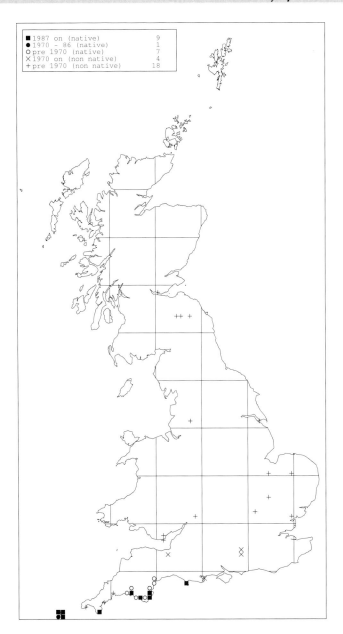

■ 1987 on (native)	9
● 1970 - 86 (native)	1
O pre 1970 (native)	7
× 1970 on (non native)	4
+ pre 1970 (non native)	18

As a native species, *P. tetraphyllum* is found in southern
and western Europe and the Middle East, northwards to
Britain and the Channel Islands and eastwards to Turkey,
Georgia, Syria, northern Iran, Arabia and the Sinai
peninsula. It is abundant around the Mediterranean.
Elsewhere, it is a widely naturalised weed species,
occurring in Asia, Africa, Australia and South America
(Davis 1967; Chater & Akeroyd 1993).

A.J. Byfield

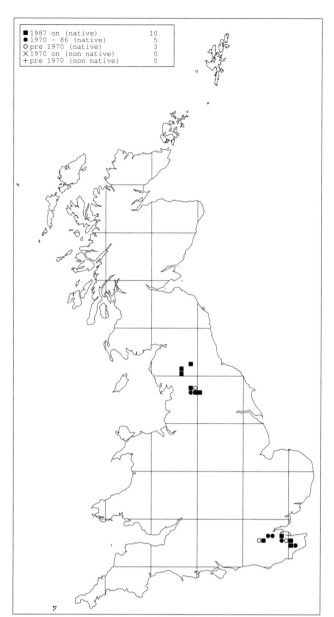

■ 1987 on (native)	10
● 1970 – 86 (native)	5
○ pre 1970 (native)	3
× 1970 on (non native)	0
+ pre 1970 (non native)	0

Polygala amarella Crantz (Polygalaceae)

P. amara auct., including *P. austriaca* Crantz
Dwarf milkwort
Status in Britain: VULNERABLE.
Status in Europe: Not threatened. Endemic.

An extreme calcicole, *P. amarella* grows in open sites in short turf overlying calcareous rocks, often on sloping banks where the drainage is good. All its sites in northern England are in areas of heavy rainfall but its habitats are not prone to waterlogging. Where this species occurs in calcareous mires, as near Malham, it grows on tussocks raised above the wetter surrounding areas. It ascends to 530 m on Cronkley Fell. In southern England, *P. amarella* favours open chalk grassland in the absence of vigorous competition. Associated species include *Carex flacca*, *C. panicea*, *Festuca ovina*, *Helianthemum nummularium*, *Sanguisorba minor* and, in the north of England, *Sesleria caerulea*. *Polygala vulgaris* is also often present, and is a potential source of mistaken identity.

It is a small, often prostrate, herbaceous short-lived perennial. Flowering usually begins in May and can continue until mid-July. Seed germinates freely. Variation in *P. amarella* is discussed in Fearn (1975).

P. amarella occurs in four disjunct areas in England: Kent, the Craven district of Yorkshire, the limestone districts near Orton, Cumbria, and the 'sugar' limestones of Upper Teesdale. Some of the sites have very few plants, but one of the Craven populations may exceed 1,000 plants, and on Widdybank, more than 100 plants were counted in June 1994. Only small populations of fewer than twenty plants occur at some sites. *P. amarella* seems to have declined markedly in Craven. It was not found at six of the eight sites surveyed in 1994 and 1995, despite prolonged searching. It seems likely that declines and extinctions can be attributed to recent intensification of farming practices, particularly increased stocking rates. Since it is a short-lived species the removal of flowers and fruit by grazing animals could quickly lead to permanent losses. It may be significant that the two sites where it was found in 1994/5 are ungrazed banks. The Kentish sites suffer the additional threats of urban development, the ploughing up of chalk grassland and, in some cases, inadequate grazing leading to the growth of scrub or coarse grass swards. Since the soils on which *P. amarella* grows are extremely skeletal with very low humus content, Kentish populations in particular may also be liable to summer drought (F. Rose and J. Pitt pers. comm.). Most of the northern sites and some of the Kent sites are within nature reserves or SSSIs, but recent records from Kent are incomplete and need updating as a matter of urgency.

Two distinct species, *Polygala amara* in the north and *P. austriaca* in the south, were once differentiated on the basis of morphological differences and geographical location. Most authorities now consider that they are best treated as different races of a single polymorphic species, *P. amarella*.

P. amarella has a wide distribution in Europe, encompassing discrete populations in the Pyrenees, northern France, the French and Swiss Alps, central and eastern Europe and Scandinavia.

M.S. Porter

Polygonatum verticillatum (L.) All. (Liliaceae)

Whorled Solomon's seal, Sêl Selyf Culddail
Status in Britain: VULNERABLE. WCA Schedule 8.
Status in Europe: Not threatened.

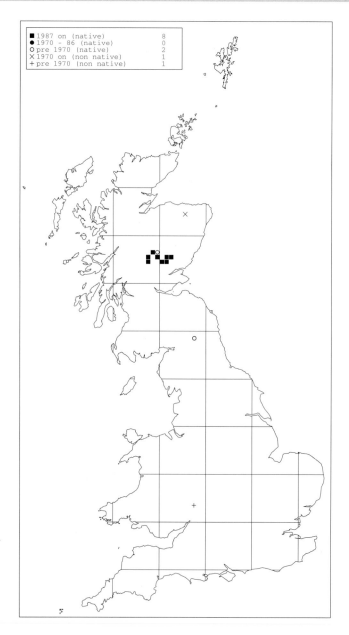

All sites for *P. verticillatum* are in thickly wooded gorges associated with ancient woodland, the one exception being a narrow strip of woodland along a level river bank. In some places it occurs on wet unstable slopes that are highly vulnerable to disturbance and trampling. High soil moisture and air humidity are the principal habitat requirements, and base-rich substrates and fertile soils are favoured. In areas of acidic geology additional nutrients are received from flushing, snow-melt, mineral downwash and periodic flooding. It rarely occurs in coniferous woodland, except where conifers have been planted on sites of former broad-leaved woods. The diversity of woodland types in which it occurs is reflected in the wide range of associates. At some sites, accompanying species include *Deschampsia cespitosa, Dryopteris filix-mas, Holcus mollis, Luzula sylvatica, Oxalis acetosella* and *Phegopteris connectilis*; elsewhere it may occur with *Ajuga reptans, Allium ursinum, Crepis paludosa, Geum rivale, Melica nutans* and *Mercurialis perennis*. There is often a good cover of bryophytes, including *Eurhynchium praelongum, E. striatum* and *Thuidium tamariscinum*.

P. verticillatum is a clonal, summer-green rhizomatous herb. Shoots emerge in late April and May with flower buds opening in late May to early June. Flowering is recorded in half the Scottish populations, but the proportion of flowering shoots ranges from less than 0.5% to 50%. The incidence of flowering is related to light and soil conditions. In nutrient-poor sites, shoots are small and mostly sterile; in nutrient-rich sites they are large. The plant flowers well only in open or semi-shaded places. It is partially self-compatible, but spontaneous selfing is rare. Bumblebees appear to be the most important pollinators. Fruiting is very poor but variable, a shortage of cross-pollen being one reason, and the number of seeds within fruits is well below the observed number of ovules. Seedlings have been detected in wild populations, but recruitment from seed appears to be very infrequent in established populations (Wright *et al.* 1993).

This species was discovered in the wild in 1790 by George Don but was grown as a medicinal herb in Britain possibly as early as the fifteenth century. A total of 16 discrete populations are currently known, but from only nine sites, all in the Tay catchment in Perthshire, though it formerly occurred in Northumberland and at least three other sites in Perthshire. Losses have been due to habitat destruction, erosion and collecting. The 16 known populations held between two and about 2,000 shoots in 1995. However, since several populations may comprise a single or few clones, the genetic stock might be very limited.

All but two sites have statutory protection, though all colonies are threatened. Erosion of stream banks and soil slippage are the most immediate threats at two sites, one of which has been progressively undercut in recent years such that rhizomes are now exposed. One population

seems to be suffering reduced vegetative vigour owing to reduced fertility under spruce, and at least three populations are at risk because of their small size. Roe deer cause some damage by trampling and soil displacement but appear rarely to graze the shoots. All populations now lie close to the woodland edge, where they are more vulnerable to various disturbances and disruptive processes, including windthrow, soil erosion, agricultural run-off, reduction in air humidity, summer drought, human disturbance, and competition from more light-demanding plants. Flowering is particularly sensitive to summer drought, and warmer winters may be detrimental to species with endogenous winter dormancy. The present isolation of populations may be restricting the chances of cross-pollination.

P. verticillatum has a sub-oceanic distribution extending from 70°N in Norway through western Scandinavia, Denmark and central Europe to the Pyrenees, Apennines,

Carpathians and Caucasus. From Russia it extends eastwards to the Himalayas. It ranges from sea level in Arctic Norway to the alpine zone but is predominantly sub-montane. In Europe, it is mostly confined to deciduous forest, but it also occurs in sub-alpine birch-willow scrub, species-rich tall-herb vegetation and wooded meadows (Carlsson 1991).

J. Wright

Polygonum maritimum L. (Polygonaceae)

Sea knotgrass, Canclwm Arfor
Status in Britain: ENDANGERED. WCA Schedule 8.
Status in Europe: Not threatened.

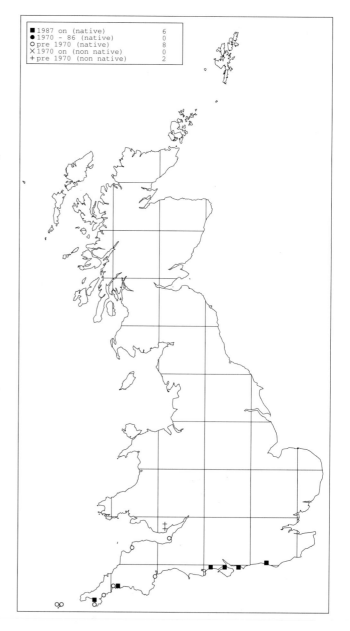

This species occurs on beaches above the high spring-tide mark in a generally very open community, mainly on fine shingle and coarse sand. In some places it may occur without close associates, but in others it accompanies typical plants of this habitat, including *Atriplex glabriuscula, A. laciniata, Beta vulgaris, Cakile maritima, Calystegia soldanella, Catapodium marinum, Glaucium flavum, Honkenya peploides, Polygonum oxyspermum* ssp. *raii, Rumex crispus, Salsola kali, Senecio vulgaris* and *Tripleurospermum maritimum*.

P. maritimum is a procumbent perennial. The stems become woody with age, but in mainland Britain most plants do not survive long enough to become large and woody. Flowering extends mainly from June to October, though a few flowers may persist until late November. It seems likely that seed can remain viable for many years, as shown by the reappearance of plants in known sites after long absences.

This species has apparently always been rare in Britain, often with a pattern of sporadic occurrences followed by periods when it appears to be absent (e.g. Clapham *et al.* 1987). It has occurred only on the south coast of England, having been first recorded at Christchurch in 1836 by W. Borrer. It has made brief appearances in Devon and Somerset, but currently seems to be absent in these counties, as also in the Isles of Scilly, where it was first recorded in 1852. Populations are currently extant in Cornwall, Dorset (in vice-county 11), Hampshire, the Isle of Wight, West Sussex and East Sussex (Harmes & Spiers 1993). The largest recent counts are at Gunwalloe (250 plants in 1993) and at Hengistbury Head (144 plants in 1993, though only 15 in 1996). Most populations are, however, much smaller. Only 14 plants were recorded, for example, in East Sussex in 1993, and 15 plants at Sandy Point, Hayling Island in 1995.

The fate of most populations seems likely to depend largely on natural events such as storms shifting beach material. For instance, at Hengistbury Head, where *P. maritimum* reappeared in 1990 after an absence of 88 years, it spread to a beach built up by storms only two years before. Conversely, it was not seen in 1995 at Lantic Beach, Cornwall, though it could reappear there. However, human activity may be significant, for example, in East Sussex where plants appeared on a beach created by recent 'engineering' works. Accidental uprooting of plants on popular beaches such as Gunwalloe and Brighton is likely, though this may constitute only a minor threat. On the other hand, over-enthusiastic beach maintenance by local authorities could be much more damaging.

P. maritimum is found throughout Europe on Atlantic, Mediterranean and Black Sea coasts, extending northwards to the Channel Islands, England and Belgium.

R.M. Walls and M.J. Wigginton

Potamogeton acutifolius Link
(Potamogetonaceae)

Sharp-leaved pondweed, Dyfrllys Meinddail
Status in Britain: VULNERABLE.
Status in Europe: Not threatened.

P. acutifolius has a narrower habitat range than most
aquatics in Britain, as it is virtually restricted to shallow,
species-rich drainage ditches in lowland grazing marshes.
It is found in calcareous, mesotrophic or meso-eutrophic
water, where typical associates include *Elodea canadensis*,
Hottonia palustris, *Hydrocharis morsus-ranae*, *Lemna minor*,
L. trisulca, *Myriophyllum verticillatum*, *Potamogeton natans*,
Ranunculus circinatus, *Sagittaria sagittifolia* and *Spirodela
polyrhiza*. It is also recorded from a pond in Middlesex and
there are single records from the Hereford & Gloucester
and Oxford Canals. There is, however, no evidence that
this species ever became established in the canal system,
nor is it known from other newly available habitats such
as gravel pits.

P. acutifolius flowers and fruits more freely than some
other linear-leaved pondweeds, including the closely
related *P. compressus*. Turions are apparently produced in
smaller quantity than in some other species, but they can
usually be found from August onwards. The life history
and reproduction of the species require detailed study.

There has been a substantial decline. It has not been
seen in Yorkshire and Lincolnshire since the late 17th
century. It was last seen in Essex in 1700, Herefordshire in
1846, Warwickshire in *c.* 1859, Gloucestershire in 1870,
Hampshire in 1898 and Northamptonshire in 1910. In
many of these counties the only evidence for its occurrence
is a single herbarium specimen. It may be extinct in Surrey,
where it was last seen in 1965, and if it still survives at its
sole site in Middlesex it does so very precariously.
Although it persists in the Wareham area, its distribution
there has contracted since the last century and it now
occurs in small quantity in one small area; it has also been
lost from some of its sites in Norfolk. The reasons for the
decline appear to be habitat destruction, the conversion of
grazing marshes for arable use, and eutrophication. The
headquarters of this species is now in Sussex, where it
occurs in abundance at Amberley Wild Brooks and has
also been recorded recently at other localities. In Norfolk it
has been successfully introduced to the RSPB reserve at
Strumpshaw.

P. acutifolius has a restricted world distribution, being
confined to temperate regions of Europe. It is found from
southern Sweden southwards to France, Italy and the
Balkans, but it is absent from the Mediterranean region.

C.D. Preston, adapted from an account in Preston &
Croft (1997)

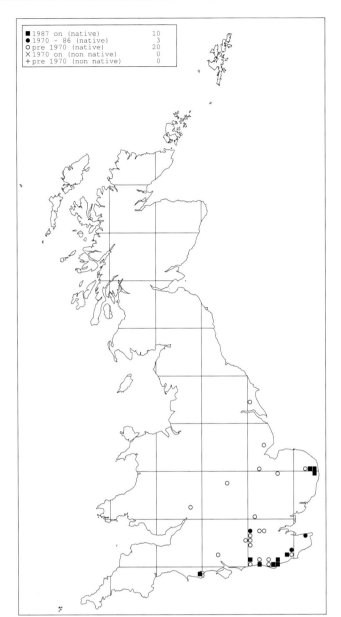

■ 1987 on (native)	10
● 1970 – 86 (native)	3
○ pre 1970 (native)	20
✗ 1970 on (non native)	0
+ pre 1970 (non native)	0

Potamogeton epihydrus Raf.
(Potamogetonaceae)
American pondweed, Lìobhag Aimeireaganach
Status in Britain: VULNERABLE.
Status in Europe: Occurs only in Britain

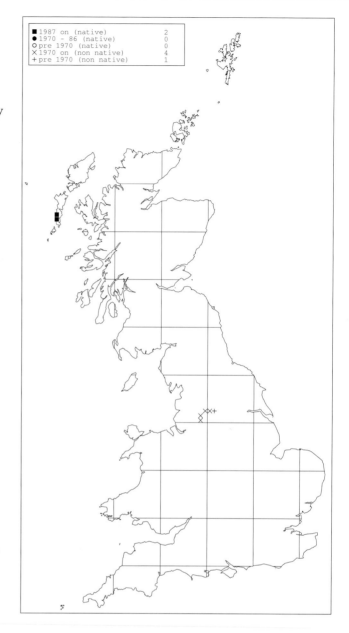

In the Outer Hebrides *P. epihydrus* grows in a few peaty lochans near Loch Ceann a'Bhaigh, where it is found in oligotrophic and base-poor water less than a metre deep. Other species in these lochans include *Eleogiton fluitans*, *Equisetum fluviatile*, the aquatic form of *Juncus bulbosus*, *Littorella uniflora*, *Lobelia dortmanna*, *Nymphaea alba*, *Potamogeton natans*, *P. polygonifolius*, *Sparganium angustifolium* and *Utricularia vulgaris sensu lato*. There is also a record from Loch an Duin. Submerged leaves that appear to belong to this species are washed up in late summer at the edge of a large, oligotrophic loch on Skye, but rooted plants have never been found here and the identification remains unconfirmed. However, *P. epihydrus* is locally plentiful in the mesotrophic water of the Rochdale Canal and the Calder & Hebble Navigation, where it grows with *Elodea nuttallii*, *Lemna minor*, *L. trisulca*, *Luronium natans*, *Potamogeton crispus*, *Sparganium emersum* and *Nitella mucronata* in water 0.4-1.2 m deep. All the known sites are lowland.

P. epihydrus is a rhizomatous perennial. It has both submerged, linear leaves and broader floating leaves, although floating leaves may not develop on plants in relatively deep water. The species is self compatible, and self-pollination occurs once the anthers dehisce, if cross-pollination has not taken place (Philbrick 1983). In the American population studied by Philbrick all ovules developed into fruits, and plants with floating leaves flower and fruit freely in both the Outer Hebrides and northern England. In addition, short stolons bearing buds or small fascicles of leaves can be found in the leaf axils in September. They are easily detached and act as vegetative propagules.

This species was originally discovered in Britain in 1907, at Salterhebble Bridge near Halifax (Bennett 1908). It is well-established in canals on both sides of the Pennines, but it has not been seen recently in the Calder, where it was collected in 1942. The English populations are believed to originate from an introduction, although it is difficult to imagine the source of the material. The native populations in South Uist were discovered by W.A. Clark and J.W. Heslop-Harrison in 1943 and 1944 (Heslop-Harrison 1949, 1950).

P. epihydrus is widespread in North America, where it is found in lakes, pools and streams from southern Alaska and Labrador south to northern California and Tennessee; it is restricted to mountains in the southern edge of its

range (Fernald 1932). The British records are the only known occurrences in Europe.

There are similarities between this species and *Luronium natans*: both have native sites where the water is oligotrophic and also grow in mesotrophic canals. However, *L. natans* spread into the canal system from its native sites, whereas the canal populations of *P. epihydrus* presumably originated independently.

C.D. Preston, adapted from an account in Preston & Croft (1997)

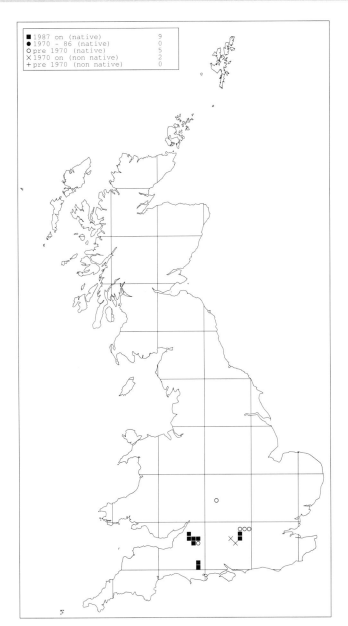

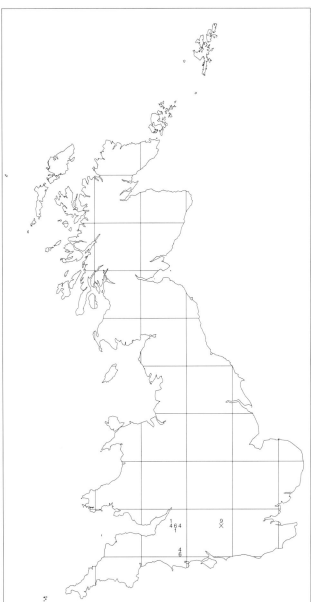

Potamogeton nodosus Poiret
(Potamogetonaceae)
Loddon pondweed, Dyfrllys Rhwydog
Status in Britain: LOWER RISK - Near Threatened.
Status in Europe: Not threatened.

This species is confined to a few calcareous and moderately but not excessively eutrophic rivers in lowland England. It is found in both shallow and relatively deep water, growing in stretches of moderately rapid flow with species such as *Elodea nuttallii, Lemna minor, Myriophyllum spicatum, Nuphar lutea, Potamogeton crispus, P. pectinatus, Sagittaria sagittifolia, Schoenoplectus lacustris* and *Sparganium emersum*. In the Loddon it is often particularly abundant in well aerated stretches below weirs and sluices (Archer 1987). In Dorset it grows in fairly shallow water over a gravelly substrate and cannot be found where there are soft clay sediments.

P. nodosus is a rhizomatous perennial that seems to vary in abundance from year to year. It flowers freely in summer when growing in shallow water. Plants do not appear to set fruit in the wild but will do so if cultivated in small containers, suggesting that the apparent sterility is caused by environmental factors. Buds develop on short stolons in the leaf axils in late summer and act as vegetative propagules. Established populations overwinter as buds on the rhizome, which develop in response to short days (Spencer & Anderson 1987). In southern Europe this species grows in a much wider range of habitats than it does in England, and fruits freely.

This species was first discovered in England in the Loddon by G.C.Druce in 1893 and was named *P. drucei* by Fryer in 1898 (Preston 1988). It was subsequently discovered in the Bristol Avon, Stour (Dorset) and Thames. However, its correct identity continued to puzzle British botanists until Dandy & Taylor (1939) established that it was the widespread continental species *P. nodosus*. The species has apparently become extinct in the Thames, where it was once locally frequent (Lousley 1944a) and perhaps disappeared as a result of eutrophication and increasing boat traffic in the 1950s. Recent reports of its rediscovery in the Thames have not been confirmed. It cannot now be found in the Warwickshire Stour at Alderminster, where it is known from a single 19th century specimen. However, it survives as vigorous stands in the Loddon and the Dorset Stour, and has recently been planted at additional sites in the Loddon and the nearby Whitewater and Blackwater.

This is the most frequent broad-leaved *Potamogeton* in the Mediterranean region, and extends northwards in Europe to England, the Netherlands and Estonia. It also occurs in Africa south to Angola and Madagascar, temperate and tropical Asia, and North and South America.

C.D. Preston, adapted from an account in Preston & Croft (1997).

Potamogeton rutilus Wolfg. (Potamogetonaceae)
Shetland pondweed, Lìobhag Ruadh
Status in Britain: LOWER RISK - Near Threatened.
Status in Europe: Not threatened. Endemic.

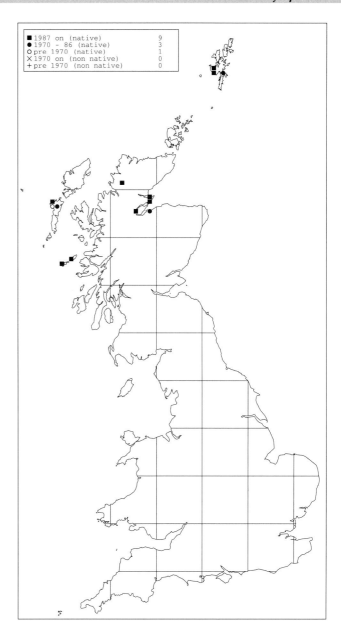

This species grows in unpolluted, lowland, mesotrophic or eutrophic lochs that receive base-enrichment; it is also found in adjoining streams. In Shetland it grows in water 1-2 m deep in mesotrophic lochs that receive drainage from limestone rocks. In the Hebrides the sites are machair lochs at the junction of acidic rock and calcareous dune sand. The habitat of *P. rutilus* at its remote British sites has not been studied in detail, partly because it is difficult to find in the deep water in which it often grows, and plants are normally detected after they have become detached and been blown to the side of a loch. In a dry summer when water levels were exceptionally low it was recorded in water only 0.5 m deep in Loch Ballyhaugh, Coll, growing in a substrate of fine silt with *Littorella uniflora, Najas flexilis, Potamogeton gramineus* and abundant *Chara aspera.*

Although flowering material of this species has been collected in Britain, fruiting plants have never been found. Plants overwinter as turions, which develop in the autumn and also provide a means of dispersal within a lake.

P. rutilus was completely misunderstood by British botanists until Dandy & Taylor (1938) revised the existing records and showed that the only correctly determined specimens were from Shetland. *P. rutilus* was discovered in the Outer Hebrides by Clark (1943). All sites known to Perring & Walters (1962) were in these two archipelagos, but this species has subsequently been found in an increasing number of sites as a result of the more systematic recording of Scottish lochs in recent years. The only site where it is thought to be extinct is Loch Flemington, where it was discovered in 1975 but appears since to have been eliminated by pollution from silage effluent and other sources.

P. rutilus is endemic to northern Europe, where it is found from the Arctic Circle south to northern France, Germany and Poland. It is a local plant throughout its range, and it may be the rarest of the European pondweeds on a world scale.

C.D. Preston, adapted from an account in Preston & Croft (1997).

Potentilla fruticosa L. (Rosaceae)

Shrubby cinquefoil
Status in Britain: LOWER RISK - Near Threatened.
Status in Europe: Not threatened.

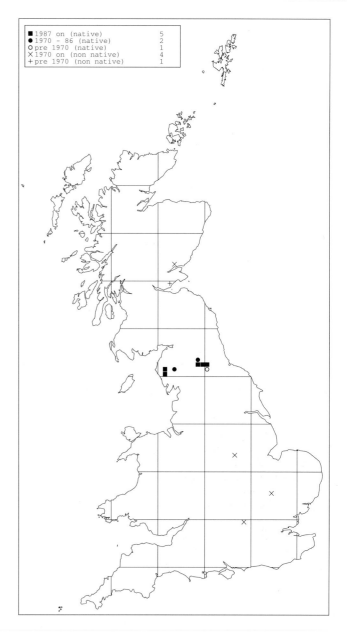

In Britain, this species is found only in Upper Teesdale and in Cumbria, occurring in contrasting habitats in these two areas. In Teesdale, it grows in the cracks between boulders and in the Whin Sill bedrock of river banks and islands in the Tees, and on silty stabilised shingle banks and islands. In both habitats, the scouring of the river at times of spate effectively removes both its less securely anchored competitors and the soil in which they grow. Constant associates are *Centaurea nigra, Galium boreale* and *Prunella vulgaris*, together with a wide variety of other species, including *Briza media, Carex flacca, Helictotrichon pratense, Leontodon hispidus, Linum catharticum, Nardus stricta* and *Thymus polytrichus*. In Cumbria, it is a montane plant, occurring on screes and cliffs on a few mountains in central Lakeland. The sites are on loose and crumbly rock, well supplied with trickling water and frequently soaked by heavy rain and low cloud. The largest plants tend to occur on shelving ungrazed ledges where there is some accumulation of soil. A few small plants also grow directly from cracks in the bed-rock. Associates include many montane species such as *Alchemilla alpina, Oxyria digyna, Saxifraga aizoides, S. hypnoides, S. oppositifolia* and *Sedum rosea*, together with more widespread species such as *Filipendula ulmaria, Geranium sylvaticum, Geum rivale, Silene dioica* and *Succisa pratensis*.

P. fruticosa is dioecious and appears to flower and set seed freely at all its sites. The flowering season is long, starting in May and often continuing well into autumn. For instance, in 1995, following a long but dry summer, numerous Teesdale plants were still in good flower at the end of September. Seedling establishment seems to be rare at present, but plants can spread by layering of their sprawling stems. Plants can grow to a metre tall but, where grazed, may flower when only a few centimetres tall.

In Teesdale, G.G. & P.S. Graham (1993b) found colonies at eleven sites along 24 km of the Tees from under Cronkley Scar at 390 m altitude down to Eggleston at 180 m. The total population was estimated at more than 3,500 plants, in colonies varying in size from a single old plant to about 1,900 plants. In Cumbria, it occurs in gullies on the Wastwater Screes at 395-450 m and on Pillar at about 610 m, both sites on the south-western side of the mountain. There is an old record from Helvellyn and a recent one from Fairfield. Populations in Cumbrian sites are very small, in contrast to the much larger ones in Co. Durham. There is fossil evidence that *P. fruticosa* occurred much more widely in Britain in late-glacial times, being now reduced to a relict scatter of populations.

It is not known whether the damming of the Tees at Cauldron Snout, upstream from all the *P. fruticosa* sites, has adversely affected the populations. Certainly, moribund plants are now apparent in most colonies, but the cause is unclear. Natural erosion seems to occur only slowly along the banks, and many colonies are probably of great age. Some parts of the river, such as the neighbourhood of Low Force and High Force, are now much eroded by human

trampling, and some plants may have been lost as a result. Most sites are ungrazed by livestock, though grazing may be beneficial in suppressing potential competitors. *P. fruticosa* may be unpalatable and grazed only as a last resort (Elkington & Woodell 1963). In Cumbria, there seem to be few threats to its sites, apart from rock-falls. Indeed, their inaccessibility confers protection, those in Wasdale being especially awkward to approach.

P. fruticosa has a disjunct range elsewhere in Europe, occurring in Ireland, the Pyrenees, the Maritime Alps and the Rhodope mountains, and in the Baltic region. It ranges eastwards to northern and central Asia, the Himalayas and Japan, and also occurs in Greenland and North America south to Mexico.

Details of the ecology and distribution of *P. fruticosa* in Britain are given in Pigott (1956) and Clapham (1978).

M.S. Porter and F.J. Roberts

Potentilla rupestris L. (Rosaceae)
Rock cinquefoil, Pumdalen y Graig
Status in Britain: VULNERABLE. WCA Schedule 8.
Status in Europe: Not threatened.

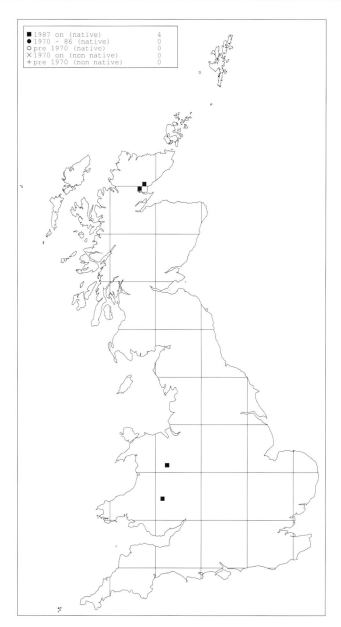

P. rupestris occurs at four lowland sites in Britain. Two are in Wales (Turner 1988), both of them small and severely threatened, and two, also small, are in Scotland. All British sites receive relatively high rainfall, but most are freely draining and subject to summer drought (e.g. Ratcliffe 1962). It chiefly occurs on soils of pH 5.2-6.7 derived from base-rich igneous rocks, and though typically found in open habitats, it appears to be tolerant of shade.

On Craig Breidden, Montgomeryshire, where it has been known since the late 17th century, populations are established on thin skeletal soil overlying a calcite-rich dolerite (Jarvis 1971). Associated species include those characteristic of basic rocks, amongst them *Festuca ovina, Geranium sanguineum, Helianthemum nummularium, Inula conyza, Scabiosa columbaria* and *Thymus polytrichus* (Wilson *et al.* 1995), together with *Calluna vulgaris, Campanula rotundifolia, Erica cinerea* and *Teucrium scorodonia. Lychnis viscaria, Sorbus leptophylla* and *Veronica spicata* grow nearby. Collecting and quarrying reduced the formerly large population to only six individuals, though in recent years there have been repeated attempts to bolster the population by *ex situ* propagation and transplantation. Currently there are about 60 transplants on the South Crags and about 200 within Criggion Quarry. It is assumed, without proof, that all the extant plants originated from Breidden stock, and it is uncertain whether the population is yet self-sustaining. The other Welsh site is on the banks of the Wye, but there is some doubt as to whether this is the same as that described by Ley (1887). The habitat is untypical, the 20 or so plants of *P. rupestris* growing on a river bank near a disused railway, in regenerating woodland, where they persist only because of the regular cutting back of branches. It may have been deliberately planted in this well botanised site.

The two Scottish sites are in East Sutherland, where plants grow on base-enriched soils on precipitous, generally south-facing granodiorite crags difficult of access. Such inaccessibility has made accurate population counts difficult, and estimates have varied widely. At the largest site, Ratcliffe (1962) reported tall clumps "thickly scattered" over the cliffs, but other estimates have ranged from only 12 plants in 1981 to 200 in 1985. However, it is not certain that all surveys covered the same area. It is likely that counts have been lower in poor flowering years, with non-flowering rosettes liable to be overlooked. The other Sutherland population, also on steep crags, numbered about 50 plants in 1985 and 1994 (P. Lusby pers. comm.). Accompanying species at Scottish sites include *Anthoxanthum odoratum, Anthyllis vulneraria, Brachypodium sylvaticum, Festuca ovina, F. rubra, Galium boreale, Helianthemum nummularium, Teucrium scorodonia, Thymus polytrichus* and *Ulex europaeus*.

It is a perennial herb with a woody branching stock that reaches deeply into crevices. Individual plants may be long-lived; some on Craig Breidden are considered to be at least 30 years old (Jones 1993). Flowers appear from May to July and are visited by bees, ants and flies. Seed is freely produced, but there seems to be no special mechanism for seed dispersal and seed is often retained in the seed-head until the following season. Germination is mostly in spring.

The plant is intolerant of heavy grazing by rabbits and sheep. Encroaching trees and scrub threaten at least the Wye population, and a limited amount of clearance has been done at Criggion Quarry on Breidden Hill. Plants were avidly collected from 1830-1930, with many specimens residing in herbaria (and doubtless many more transplanted in gardens), but since the plant seeds prolifically, any adverse effect on native populations from collecting may be more apparent than real. Habitat change is much the more likely cause.

In Europe, the range of *P. rupestris* extends from the Iberian peninsula and North Africa northwards to Britain and Sweden, and eastwards to Greece, and the Caucasus, Urals and Altai mountains. Plants in North America are now regarded as *P. glandulosa* var. *pseudorupestris*.

W.J. Whittington and M.J. Wigginton

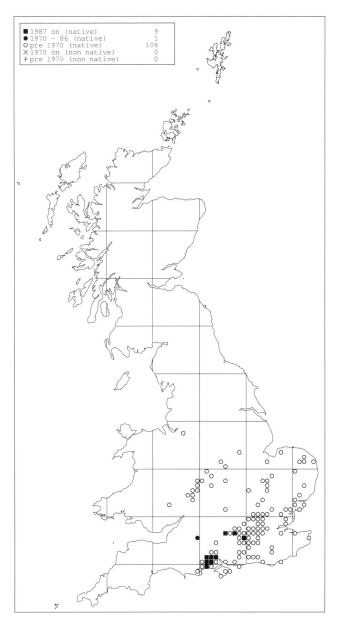

■ 1987 on (native)   9
● 1970 - 86 (native)   1
○ pre 1970 (native)   106
✕ 1970 on (non native)   0
+ pre 1970 (non native)   0

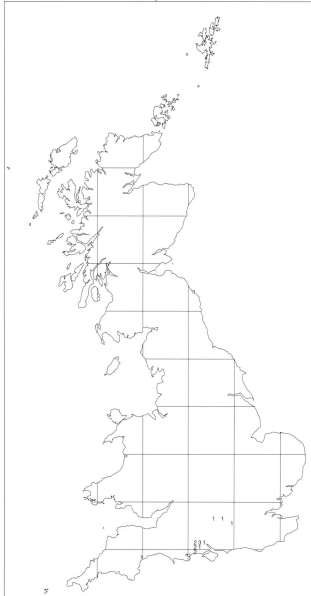

Pulicaria vulgaris Gaertner (Asteraceae)

Lesser fleabane, Cedowydd Bach
Status in Britain: VULNERABLE. WCA Schedule 8.
Status in Europe: Vulnerable.

P. vulgaris is a plant of winter-flooded hollows in grassy places (Hare 1991), an unostentatious plant and a victim of the changes brought to the English countryside during the twentieth century. In the past, typical sites for *P. vulgaris* were village greens and cart tracks, in damp ground that was well-hoofed, well-grazed and fertilised by animals. It has declined drastically, except in the New Forest, and is elsewhere exceptionally rare and localised.

P. vulgaris has always had a stronghold in the New Forest, where it grows mainly in depressions in the grassy lawns where ponies graze along with smaller numbers of other livestock including geese. These irregular depressions hold water through the winter and dry out in spring. Their origin is obscure, though they may be old

wallows dug out by pigs, which were once turned out into the Forest in their thousands (Tubbs 1986). Other haunts of *P. vulgaris* include cart-tracks and ditch-edges. In the New Forest it grows with an interesting assemblage of species, which give a tantalising glimpse of what the greens and waysides of England must have been like a century or more ago. The closest associates are *Alopecurus geniculatus*, *Bidens tripartita*, *Gnaphalium uliginosum*, *Lythrum portula*, *Polygonum aviculare*, *Potentilla anserina* and *Senecio aquaticus*, and it occurs occasionally with the much rarer *Chamaemelum nobile*, *Polygonum minus* and *Mentha pulegium*.

P. vulgaris is an annual. Germination takes place chiefly in spring, in areas of open mud or sandy soil exposed by falling water levels, though a few plants probably overwinter, having germinated in late autumn. The yellow disc-like capitula appear in August and September. Seeds are unlikely to be blown far by the wind but may get carried on the feet of domestic animals. In the era of unmetalled roads, dispersal by these means was probably common, but now such opportunities are few. Its seeds are long-lived, as experiments with buried seed have demonstrated (Hare 1986).

Outside the New Forest, *P. vulgaris* has recently occurred in just one site in Surrey (a sandy-gravelly pond edge), and two in North Hampshire, though it has not been seen in one of the latter for several years. Fewer than 10 plants may occur in these sites. The New Forest population is spread between some 10 more-or-less discrete sites and consists of thousands of plants, though even here it seems to have declined in recent years. A review of populations in 1985 suggested that more than 100,000 plants occurred in the New Forest, but in 1990 the total decreased to about 10,000. This national decline in numbers was, however, mainly because of changes in a single sub-site. In 1995, a comprehensive survey revealed a total of 28,000 plants, four populations having increased significantly (FitzGerald *et al.* 1997). *P. vulgaris* was formerly widespread across southern England and the Midlands and must once have been frequent or common around London, where working horses were fed and rested. There are many old records from Middlesex and Surrey, especially to the south-west of London, and it was

also found in such places as Hampstead Heath and Islington, improbable locations indeed today. However, because of the long viability of buried seed, reappearances at old sites have occurred. For instance, at the present Surrey site it was rediscovered in 1979 after a gap of about seven years. Heavy poaching by livestock had restored the quite open, disturbed conditions that favour the plant, and presumably had brought seeds to the surface.

Current management of sites in the New Forest and south-east England seems to be suitable, though habitat restoration or selective reintroduction may locally give the species more security. A full account of the species is given in Prince & Hare (1981), and other details are given in Chatters (1991).

P. vulgaris is distributed throughout Europe from England, southern Sweden (now extinct) and central Russia southwards to the Mediterranean and Turkey. It is declining rapidly across much of its range, and the New Forest populations are probably the most important in western Europe.

A.D.R. Hare and F. Rose

Pulmonaria obscura Dumortier (Boraginaceae)

Unspotted lungwort
Status in Britain: VULNERABLE.
Status in Europe: Not threatened.

P. obscura grows in three adjacent woods in East Suffolk. All three are ancient woods with a history of management as coppice with standards. It is found in areas with poorly drained, fertile soil where the canopy and shrub layer are dominated by ash, field-maple and hazel. Frequent associates include *Ajuga reptans, Anemone nemorosa, Arum maculatum, Carex sylvatica, Circaea lutetiana, Geum urbanum, Glechoma hederacea, Lamiastrum galeobdolon, Listera ovata, Mercurialis perennis, Orchis mascula, Poa trivialis, Rubus fruticosus, Urtica dioica* and *Viola reichenbachiana*. Although it can persist in the shade of a dense woodland canopy, it flowers more freely, produces larger leaves, and grows more vigorously in light shade such as is provided by some stages of the coppice cycle (Birkinshaw & Sanford 1996).

This species is a perennial that spreads vegetatively by creeping rhizomes. It is similar in appearance to *P. officinalis*, except that its leaves are unspotted and do not overwinter (Bolliger 1978). It flowers from late February to early May and sheds its seeds in June. Both 'pin' and 'thrum' flower types occur, this adaptation being considered likely to promote cross-pollination. The discovery in 1993 of two large and apparently new clumps of lungwort growing on recently created woodland rides suggests that seed may remain viable in the soil for many years.

P. obscura was first recorded by C.J. Ashfield in 1842, but as *P. officinalis*, under which name it is described also in Simpson (1982). It has since been found in three additional sites. However, one population was destroyed in 1970 when electricity pylons were erected through the wood. The populations at two of the other sites have apparently declined since the 1930s (Simpson 1982), probably because coppicing has ceased. The third population was first recorded in 1993. In 1994, the lungwort populations covered in total some 18 square metres and produced about 600 flowering stems.

Several *Pulmonaria* species are grown in gardens and have subsequently become established locally in semi-natural habitats. However, *P. obscura* may be truly native in Suffolk, since the habitat and associated species there closely match those of native sites in Belgium and northern France. Ideally, sites for this species should be managed in coppice rotation, and this has resumed in one of its three sites.

In Europe, *P. obscura* ranges from southern Sweden east to the Urals, south to the Alps and Ukraine, and westwards to north-east France and the Ardennes region of Belgium (Bolliger 1978).

C. R. Birkinshaw and M.N. Sanford

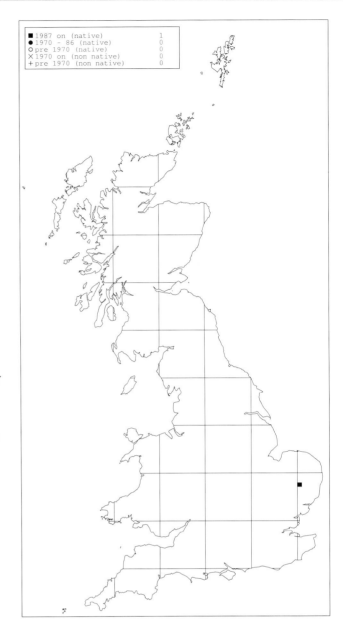

■ 1987 on (native)	1
● 1970 – 86 (native)	0
O pre 1970 (native)	0
X 1970 on (non native)	0
+ pre 1970 (non native)	0

Pyrus cordata Desv. (Rosaceae)
Plymouth pear, Gellygen Plymouth
Status in Britain: VULNERABLE. WCA Schedule 8.
Status in Europe: Unknown

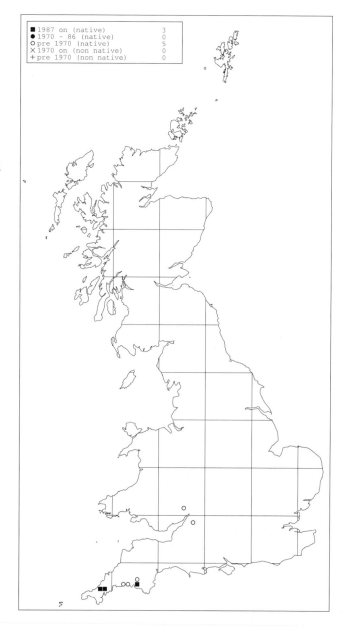

■ 1987 on (native)	3
● 1970 – 86 (native)	0
○ pre 1970 (native)	5
✕ 1970 on (non native)	0
+ pre 1970 (non native)	0

P. cordata is a small hedgerow tree up to 10 m high. It is known from just two localities in Britain, both in south-west England. The once rural hedge bank in Plymouth, where it was first discovered in 1865 (Archer Briggs 1880), is now subsumed within a light-industrial estate, though a few *P. cordata* trees remain. The second population, verified near Truro in 1989, is within a landscape of small agricultural enclosures. From these intensively managed British locations it is impossible to determine its favoured habitat. However, in Brittany, in addition to hedgerows it is a component of mixed deciduous woodland with oak, holly, beech, hornbeam, birch, blackthorn, hazel and wild service tree. In densely shaded woods it appears to reproduce asexually by suckers only; flowering and fruiting seem restricted to glades and woodland edges.

Because *P. cordata* is able to produce abundant suckers, it is often difficult to identify individual plants. The two British populations exist on seven sites, six of which are protected as SSSIs. The Cornish sites are separated by up to 1.9 km (Tonkin 1993) and the two in Plymouth by about 1.6 km. In total, over 120 larger stems (over 1 cm in diameter) have been uniquely numbered and more than 460 other small plants recorded and mapped. There is considerable morphological variation between the two populations, including the habit, the shape of the leaf, flowering time and the number of carpels. Molecular research has confirmed a genetic difference between the Truro and Plymouth plants but no genetic variation has been found within either population (Jackson *et al.* 1997).

The most significant problem for the long-term survival of *P. cordata* in Britain is likely to be its poor ability to cope with environmental change, which is restricted by poor sexual reproduction of both populations owing to the lack of genetic variation and resultant self-incompatibility. For example, following the best recorded fruit-set during 1992, in Plymouth, 1,178 fruits yielded only nineteen seeds which were either inbred offspring or the result of hybridisation with local domesticated pears. Research on *P. cordata* has considered its taxonomy, genetics, reproductive biology, ecology and horticulture (Jackson 1995). Since there are only two genotypes, the carefully designed breeding and propagation programme seeks to enhance its genetic diversity. Planting of both genotypes and seedlings raised from controlled hybridisations between them began in November 1995. This programme should be completed by 1999, with plants starting sexual reproduction by about 2010.

The removal of hedgerows and their unsympathetic management have been significant problems. Other threats to the wild populations include pathogens such as fireblight (*Erwinia amylovora*) and the importation of European genetic material. Theft of experimental native stock has also occurred. *Ex situ* measures to offset these threats include the cultivation of clonal material in botanic gardens and arboreta and the long-term storage of seed. In Plymouth, 28 suckers of local origin have been transplanted to semi-natural areas managed as Local Nature Reserves. Three years after planting all have become established and some have started to flower.

The British populations are the most northerly of its largely western European distribution. However, a recent study of European pears (Aldasoro *et al.* 1996) has expanded the range of *P. cordata* to North Africa and the Near East. They also identified a herbarium specimen from a Gloucestershire site as *P. cordata*. Field surveys may yet reveal its presence at further sites in southern England.

A. Jackson

Ranunculus ophioglossifolius Villars
(Ranunculaceae)

Adder's-tongue spearwort, Llafnlys Tafod y Neidr
Status in Britain: ENDANGERED. WCA Schedule 8.
Status in Europe: Not threatened.

In recent years, *R. ophioglossifolius* occurs in only two
sites in Britain, at Badgeworth and at Inglestone Common,
both in Gloucestershire. It is a plant of semi-permanent
ponds and marsh. At its main site, Badgeworth, the plants
grow in a depression in surrounding flat meadowland, on
impervious Lias clay, which fills up seasonally with rain
water. Periods of heavy rain can lead to dramatic rises in
water level. Associated species include *Alopecurus
geniculatus*, *Eleocharis palustris*, *Glyceria fluitans* and *Myosotis
scorpioides*. The site at Inglestone is also a pond on Lias
clay, though there is at present almost no open water.
Species occurring in the swampy grassland include *Carex
disticha*, *Glyceria fluitans*, *Juncus effusus*, *J. inflexus*,
Ranunculus flammula, *R. repens*, *R. trichophyllus* and *Senecio
aquatilis*.

It is an overwintering annual. For optimal growth it
requires bare, moist soil from August to October, with an
absence of frost for germination and seedling growth,
sufficient winter rain to submerge the plants, and a
reduction of water level in spring (Dring & Frost 1971;
Holland 1977). The germination stage is particularly
sensitive, and autumn frosts may destroy every seedling.
Seeds can also germinate the following spring, though the
resultant plants are stunted and may not exceed 3 cm in
height, whereas autumn-germinated plants generally grow
to 40-50 cm (Dring & Frost 1971; Holland 1977; Jones 1978).
Seeds can remain viable for at least 30 years but are killed
by prolonged desiccation (Jones 1978; Toase 1992).
Populations vary greatly in response to climatic conditions,
and during peak years at Badgeworth, more than a
thousand plants may flower, though in poor years there
may be none (Holland *et al.* 1986; Doe 1993). At Inglestone
Common, about 100 plants flowered in 1966 following
vegetation clearance, but in recent years there have been
few plants, and often none at all (Rich 1993b; Kitchen *et al.*
1995). In 1993-5, only one, five, and two plants were
recorded, but in 1996 cattle were excluded during the
flowering and fruiting season and 36 plants were seen
inside the fence, with nine outside it (Rich & Davis 1996).

It was formerly known from two other sites in Britain,
in Hampshire (one record, 1878) and Dorset, but both were
lost to drainage and development by the early part of this
century. It has been protected at Badgeworth since the
establishment of the reserve (now an NNR) in 1933, and
the reserve is further protected within a larger SSSI. Since
1962, the area has been actively managed for
R. ophioglossifolius, which crucially includes each year the
creation of patches of bare ground for germination
(Holland 1977; Frost 1981). Within a small area of the
reserve, grazing by cattle and horses between mid-July and
late September has also been beneficial, since this also
provides the necessary disturbance and control of
competing species, particularly *Glyceria fluitans*. The site at

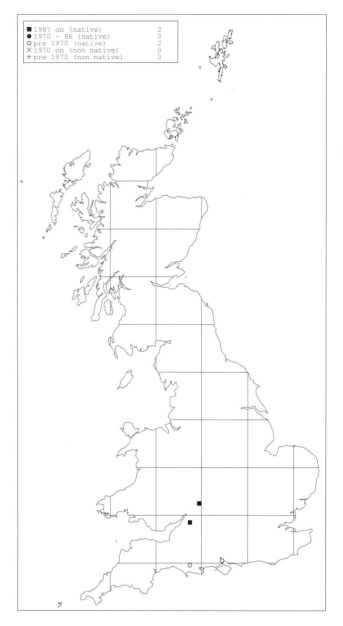

■ 1987 on (native)	2
● 1970 – 86 (native)	0
○ pre 1970 (native)	2
× 1970 on (non native)	0
+ pre 1970 (non native)	0

Inglestone Common is within the Lower Woods SSSI. Some
beneficial management was carried out in past years, and
following a period of neglect, habitat management has
resumed (Rich 1993b).

R. ophioglossifolius is widespread in southern Europe,
from Spain to Greece and the Crimea, with disjunct
populations in England and Gotland (Holland 1977). It also
occurs in North Africa and West Asia. It is extinct in
Jersey. The Badgeworth and Inglestone plants are
morphologically similar to the large-flowered var. *genuinus*,
but a taxonomic review of the species is required before
any firm conclusions can be drawn (Dring & Frost 1971;
Frost 1981). Nonetheless, plants from Badgeworth have
been shown to be genetically distinct from Portuguese
material (Frost 1981).

D.A. Callaghan

Ranunculus reptans L. (Ranunculaceae)
Creeping spearwort, Glaisleun Ealaidheach
Status in Britain: ENDANGERED.
Status in Europe: Not threatened.

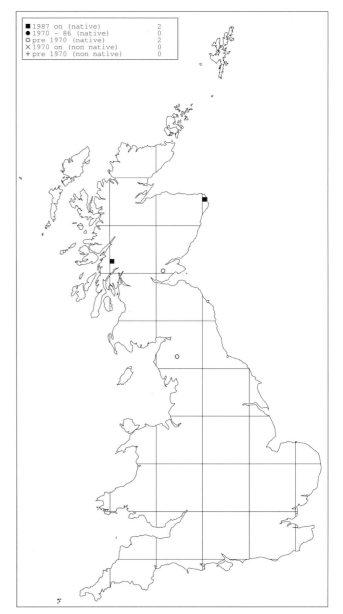

Plants that are referable to *R. reptans* are known from four localities in Britain: the gravelly shores of Ullswater and Loch of Strathbeg, where they have been recorded with *R. x levenensis*, the shore of Loch Leven, where they were almost certainly associated with the same hybrid, and Loch Awe, where they have recently been found on silty sand near the top of the shore. At the Loch of Strathbeg, it occurs on the gently-sloping shore in locally flushed, open vegetation of *Eleocharis palustris*, in company with such species as *Agrostis stolonifera, Caltha palustris, Juncus articulatus, J. bufonius, Mentha x verticillata, Persicaria amphibia* and the mosses *Cratoneuron filicinum* and *Drepanocladus aduncus*. At Loch Awe associated species include *Juncus bulbosus, Littorella uniflora* and *Lobelia dortmanna*.

This species is a stoloniferous perennial, which creeps and roots at the nodes. Flowers have been recorded from June to September. It is partially self-compatible.

R. reptans appears to be a transient member of the British flora. It was collected at Ullswater in 1887 and from 1911 to 1917 and at Loch Leven in 1869, 1896 and 1935. It is currently known from Loch of Strathbeg, where it was collected in 1876, 1900 and 1989 and was still present in 1996, and from Loch Awe, where it was collected in 1992. Gornall (1987) suggests that it is introduced repeatedly by waterfowl migrating in autumn from Iceland, Scandinavia or northern Russia.

Plants of *R. reptans* hybridise with *R. flammula* and give rise to persistent populations of the hybrid *R. x levenensis*. The reason that *R. reptans* itself fails to survive is not clear, but Gibbs & Gornall (1976) have shown that it is less phenotypically plastic than the hybrid and may therefore be at a disadvantage on lake shores where the water-level fluctuates. *R. x levenensis* has been recorded from sites in north-east Scotland and northern Ireland where *R. reptans* has never been reliably recorded but where it was presumably once present.

R. reptans has a circumpolar distribution, being widespread in the boreal zone and extending southwards in the mountains of Europe, Asia and North America. In Europe its main range is in Scandinavia and the north-east, but it is found at scattered sites south to the Alps, central Italy and Bulgaria (Jalas & Suominen 1989). There are records from Iceland (where it is common in, for example, shallow roadside pools) and Greenland.

The distinction between *R. reptans* and the variable hybrid *R. x levenensis* is difficult and somewhat arbitrary. This account is based on specimens determined by R.J. Gornall and P.D. Sell. Further details of *R. reptans* at the Loch of Strathbeg are given in Birse (1997).

C.D. Preston, adapted from an account in Preston & Croft (1997).

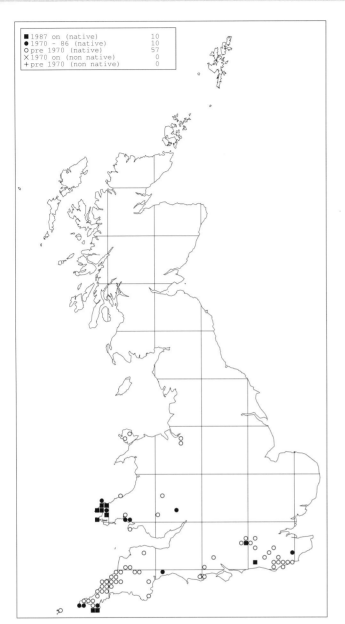

■ 1987 on (native)   10
● 1970 - 86 (native)   10
○ pre 1970 (native)   57
✕ 1970 on (non native)   0
+ pre 1970 (non native)   0

Ranunculus tripartitus DC. (Ranunculaceae)

Three-lobed crowfoot, Crafanc Trillob
Status in Britain: VULNERABLE.
Status in Europe: Not threatened.

R. tripartitus occurs in shallow but seasonal bodies of water among heaths or related communities, especially shallow ditches and ponds, cart-tracks and gate-ways, wet in winter and spring but dry by summer. The soil base-status and pH are moderately high. Associates include *Apium inundatum*, *Cicendia filiformis*, *Eleogiton fluitans*, *Juncus pygmaeus* (Lizard peninsula), *Lythrum portula*, *Pilularia globulifera*, *Potamogeton polygonifolius* and a range of charophytes. Like other pond and trackway species of heaths and commons, *R. tripartitus* is intolerant of competition from other plant species, needing open sites maintained by fluctuating water levels, grazing and poaching by livestock, and disturbance by traffic.

It is typically a winter annual, though it may sometimes perennate, germinating in autumn and flowering in April and May, or in mild winters as early as February - earlier than related species. Its sites often dry out completely by June. Wider dispersal of seed is likely be on the feet of birds and other animals.

R. tripartitus may be under-recorded, partly because of its early flowering, partly from problems of identification. *R. tripartitus* is very local and diminishing because of the destruction of heaths, the drainage or infilling of its habitats, and the cessation of grazing and disturbance, allowing the uncontrolled growth of competitors and eventually *Salix aurita* and *S. cinerea* carr. However, the long viability of its seed allows it to recover if overgrown sites are cleared, provided drainage has not been too severe. *R. tripartitus* is frequent only in parts of the Lizard and in western Pembrokeshire. Some plants recorded as '*R. lutarius*' in the New Forest are pentaploid hybrids

311

between *R. omiophyllus* and *R. tripartitus (R.* x *novae-forestae*) and are frequently very similar in form to *R. tripartitus* (Cook 1966; Webster 1990). However, recent isoenzyme studies by Kay & John (1995) suggest that some New Forest populations should be regarded as the true species.

R. tripartitus is an oceanic-western European species, occurring from south-west Spain to northern Germany, and is also reported from Greece and Morocco (Jalas & Suominen 1989). It is declining throughout the northern part of its range.

A.J. Byfield, in Stewart *et al.* (1994)

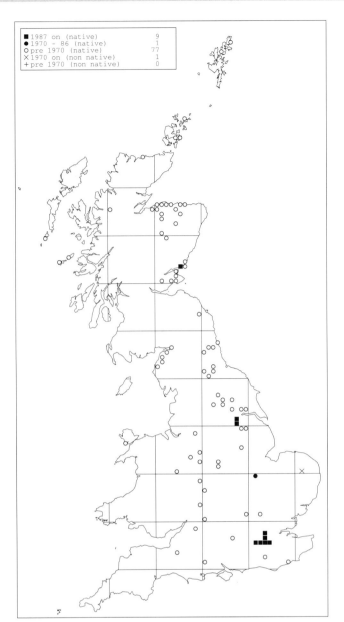

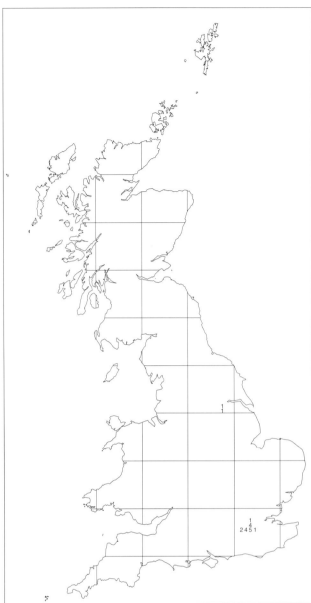

Rhinanthus angustifolius C. Gmelin
(Scrophulariaceae)
R. serotinus (Schönheit) Oborny
Greater yellow-rattle, Cribell Felen Fawr
Status in Britain: VULNERABLE. WCA Schedule 8.
Status in Europe: Not threatened.

R. angustifolius grows in a range of lowland grassy habitats. It owes its wide ecological tolerance to its semi-parasitic habit, its main requirement being for a sufficient density of host roots through which it can obtain water and nutrients. It has been recorded widely across Britain, although robust forms of *R. minor* may sometimes have been mistaken for it, and the map may contain some errors. Nevertheless, there has been a widespread decline, the reasons for which are obscure, and it is now found in only three areas - in Surrey, North Lincolnshire and Angus.

The only area where this species is at all frequent is a small section of the North Downs on the Surrey/Greater London borders, especially Happy Valley and Farthing Downs, where it occurs in 18 1 km squares in six hectads. There it is now mainly a plant of meadows on the chalk, where it grows with such species as *Centaurea scabiosa*, *C. nigra*, *Festuca pratensis*, *Helictotrichon pratense*, *Lathyrus pratensis* and *Onobrychis viciifolia*. In Lincolnshire, it still occurs in two localities: on Crowle Moors, two small populations on a trackside and one close-by in an area of cleared bracken, and near Belton, in species-rich rank grassland between an old railway and arable land. At the latter site, associated species include *Arrhenatherum elatius*, *Centaurea nigra*, *Dactylis glomerata*, *Heracleum sphondylium*, *Lathyrus pratensis*, *Leucanthemum vulgare*, *Leontodon hispidus*, *Phragmites australis*, *Poa trivialis*, *Sanguisorba officinalis* and *Trifolium medium*. In Angus, a single tiny colony just survives in sandy coastal grassland near East Haven.

This species is an annual hemi-parasite. Flowers generally appear in late June and July, but may extend to September. They are pollinated by bumble-bees (Kwak 1978). Flowering times overlap with *R. minor* which, however, generally starts flowering earlier. Hybridisation is suspected, as indicated by the intermediate characters of some individuals, and it is possible that some colonies of pure *R. angustifolius* have disappeared through introgression. Seed germinates in spring, and plants may establish themselves in a more or less closed sward, unlike most annuals, which require patches of open ground. Studies of related species suggest that *R. angustifolius* can probably accept a wide range of host plants, though particular species are favoured in certain situations, such as legumes in low nitrogen soils. However, it can survive without a host, even to produce seed, though such plants are depauperate. The quality of the host plant also seems to affect the morphology of the *R. angustifolius*, and this has contributed to taxonomic confusion, with forms variously described as species, subspecies, and seasonal or altitudinal ecotypes.

The ideal management for *R. angustifolius* is a late hay-cut after seed has set. Seed has little longevity beyond the following spring, and therefore any break in continuity of seed production can quickly lead to the loss of populations. Aftermath grazing is acceptable, as is light grazing earlier in the season. The plant seems to be unpalatable to livestock. Given that distinctive forms have evolved in response to particular mowing regimes, it is highly desirable to retain the local management pattern to ensure seed production is in synchrony with mowing. At the Surrey sites, it has increased following scrub clearance and late hay cutting. It also seems to have extended its range, perhaps through seeds being inadvertently carried to new sites by mowing machinery. However, some new colonies are suspected of having originated from continental seed, as was the alien *R. alectorolophum* in Warwickshire.

Elsewhere, its distribution extends across much of central Europe, northwards to Scandinavia and southwards to Turkey and Siberia. It is declining in the west, but is perhaps stable further east, especially where it retains a strong presence as an arable weed.

C. Gibson

Romulea columnae Seb. & Mauri (Iridaceae)

Sand crocus
Status in Britain: VULNERABLE. WCA Schedule 8.
Status in Europe: Not threatened.

In Britain, this species is restricted to Dawlish Warren, where it was first recorded in 1834. It grows on well-drained, leached, sandy or gravelly soils in established sand-dune grassland. Most of the population lies within a links golf course, where it grows best in shortly grazed or mown turf. Associated herbs in the sometimes rich sward include *Aphanes arvensis, Montia fontana, Myosotis ramosissima, Ornithopus perpusillus, Plantago coronopus, Rumex acetosella, Sedum acre* and *Veronica arvensis*, together with the less common *Moenchia erecta, Poa infirma* and *Teesdalia nudicaulis*. It may be able to hold its own for a while in taller vegetation up to 10 cm high, but seedlings do not establish readily and plants will eventually be crowded out.

It is a perennial species of about 5-6 years' longevity. The leaves emerge in autumn and remain green in winter. Flowers appear between late March and early May, the larger plants producing three or more flowers. By early summer the leaves have shrivelled and only the seed-pods remain. Seed is produced in quantity and is set before the soil becomes desiccated during the summer months. Reproduction from seed appears to be much more important than by division of the corm, and open soil is required for seedlings to become established.

At Dawlish, there are many colonies scattered over an area of about 30 hectares of golf course (De Lemos 1992), where it benefits from frequent mowing and rabbit-grazing, which keep the vegetation short. However, fertilisers and irrigation may be detrimental where they encourage the vigorous growth of grass. In 1981, sections of turf containing corms of *R. columnae* were successfully transplanted from an area earmarked for a car park, and the new colony continues to thrive. It can occur at high densities, up to 500 plants per square metre. *R. columnae* was reported from Cornwall in 1879 and 1881 (Davey 1909) but has not been reported since. It is still frequent on some cliff tops in the Channel Islands.

R. columnae is a Mediterranean-Atlantic species that extends from England to the Azores and eastwards to Turkey. Throughout its range it occurs mainly near the coast. The British plant is ssp. *Columnae*, which occurs throughout the range of the species. Many varieties have been named, and our plant has been referred to var. *occidentale* (Sell & Murrell 1996).

N.F. Stewart

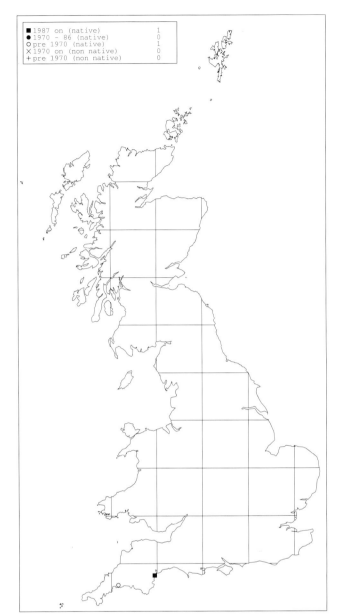

■ 1987 on (native)	1
● 1970 – 86 (native)	0
○ pre 1970 (native)	1
✕ 1970 on (non native)	0
+ pre 1970 (non native)	0

Rorippa islandica (Oeder ex Murray) Borbás (Brassicaceae)

Northern yellowcress, Biolair Bhuidhe Thuathach, Berwr Melyn y Gogledd
Status in Britain: LOWER RISK - Nationally Scarce.
Status in Europe: Not threatened.

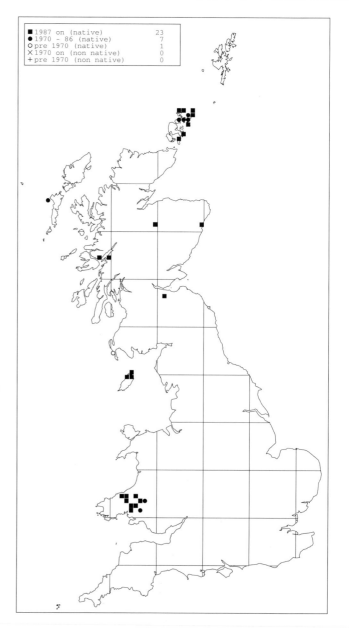

R. islandica is mainly, but not exclusively, a plant of wetland habitats. In Wales, it occurs in the muddy, open vegetation on the edges of ponds and depressions in the flood plains of rivers. In mainland Scotland it occurs in small ponds, on damp ground, and on the margins of loughs and reservoirs, and in Orkney also around the margins of old kelp burning pits. It is found occasionally on waste ground and tips and occurs at one site on a rock outcrop. The substrate is usually above about pH 6.0 and the mud may be rich or poor in nutrients. In Wales it is associated with *Callitriche brutia*, *Gnaphalium uliginosum* and *Persicaria hydropiper* and elsewhere often with *Juncus bufonius* and *Potentilla anserina*, but only rarely with *Rorippa palustris*.

This species is probably annual or biennial in the wild, matching the temporal nature of its habitats, but it may also be a short-lived perennial. Colonies range in size from a few individuals to many thousands of plants, which in the latter case can sometimes turn the soil orange with ripe seeds. It is dispersed locally by water and in mud and probably over longer distances by wildfowl.

It was at first thought to be nationally rare in Britain and was initially known from only five localities (Randall 1974). However, recent surveys have revealed it to be rather more frequent. The plant has subsequently been found in scattered localities in Scotland (at present mainly in Orkney) and somewhat erratically in the Isle of Man (Rich 1991), the Republic of Ireland (Goodwillie 1995) and the Afon Teifi valley and associated catchments in South Wales (Chater & Rich 1995). It is undoubtedly more widespread, though localised in distribution. The main threats are river works, the loss of ponds, the removal of grazing livestock, and agricultural improvement.

R. islandica was first recognised as distinct from its common relative *R. palustris* by Jonsell (1968). Formerly, both species were included under one taxon '*R. islandica*', which still causes some confusion with old records. The characters distinguishing the two species have been reviewed by Chater & Rich (1995); new records for *R. islandica* should be supported by a voucher specimen which need consist of only a few fruits with ripe seeds.

The European distribution is mapped in Jalas & Suominen (1994). It is rare around the Atlantic coasts of northern Europe, occurring also in Iceland and Greenland, and around alpine lakes in central Europe. The Afon Teifi and the Irish turloughs appear to be the major stronghold for the species in north-west Europe. Habitats in Norway include coastal pools and potato fields, and in Iceland it grows also around hot springs. British plants are referable to the European ssp. *islandica*. A different sub-taxon (ssp. *dogadovae*) occurs in Russia.

T.C.G. Rich

Rosa agrestis Savi (Rosaceae)

Small-leaved sweet-briar, Miaren Gulddail
Status in Britain: LOWER RISK - Nationally Scarce.
Status in Europe: Not threatened.

The principal habitat of *R. agrestis* is open scrub on dry calcareous grassland, which is also the habitat of a number of other *Rosa* species. *R. agrestis* is a free-standing, erect shrub with slightly flexuous stems. It is rather inconspicuous, with no striking features to make it obvious from a distance, and is apt to occur as single isolated individuals rather than in large stands. For these reasons it can be easily overlooked in the presence of other *Rosa* species, which may sometimes be abundant in this habitat, especially *R. canina*, which when young and growing free-standing may have a habit very similar to that of *R. agrestis*.

Only recently has the taxonomy of the British *Rosa* species been placed on a sound basis (Stace 1991; Graham & Primavesi 1993). Before this, few people took a serious interest in the genus, and therefore authenticated records for even the commonest taxa are few. Our knowledge of the status of *R. agrestis* is consequently incomplete. Since 1970, it has been recorded from only about 13 sites on the chalk in southern and south-east England, with outlying records from Somerset, Dorset, Worcestershire, Norfolk and Caernarvonshire. Older records, supported by herbarium vouchers, confirm its mainly southern distribution, though there is a single apparently anomalous record from Northumberland.

The future of *R. agrestis* depends largely upon the permanence of its habitat. Much of this type of open scrub has recently gone under the plough or has been diverted to other uses. Furthermore, it is a stage in succession leading to a climax vegetation, usually woodland, and its stability depends upon the activities of man or grazing animals. Total neglect and lack of grazing would, in time, be detrimental. The first great epidemic of myxomatosis, with a severe reduction in the rabbit population, showed how rapidly scrub became dense and impenetrable, in which state *R. agrestis* would ultimately be shaded out. Notwithstanding the need for grazing to control scrub, young plants of *R. agrestis* must survive the predation of grazing animals. The climbing species of *Rosa* are probably less susceptible to this predation because they are partially concealed and protected in their seedling stages by the plants up which they climb. The rather blunt prickles of *R. agrestis* may provide less of a defence than do the fiercely-hooked prickles of *R. rubiginosa*, which is also characteristic of open scrub.

R. agrestis is widespread in Europe, becoming rarer northwards and being absent in the extreme north (Klástersky 1968). It is likely that careful survey in Britain would reveal further sites within its historic range, though

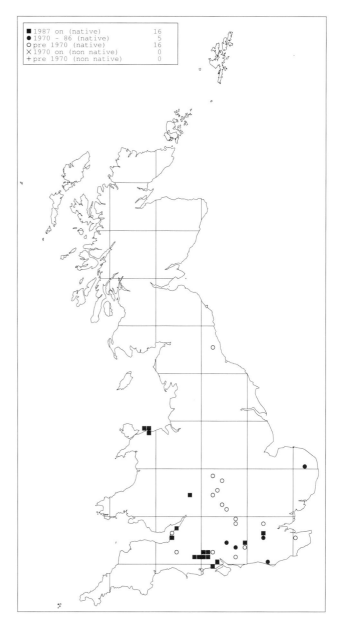

it is probably a truly uncommon species. It may be significant that its apparent greater frequency in similar habitats in central Ireland reflects the thoroughness with which a keen local botanist has scoured that country with a critical eye.

Because of the comparative rarity of *R. agrestis*, its hybrids with other species appear to be correspondingly rare. Authenticated records, mostly from old herbarium material, are only about 10 in number, including hybrids with *R. stylosa, R. canina, R. sherardii* and *R. micrantha*.

A.L. Primavesi

Rubus fruticosus L. agg. (Rosaceae)
Bramble, Dris, Mwyaren Ddu

Taxa within the *R. fruticosus* complex can be grouped according to morphological characters, including the relative density and arrangement of stem armament and indumentum, panicle architecture and growth habit. Relative distribution and frequency of European examples are, however, best illustrated on the basis of geographical range - allowing for the fact that since many microspecies are more or less apomictic, a high degree of endemism is exhibited. Mapping of *Rubi* in north-west Europe has been undertaken intensively over the last 20 years and it is fair to say that in the British Isles, Germany, Denmark, Belgium and the Netherlands, as well as some countries of central and eastern Europe, their distribution is reasonably well understood. Three categories of range are generally recognised by European batologists - Widespread (diameter of range greater than 400 km), Regional (50-400 km) and Local (less than 50 km). Edees & Newton (1988) describe the application of these categories and comment further on the distribution of bramble assemblages.

The British *Rubus* list currently includes 310 named microspecies. In addition to six introduced taxa (three of which are actively spreading into the wild from horticultural or ornamental origins), it includes the diploid out-breeding *Rubus ulmifolius* and the tetraploid pollen-fertile *R. caesius* (Dewberry) and three extinct taxa. Recent hybrids can be encountered where stable long-established communities of *Rubus* are met by the intrusive and aggressive microspecies *R. ulmifolius, R. caesius* and some others of open ground. Such hybrids are particularly frequent in south Wales and in southern and south-west England on the coastal fringes and estuary margins. Most of these hybrids are infertile but many are capable of spreading vegetatively.

Of the named microspecies, many owe their recognition to chance factors, such as the interest of a local 19th century batologist and subsequent collection and correlation through the Botanical Exchange Club, culminating in separate recognition by monographers such as H. Sudre, W.M. Rogers and W. Watson, or county flora writers such as A. Ley and W.R. Linton. A number of 'local' microspecies have thus been coined which, under current conventions, would not have qualified for description. At the moment, presence in at least four hectads and in two Watsonian vice-counties, or a conspicuous notice in the literature, are deemed to be *de rigueur* for such recognition.

Most British and Irish vice-counties possess a number of undescribed local taxa, varying in quantity from about 20 in southern England through about 10 in the Midlands to a small number (if any) further north. To some extent therefore the selection of a list of the 'rarest' brambles is tendentious, favouring named local endemism at the expense inevitably of the undescribed. If such a list is to be attempted it should include the following, where the known current range is restricted to very few hectads.

R. adenoleucus - Surrey
R. andegavensis - Kent
R. arrhenii - S. Wiltshire
R. bucknallii - W. Gloucestershire
R. crespignyanus - W. Kent, Surrey
R. dasycoccus - W. Glouc., Monmouthshire
R. diversiarmatus - N. Somerset
R. dobuniensis - W. Gloucestershire
R. herefordensis - Herefordshire
R. hirsutissimus - Herefordshire, Monmouthshire
R. hyposericeus - Herefordshire
R. iodnephes - Surrey
R. laxatifrons - Herefordshire
R. longifrons - E. Sussex, W. Kent
R. patuliformis - E. Sussex, W. Kent
R. permundus - Surrey
R. pervalidus - Surrey, S. Essex
R. pliocenicus - Kent
R. powellii - S. Essex
R. pseudoplinthostylus - Dorset
R. putniensis - Surrey
R. regillus - Herefordshire
R. sagittarius - S. Devon
R. salteri - Isle of Wight
R. spadix - W. Kent
R. trelleckensis - Monmouthshire
R. tresidderi - Cornwall
R. wolley-dodii Cheshire

A further 30 species, whilst falling within the 'regional' category, are decidedly scarce and might also warrant inclusion in an expanded list.

In general, the greatest diversity of brambles is to be found in areas that have proved difficult or uneconomic to develop and where the habitat is less intensively managed. Major threats to the continuing survival of scarce and local taxa consist in the felling and grubbing up of woods, ploughing of heathland margins and uprooting of hedges, particularly in districts where complex bramble communities have developed over the centuries on congenial terrain.

A. Newton

Rumex aquaticus L. (Polygonaceae)
Scottish dock, Copag Albannach
Status in Britain: VULNERABLE.
Status in Europe: Not threatened.

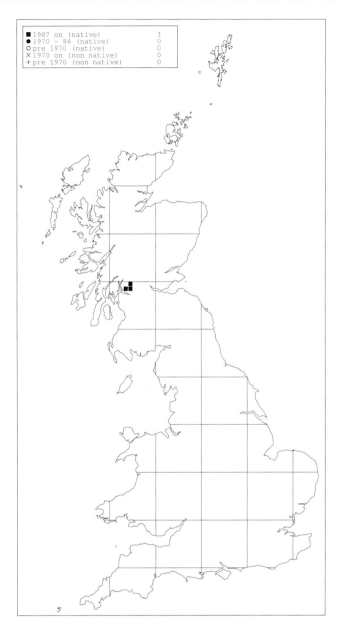

R. aquaticus in known in Britain only in southern Loch Lomondside. It occupies a variety of mesotrophic wetland niches within its restricted range: alluvial and gravelly loch shores, river banks, the edges of ditches and streams, marshes and overgrown ox-bow ponds, clearings in wet woods, damp meadowland and even rush-infested abandoned arable ground where the drainage system has broken down. The size of colonies is also very variable, from isolated individuals to impressive assemblages of 200 or more plants. Amongst the most frequent associates are widespread species such as *Carex vesicaria, Deschampsia cespitosa, Equisetum fluviatile, Filipendula ulmaria, Phalaris arundinacea* and *Sparganium erectum.* At three sites *R. aquaticus* can be found in the more distinguished company of *Carex elongata.*

This species is perennial, growing up to 2 m tall with large basal leaves up to 0.5 m long. Flowering is in July and August. The valves of the fruit are more or less acute and longer than wide, which immediately distinguishes it from the superficially similar *R. longifolius,* in which the valves are rounded.

Despite being a conspicuous plant, *R. aquaticus* was not discovered in Britain until 1935, when R. Mackechnie collected specimens near the mouth of the River Endrick at the south-eastern corner of Loch Lomond (Lousley 1939). Subsequent survey work in the surrounding area by Lousley (1944b), Idle (1968) and others showed *R. aquaticus* to be well-established between Ross Priory and Balmaha, with a few outlying plants at Arden-Midross and two expanding colonies at Rossdhu on the west wide of the loch. Away from the immediate vicinity of Loch Lomond, an exceptionally strong colony (300 plants in 1995) occurs beside the Endrick above Croftamie, 12 km upstream.

Since 1967, periodic monitoring of *R. aquaticus* colonies in the lower flood plain of the Endrick has highlighted both gains and losses. On the positive side, colonising plants quickly became established at one locality where wetland restoration successfully reversed a falling water table. The most serious single loss followed the excavation of an unauthorised deep drainage ditch right through the largest known colony of some 250 plants. Far less obvious, but possibly the greatest threat to *R. aquaticus* in the long term, is introgressive hybridisation with *R. obtusifolius.* The hybrid (*R.* x *platyphyllos*) is all too common wherever *R. aquaticus* is found, and there are several instances where this fertile hybrid and probable back-crosses with the parent *R. obtusifolius* have gradually replaced former pure stands of *R. aquaticus.* Removal of *R. obtusifolius* from the vicinity of some *R. aquaticus* colonies has been carried out in an attempt to reduce the incidence of hybridisation (Lusby & Wright 1996). Other associated *Rumex* species would seem to pose much less of a problem. *R. aquaticus* x *R. crispus* (*R.* x *conspersus*) has been recorded only about six times and *R. aquaticus* x *R. sanguineus* (*R.* x *dumulosus*) once or twice in the last 60 years.

R. aquaticus is widely distributed in central and northern Europe, and in northern Asia, extending well beyond the Arctic Circle.

J. Mitchell

Rumex rupestris Le Gall (Polygonaceae)

Shore dock, Tafolen y Traeth
Status in Britain: ENDANGERED. WCA Schedule 8. EC
Habitats & Species Directive, Annexes II and IV.
Status in Europe: Vulnerable. Endemic.

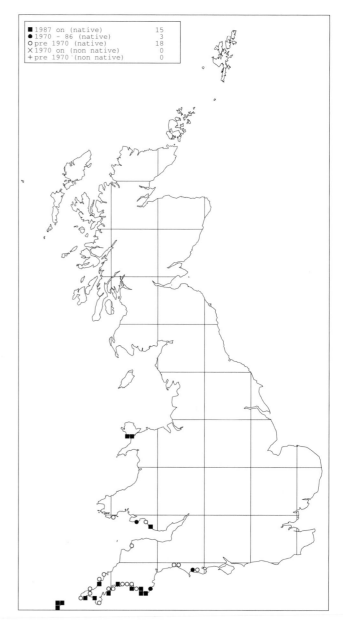

■ 1987 on (native)	15
● 1970 – 86 (native)	3
○ pre 1970 (native)	18
✕ 1970 on (non native)	0
+ pre 1970 (non native)	0

R. rupestris is a strictly maritime species, typically
occurring on rocky or sandy shores and the lower slopes
of cliffs, and more rarely on wet cliff-ledges, on strandlines
and in wet slacks in sand-dune systems. It seems to
require a constant supply of freshwater, so is commonly
found growing where streams debouch onto the shore, on
oozing soft-rock cliffs, and on hard-rock cliffs in clefts and
gullies, especially where freshwater trickles onto wave-cut
platforms. Plants can be found in a range of communities,
including stands of tall-herb perennial plants at the base of
flushed cliffs, or may occur alone on the strand line.
Associates are varied and include *Agrostis stolonifera,
Atriplex prostrata, Beta vulgaris, Cakile maritima, Carex
distans, C. otrubae, Eupatorium cannabinum, Hydrocotyle
vulgaris, Plantago maritima, Potentilla anserina, Phragmites
australis, Pulicaria dysenterica, Rumex crispus* var. *littoreus*
and *Tripleurospermum maritimum*. In Glamorgan,
R. rupestris grows on steeply-sloping densely-vegetated cliff
ledges, 8-12 m above high water mark, together with
*Calamagrostis epigejos, Festuca rubra, Filipendula ulmaria,
Holcus lanatus* and *Samolus valerandi*.

R. rupestris is perennial. Some individuals can live to at
least ten years, producing large, woody rootstocks and
many flower spikes. However, most succumb to winter
storms before their full span. Germination may occur
throughout the year, though it is probable that seedlings
emerging later in the year are less likely to survive the
winter. Flowering probably depends on the size of the
crown or rosette, and most plants will reach their
flowering size in the second year. Flowering is in
midsummer, with fruiting in August and September. It was
assumed that flowering is biennial, but it is now thought
that plants are able to flower in successive years in suitable
conditions. Seed viability varies greatly between
populations (Daniels *et al.* 1996). Since fruits can remain
buoyant for relatively long periods in seawater, dispersal is
likely to be at least in part by water. Experiments carried
out in 1938/9 by J.E. Lousley showed 56% survival of seed
after 180 days in seawater, and Q.O.N. Kay (1996)
demonstrated 100% flotation and 80% viability after
30 days.

In Britain, this species occurs only in the Isles of Scilly,
Cornwall, South Devon, Glamorgan and Anglesey. It is
presumed extinct in North Devon and Dorset. Detailed
surveys in 1994 (Parslow & Colston 1994; McDonnell
1995b) revealed 22 sites holding a total population of 290-
320 plants. The main stronghold is in the Isles of Scilly,
where, in 1994, 165 plants were counted in ten sites on five
islands (Parslow & Colston 1994). Except in the Isles of
Scilly, only three colonies exceeded 20 mature plants in
1994; most comprised fewer than 10 individuals, indicating
a continuing decline. In the late 1980s, the largest colonies,
near Plymouth and Perranporth, supported 50-60 plants
(King 1989). In 1996, a colony of at least 21 plants was

found in Glamorgan at a site that may not have been
visited since *R. rupestris* was discovered there in 1934
(Q.O.N. Kay 1996).

About 17 of the known colonies lie within SSSIs or are
on land owned by the National Trust. Populations
fluctuate according to the severity of winter storms, and
several small colonies have gone in recent years as a result
of natural erosion and accretion, plants being uprooted
during storms or buried under slumping cliffs or by fresh
accumulations of sand and drift litter. At many old
localities suitable habitat has been lost through the
construction of coastal defences, the culverting of streams
and the construction of boat ramps. Visitor pressure
appears to be a significant factor in the decline of
R. rupestris in the Isles of Scilly, including the use of
dried inflorescences (in some cases whole plants) for
beach bonfires.

The plight of *R. rupestris* is now widely recognised, and is the subject of much conservation effort. It is the subject of 'recovery' programmes, and studies of its autecology throughout its range are underway. Reintroductions are in progress at three of its former sites, two in Cornwall and one in South Devon, with further introductions planned.

R. rupestris is threatened not only in Britain, but throughout its world range. It has a mainly lusitanic distribution, and is restricted to the coastal margins of Britain, the Channel Islands, France (Normandy, Brittany and Vendée) and Spain (Galicia). In the Channel Islands, three sites in Jersey held 23 plants in 1994, but there appears to be no recent information for the other islands. Three sites were known in Guernsey (McClintock 1984), and older records exist for Lethou, Herm and Alderney.

M.P. King, E.J. McDonnell, S.J. Leach and M.J. Wigginton

Sagina nivalis (Lindblad) Fries (Caryophyllaceae)

Sagina intermedia Fenzl
Snow pearlwort
Status in Britain: VULNERABLE.
Status in Europe: Not threatened.

S. *nivalis* is probably one of the least known of British montane plants and often eludes even skilled botanists, presumably because of its diminutive and inconspicuous form. It was probably first collected by H.H. Harvey in 1825 but mistaken for *Minuartia rubella* (Payne 1981c). White (1898) attributes the earliest record to Balfour in 1847 (but states that it was "probably first gathered in Britain by Greville before 1840"), but its unequivocal recognition as a British species was published in 1863 and 1864 (Payne 1981c).

S. *nivalis* is found at altitudes between 840 and 1,190 m, but mostly above 900 m, usually in open, more or less unstable habitats with little vascular plant cover. It frequently grows on a surface mat of bryophytes above a gravelly substrate or rock face, at any angle from horizontal to vertical. Drier sites are usually subject to frost-heave and occasional sloughing off or landslip, the wetter sites being subject to extreme frost-heave and frequent or periodic flushing. Many sites are among the first areas to be covered with snow, but they are not subjected to persistent late snow cover. The soils are calcareous, with pH in the range 5.7-6.9 (Evans 1982). Frequently associated species on Ben Lawers are *Cochlearia micacea, Sagina saginoides, Salix herbacea* and *Saxifraga oppositifolia*; in wet flushes *Blindia acuta* is almost invariably present, frequently with *Juncus triglumis, Saxifraga stellaris* and, often, *Juncus biglumis*; on drier sites it may accompany *Alchemilla alpina, Festuca vivipara, Minuartia sedoides, Silene acaulis* and sometimes *Saxifraga nivalis*.

It is a dwarf, tufted, short-lived perennial, usually forming a small cushion rarely more than 40 mm across. The tap-root provides anchorage in unstable gravelly soil, and the woody stock is usually also submerged in the ground. Branching stems spread from the central shoot, sometimes through the bryophyte mat. It flowers freely between July to September (mainly in August), the flowers opening only on warm or sunny days. It is homogamous and self-pollinated. Seedling establishment is extremely variable, from negligible to prolific, but very few survive to maturity. Clusters of seedlings often develop adjacent to parent plants, but seeds may fall and establish well below the parent plants on steep cliffs. Fixed quadrats within 10 colonies have been monitored over 14 years and have shown marked fluctuations in numbers. These reflect natural catastrophic events such as landslip, or being inundated from above. Colonies may be extinguished in this way, but at the same time new ground becomes available for colonisation.

In Britain, S. *nivalis* is confined to the Breadalbane mountains in Perthshire (Payne 1981c). Since 1980, this species has been recorded from 14 1 km squares in four hectads, with most of the population and most of the sites encompassed within Ben Lawers NNR. Some locations recorded in Breadalbane prior to 1980 have not been confirmed recently. The time required to find and effectively

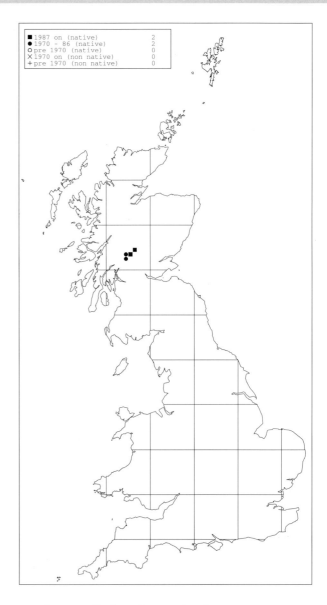

count these small plants has rendered an accurate measure of population size impracticable. However, the distribution and numbers recorded by Payne (1981c) were such that the plant was not considered to be endangered, and more recent monitoring suggests that this is still the case.

The entire range of this plant in Britain is subject to sheep-grazing, with most sites accessible to sheep. The fragile habitat is easily damaged by sheep hooves and this occurs regularly, probably causing the extinction of some colonies, but perhaps also providing open ground into which the plant can seed. Trampling by humans is totally destructive, but although many colonies on Ben Lawers NNR are close to walking routes, they are mostly bypassed. The main human threat is from botanists searching for or photographing rare plants at sites of S. *nivalis*.

S. *nivalis* has an arctic circumpolar distribution, occurring in North America from Alaska to Labrador, in Greenland, Fennoscandia, arctic Russia, the western European Alps and the arctic coast of Asia.

D.K. Mardon

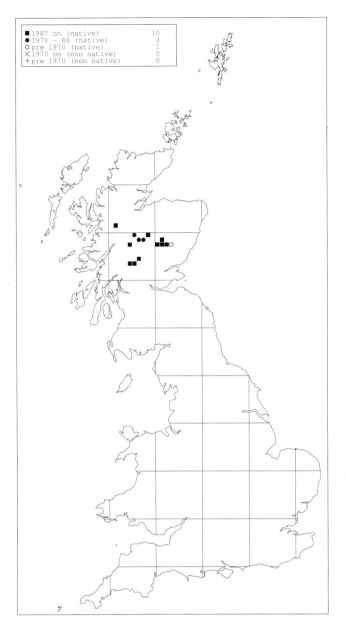

Salix lanata L. (Salicaceae)

Woolly willow, Seileach Clòimheach
Status in Britain: VULNERABLE.
Status in Europe: Not threatened.

S. *lanata* is a montane willow inhabiting damp situations on basic rocks (usually schists or, in one case, limestone) in north-facing corries. It is almost confined to ungrazed ledges and crags and is often associated with late snow, some bushes emerging from under the snow as late as July. Although the altitude range is 620-1,036 m, most populations are found between 700 and 900 m. S. *lanata* often grows with other montane willows, especially S. *lapponum*. The habitat also supports herb-rich communities including *Alchemilla alpina, Luzula sylvatica, Oxyria digyna, Saxifraga aizoides, S. oppositifolia, Sedum rosea, Vaccinium myrtillus* and *Racomitrium* and *Sphagnum* species.

It is a dioecious species pollinated by bumble-bees, and thus a population that becomes fragmented is likely to suffer reduced seed-set. The short-lived seeds are wind dispersed. The bushes, some of which may be up to a metre tall, are capable of horizontal growth, rooting in the mossy substrate. The tallest are found in Coire Fee at an altitude of 750-770 m, whereas those at the highest altitudes (on Geal Charn) were only 20 cm high.

In Britain the species is confined to Scotland, where it was first recorded in Glen Callater in 1812 (Meikle 1984). It has always been very localised, but a new colony was found in Gleann na Ciche in 1991. In a survey in 1994 of 14 known sites, five were considered to hold viable populations of from 30 to almost 1,000 plants. Three other sites held populations of up to 30 plants and each was reckoned to be at risk, and at a further three sites only a single, in each case female, bush remained. At the remaining three sites no sign of the species was found, which may be because of extinction, past misidentification or erroneous map references attached to old records.

The more intensive grazing by increasing numbers of deer and sheep threaten this species. Rockfall and unstable slopes pose a threat to individual bushes whilst at the same time providing sites for seedlings. The three sites where *S. lanata* is reduced to a single female plant are clearly doomed. At Coire Fee, one of the major sites for the species, the whole coire has been fenced against the larger grazing animals, and it will be interesting to see whether this and other species can spread naturally off the crags on to more accessible ground. However, if the total population and the number of localities continue to decline, then this species will become endangered in the near future.

S. lanata has an arctic and subarctic distribution, and is found from Iceland and the Faeroe Islands eastwards through Scandinavia and Finland, Russia and eastern Siberia. Related taxa (perhaps subspecies) occur in North America (Meikle 1984).

R.W. Marriot

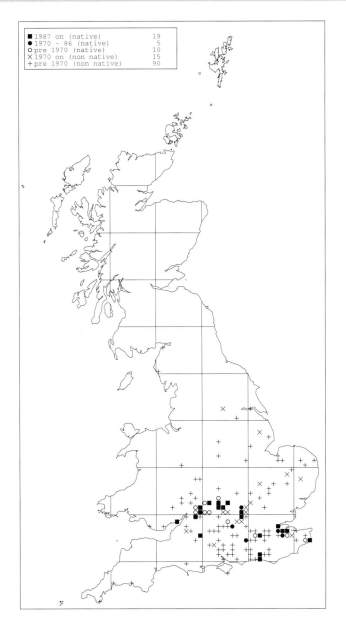

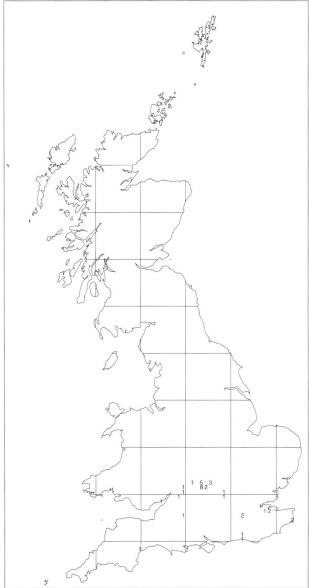

Salvia pratensis L. (Lamiaceae)

Meadow clary, Saets y Waun
Status in Britain: LOWER RISK - Nationally Scarce. WCA
Schedule 8.
Status in Europe: Not threatened.

This species grows on shallow, well-drained calcareous soil derived from chalk or oolitic limestone, most often in herb-rich unimproved grassland on flat to steeply-sloping sites with a southerly to westerly aspect. Grazed pastures and meadows are the usual habitats, but it also occurs on grassy lane-sides, road verges, and on disturbed ground. Managed grasslands are preferred. It occurs in short-grazed chalk and limestone grasslands in which *Briza media, Festuca ovina, Helictotrichon pratense, Leontodon hispidus, Linum catharticum, Scabiosa columbaria* and *Thymus polytrichus* are prominent. It is also found in ranker grasslands dominated by *Bromopsis erecta, Brachypodium pinnatum* or *Arrhenatherum elatius* (perhaps indicating a relaxation of grazing), amongst scrub, and in woodland edges.

S. pratensis is an imposing long-lived perennial with stems up to a metre tall bearing heads of large violet-blue flowers. Populations often contain a small percentage of male-sterile plants in which stamens are aborted, and those plants have much smaller flowers. The peak flowering period is from late May to early July. The flowers are visited mainly by bumble-bees. A high proportion of seed is set, even in late-flowering male-sterile plants, suggesting that hermaphrodite plants are substantially outcrossed. Seedlings establish freely where disturbance or stress creates gaps in the sward. Vegetative reproduction appears to be good at some sites, and rooting from the nodes of decumbent stems has been observed. Dormant buds on the rootstock develop rapidly if the main stem is damaged by grazing or cutting.

S. pratensis is accepted as native or probably native in Kent, Surrey, Sussex, Oxfordshire, Buckinghamshire, Wiltshire, Gloucestershire, Monmouthshire and Berkshire, but is no longer present in the last county. Since 1990 it has been recorded at 27 sites, 11 of which are accepted as native, with a further nine possibly native (Rich 1996b).

The four largest British populations are in Oxfordshire (comprising 4,000-5,000, 1,300, 400, and 300 plants), and that county remains its stronghold. Three other populations (in Gloucestershire, Oxfordshire and Sussex) hold more than 100 plants, but all the others are dismally small, ranging from a few tens of plants down to a singleton. However, single isolated plants can survive for long periods, perhaps as long as 30 years (Rich, Lambrick & McNab in press). It is widely scattered in southern England as a probable or certain introduction (Rich 1995d).

Short swards maintained by rabbit-grazing provide favourable conditions for seedling establishment, and suitable niches include rabbit scrapes and small mammal runs. Establishment is generally unsuccessful in rank grassland where litter or moss has accumulated (for instance under *Brachypodium pinnatum* or scrub) or where plants are cut or intensively grazed before seed-set. Grassland management holds the key to the survival of this species, but an ideal regime may not be easy to achieve. Sites should preferably be grazed at high stocking density in spring, with livestock withdrawn in early May to allow flowering shoots to develop then reintroduced in late July or early August after seed has ripened. Spring grazing is especially important to control aggressive grasses, especially *Brachypodium pinnatum*. Hay meadows

should be cut in late July or early August to allow a good proportion of seed to ripen. It appears to be able to survive vegetatively for many years, and so can tolerate an unsympathetic regime for a while. But if regular cutting or grazing prevents the production of seed, in the absence of vegetative reproduction populations can be expected to decline as plants age. Likewise, plants can survive in rank unmanaged grassland but in the longer term are unlikely to remain. Regular scrub control is necessary at many sites. Detailed studies of its ecology are underway as part of 'recovery' programmes.

It is rare in northern Europe and Scandinavia, and investigations of populations in these areas (Ouborg & van Treuren 1994, 1995) have indicated some genetic erosion there

It is a southern-continental species, reaching its north-western limit in Britain and extending from Morocco and the Pyrenees to Turkey and the Urals. It is rare in northern Europe and Scandinavia, and investigations of populations in these areas (Ouborg & van Treuren 1994, 1995) have indicated some genetic erosion there. Its ecology and conservation are discussed in Scott (1989).

M.J. Wigginton

Saxifraga cernua L. (Saxifragaceae)

Drooping saxifrage, Lus Bheinn Labhair
Status in Britain: VULNERABLE. WCA Schedule 8.
Status in Europe: Not threatened.

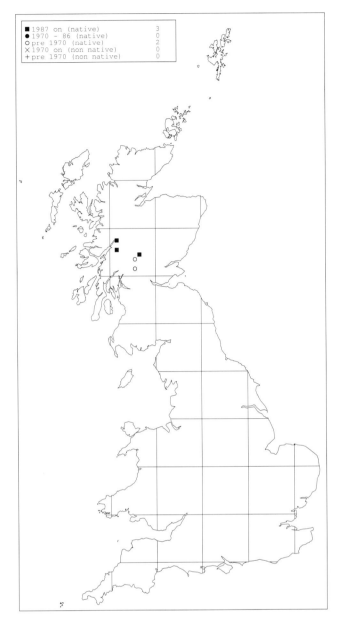

```
■ 1987 on  (native)        3
● 1970 - 86 (native)       0
○ pre 1970  (native)       2
✕ 1970 on   (non native)   0
+ pre 1970  (non native)   0
```

S. cernua is a plant of basic rocks, occurring at its Scottish sites at an altitude of 950-1,190 m. The plant usually occupies crevices or ledges below overhangs rather than very exposed sites, although this confinement might reflect the effects of grazing. The sites are all subject to late snow-lie and, when free of snow, are often humid because of prolonged cloud cover. *S. cernua* usually grows among bryophytes but with little vascular plant cover. Species loosely associated on Ben Lawers include *Cerastium alpinum, Cochlearia micacea, Draba norvegica, Festuca vivipara, Minuartia sedoides, Oxyria digyna, Persicaria vivipara, Saxifraga hypnoides, S. nivalis, S. oppositifolia* and *Silene acaulis*.

This perennial herb flowers infrequently in Britain and does not set seed (Raven & Walters 1956), apparently because viable pollen is not formed: experimental crossing using pollen from arctic Norway has produced seed (Godfree 1979). Reproduction is vegetative, by axillary bulbils. The bright red bulbils are produced annually in moderate numbers, up to 40 per stem (Batty & Batty 1978). Up to about 50 flowering stems per plant have been recorded at Ben Lawers. Incomplete flowers, with only a calyx visible, develop in smaller numbers more frequently. The proportion of plants with bulbil-bearing stems has been estimated at 50% in Glen Coe, and 10-25% at Ben Lawers.

The recorded range of *S. cernua* is within three hectads; there are records since 1970 in two 1 km squares on the Ben Nevis range, two in Glen Coe, and a single 1 km square on the Ben Lawers range. Population counts have varied considerably and have proved difficult to interpret because of the difficulty in recognising individual plants. However, on Ben Lawers, the number of mature, stemmed plants is probably within the range 100-300, and the total number of plants in the range 500-1,000. Counts of separate populations at Glen Coe, although not all done in the same year, suggest a total population exceeding 700, the largest group being estimated to be a few hundreds. On the Ben Nevis range, numbers at two separate colonies have varied, but both probably hold about 100 plants. In the absence of sexual reproduction, Scottish populations probably each consist of a single clone.

Some evidence of grazing has been observed at Ben Lawers, and plants may be trampled both there and on Ben Nevis, whereas the Glen Coe colonies probably escape this at the present sites. Collecting of individual plants has been recorded several times at Ben Lawers during the 1970s and 1980s, usually of good, flowering specimens.

Effective protection is hardly possible at any of the present sites, though long-term survival is not considered seriously threatened. However, since plants occur in so few places, the species must be regarded as vulnerable.

S. cernua is an arctic-alpine circumpolar species. It occurs in the mountains of central and eastern Europe, in Scandinavia, the Urals, arctic and central Asia eastwards to Kamchatka and Japan, North America from Alaska to Labrador, southwards in the Rocky Mountains to New Mexico, and in Greenland (Benum 1958).

D.K. Mardon

Saxifraga cespitosa L. (Saxifragaceae)
Tufted saxifrage, Clach-bhriseach Bhuidhe, Tormaen Siobynnog
Status in Britain: VULNERABLE.
Status in Europe: Not threatened.

This montane species occurs on well-drained base-rich substrates at altitudes between 520 and 1,065 m, on sheltered or exposed rock ledges (often moss-covered), in crevices and on boulder-scree slopes. Tufts are often isolated or grow in a mat of *Racomitrium lanuginosum* or other bryophytes, but species recorded growing nearby include *Asplenium viride, Cerastium arcticum, Cochlearia micacea, Cystopteris fragilis, Festuca vivipara, Oxyria digyna, Saxifraga nivalis, S. oppositifolia, S. stellaris, Sedum rosea* and *Silene acaulis.*

S. cespitosa is a tussock-forming stoloniferous perennial. Plants are long-lived, and their population structure often becomes shifted into small numbers of large plants on small sites; only on large sites does active recruitment provide a more diverse population structure. Flowering is between May and July, and the flowers are visited by flies. Generally only a proportion of plants flower in a population, but viable seed is freely produced. Little appears to be known of the reproductive strategies in British populations, though Webb (1950) considers that vegetative reproduction is insignificant. *S. cespitosa* has been reported as preferentially selfing, but protogyny has also been described in British populations (Webb 1950), suggesting that outcrossing may occur. A study of populations in northern Sweden (Molau & Prentice 1992) has shown that many hermaphrodite flowers are male-sterile, but it is not known if this is the case in British populations.

In Scotland, *S. cespitosa* has apparently been recorded on only seven mountains since 1980, and on three others in the 1970s, where, however, it may still occur. In the Cairngorms, it has been recorded recently from only one site - a small colony of about six plants on Ben Avon in 1995. All but one of the recently recorded populations are small, consisting of 2-30 plants. On Ben Nevis, a large population of 108 plants was recorded in 1985, but its present size is not known. Trends are not easily discerned because monitoring of populations has been sporadic, but there is some indication of a decline at some sites. In Wales, it has been recorded only on base-rich rocks in Cwm Idwal. The population declined to just four plants by 1975, but small colonies were then established at several nearby sites from plants grown *ex situ* from seed collected from the original colony. Seed was also distributed at suitable sites. However, the population of perhaps 70-80 plants subsequently declined, and only about 10-20 plants in two colonies remained in 1996 (H. Roberts pers. comm.).

The small size of the populations makes this species very vulnerable to natural events such as erosion and rock-fall, and locally also to trampling and accidental

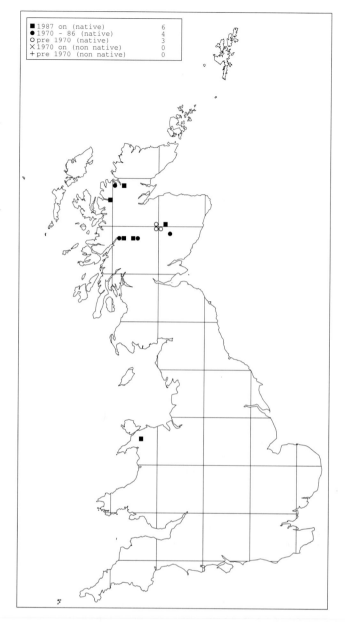

■ 1987 on (native)	6
● 1970 - 86 (native)	4
O pre 1970 (native)	3
X 1970 on (non native)	0
+ pre 1970 (non native)	0

disturbance. Extremes of climate may also play a part; for instance, one of the North Wales populations succumbed to spring and summer drought in 1976. All British populations are protected within SSSIs and some in NNRs.

S. cespitosa has an arctic and sub-arctic circumpolar distribution. In Europe it extends southwards to about 59N in Norway and the Urals, and further south in Scotland and Wales. It extends across North America from Alaska to Labrador, with other subspecies extending southwards in the Rockies to California and New Mexico (Webb & Gornall 1989).

M.J. Wigginton

Saxifraga hirculus L. (Saxifragaceae)

Yellow marsh saxifrage, Moran Rèisg
Status in Britain: VULNERABLE. WCA Schedule 8. EC
Habitats & Species Directive, Annexes II, IV.
Status in Europe: Vulnerable.

S. hirculus is a plant of base-rich flushes and mires.
Though at one time occurring from near sea-level, all its
lowland habitats have been lost, and it is now a montane
plant, ascending to 800 m. In the species-rich flush and
mire communities where *S. hirculus* occurs, sedges such as
Carex echinata, C. flacca, C. panicea or *C. viridula* ssp.
oedocarpa are usually prominent, together with a wide array
of other associates including *Holcus lanatus, Lychnis flos-
cuculi, Montia fontana, Parnassia palustris, Pedicularis sylvatica,
Ranunculus flammula, Sagina nodosa, Sedum villosum,
Selaginella selaginoides, Triglochin palustre, Viola palustris,* and
robust mosses such as *Bryum pseudotriquetrum,
Drepanocladus revolvens, Palustriella commutata* and *Philonotis
fontana* (Eddy *et al.* 1969). It also grows in taller, open
mesotrophic mires in which *Carex rostrata, Juncus acutiflorus*
and *Sphagnum* species may be common.

It is a rhizomatous perennial, producing slender
upright flowering shoots and creeping runners. Some
rhizomes do not produce flowering shoots in some years,
but there are normally between one and five runners.
Flowering is relatively late in the season, extending from
July to September. The flowers are bright-yellow, insect-
pollinated and normally protandrous. In studies of Danish
populations, most insect visitors were hoverflies. A high
proportion of seeds are viable.

In Britain, *S. hirculus* was known in 12 vice-counties in
the 19th century, eight in Scotland and four in northern
England. It has become extinct in several vice-counties,
mostly in Scotland, and its main stronghold is now a
50 km-long section of the northern Pennines between
Stainforth and South Tyne and Weardale. Within this block
the total population probably exceeds 50,000 plants, and
several localities have thriving populations of more than
1,000 flowering shoots (Taylor 1987a). Together these
represent 80-90% of the British population. It may still
occur at some Pennine sites for which there are only pre-
1987 records. In Scotland four sites remain, one in
Midlothian and three in Aberdeenshire (Welch 1970, 1992).
The two largest populations (in the latter county) held
nearly 300 and 485 plants in 1995. Further details on its
habitats and status in Scotland are given in Welch (1995).

Its sites are often heavily grazed, and, probably because
of its low stature, *S. hirculus* seems unable to survive in
fens and mires with tall vegetation. Sheep graze the
flowers and reduce seed output, but they check other
plants, so moderate levels of grazing are probably
beneficial. In Scotland, some sites where *S. hirculus* has
disappeared still seem suitable and the reasons for its
disappearance are unclear. Drainage, afforestation and land
claim for intensive agriculture are the main threats to
existing populations. Flowers are often scarce because of
grazing, and the plant is often a shy flowerer. This,
together with the similarity in the colour of the flowers to

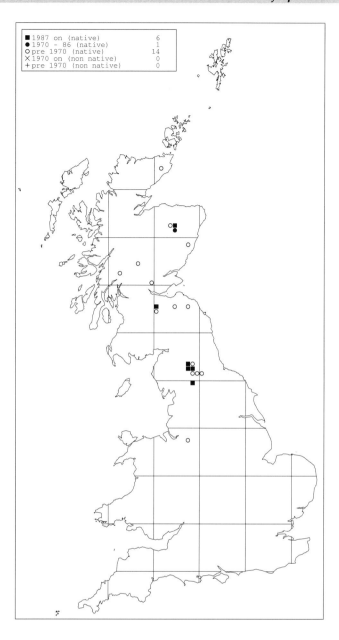

■ 1987 on (native)	6
● 1970 - 86 (native)	1
○ pre 1970 (native)	14
✕ 1970 on (non native)	0
+ pre 1970 (non native)	0

Leontodon autumnalis and *Ranunculus flammula,* which are
often abundant in flushes, has given *S. hirculus* a
reputation of being elusive. Furthermore, the entire leaves
can be difficult to distinguish from those of *Ranunculus
flammula* and *Epilobium palustre* without close scrutiny.

S. hirculus varies markedly in morphology across its
range, and four subspecies are now recognised (Hedberg
1992). As far as it is known, only ssp. *hirculus*, the most
widely-distributed subspecies, occurs in Britain.

This plant is widely distributed in Europe but is
declining or threatened in most countries. It is endangered
in Germany, Switzerland, Ireland, Poland and Lithuania,
vulnerable in Estonia, Latvia, Norway and Denmark, and
rare in Turkey. It also occurs in the Caucasus and
Himalayas and has a circumpolar range in Asia and North
America.

D. Welch

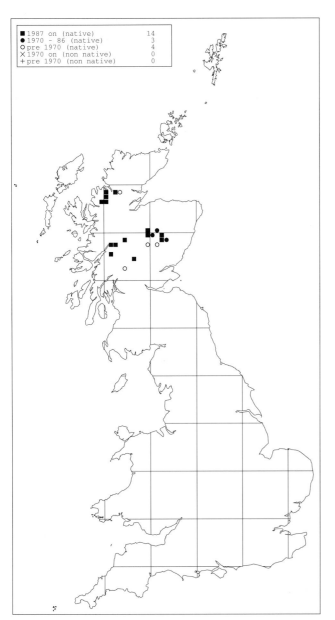

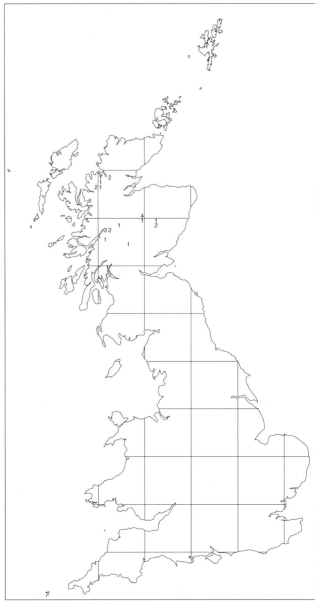

Saxifraga rivularis L. (Saxifragaceae)

Highland saxifrage, Clach-bhriseach t-Slèibhe
Status in Britain: LOWER RISK - Near Threatened.
Status in Europe: Not threatened.

In Scotland, *S. rivularis* is a plant of areas of late snow-lie and springs that are derived from meltwater. Thus, its occurrence is limited to approximately 20 areas where significant snow patches persist into the summer months. It descends to about 820 m on Beinn Dearg and ascends to its highest altitude of 1,200 m on Ben Nevis. All sites have a northerly, north-easterly or easterly aspect and occur on steep, rocky slopes, gullies and crags, though some flush sites are locally much less steep. The plant usually occurs in a matrix of mosses, organic material and sand or gravel in a regularly irrigated or permanently wet skeletal soil, over block scree, on ledges or crevices of crags or in bryophyte flushes. The most constant associates are mosses such as *Hygrohypnum ochraceum, Philonotis fontana, Pohlia ludwigii,* and *P. wahlenbergii* var. *glacialis,* but frequent

vascular plant associates include *Cerastium arcticum, C. cerastoides, Chrysosplenium oppositifolium, Cochlearia officinalis, Epilobium anagallidifolium, Saxifraga stellaris* and *Stellaria uliginosa.* On a few sites where the substrate is more base-rich, it is associated with *Saxifraga cernua, S. cespitosa* and *S. nivalis.*

S. rivularis is a small, stoloniferous perennial forming rather fragile, loose tufts. Bulbils are developed in the axils of the basal leaves during summer and autumn and these germinate before the flowering season, giving rise to slender runners from which develop new plants. The flowering period is dependent on the amount of accumulated snow and the pattern of snow melt; in the same season it is possible to find seed-heads in some populations in July and find flowers in other populations at the end of August. In a poor summer; some populations may not set seed. Most populations consist of a relatively small number of mature plants, many of which produce flowers in a normal year, and a larger number of juveniles, often in a dense tuft, developed from the previous year's bulbils.

Most populations are in the central Highlands, from Glen Coe in the west to Lochnagar in the east, but there are isolated sites from Ben Lawers in the south to the Torridon Mountains and Beinn Dearg in the north. Estimating population size presents considerable problems because of the large numbers of juveniles, but the largest populations in Glen Coe and the Cairngorms exceed 1,000 plants. More isolated populations are usually much smaller, often fewer than 50 plants. Population size varies to an extent because of the inherent instability of many of the sites, but observations of the more accessible populations suggests that they are constant in occurrence, though the numbers of plants may fluctuate. Most of the populations are on SSSIs or NNRs and so are safeguarded to some degree, and many sites are either remote or difficult of access, which provides further protection. A recent survey of several sites revealed some damage from grazing and trampling by sheep and deer. At least two sites are threatened by erosion caused by increased recreational activity, at least partly owing to the provision of easier access for skiers.

S. rivularis is a widespread species in the arctic extending southwards to Iceland and southern Norway in Europe, to Kamchatka in Asia and in North America to Montana, Newfoundland and southern Greenland. Outlying stations are in California, Colorado and in the Scottish Highlands (Webb & Gornall 1989).

G.P. Rothero

Scheuchzeria palustris L. (Scheuchzeriaceae)
Rannoch rush, Luachair Rainich, Brwynen Rannoch
Status in Britain: VULNERABLE.
Status in Europe: Not threatened.

S. palustris is now restricted to scattered localities on Rannoch Moor where it was first recorded in 1910 by Scarth. It occurs there in mid-Perth and Argyll but, oddly, has not been detected in the adjacent West Inverness part of the Moor. The altitude is about 300 m. Its habitat is typically acid runnels, pools or very wet *Sphagnum* mire. It has few associates in runnels; the commonest associates elsewhere include *Carex lasiocarpa, C. limosa, C. pauciflora, C. rostrata, Drosera rotundifolia, Eriophorum angustifolium, Menyanthes trifoliata, Molinia caerulea, Myrica gale, Narthecium ossifragum, Trichophorum cespitosum, Vaccinium oxycoccos* and calcifugous *Sphagnum* species.

S. palustris is a perennial herb with scattered stems 10-20 cm tall spaced along a creeping rhizome. The leaves bear a conspicuous pore at the tip. Often the vast majority of stems are merely vegetative, and the lax 3-10 bisexual-flowered inflorescences are decidedly infrequent. Flowering occurs from June to August, and the species is wind-pollinated. However, there is perhaps little reproduction by seed, and propagation is likely to be mostly by its spreading rhizomes.

Many more sites for *S. palustris* have been found in recent years, and it is doubtful whether any have been lost this century. Although its sites on blanket mires are potentially threatened by afforestation, and to a lesser extent by burning in dry conditions and by agricultural improvement (drainage and fertilisation), virtually all sites lie within NNRs or SSSIs and so are safeguarded to some degree. There are more than 20 locations for *S. palustris* known at present on the Moor or adjacent areas, some with several hundreds, if not thousands, of plants or shoots. The shoots are grazed by red deer.

Last century, *S. palustris* is known to have occurred at four lowland sites in England, and at one site in the mid-Perth lowlands. The first record of the species in Britain was in 1807 from Leckby Carr, near Boroughbridge, and this was followed by discoveries in Shropshire, Cheshire and south-east Yorkshire, and at the White Myre, Methven, in mid-Perth. All sites were acidic mires. The loss of all the English sites before 1900 is attributed to drainage coupled with peat cutting. By contrast, the plant disappeared from Methven because of rapacious collecting, combined with the flooding of the mire and nutrient enrichment from a gull colony there. In Ireland, it was discovered in Co. Offaly in 1951, but is no longer extant, despite attempts at transplantation to other sites (Curtis & McGough 1988). Further details of its distribution and ecology in Britain are to be found in Sledge (1949).

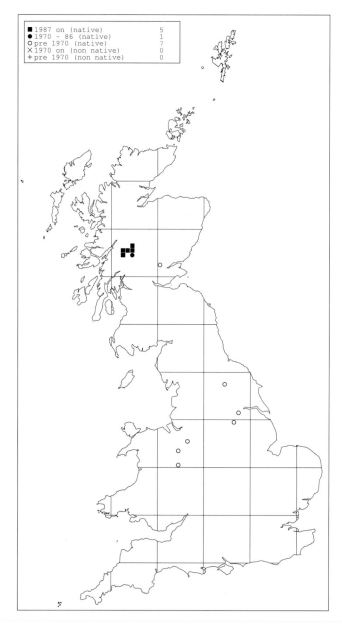

Elsewhere, *S. palustris* has a circumpolar distribution, mainly occurring between 40°N and 60°N, but is absent from Greenland, Iceland, much of continental North America and a few parts of eastern Asia. The American plants are considered morphologically distinct but are apparently similar ecologically. It seems to be in general decline over much of Europe and is vulnerable in several countries.

R.A.H. Smith

Schoenoplectus triqueter (L.) Palla (Cyperaceae)
Scirpus triqueter L.
Triangular club-rush
Status in Britain: CRITICALLY ENDANGERED. WCA
Schedule 8.
Status in Europe: Not threatened.

S. triqueter grows on mud banks along the lower
reaches of tidal rivers. It forms clumps on the riverside
edge of fringing beds of *Phragmites australis,* along with
other *Schoenoplectus* species. The mud on which it grows is
exposed at low tide, but the plants may become
completely submerged, at least by the highest tides. It is
tolerant of brackish conditions, but is not found in
association with any true halophytes. It may be favoured
by freshwater seepage around its roots; by the Tamar,
Agrostis stolonifera and *Oenanthe crocata* grow nearby.

This species is a perennial, flowering from July to
September, about one month later than other *Schoenoplectus*
species and somewhat later than the hybrids. However, the
last remaining plant of this species in Britain is reported to
flower poorly and erratically, and fruit-set has not been
observed in the wild in Britain or Ireland.

At one time this species occurred in the Tamar (Devon
and Cornwall), Arun (West Sussex), Medway (Kent) and
Thames (Greater London). It is now restricted to the
Tamar, Devon, but even here it appears to be on the verge
of extinction. In 1985, surveys revealed six patches
covering about 6 square metres, but in 1989 and 1994, only
four clumps covering 2 square metres were detected. It is
believed that, in 1996, only a single unhybridised clump
remained, in which case this may now be the rarest
vascular plant in Britain. The hybrid with
S. tabernaemontani (*S.* x *kuekenthalianus*) is still very locally
frequent on the Tamar and occurs in small quantity on
the Arun and Medway. The hybrid with *S. lacustris*
(*S.* x *carinatus*) has also been recorded from the Tamar,
and also from the Thames (Jackson & Domin 1908; Lousley
1931), but is now very rare or extinct (Stace 1997).

Both *S. triqueter* and *S.* x *carinatus* were once locally
frequent by the Thames. It disappeared gradually over a
period of at least 50 years as a result of land claim and
work on river embankments and was last seen in about
1946 (Lousley 1976). Similarly, the Medway population of
S. triqueter disappeared in about 1940, following works to
make the river more navigable, and records of it in recent
years are erroneous (only *S.* x *kuekenthalianus* remains).
River engineering contributed to its extinction on the Arun,
but mudbanks are still common by the Tamar and reasons
for the decline of *S. triqueter* there are unclear. The hybrid
S. x *kuekenthalianus* has been present at least since the end of
the 19th century and there is no evidence to suggest that
there is significant competition between it and *S. triqueter.*

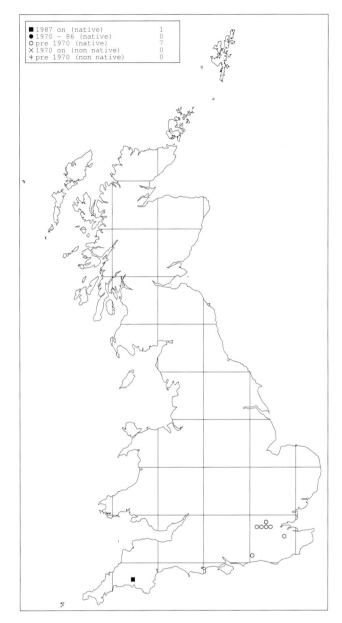

■ 1987 on (native)	1
● 1970 – 86 (native)	0
O pre 1970 (native)	7
X 1970 on (non native)	0
+ pre 1970 (non native)	0

This species is widely distributed across central and
southern Europe, extending to western Asia and North
Africa, and occurring also in North America and South
Africa. It becomes local in the northern parts of its range
and is absent from Scandinavia. It seeds readily in warmer
climates, but does not appear to do so in the wild in the
north. It may also be less able to compete vegetatively in
northern areas because of the cooler climate, and in these
areas it is normally restricted to habitats where
competition with other swamp species is reduced.

N.F. Stewart

333

Schoenus ferrugineus L. (Cyperaceae)

Brown bog-rush, Sèimhean Ruadh
Status in Britain: VULNERABLE.
Status in Europe: Not threatened. Endemic

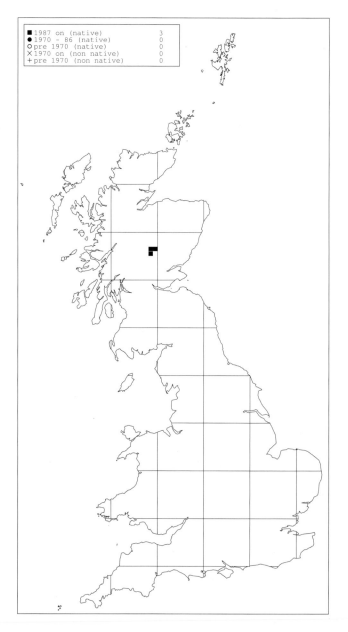

S. ferrugineus is found in base-rich flushes within calcareous grassland, usually adjacent to unimproved heathland. The flushes have high concentrations of calcium (pH 6.2 - 7.3) and low concentrations of nitrogen and phosphorus. It is tussock-forming except when intensely grazed. Commonly associated species include *Carex flacca, C. panicea, Eleocharis quinqueflora, Eriophorum latifolium, Juncus articulatus, Molinia caerulea, Saxifraga aizoides* and *Selaginella selaginoides,* and the mosses *Campylium stellatum, Ctenidium molluscum, Drepanocladus revolvens* and *Scorpidium scorpioides.*

It is a perennial plant. New shoots often emerge in autumn, grow a few centimetres, then overwinter, during which time they remain bright green near the base, unlike similar species in the habitat. Growth of all shoots resumes as early as March, with full shoot extension usually by July, but flowers appear before shoots are fully extended. Seed production is low, never exceeding six seeds per inflorescence, with a mean of 0.4-1.9 in Scottish populations in 1993. This contrasts with a mean of 3.4 from Swedish samples. Long-distance dispersal is restricted by retention of much of the seed within the inflorescence for up to a year after flowering. Retained seed germinates within the dead flower when it makes contact with the moist soil surface. Tussocks appear to establish by seed and subsequently expand vegetatively.

In Britain, *S. ferrugineus* is known only in Scotland, where it was first discovered in 1884 on the shores of Loch Tummel. The water-level of the loch was raised in 1950 for the generation of hydro-electricity, and attempts to establish plants from the threatened population at nearby sites were unsuccessful (Brookes 1981). However, it had been introduced to sites on Ben Vrackie in 1945, and two small but increasing populations are still extant there. Other extant populations were discovered during routine SSSI boundary survey (Smith 1980): 10 populations are currently known at six localities, all near Blair Atholl, Perthshire. In all, there are about 12,000 plants, the largest population holding about 80% of the total. All populations occur at a relatively low altitude of 200-400 m.

Most populations are close to improved ground, and appear vulnerable to pasture improvement, drainage, afforestation and, for one population, mining. All populations are, however, within SSSIs, providing at least a measure of protection. Excessive grazing by domestic animals is detrimental, stunting growth and inhibiting seed production. Recent studies have shown that, in two populations (including the largest), the regeneration potential is much lower and the mortality much greater than in the others. This seems likely to be because of increased stocking, especially in winter (Cowie & Sydes

1995). By contrast, in a flushed area where grazing has been excluded for years, the tussocks of *S. ferrugineus* are large and luxuriant, with good seed production. In this area, the flushes remain open naturally, thus providing new ground for colonisation.

S. ferrugineus is restricted to Europe, where it is found from south-east France northwards to Scandinavia, and eastwards to Russia and the Balkans. It is most abundant in the Baltic region and in the foothills of the Alps (Hultén & Fries 1986), but is very local over much of its range. In Europe it occurs in similar vegetation to that in Scotland, but is also reported from the field layer of fens (Wheeler *et al.* 1983).

C. Sydes

Scirpoides holoschoenus (L.) Sojak (Cyperaceae)

Scirpus holoschoenus L., *Holoschoenus vulgaris* Link
Round-headed club-rush, Clwbfrwynen Bengrwn
Status in Britain: VULNERABLE.
Status in Europe: Not threatened.

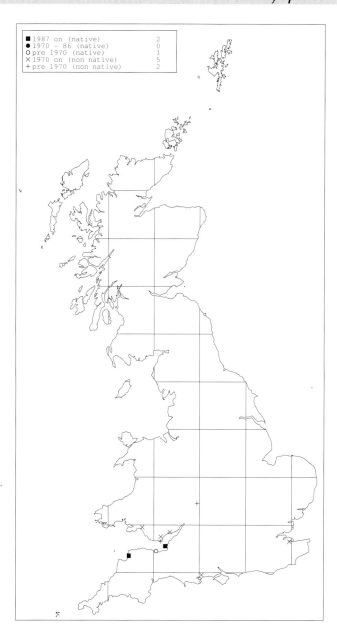

In Britain, native populations of *S. holoschoenus* occur at
Braunton Burrows, North Devon, and Berrow Dunes,
North Somerset, both in coastal dunes. Populations at
Braunton Burrows occur in damp dune-slacks and in
adjacent very low dunes. Frequent associates in the slacks
include *Agrostis stolonifera*, *Anagallis tenella*, *Juncus acutus*,
Mentha aquatica, *Pulicaria dysenterica* and *Salix repens*, whilst
on the low dunes they include *Carex arenaria*, *Festuca rubra*,
Hypochaeris radicata, *Ligustrum vulgare*, *Ononis repens*, *Poa
humilis* and *Rubus fruticosus*. At Berrow, the single tiny
patch occurs in a damp sandy hollow on a coastal golf
course, together with *Carex disticha*, *Festuca arundinacea*,
Hydrocotyle vulgaris, *Juncus inflexus* and *Potentilla anserina*.

S. holoschoenus is a densely tufted rhizomatous
perennial that grows up to 150 cm tall but is usually about
100 cm. Flowering occurs in August and September, and
only after a long hot summer do fruits containing viable
seeds appear to form and ripen in quantity. After a
favourable summer the seed bank may be substantial and
short-term persistent, lasting for at least one year but
fewer than five years (Thompson *et al.* 1997). Seedling
establishment typically occurs at sites with bare ground
within reach of the water table, but regeneration by
seedlings has never occurred at Berrow (A.J. Willis pers.
comm.). The plant may form 'dunelets' by trapping and
growing up through sand (Willis 1985), and this probably
explains its occurrence in very low dunes at Braunton
Burrows.

The stronghold for this species is at Braunton Burrows,
where the number of clumps has been estimated to exceed
3,000. *S. holoschoenus* also occurs sporadically as an
introduction, particularly in the vicinity of docks and
industrial areas. A single clump in Dorset, presumed to be
an introduction, grows just above the high tide line in
Poole Harbour, where associates include *Elytrigia atherica*,
Picris echioides and *Plantago lanceolata*. Other presumed
introductions have been reported from docks in Cardiff
and Newport and from a brickyard in Kent. It is unclear
whether any of these are extant: a colony at Newport was
lost to development, though some plants may still survive
on industrial land nearby.

The areas of Braunton Burrows where *S. holoschoenus*
occurs currently receive no grazing other than by rabbits,
and scrub encroachment is a continual threat. Rabbit viral
haemorrhagic fever has recently been confirmed there, and
this may pose a threat to the regeneration of *S. holoschoenus*
should the level of rabbit-grazing decrease substantially.
Scrub is kept in check by annual mowing in the months
between November and March, and this management
appears to be maintaining the population of
S. holoschoenus. The recent de-declaration of Braunton

Burrows as an NNR may put the species at further risk
there. Plants occurring near industrial areas are liable to be
lost when waste ground is developed or tidied, and such
losses have increased in recent years. The species is
tolerant of cold, but its range in Britain is probably limited
by its requirement for high summer temperatures for
regeneration by seed (Willis 1985).

S. holoschoenus occurs widely in Europe, from Iberia
northwards to Britain, and eastwards to central Russia and
Greece. It also occurs in the Canary Islands, north-west
Africa and south-west and central Asia. It is often common
in the southern parts of its range, including all the main
Mediterranean islands, and occurs in a wide variety of
open habitats.

J.H.S. Cox

Scleranthus perennis ssp. *perennis* P.D. Sell (Caryophyllaceae)

Perennial knawel, Dinodd Parhaol
Status in Britain: CRITICALLY ENDANGERED. WCA
Schedule 8.
Status in Europe: Not threatened.

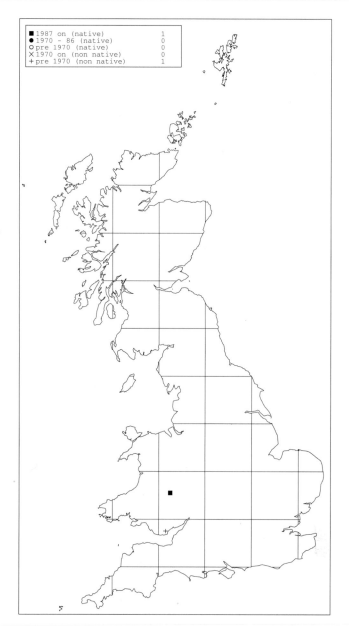

■ 1987 on (native)	1
● 1970 - 86 (native)	0
O pre 1970 (native)	0
X 1970 on (non native)	0
+ pre 1970 (non native)	1

In Britain, *S. perennis* ssp. *perennis* is restricted to an area of rock only a few square metres in extent on the dolerite and gabbro cliffs of Stanner Rocks, Radnorshire, where it was first confirmed in 1850. It grows in shallow pockets of soil that are vulnerable to summer drought, growing in association with *Aphanes arvensis*, *Erophila verna*, *Festuca ovina*, *Rumex acetosella*, *Saxifraga tridactylites*, *Teesdalia nudicaulis* and the cryptogams *Dicranum scoparium*, *Hedwigia stellata* and *Cladonia* species. Within a few metres the microtopography changes and species such as *Allium vineale*, *Erodium cicutarium*, *Hypochaeris radicata*, *Helianthemum nummularium*, *Jasione montana*, *Ornithopus perpusillus*, *Sedum forsterianum* and occasional *Gagea bohemica* occur.

Despite its name, *S. perennis* ssp. *perennis* is normally a biennial, flowering between June and August. The plants are variable in size and in the number of flowering stems, which can range from 3-24 per plant. Seed production is variable.

The population at Stanner fluctuates considerably, as also does the incidence of flowering. Annual monitoring between 1987 and 1994 showed that between eight and 135 plants occurred in any year, but usually only a small proportion flowered (up to a few tens of plants, but sometimes none at all). The plant has probably never been abundant at Stanner. For instance, on a visit by the Woolhope Club in 1888, it was noted that "... the other Stanner rarity, *Scleranthus perennis* ... was hunted for today in vain."

There are two main threats to the plant. Shading by taller plants can diminish the effects of the competition-reducing summer drought, and grazing by molluscs leave little foliage undamaged, so that ungrazed pot-grown plants look very different from those growing *in situ*. Encroaching bramble and ivy are being kept in check, and shading by trees is controlled by judicious pruning. Seeds are collected and scattered elsewhere on the site, but it seems that they do not germinate in their first year. However, seeds readily germinate in pots, and some stock plants are kept *ex situ* as a back-up. It will also grow readily from even the smallest cutting. It may be confined to its small area by grazing pressure (Woods 1993).

S. perennis ssp. *perennis* is widespread throughout central and southern Europe, extending northwards to Britain, Norway and Poland, and eastwards from Iberia to Greece and the Crimea; it is also in Turkey. It is declining in north-west Europe.

F.M. Slater

Scleranthus perennis ssp. *prostratus* Sell
(Caryophyllaceae)
Perennial knawel
Status in Britain: ENDANGERED. ENDEMIC. WCA
Schedule 8.

S. perennis ssp. *prostratus* is endemic to the East Anglian
Breckland. It grows on open, acidic, sandy soils of pH 4.9-
6.8 (Watt 1971) but is also found in semi-closed, very short
permanent grassland and on compacted tracks. It has also
occurred in disturbed sands in abandoned arable land and
in open fallow (Trist 1979). Associated species include
*Agrostis capillaris, Aira praecox, Filago minima, Jasione
montana, Ornithopus perpusillus, Rumex acetosella, Silene otites*
and *Teesdalia nudicaulis.* At one site it is found alongside
Thymus serpyllum and *Veronica verna,* two other Breckland
rarities tolerant of more acidic ground.

S. perennis ssp. *prostratus* is a biennial or short-lived
perennial. Its slightly woody stems with short internodes
are normally procumbent but occasionally ascending. The
usual flowering period is June to September, but recent
observations have shown that flowers may occur
throughout the year. Open soil is required for seedling
establishment, and this appears to be optimal where there
is 35-50% bare ground in the turf. It is a poor competitor
(Watt 1971).

This taxon has never been plentiful. Only 25 sites have
been recorded and never more than six at any one time
(G. Crompton 1974-86). Latterly, appearances of the plant
have been sporadic, most being in the southern part of
Breckland. It became extinct in Norfolk in 1961. Several of
its former sites have been lost to afforestation, agricultural
intensification and housing development, and it is now
restricted to three native sites, all SSSIs in the Suffolk
Breckland. By far the largest colony, of several thousand
plants, flourishes on the airfield at RAF Lakenheath, where
the sandy soil and regular grass cutting provide a suitable
habitat. In 1980, this plant was successfully established on
a Suffolk NNR, and attempts have been made to introduce
it to other sites.

A long-term strategy for the conservation of *S. perennis*
ssp. *prostratus* is being developed through 'recovery'
programmes. Plants are now in cultivation, and the
intention is to reintroduce the species to former sites.

Y. Leonard

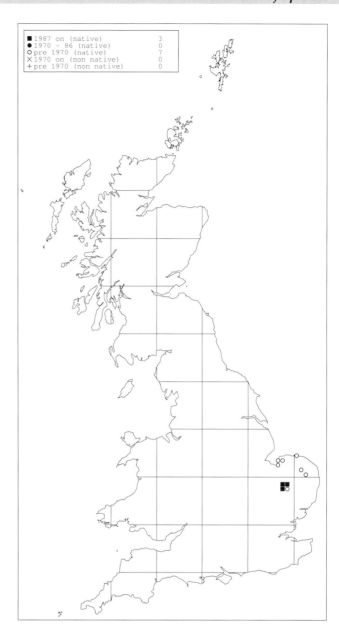

■ 1987 on (native)   3
● 1970 - 86 (native)   0
○ pre 1970 (native)   7
✕ 1970 on (non native)   0
+ pre 1970 (non native)   0

Scorzonera humilis L. (Asteraceae)

Viper's grass, Llys y Wiber
Status in Britain: VULNERABLE. WCA Schedule 8.
Status in Europe: Not threatened.

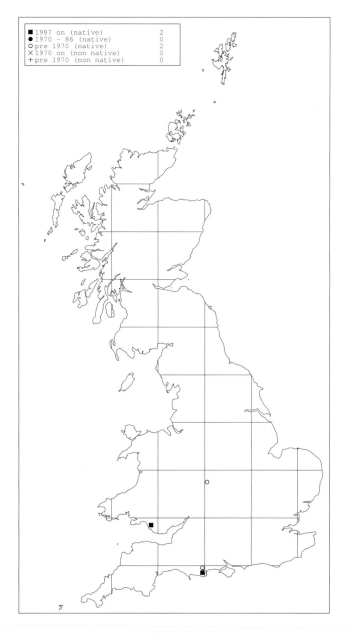

In Britain, native populations of *S. humilis* are known only at sites in Dorset and Glamorgan. It is a plant of damp lowland unimproved grassland or fen-meadow, sometimes occurring in a species-rich sward. In Glamorgan, its common associates include *Carex hostiana, C. panicea, C. pulicaris, Cirsium dissectum, Juncus acutiflorus, Molinia caerulea, Potentilla erecta, Succisa pratensis, Calliergonella cuspidata* and *Rhytidiadelphus squarrosus*. In Dorset the core of the population occurs in damp rank grassland dominated by *Festuca rubra*, other prominent species including *Anthoxanthum odoratum, Carex nigra, Carum verticillatum, Cirsium dissectum, Holcus lanatus, Lotus corniculatus* and *Succisa pratensis*, with sparse *Molinia caerulea*.

S. humilis is a perennial herb with a stout, black rootstock, broad basal leaves and one or a few flowering stems. The main flowering period is from mid-May to late June, with a few stragglers later in the year. Flowers are visited by bees and other insects, which presumably pollinate them, but flowers may also be self-pollinated. Seed production is good: 70-80% viable seed was recorded in a sample of fruiting capitula from Glamorgan, but no seedlings were found during a search in autumn. By late August 1996 there was little evidence that *S. humilis* occurred in such abundance at the South Wales site, since much of the above-ground parts had disappeared through die-back and grazing. Seedling establishment is unlikely to have occurred for a number of years at the Dorset site because of the rank vegetation.

S. humilis was recorded first in 1914, at Ridge, Dorset, where it still occurs in populations numbering many thousands of plants. In addition, two plants occur in a roadside ditch about 1.5 km away. At the Glamorgan site, discovered in June 1996, many thousands of plants occur over several hectares of marshy pasture enclosed in five fields near Bridgend. *S. humilis* was reported by L.B. Hall in the Poole Harbour area in 1927 (Druce 1929), but this site has not been refound, unless it is the same as the present roadside site. Five plants were found in 1954 in a damp meadow in the Earlswood district of Warwickshire (Hawkes & Phipps 1954), but they had gone by 1967.

The extent of the population at Ridge may have become somewhat reduced since first recorded. The site has not been grazed or cut for several years, and rank *Molinia caerulea* now covers about a third of the field (together with invading *Carex acutiformis* and *Phragmites australis*), with most of the *S. humilis* occurring in the adjacent rank *Festuca rubra-Anthoxanthum odoratum* dominated grassland. The Glamorgan site is extensively grazed by cattle, and rank weeds are topped (if weather permits) late in the year.

The status of *S. humilis* in Britain has been questioned, but its habitats here closely match those on the continental mainland (Hansen 1976; Wolton & Trowbridge 1990, 1992) and most authorities accept it as a true native (e.g. Druce 1929; Stace 1997).

S. humilis is of widespread occurrence in Europe, ranging from Iberia to the Balkans, southern Russia and the Caucasus, northwards to Denmark, southern Sweden, Karelia and central Russia (Hultén & Fries 1986).

J. Woodman & R.M. Walls

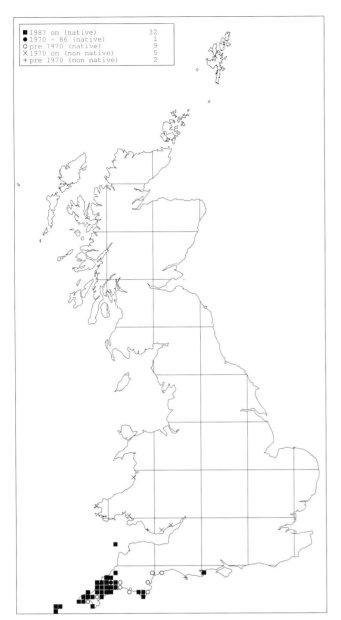

Scrophularia scorodonia L. (Scrophulariaceae)
Balm-leaved figwort, Gornerth Gwenynddail
Status in Britain: LOWER RISK - Nationally Scarce.
Status in Europe: Not threatened. Near-endemic.

In Britain *S. scorodonia* typically occurs in rather scruffy coastal or near-coastal habitats, including rough grassy or scrubby field margins, hedgebanks, roadsides, old walls, deserted quarries, disused railway lines and waste ground, often in the vicinity of ports and estuaries. Its associated species are unremarkable and include *Ballota nigra*, *Brachypodium sylvaticum*, *Dactylis glomerata*, *Hedera helix*, *Prunus spinosa*, *Rubus fruticosus*, *Smyrnium olusatrum* and *Urtica dioica*.

It is a robust and long-lived perennial, generally 60-100 cm high but sometimes taller when growing in scrub. It flowers from May to October and, like *S. nodosa*, appears to be mainly pollinated by Hymenoptera. Plants can produce copious amounts of seed, but nothing is known

about seed viability or longevity, and patches of *S. scorodonia* may sometimes enlarge by vegetative expansion or fragmentation rather than by recruitment of new plants from seed.

In Britain, *S. scorodonia* is restricted as a presumed native to the Isles of Scilly, Cornwall, Devon and Dorset. It has also been recorded as an introduction in South and West Wales. In the Isles of Scilly it occurs on all the inhabited islands except St Agnes, from where it was last recorded in the 1970s. It is most frequent, however, on two uninhabited islands, Samson and St Helen's. In mainland Cornwall it is widely scattered but local, being abundant only around the Gannel Estuary and Newquay. In Devon it is found on Lundy and in the South Hams around Salcombe and the Kingsbridge Estuary. The latter area is probably now one of its chief strongholds in mainland Britain, with many new populations found in recent years, for example at Batson Creek, East Portlemouth, South Pool, Frogmore and Kingsbridge. It is uncertain whether the species is spreading in South Devon, since it may merely have been under-recorded in the past.

S. scorodonia does not seem to be in any immediate threat, and there is no evidence of decline. Nevertheless, hedgerow removal and the herbiciding or tidying up of scrubby field margins and waste ground might put some populations at risk. At several hedgebank sites the plant seems to have been smothered in dense stands of *Galium aparine* and *Urtica dioica*, which have developed possibly as a result of nutrient enrichment from the use of fertilisers on neighbouring farmland. Conversely, it thrives where vegetation has been cut back, and appears also to benefit from hedge maintenance as currently practised.

S. scorodonia is an Oceanic-Western species, confined in Europe to western France, Spain, Portugal, Madeira and the Azores. It also occurs in Morocco and perhaps in Tunisia. In the Channel Islands, it is the commonest *Scrophularia* species. *S. scorodonia* is regarded as native in Britain (e.g. Clapham *et al.* 1987; Stace 1997), but, as with *Spergularia bocconei*, the long history of shipping trade between Iberian and western English ports hints at the possibility of it being an ancient introduction.

Further details of the history and distribution of the species in Britain are given in Meredith (1994).

H.M. Meredith

Selinum carvifolia L. (Apiaceae)
Cambridge milk-parsley
Status in Britain: LOWER RISK - Near Threatened.
Status in Europe: Not threatened.

In Britain, *S. carvifolia* is restricted to calcareous fens
and meadows in eastern England. It grows on shallow fen
peat overlying deposits of chalk. The substrate pH is in the
range 7.0-7.7 (Grimshaw 1991). It does not grow in the
wettest parts of the fen but is mainly confined to the
slightly drier fringe areas and low, freely drained, raised
banks. Among the most frequent associates in the rich and
diverse communities are *Carex hostiana, C. panicea, Cirsium
dissectum, Festuca rubra, Juncus subnodulosus, Molinia
caerulea, Potentilla erecta, Serratula tinctoria, Succisa pratensis,
Valeriana dioica* and *Calliergonella cuspidata.*

S. carvifolia is an erect glabrous perennial usually up to
60-100 cm, but occasionally taller. Flowers are produced
from mid-July to September and the usually abundant fruit
in October. However, the seeds appear to be short-lived
(Thompson *et al.* 1997). The bracts are often shed after
flowering. The plant does not flower until it is at least two
or three years old, and there are normally many non-
flowering plants in the population.

This species has only ever been recorded at six sites in
Britain, four in Cambridgeshire and one each in
Lincolnshire and Nottinghamshire (Walters 1956). It now
remains at just three sites, all in Cambridgeshire:
Chippenham Fen, Snailwell Meadows and Sawston Hall
Meadows, all holding thriving populations. The largest
population is at Chippenham Fen, where it grows fairly
plentifully at sites well scattered over the 117 hectare
reserve. A careful census in 1990 revealed a total of about
960 flowering plants there, together with many non-
flowering plants. A later survey of the largest colony (in
the North Meadows) showed that there were about six
vegetative plants for every one flowering, averaged over
six years (O'Leary 1995). In August 1996, more than 17,000
plants flowered in the whole reserve, and it is considered
that this represents a real increase in the population. At
Snailwell Meadows *S. carvifolia* grows in damp, low-lying
cattle-grazed meadows beside the River Snail. In 1990,
some 480 flowering plants in two main colonies were
counted there and just over 600 in 1995. Sawston Meadows
hold a few hundred plants.

All three extant sites have some measure of protection.
Chippenham Fen has been an NNR since 1963 and both
Snailwell Meadows and Sawston Hall Meadows are SSSIs.
Conservation management includes cattle-grazing during
the autumn and winter and cutting and removal of fen
vegetation during the same period.

This species occurs over much of Europe, as far north
as Finland and east to the Urals, but is absent from most
of the Mediterranean basin. In continental Europe
S. carvifolia has been recorded from a wider range of
habitats than in Britain, including oakwoods in Poland and
on hot dry limestone in Bosnia and Croatia.

M. Wright

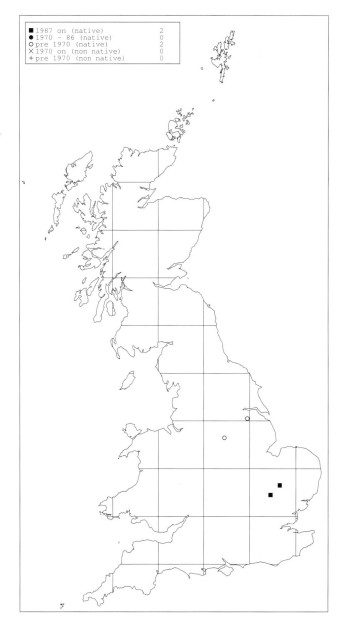

■ 1987 on (native)   2
● 1970 - 86 (native)   0
○ pre 1970 (native)   2
✕ 1970 on (non native)   0
+ pre 1970 (non native)   0

Senecio cambrensis Rosser (Asteraceae)
Welsh groundsel, Creulys Cymreig
Status: LOWER RISK - Near Threatened. ENDEMIC.

S. cambrensis is an ephemeral species inhabiting disturbed sites such as rubble, gardens and thin soils alongside roads and footpaths. *S. squalidus, S. vulgaris, Sonchus oleraceus* and *Taraxacum* species are commonly associates.

S. cambrensis is typically an annual, germinating in spring or autumn and flowering in May through to October, with particular flushes of flowering occurring in June and September. Some individuals are short-lived perennials, with mature individuals surviving the winter to flower again in the following spring. Such individuals invariably develop woody stem bases. The longer-lived individuals are almost always found growing amongst the cracks in walls, and it is possible that such less vulnerable habitats have been important in the establishment of the species following its evolution. Certainly the only persistent Leith population is found in such a habitat.

S. cambrensis was initially detected by H.E. Green at Cefn-y-Bedd in 1948 and was described by Effie Rosser (1955) from a specimen collected at Ffrith, near Wrexham. The species is still extant at Ffrith and other sites around Wrexham, including Ruabon, and Chirk to the south. In addition, it is found at Mochdre, near Colwyn Bay, and at Leith, near Edinburgh. A population was known at Ludlow (Herefordshire) until 1980, and one at Ness, on the Wirral, now extinct, had been introduced from Wrexham stock.

The Wrexham population had spread only slightly from Ffrith between 1955 and 1972. However, over the next ten years, it became well-established further afield, with notable populations up to 100 plants regularly recorded at Ffrith, Brymbo, Southsea and Rhostyllen. Population numbers in 1994 and 1995 around Wrexham were lower than in previous years. However, whether this is a real decline or part of the cycle of large yearly fluctuations related to the ephemeral nature of the habitat remains to be seen. The population from Mochdre, discovered in 1966, occupies 100 m of roadside verge and has typically supported over 100 individuals each year over the last ten years. It has, however, shown no sign of spreading. The Leith population (several colonies) has persisted since its discovery by 1982. Redevelopment in Leith during the 1980s has greatly reduced available habitat, and the population of over 100 individuals in 1983 (Abbott *et al.* 1983) has fallen to fewer than five individuals in the last few years. This population may, therefore, shortly become extinct.

The species is most likely to have evolved when an infertile hybrid resulting from a cross between *S. vulgaris* and *S. squalidus* underwent chromosome doubling. This process of allopolyploidy is extremely important in plant evolution, although it is rarely observed or documented. Analysis of protein variation within *S. cambrensis* and its parental species from its various sites showed that the Welsh populations had arisen independently from the Scottish material (Ashton & Abbott 1992). This was

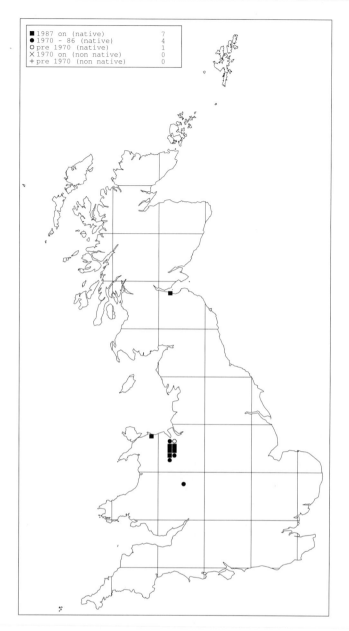

confirmed by DNA analysis (Harris & Ingram 1992). Moreover, protein variation suggests that there have been two separate origins within Wales (at Wrexham and Mochdre). The proximity of *S. vulgaris* and *S. squalidus* throughout most of Britain raises the possibility of *S. cambrensis* evolving at other sites.

The conservation of such ephemeral species is problematical. The suitable disturbed habitat that results from mowing or herbicide application also leads to loss of individuals. Within the Wrexham area a staggered regime of roadside management may be appropriate. This may facilitate recolonisation from local sources if any populations are lost. Seed from seed-banks may help offset any catastrophic losses.

Further details are in Ashton (1990) and Ingram & Noltie (1995).

P.A. Ashton

Senecio paludosus L. (Asteraceae)
Fen ragwort
Status in Britain: CRITICALLY ENDANGERED.
Status in Europe: Not threatened.

S. paludosus is currently confined as a native species to a single roadside ditch running alongside arable land near Ely. The ditch is usually flooded in winter but dries out in summer. The associated species at this site are unremarkable and include *Arrhenatherum elatius*, *Calystegia sepium* and *Elytrigia repens*.

S. paludosus is a long-lived perennial. Plants die down during the winter. In spring robust shoots grow from the rootstock and develop into tall stems with a loose panicle of large flowers. At the only native site seed-set is poor, but material from this site that has been cultivated in Cambridge and Abbots Ripton shows improved seed-set, as do plants that have been introduced to Wicken and Woodwalton Fens. This suggests that the poor reproductive performance can be attributed in part to the environmental conditions at the Ely site, where pollinating insects may be scarce and the inflorescences are regularly buffeted in the slipstream of passing traffic. Seedlings have not been seen at the Ely site.

Between 1660 and 1860 this species was recorded at scattered localities in the fens of Cambridgeshire, Lincolnshire, Norfolk and Suffolk. It appears to have been most frequent in Cambridgeshire, where Ray (1660) recorded it "in many places about the Fens, as by a great ditch side near Stretham Ferry". There were few recorded localities in the other counties. The species was eliminated from most of its sites by fenland drainage. C.C. Babington found a single plant at Wicken Fen in 1857 and this was the last definite record to be made in the 19th century. Bennett (1899) commented that a few plants were found after 1857, but this claim, although not improbable, has never been substantiated by precise records. For many years *S. paludosus* was regarded as extinct in Britain, but it was discovered in 1972 at its present site near Ely by T.W.J.D. Duprée (Walters 1974). The precise origin of the plants discovered in 1972 is uncertain: the ditch in which they occur was constructed in 1968 and it is possible that the plants grew soon afterwards from dormant seed.

The number of vegetative and flowering stems in the Ely population increased after its discovery, but has decreased in recent years. Even at its maximum extent the colony has never been extensive, and it could easily be damaged by the tipping of rubbish or other damaging events. The species was therefore adopted by English Nature's 'recovery' programme, and plants grown from seed of English origin have been introduced to a ditch near the existing site and to other places including Wicken and Woodwalton Fens. Populations are thriving at both these sites, flowering well and producing the occasional viable

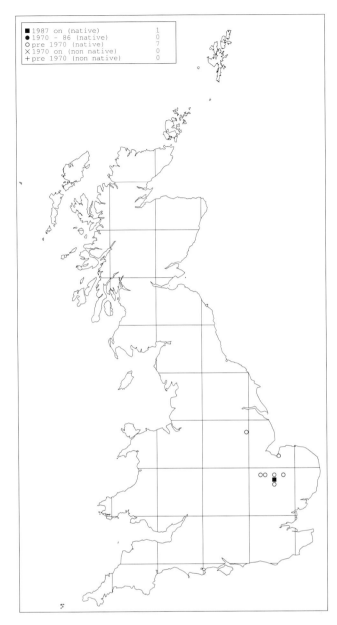

■ 1987 on (native)	1
● 1970 – 86 (native)	0
○ pre 1970 (native)	7
✕ 1970 on (non native)	0
+ pre 1970 (non native)	0

seed. Of the 50 originally planted, about 30% still survive at Wicken, and about 60% at Woodwalton.

S. paludosus occurs in central and eastern Europe and western Asia (Hultén & Fries 1986). It is regarded as extinct in Denmark and has decreased in frequency in north-west Germany. In the Netherlands and southern Germany, as in Britain, populations tend to be small and consist of mature plants but no seedlings; it is not known whether this is a cause for concern or whether natural regeneration has always taken place only infrequently.

C.D. Preston and T.C.E. Wells

Seseli libanotis (L.) Koch (Apiaceae)
Moon carrot
Status in Britain: VULNERABLE.
Status in Europe: Not threatened.

In Britain, *S. libanotis* is restricted to a very few localities in southern and eastern England, where it is mainly a plant of chalk grassland. It most often occurs in a sward containing *Briza media*, *Bromopsis erecta* and *Festuca ovina*, together with a wide variety of common calcicolous herbs such as *Asperula cynanchica*, *Campanula glomerata*, *Cirsium acaule*, *Daucus carota*, *Linum catharticum*, *Pimpinella saxifraga*, *Plantago media* and *Sanguisorba minor*. Several uncommon species, including *Hypochaeris maculata*, *Phyteuma orbiculare*, *Pulsatilla vulgaris* and *Tephroseris integrifolia* ssp. *integrifolia* occur with it at some its sites. In Cambridgeshire, *S. libanotis* also grows on chalky roadside banks, and the ledges of abandoned chalk quarries.

This plant is usually biennial, though is sometimes a short-lived perennial. It flowers in late summer and is generally in full flower in mid-August. Seed-set and germination are presumed to be good, at least in those populations that are maintaining their numbers.

It occurs at six sites in Bedfordshire, Cambridgeshire and East Sussex, and formerly grew at one other in Hertfordshire. All extant populations lie within SSSIs or NNRs and thus have a degree of protection. There are two large populations near Beachy Head, East Sussex, each of which holds more than 3,000 plants. It is possible that *S. libanotis* is still extant at a third site nearby (the edge of a golf-course), though only a single plant was recorded in 1986 (Everett 1988). The other large British population is at Knocking Hoe, Bedfordshire, where up to 12,000 plants have been recorded (Crompton 1974-1986). Both populations in Cambridgeshire are small. On the Gog Magog Hills, it appears to be in decline, perhaps because of the lack of suitable grassland management. Hundreds of plants were recorded there in 1986, but only 70 in 1988 and 12 in 1993, though the later surveys might have been partial. Only 23 plants were counted at the other Cambridgeshire site, at Cherry Hinton, in 1991.

In most populations, the number of flowering stems varies from year to year according to the intensity of rabbit-grazing. The largest populations seem to be more or less stable at the present time, or are, perhaps, slightly increasing. This encouraging picture suggests that current management is suitable, comprising sheep- and cattle-grazing early in the year, together with occasional 'topping' of the sward after seed-set to prevent it becoming too rank.

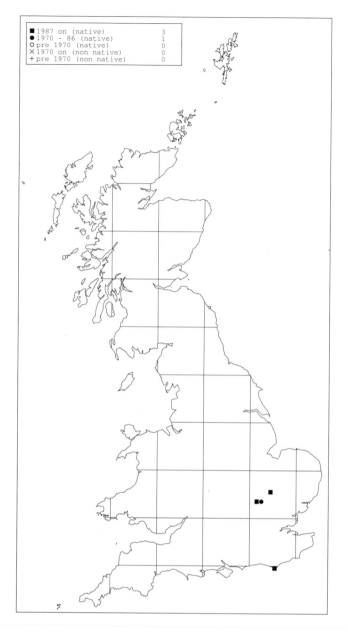

■ 1987 on (native)	3
● 1970 - 86 (native)	1
○ pre 1970 (native)	0
✕ 1970 on (non native)	0
+ pre 1970 (non native)	0

S. libanotis occurs widely in Europe, from Spain northwards to Britain, the Benelux countries and Fennoscandia and eastwards to Poland, Russia and Bulgaria. Its also occurs in North Africa and in south-west Asia. Elevated plateaux and montane meadows are its usual habitats in continental Europe. A number of closely-related species occur in eastern Europe and Asia. *S. libanotis* has also been recorded from the eastern United States of America, although it is adventive there.

P.A. Harmes

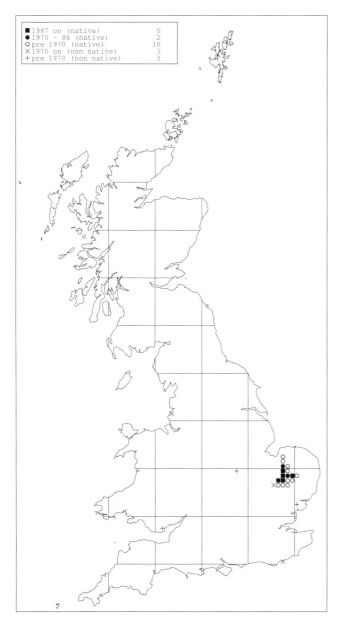

Silene otites (L.) Wibel (Caryophyllaceae)
Spanish catchfly
Status in Britain: LOWER RISK - Near Threatened.
Status in Europe: Not threatened.

Native sites for *S. otites* in Britain are now confined to the East Anglian Breckland. It grows on shallow, light, calcareous soils on grass-heaths and roadside verges, where it may be locally abundant. It is favoured by low, open vegetation where disturbance produces plenty of open ground for seed germination and seedling development. Associates include *Festuca ovina, Galium verum, Koeleria macrantha* and *Plantago lanceolata* and, frequently, Breckland specialities including *Medicago minima, M. sativa* ssp. *falcata, Phleum phleoides, Silene conica* and *Veronica verna*.

It is a more or less dioecious perennial, female plants lacking stamens and male plants sometimes bearing a few hermaphrodite flowers, though these usually these have only a vestigial ovary. The loose spikes of pale greenish-yellow flowers appear from June to September. Mature plants are deeply rooted and can withstand the low summer rainfall, but spring droughts frequently kill young seedlings. Plants may survive for a time in taller swards by reproducing vegetatively but ultimately succumb if competition becomes too intense.

S. otites has declined significantly in both range and abundance since the 1920s, when W.G. Clarke wrote "the greenish spikes of the Spanish catchfly are about as thick as a hay crop on some heaths." Many of those heaths have been ploughed up or afforested. The range has further contracted since 1960, but the population now seems more or less stable overall. The Suffolk Breck remains the stronghold for this species, with about 21 sites (Leonard 1993), but elsewhere there are only five other extant native sites, three in Norfolk and two in Cambridgeshire. Casual plants are recorded occasionally from other parts of the country, and one female plant at Ainsdale NNR in

Lancashire has persisted for several years. All its Breckland sites are on SSSIs, County Wildlife Sites or Protected Roadside Verges and thus have a degree of security so long as they are appropriately managed.

This species is widespread throughout Europe, extending from Iberia northwards to Britain and eastwards to Poland, the Caucasus, northern Iran and Siberia. It is common in eastern, central and southern Europe, but is very local in the northern parts of its range. It is a variable species that has been divided by some authorities into a number of separate taxa.

S. otites is the sole larval food plant of the viper's bugloss moth *Hadena irregularis* which, in Britain, was restricted to Breckland (Haggett 1952). It has not been seen since 1977, despite much searching during systematic surveys, and is presumed to be extinct.

G. Beckett

Sorbus L. (Rosaceae)
Whitebeam

Sorbus in Britain consists of three widespread, variable, sexually-reproducing species, *S. aria*, *S. aucuparia* and *S. torminalis*, one extremely rare sexual species (*S. domestica*) and some 20 apomictic microspecies that are genetically more or less uniform and form populations of essentially identical individuals, often with very restricted distributions. These apomicts fall into three groups, which appear to have arisen respectively from parent stocks of *S. aria sensu lato* (*S. aria* agg.), from hybridisation between *S. aria sensu lato* and *S. aucuparia* ('*S. intermedia* agg.'), or from hybridisation between *S. aria* agg. and *S. torminalis* (*S. latifolia* agg.).

Stace (1991) recognises 16 native apomictic microspecies in Britain. There is a further well-characterised microspecies forming substantial populations from Cwm Clydach to the Black Mountains in South Wales (Proctor & Groenhof 1992) and another (mentioned in the account of *S. vexans*) on the Bristol Channel coast of Devon and Somerset (Proctor *et al.* 1989). There appear to be other local uniform apomictic populations, comparable to some of those that have been formally described and named, in the Avon Gorge and in the Wye valley below Symonds Yat. Intensive study would undoubtedly disclose more of them. We should neither belittle the biological interest of the *Sorbus* (and other) apomicts, nor get them out of proportion. They are not comparable with 'Linnaean' sexually-reproducing species, and individually can hardly be considered to have the same intrinsic value. Their greatest interest is perhaps in relation to what they can tell us about biological processes - about the ways that different groups of plants deploy their stock of genetic variation, and how these have evolved. Apomixis allows a plant to multiply and disperse a well-adapted genotype - to clone itself by seed - but stands in the way of the openness to change conferred by (sexual) genetic recombination. What conditions favour apomixis? Do apomictic microspecies have (geologically) short life-spans? More research is needed before we shall really understand the ecological and evolutionary genetics of apomictic plant groups. It is likely to be of much more value to conserve sites with large and diverse apomictic populations (e.g. Craig-y-Cilau, Symonds Yat area, Avon Gorge, West Somerset - North Devon coast) than isolated outlying occurrences of taxa with their main area elsewhere.

M.C.F. Proctor

347

Sorbus anglica Hedlund (Rosaceae)

Whitebeam, Cerddinen Seisnig
Status in Britain: VULNERABLE.
Status in Europe: Vulnerable. Endemic.

This rare shrub or small tree up to 3 m high occurs locally in small populations, usually on Carboniferous limestone, in Devon, the Mendips, the Avon Gorge and Wye valley, Shropshire, Breconshire, Montgomeryshire and Denbighshire, and in Kerry. It is light-demanding, often forming part of the woodland edge in rocky woods and also occurring in scrub on crags and cliffs, in association with *Crataegus monogyna, Fraxinus excelsior, Ilex aquifolium, Sorbus aucuparia* and *Taxus baccata*. The largest populations are on Eglwyseg, Denbighshire (about 240 trees), at Breidden and at Cwm Clydach, where more than 80 plants occur, and at Llanymynech Hill, which supports more than 50 plants. The national population comprises about 600 plants.

It is apomictic, although - atypically for non-sexual species - the populations show some variability. Despite variations in leaf shape, samples from Mynydd Llangattock, Wye valley, Avon Gorge, Cheddar, Breidden and Llangollen all showed the same peroxidase phenotypes, but specimens examined from Llangollen may be a different and undescribed species (Proctor & Groenhof 1992). *S. anglica* is generally thought to have originated from *S. aucuparia* and *S. rupicola*, although it is possible that *S. porrigentiformis* may also have been involved along the way. Closely allied species occur in Norway and the Alps, Pyrenees, Carpathians, and the Balkan peninsula.

Populations of *S. anglica* seem to be relatively stable. Large populations occur in NNRs, SSSIs and other protected areas. In the Avon Gorge, clearance of invasive plants including *Quercus ilex* and *Cotoneaster* species is being carried out with the aim of increasing good *Sorbus* habitat. The *Sorbi* of the Avon Gorge seem unaffected by the toxic waste from the shot-blasting of the Clifton Suspension Bridge in 1995 (P.J.M. Nethercott pers. comm.). Generally, *S. anglica* is not under great threat, partly because of its occurrence in many protected sites, and partly because many of its sites are relatively inaccessible to public access. However, because there are fewer than 1,000 trees nationally, it must be placed in the Vulnerable threat category.

Outside Britain it is recorded only from Co. Kerry, Ireland.

C.S. Crook

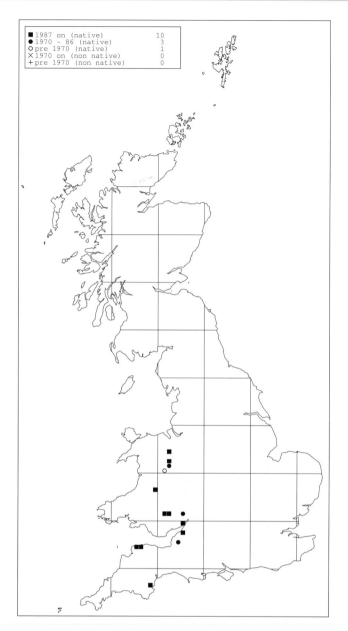

■ 1987 on (native)	10
● 1970 - 86 (native)	3
O pre 1970 (native)	1
X 1970 on (non native)	0
+ pre 1970 (non native)	0

Sorbus arranensis Hedlund (Rosaceae)

Arran whitebeam

Status in Britain: VULNERABLE. ENDEMIC.

S. arranensis is restricted to a few glens in northern Arran, where it occurs mainly as scattered, small, often stunted trees, in fragmentary woodland. The plants usually grow on or near the almost vertical edges of the gorges of streams, where these are cutting back into the upper slopes, often in cracks between rocks (Hull & Smart 1984). It requires sharp drainage and is intolerant of water-retentive soils, including peats. It is usually to be found with *Betula pubescens* and *S. aucuparia* and, at some sites, with its putative hybrid offspring *S. pseudofennica*. The ground flora is typical of that found in the area on acid to very acid soil, usually with abundant *Calluna vulgaris* and *Molinia caerulea*; other associates include *Deschampsia flexuosa, Erica cinerea, Galium saxatile, Pteridium aquilinum* and, more rarely, low-growing shrubs of prostrate *Juniperus communis* and *Arctostaphylos uva-ursi*. *S. arranensis* and *S. pseudofennica* are the only *Sorbus* species endemic in Britain that occur on acid substrates.

It is a small slender tree growing to 7-8 m, though it would probably grow taller and more robust if a restriction in grazing allowed it to become established in fertile, sheltered lowland sites. It is an apomictic, fully fertile hybrid derived from *S. aucuparia* and *S. rupicola*. The latter species does not occur in the locality, but probably did so in the past (another possibility is that the primary hybrid event occurred elsewhere and seeds were carried to the current area). Elsewhere on Arran *S. rupicola* is rare, but there is a sizeable colony on Holy Isle some 20 km to the south-east, with further sites to the north in Argyll. At some sites, populations of *S. arranensis* include young trees, but there has been no systematic study of the extent and success of natural regeneration. Isozyme studies are underway to determine the genetic make-up of populations.

Because of the precipitous nature of its sites, the population is difficult to count, and a regular comprehensive census has not been carried out. Furthermore, there are taxonomic difficulties for the observer, some trees appearing to be intermediate in leaf morphology between *S. arranensis* and *S. pseudofennica*, suggesting that there may have been multi-hybridisation events (Hull & Smart 1984). The last complete census was in 1980/81, which showed that there were about 400 trees of *S. arranensis* of all ages at the several sites within Glen Catacol, Glen Easan Biorach, Glen Iorsa and their associated tributary valleys. A partial census was carried out in 1970/71 and in 1986, with one new site (of eight trees) discovered in the latter year. There have been no population counts since then. The largest concentration of *S. arranensis* is to be found along the gorge of the Glen Diomhan stream, part of which was established as an NNR in an attempt to improve the future prospects of the species, but there has apparently been no population census within the NNR since 1970. A section of the reserve is enclosed by a deer- and sheep-proof fence to control browsing, and there has been some planting within the

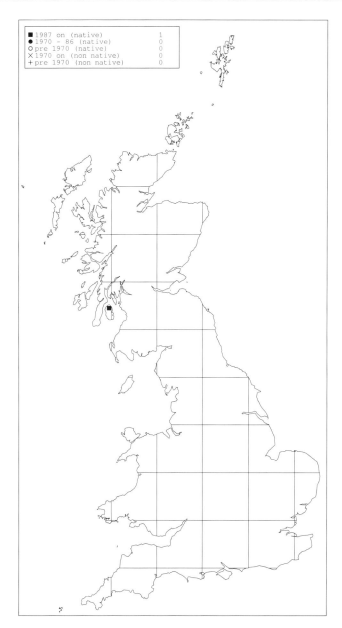

■ 1987 on (native)	1
● 1970 – 86 (native)	0
○ pre 1970 (native)	0
✕ 1970 on (non native)	0
+ pre 1970 (non native)	0

enclosure of saplings grown *ex situ*. This has been partially successful, with about 30 trees of both species surviving on better drained ground adjacent to the top of the gorge. Failures may have been the result of planting saplings in waterlogged ground or because of browsing during a period when the fence became dilapidated (the fence has since been renewed). There has been much speculation as to whether *S. arranensis* occurred more widely in the past when the surrounding moorland may have been tree-covered, but pollen analysis from near the largest extant colony in Glen Diomhan suggests that for the last 8,000 years the woodland in the gorge has never been much more extensive (Steven & Dickson 1991).

The Arran whitebeams were considered by McVean (1954) to be closer to extinction than any other tree or shrub in Scotland with the exception of *Salix lanata* (Bignal 1980). Despite some speculation that the total population is in decline, there is no positive evidence for this. On the basis of available evidence, it is not possible to determine whether the population size and structure have changed since 1970. The enclosed area of the NNR is only a relatively small area of the total extant distribution and it is more likely that there would be a significant increase in the population only if the numbers of browsing red deer were much reduced in northern Arran, allowing the possibility of regeneration on suitable well-drained but currently accessible ground.

This taxon is available commercially in a few nurseries (S.J. Leach pers. comm.). Further information on the endemic Arran whitebeams is given in Bignal (1980).

A.R. Church and C.S. Crook

Sorbus bristoliensis Wilmott (Rosaceae)
Bristol whitebeam
Status in Britain: ENDANGERED. ENDEMIC.

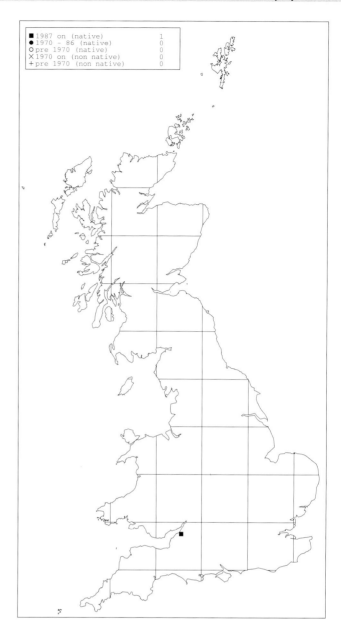

■ 1987 on (native)	1
● 1970 - 86 (native)	0
O pre 1970 (native)	0
× 1970 on (non native)	0
+ pre 1970 (non native)	0

This tree, up to about 10 m tall, is restricted to the Avon Gorge, where it occurs on both the Somerset and Gloucestershire sides of the Gorge, with most trees in the former county. It occurs mainly on Carboniferous limestone, on cliffs and crags, in woodland and scrub with *Fraxinus excelsior*, *Quercus robur* and *Tilia cordata*, and sometimes with *Sorbus anglica*, *S. eminens* and *S. wilmottiana*. The population of about 100 trees is thought to be increasing.

S. bristoliensis is apomictic and appears to regenerate moderately well, although, in common with most apomictic *Sorbus* species, fruiting can be erratic. It was once thought to have originated from hybridisation between *S. rupicola* and *S. torminalis*, but *S. rupicola* has nowhere been found in the Bristol area and there is nothing about the leaf and fruit of *S. bristoliensis* to show the influence of *S. rupicola* (P.J.M. Nethercott pers. comm.).

All known trees occur within the Avon Gorge SSSI and thus are afforded a degree of protection. The aim of conservation management has been to keep potential *Sorbus* habitat open and free from competition. To this end, some removal of woody aliens, especially *Cotoneaster* species and *Quercus ilex*, has been carried out but further clearance is required. At some of its sites, the inaccessibility of its habitat provides good protection.

This taxon is available commercially in a few nurseries (S.J. Leach pers. comm.).

C.S. Crook

Sorbus domestica L. (Rosaceae)
Service-tree, Cerddinen Ddof
Status in Britain: CRITICALLY ENDANGERED.
Status in Europe: Not threatened.

In England, *S. domestica* is found on south-facing river cliffs on Triassic mudstones and Carboniferous shales. The cliffs are relatively unstable and subject to landslip, but the immediate areas around the trees have more stability. In South Wales, it occurs on south-facing maritime limestone cliffs, where it grows as a wind-cut, gnarled shrub or semi-prostrate small tree on inaccessible ledges and in cliff scrub. Associates common to all its sites include *Brachypodium sylvaticum*, *Crataegus monogyna*, *Hedera helix*, *Rubia peregrina* and *Rubus ulmifolius*. A wide range of other associates in English sites includes *Bromopsis erecta*, *Cornus sanguinea*, *Hyacinthoides non-scripta*, *Ligustrum vulgare*, *Origanum vulgare*, *Sorbus eminens*, *Teucrium scorodonia* and, in Welsh sites, *Adiantum capillus-veneris*, *Clematis vitalba*, *Festuca rubra*, *Lithospermum purpureocaeruleum* and *Prunus spinosa*, among others.

S. domestica, a tree up to about 20 m tall but often very much shorter, has deciduous pinnate leaves; although vegetatively similar to *S. aucuparia*, it can be distinguished by its fissured bark, forked stipules and glabrous viscid buds. Its flowers, which appear in late April and early May, are pink-flushed, larger than in *S. aucuparia* and in a rather pyramidal corymb, and its fruits, which are rarely produced in Welsh populations, resemble small pears up to about 2.5 cm in diameter. It is a sexual species that is not closely related to the other *Sorbus* species that grow in Britain and does not form hybrids with them (Richards 1975; Warburg & Kárpáti 1968). The discovery of trees in England is very recent, but flowering seems to be poor, with some trees not flowering every year, and mature fruit has not yet been observed.

There are three known native sites in England: two sites by the Severn, each with a single tree, and one by the Bristol Avon, with perhaps six trees, though at the latter site suckering and problems of access make a precise count difficult.

The single trees on cliffs appear to be a great age. An extraordinary record of *S. domestica* was given by Nennius, a monk, who wrote in his Historia Brittonum in about AD 829 "by the river called Guoy (Wye), apples are found on an ash tree in the wood near the mouth of the river." The mouth refers to the Severn. There is no doubt that the tree was *S. domestica* and that he considered it to be rare at that time. In Wales, one site holds a population of about 14 trees growing on a 100 m length of low cliffs, and the other, about 4 km to the east of the first, has eight trees and about 60 saplings or sucker shoots scattered along the upper ledges of a 230 m length of cliffs. The older trees may be several hundred years old (Hampton & Kay 1995). In the past, they may have been confused with *S. aucuparia* or *Fraxinus excelsior,* which, from a distance, they quite closely resemble; it is possible that other populations of

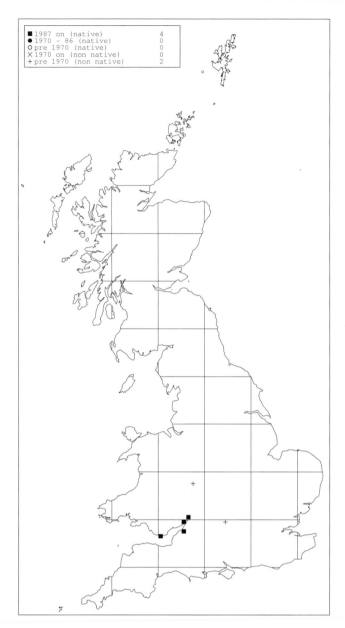

■ 1987 on (native)	4
● 1970 – 86 (native)	0
○ pre 1970 (native)	0
✕ 1970 on (non native)	0
+ pre 1970 (non native)	2

S. domestica may have been similarly overlooked elsewhere, as shown by the recent discoveries in England. Previously, it had been known in Britain only as a single tree that formerly grew in the Wyre Forest and was described in 1678. This tree no longer exists, but five trees currently occur in a woodland near Wyre Forest and, though probably deliberately planted, are thought to derive from the original tree.

The range of *S. domestica* extends from Britain, western France and southern Germany southwards to Spain and North Africa, and eastwards to the Balkans and Turkey.

Further details of its morphology, history and habitats are described in Hampton (1996).

Q.O.N. Kay, M.A.R. Kitchen & C. Kitchen

Sorbus eminens E.F. Warburg (Rosaceae)
Whitebeam, Cerddinen Mynwy
Status in Britain: VULNERABLE. ENDEMIC

S. eminens is a plant of deciduous woodland on
Carboniferous limestone, often occurring on steep slopes
and sometimes on rock scars. In the Wye valley woods,
common associates include *Anemone nemorosa,
Brachypodium sylvaticum, Carex sylvatica, Crataegus
monogyna, Euphorbia amygdaloides, Hyacinthoides non-scripta,
Mercurialis perennis, Teucrium scorodonia* and *Viola
reichenbachiana.*

This apomictic species is a shrub or small tree up to
6 m high. Fruiting tends to be erratic, as in most *Sorbus*
species, although fertile seed is produced and (in the
absence of grazing) regeneration occurs.

S. eminens occurs very locally in the Wye valley and the
Avon Gorge, with a few outlying populations in Somerset.
In the Avon Gorge, it is well represented on the Leigh
Woods (Somerset) side, where it shows high adaptability to
the stonework of the old railway (not only on
Carboniferous limestone) and the tow-path area of the
riverbank. The total population probably numbers about
250 trees, nearly half of which occur in the Avon Gorge.
However, Proctor & Groenhof (1992) have shown that
S. eminens from near Symonds Yat, Wye valley, differs
genetically from plants from the Avon Gorge and,
moreover, the difference in peroxidase phenotype suggests
that the two populations are not very closely related. There
are morphological differences in the leaves from the two
areas, as was first recognised by E.F. Warburg (1957).
Proctor & Groenhof (1992) have also shown that *S. eminens
sensu lato* from the Avon Gorge has genetic affinities with
trees at Cheddar and Weston-super-Mare, with plants at
Bangor, North Wales (which have been named erroneously
in the past as *S. porrigentiformis*), and with the widespread
S. hibernica. Further study of these plants is required.

Compared with other rare *Sorbus* species, this species
may be more tolerant of shade, though not of dense shade,
under which it may develop a weak stem having a
tendency to decay at the base. A coppice-with-standards
management regime is perhaps ideal (Taylor 1990e).
Conservation management in the Avon Gorge includes the
control of scrub and the removal of competitive alien
species, such as the invasive *Quercus ilex* and *Cotoneaster*
spp., which are increasingly occupying good *Sorbus* habitat
(Taylor 1990f). Several trees of *S. eminens* have been
pruned in order to allow public access along a tow path,
but this may not be damaging in the longer term. If the
railway track in the Gorge is converted, as seems likely, to
a passenger light railway, the construction of this and its
use will be detrimental to the present favourable status of
S. eminens (P.J.M. Nethercott pers. comm.).

M.J. Wigginton

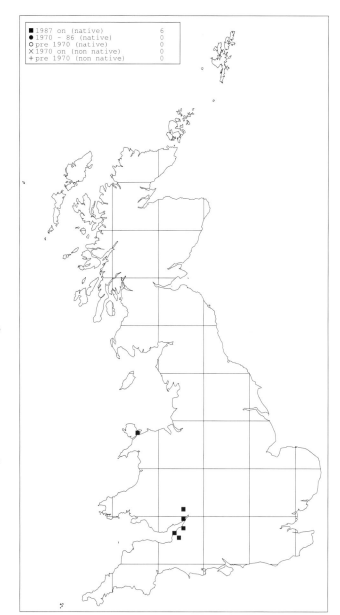

■ 1987 on (native)	6
● 1970 – 86 (native)	0
○ pre 1970 (native)	0
✕ 1970 on (non native)	0
+ pre 1970 (non native)	0

Sorbus lancastriensis E.F.Warburg (Rosaceae)
Whitebeam
Status in Britain: LOWER RISK - Near Threatened.
ENDEMIC.

This endemic small tree or shrub occurs locally within some 30 km of Morecambe Bay. It is found mostly on open Carboniferous limestone screes, crags, scars, cliffs and pavements, usually in small populations. There are, however, at least two sites where *S. lancastriensis* occurs on Silurian slate (Witherslack and Poolbank), and one where it occurs on glacial drift (Roughholme Point).

S. lancastriensis is intolerant of shade and is found mainly at woodland edges or in open woodland, particularly where the canopy is relatively low and there is regular coppicing. It is usually found in association with *Fraxinus excelsior*, *Ulmus glabra* and occasional *Taxus baccata*, with a ground flora including such species as *Anemone nemorosa*, *Helianthemum nummularium*, *Hippocrepis comosa*, *Leontodon hispidus*, *Orchis mascula*, *Potentilla neumanniana*, *Primula vulgaris*, *Thymus polytrichus* and *Viola reichenbachiana*.

It is apomictic, flowers and sets fruit freely, and where grazing pressure is low is able to regenerate successfully.

Populations of up to 50 trees occur on sea-cliffs around Morecambe Bay, together with *S. rupicola* and *S. aria*, though most other populations are smaller than this. There are thought to be about 2,000 trees in all. *S. lancastriensis* is not immediately threatened. Many of its populations are within SSSIs or other protected sites. Elsewhere, it occurs on inaccessible cliffs and steep crags which are relatively free from grazing and potentially damaging activities.

Further details are given in Rich & Baecker (1986, 1992).

C.S. Crook

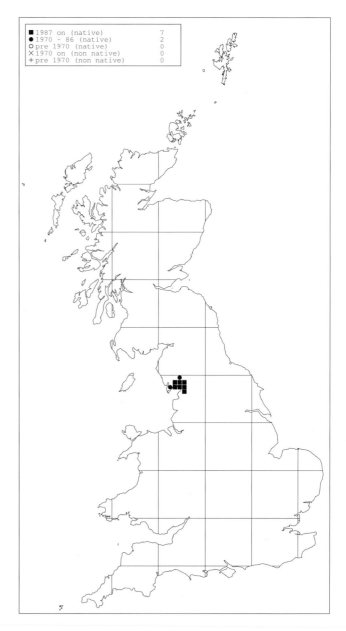

■ 1987 on (native)	7
● 1970 - 86 (native)	2
O pre 1970 (native)	0
✕ 1970 on (non native)	0
+ pre 1970 (non native)	0

Sorbus leptophylla E.F. Warburg (Rosaceae)
Whitebeam, Cerddinen Gymreig
Status in Britain: CRITICALLY ENDANGERED.
ENDEMIC.

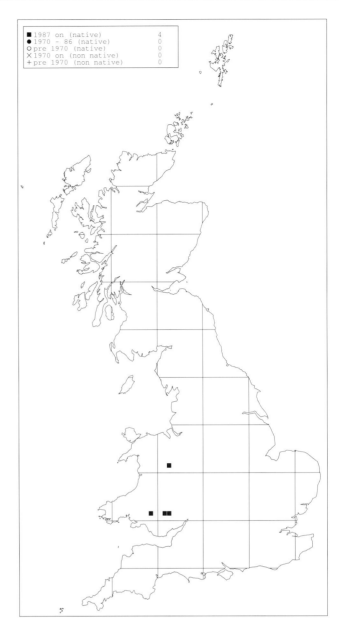

S. leptophylla typically forms a sprawling tree with long, more or less pendulous branches, and is certainly known only on shady crags of Carboniferous limestone at two sites in Breconshire. It grows there in association with related species of *Sorbus* and other trees and shrubs including *Corylus avellana, Fraxinus excelsior, Ilex aquifolium* and *Viburnum lantana*. The ground flora includes *Brachypodium sylvaticum, Campanula rotundifolia, Hedera helix, Pilosella officinarum, Succisa pratensis* and *Teucrium scorodonia*. Fewer than 50 trees of *S. leptophylla* have been recorded in Breconshire; about 16 individuals were recorded on Craig-y-Rhiwarth near Glyntawe in 1988, and about 28 on Craig-y-Cilau. E.F. Warburg (1957) reported a population of *S. leptophylla* on Craig Breidden, Montgomeryshire. Some of those plants have leaves similar in shape to *S. leptophylla*, though smaller. In 1993, P.J.M. Nethercott located two very slender-stemmed specimens, 2-3 m long, growing almost horizontally from a vertical rock face at the north end of Craig Breidden; the leaves were a little smaller than at Craig-y-Rhiwarth but a good match, the very few fruit also seeming right. In other studies (Proctor & Groenhof 1992), the one tree sampled from this population appeared to be identical in peroxidase phenotype with Breconshire *S. leptophylla*, indicating a likely relationship with that species. Further investigation of the Craig Breidden populations is desirable.

S. leptophylla appears to be genetically uniform, but varied in habit and leaf shape, depending on exposure or shading. It is probably at least partially apomictic and is apparently able to set seed freely. Flowers appear in May, with ripe fruit in September. A high percentage of well-formed pollen grains has been observed, with good germination in laboratory conditions (Proctor & Groenhof 1992).

Its two Breconshire sites are SSSIs, and one is an NNR. Most trees of *S. leptophylla* are inaccessible to grazing animals, and the population appears to be stable.

C.S. Crook

Sorbus leyana Wilmott (Rosaceae)

Ley's whitebeam, Cerddinen Darren Fach
Status in Britain: CRITICALLY ENDANGERED. ENDEMIC.

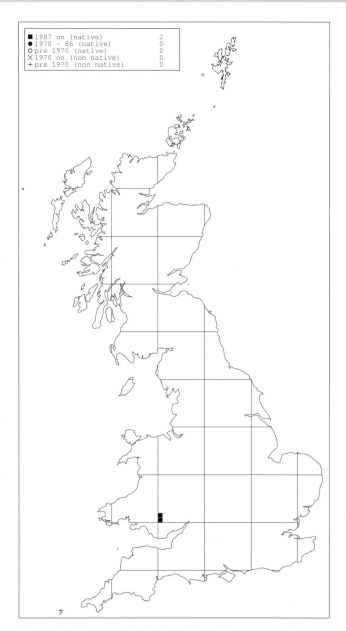

S. leyana is a distinctive, large, deciduous shrub or small tree known only from two localities on Carboniferous limestone cliffs in southern Breconshire. At the first locality, it grows in crevices on vertical or steeply sloping west-facing cliffs, with most of the few trees (13 in 1983) on the upper part of the partly wooded cliffs at an altitude of 280-305 m, above open ashwood on steep scree, and below grassy and rocky slopes rising to 450 m. Associates on the cliffs include *Ilex aquifolium, Rhamnus catharticus* and *Sorbus aucuparia*, with *Crataegus monogyna* and *Fraxinus excelsior* in the wood below. At the second site, on the opposite side of the Taff valley, only three native trees of *S. leyana* are known to survive, on limestone cliffs with an easterly aspect; however seven saplings raised from their seeds (collected in 1959) were planted on the slopes above the cliffs in 1962, and four of them were extant in the late 1980s. Its total known wild or semi-wild population thus consists of about 20 trees. Both sites are protected in SSSIs.

It is thought to be a triploid apomictic microspecies derived from diploid sexual *S. aucuparia* and the tetraploid *S. rupicola*, which is also apomictic (Warburg & Kárpáti 1968; Richards 1975). The leaves of *S. leyana* have two pairs of basal lobes that are free at least halfway to the midrib; the leaves are intermediate in appearance between those of *S. aucuparia* and *Sorbus rupicola*. The creamy-white flowers, densely massed in flat-topped corymbs, appear in late May and early June, and the red fruits follow in September, but are produced rather sparsely, as is usually the case in triploid *Sorbus* apomicts; the pollen-grains have correspondingly low (under 15%) stainability (M.C.F. Proctor pers. comm.). Germination of apparently well-set seeds is poor, with complete failure in some years, although saplings are fairly vigorous when established in cultivation from seeds produced in a 'good' year such as 1976. Morphologically, and also in isoenzyme phenotype (Proctor & Groenhof 1992), *S. leyana* appears to be essentially uniform and well delimited.

It is likely to have originated at or near its present site since the late-glacial period. *S. minima*, which appears to have the same parentage and general cytotype, but is morphologically clearly different, occurs as an endemic on north-facing Carboniferous limestone cliffs from Cwm Claisfer to Craig-y-Ciliau, about 15-20 km to the east. *S. arranensis*, endemic to the Isle of Arran, and three similar Norwegian microspecies are also believed to have the same parentage but are widely geographically separated from *S. leyana* and *S. minima*.

Some former sites for *S. leyana* may have been destroyed by quarrying. It was first recorded, at the eastern site, by Ley in 1896 but was not described as a species until 1934.

Q.O.N. Kay

Sorbus minima (A.Ley) Hedl. (Rosaceae)
Lesser whitebeam, Cerddinen Wen Leiaf
Status in Britain: VULNERABLE. ENDEMIC.

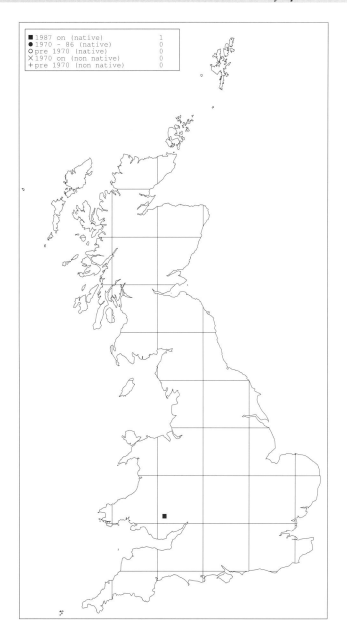

S. minima is a large deciduous shrub endemic to Carboniferous limestone cliffs near Llangattock and Llangynidr in southern Breconshire, where it was first recorded by Augustin Ley, in 1893. It grows in crevices on vertical or steeply sloping north-facing cliffs at altitudes of about 360-450 m, with *Asplenium adiantum-nigrum, Centaurea nigra, Crataegus monogyna, Festuca ovina, Fraxinus excelsior, Hedera helix, Oxalis acetosella, Pimpinella saxifraga, Polypodium vulgare, Sorbus anglica, S. aucuparia, S. leptophylla* and *S. porrigentiformis*. Although the easternmost part of the main range of cliffs on which it grows was badly affected by quarrying in the past, *S. minima* appears to have recolonised some quarried areas and also grows on many smaller unquarried outcrops above and to the west, extending overall along about 5-6 km of the limestone escarpment, often in inaccessible sites. Its total population probably numbers fewer than 350 individual plants; some 305 were counted from Craig-y-Ciliau to Craig-y-Castell in 1976 (and there has probably been little change since) and 17 in the separate western population at Cwm Claisfer in 1988.

Its creamy-white flowers appear in late May and early June and the red fruits are produced in September, more abundantly and containing seeds with higher viability than in *S. leyana*, from which it is morphologically clearly distinct. Like *S. leyana*, it is a triploid apomictic microspecies thought to be derived from *S. aucuparia* and apomictic *S. rupicola* (Warburg & Kárpáti 1968; Richards 1975) and probably originated at or near its present site since the late-glacial period.

Its leaves are smaller, narrower and markedly less divided than in other apomictic *Sorbus* species with the same presumed parentage and have no free or partly free lobes. Like *S. leyana*, it appears to be essentially uniform and well delimited, both morphologically and in peroxidase isoenzyme phenotype (Proctor & Groenhof 1992), but the two species differ quite markedly from one another in both respects. The single native locality for *S. leyana* is about 15 km to the west of the westernmost locality of *S. minima*.

Q.O.N. Kay

Sorbus pseudofennica E.F.Warburg (Rosaceae)
Arran Service-tree
Status in Britain: VULNERABLE. ENDEMIC.

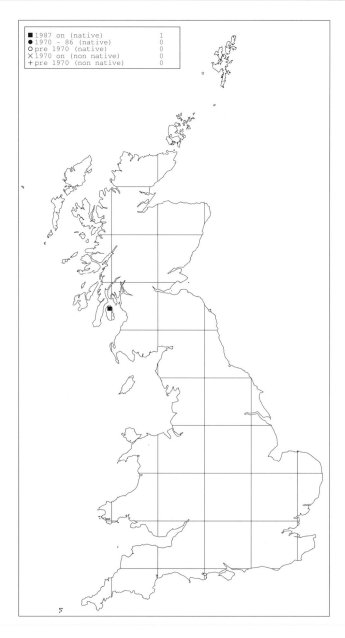

■ 1987 on (native)	1
● 1970 - 86 (native)	0
○ pre 1970 (native)	0
✕ 1970 on (non native)	0
+ pre 1970 (non native)	0

S. pseudofennica is usually found growing from cracks in rocks and on ledges on the sides of stream gorges but occasionally strays on to flatter, well drained ground by the stream bed, or on the slopes above the gorges where grazing pressure permits. Most trees are associated with granite, but there is a population of about 40 trees in the gorge of the Abhainn Beag, a small tributary of Glen Catacol, Arran, growing on metamorphic schist.
S. pseudofennica is normally associated with *S. arranensis*, *S. aucuparia*, *Betula pubescens* and typical moorland plants such as *Calluna vulgaris, Deschampsia flexuosa, Erica cinerea, Galium saxatile* and *Molinia caerulea*, with occasional prostrate *Juniperus communis* and *Salix aurita;* more rarely it is found with *Quercus* cf. x *rosacea* and *Ilex aquifolium*.

S. pseudofennica is believed to be a hybrid derivative of *S. arranensis* and *S. aucuparia*. It is somewhat variable, with some trees not clearly distinguishable morphologically from *S. arranensis* (Hull & Smart 1984). There may have been multiple hybrid events with continuing occasional gene flow. Similar sequences of hybridisation have occurred in Scandinavia, giving rise to a series of clones included under *S. hybrida*, a species commonly found in cultivation in Britain. It is apomictic and fully fertile; like *S. arranensis*, it is usually a small, sometimes multi-stemmed and often stunted tree up to a maximum of 7 m high. In winter both *S. arranensis* and *S. pseudofennica* can be distinguished from intermixed *S. aucuparia* by their more upright growth and darker, purplish twigs and bark. Natural regeneration does occur, but has not been systematically studied.

S. pseudofennica occurs as three more or less discrete colonies in Glen Catacol and two of its tributaries in north-west Arran. Population counts were carried out in 1970/71, 1979-81 and in 1986, but none covered all sites. However, an overall assessment and extrapolation of the figures suggests that the population may have numbered between 400 and 430 trees during that period, about half of which were recorded within the Glen Diomhan NNR. There have been no counts since 1986 and apparently none within the NNR since 1970. On available evidence, no conclusions can drawn on any possible changes in the population size and structure. Experimental planting of saplings within the reserve enclosure has met with limited success.

Palynological studies from the lower part of Glen Diomhan suggest that moorland outside of the gorges has never been much wooded in the last 8,000 years (Steven & Dickson 1991). Nonetheless, it is possible that populations of *S. pseudofennica* might become established on suitable well-drained ground outside the gorges if the numbers of red deer were considerably reduced. Short-tailed voles may also have a significant browsing effect on young plants, especially during the peak season of their population cycles.

Further information is to be found in the account of *S. arranensis* and in Bignal (1980).

A.R. Church and C.S. Crook

Sorbus subcuneata Wilmott (Rosaceae)
Whitebeam
Status in Britain: VULNERABLE. ENDEMIC.

S. subcuneata is very local in open, rocky *Quercus petraea* woodland on Old Red Sandstone, in which other prominent species include *Deschampsia flexuosa, Lonicera periclymenum, Luzula sylvatica* and *Vaccinium myrtillus*. It is known only in the valley of the East Lyn near Lynmouth, North Devon, in a few places on steep slopes facing the sea near Martinhoe and Trentishoe, and on North Hill, Minehead. The largest population occupies a short stretch of the East Lyn valley within a few hundred metres of Watersmeet. There is an outlying population (a dozen or more trees) in and near West Woodybay Wood, Martinhoe, and about a dozen trees in a steep fragment of woodland just above the shore east of Greenaleigh Farm, Minehead. Smaller outlying groups of trees occur at Neck Wood, Trentishoe, Invention Wood, Martinhoe, and on the outskirts of Minehead.

It is a small tree, up to 12 m tall, differing from the related *S. devoniensis*, with which it grows at several localities, in its narrower, later-flushing, more cuneate-based leaves with whiter tomentum on the underside, a more open canopy and slightly smaller, more orange fruits. It is apomictic. The constant peroxidase phenotype suggests substantial genetical uniformity, but there may be a slight difference in leaf shape between the Minehead trees and those further west. Fruiting varies from year to year, but in good seasons abundant fruit is produced. *S. subcuneata* reproduces from seed, and young trees are frequent.

The Devon sites are nearly all in SSSIs or on National Trust properties and should be reasonably safe. There seems no likely threat to the population near Greenaleigh, Minehead, but the few trees above High Town, Minehead, are potentially threatened.

M.C.F. Proctor

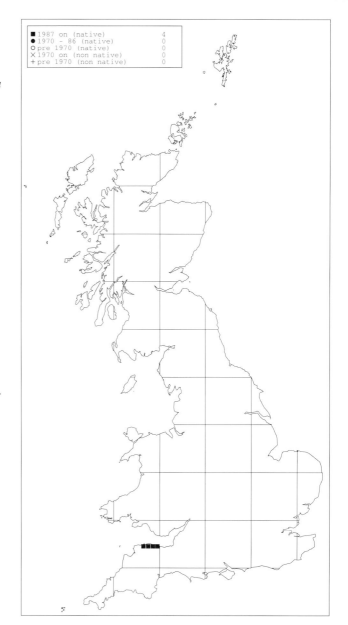

■ 1987 on (native)	4
● 1970 – 86 (native)	0
○ pre 1970 (native)	0
✕ 1970 on (non native)	0
+ pre 1970 (non native)	0

Sorbus vexans E.F. Warburg (Rosaceae)
Whitebeam
Status in Britain: VULNERABLE. ENDEMIC.

S. vexans is a small tree to about 10 m tall, often multi-stemmed from the base. It grows mainly in open *Quercus petraea* woodland on thin acid soils with a varied ground flora including such species as *Agrostis vinealis, Calluna vulgaris, Deschampsia flexuosa, Dryopteris dilatata, Lonicera periclymenum, Luzula sylvatica, Pteridium aquilinum* and *Vaccinium myrtillus*. It occurs on steep, rocky Old Red Sandstone slopes facing the sea on the coast between Culbone, Somerset, and just west of Trentishoe, Devon. There are some occurrences on open cliffs, as at Woody Bay, Martinhoe, and there is a population (on Lynton Slates) a little back from the sea in the East Lyn valley near Lynmouth.

S. vexans is apomictic and appears uniform in morphological characters and peroxidase phenotype. As in other *Sorbus* species, fruiting is erratic from year to year, but plenty of fruit is produced in favourable seasons, and small (presumably young) saplings are frequent. *S. vexans* regrows readily from the base following damage.

This species is accompanied in its coastal cliff woodland sites by another, unnamed, apomictic microspecies with similar leaves but of somewhat more vigorous habit, with a broader, darker-red fruit, and a different peroxidase phenotype. This tree ('Taxon D' of Proctor *et al*. 1989) is probably somewhat more numerous than *S. vexans*, and has a similar but slightly wider geographical range (from just east of Culbone to Combe Martin). It is commoner on non-wooded slopes than *S. vexans*, and reproduces freely from seed. There is no significant threat at present to either *S. vexans* or its companion apomict. Uninformed woodland management and excessive spread of *Rhododendron ponticum* could be potential hazards at some sites.

M.C.F. Proctor

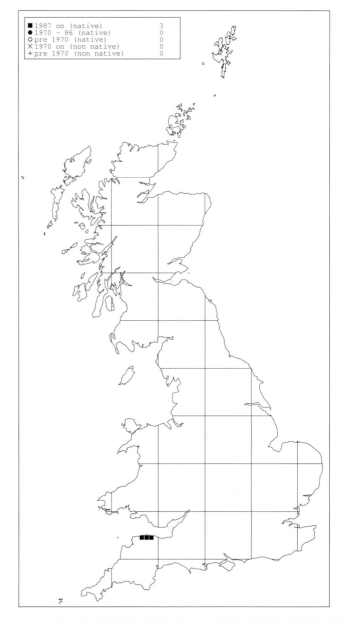

■ 1987 on (native)	3
● 1970 – 86 (native)	0
○ pre 1970 (native)	0
✕ 1970 on (non native)	0
+ pre 1970 (non native)	0

Sorbus wilmottiana E.F. Warburg (Rosaceae)
Whitebeam
Status in Britain: CRITICALLY ENDANGERED. ENDEMIC.

S. wilmottiana is one of the rarest of the *Sorbus* species endemic in Britain. It is restricted to the Avon Gorge, where it occurs on both the Gloucestershire and the Somerset sides. In the open it forms a tree up to about 8 m, but it is shorter, smaller or shrubby in less favourable sites. On woodland margins and cliff edges its associated species include *Bromopsis erecta, Carex humilis, Festuca ovina, Fraxinus excelsior, Sorbus aria* and *S. bristoliensis*.

Such is the secrecy surrounding the number and locations of extant trees that accurate information on the population cannot be given here, but it is understood that there are very few individuals. This species has the smallest population of any named native species in the Gorge and has long suffered the worst damage of any *Sorbus* there. In 1996, P.J.M. Nethercott reported possible thefts of two plants, and another two were accidentally cut down during conservation work, though both of the latter are now showing signs of regeneration.

The present management of the area is aimed primarily at the conservation of the rare plants of the Gorge, and mainly involves the removal of alien species, including *Cotoneaster* spp. and *Quercus ilex*, along with some scrub clearance to open up the habitat (Taylor 1990f). Tight management of conservation work is essential.

M.J. Wigginton & M.A.R. Kitchen

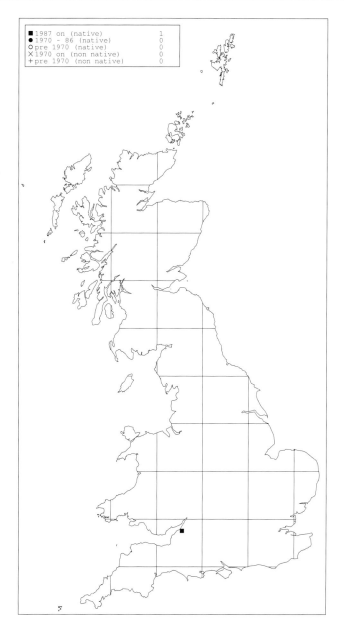

■ 1987 on (native)	1
● 1970 - 86 (native)	0
O pre 1970 (native)	0
✕ 1970 on (non native)	0
+ pre 1970 (non native)	0

Spergularia bocconei (Scheele) Graebner (Caryophyllaceae)
Greek sea-spurrey, Troellys Boccone
Status in Britain: CRITICALLY ENDANGERED.
Status in Europe: Not threatened.

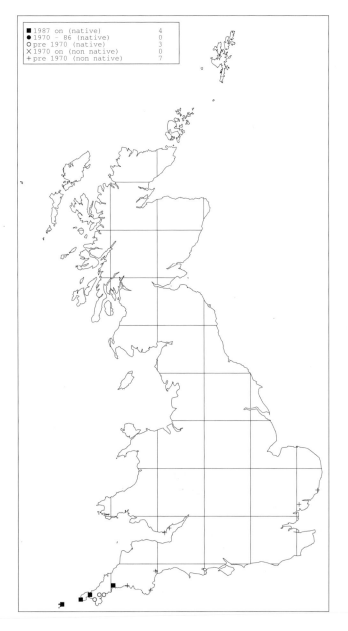

Despite some authorities having regarded this species as probably native in south-west England (e.g. Clapham *et al.* 1987), it seems more likely to be an ancient denizen. *S. bocconei* is mainly a plant of warmer climes and is particularly abundant around the Mediterranean; its British sites are all coastal, mostly at or near trading ports. In view of the long history of shipping trade between the Mediterranean and Cornish ports, *S. bocconei* could be a very ancient introduction.

Since 1950, *S. bocconei* has been recorded at about five sites on the mainland and one or two in the Isles of Scilly. It has now gone from Devon and seemingly also from the Isles of Scilly, and only two or three small populations are extant on the mainland. The largest (of about 50 plants in 1989 and 1994) is on sandy waste ground near the beach at Par, growing there in company with such species as *Agrostis stolonifera, Coronopus didymus, Lolium perenne, Plantago major* and *Polygonum aviculare*. A nearby population at Par china clay docks, which numbered in the hundreds in 1983, has been obliterated by concrete and tarmac. Only a few small plants were found in 1989, two of them right by the wall of a warehouse, but none has been seen since. In 1989, a few individuals were noted in a car-park at Land's End, in a weedy community with *Matricaria matricarioides, Poa annua* and *Polygonum aviculare* (FitzGerald 1990c), but not since. A few plants were seen on St Michael's Mount in 1990, but it is not known whether the plant still occurs there.

S. bocconei is an annual or biennial, resembling *S. rubra*, but differing in characters of the flower and inflorescence. It has a long flowering period between May and September. Little appears to be known of its biology.

Whatever its status, it is certainly one of our most threatened plants, not only because of its very few sites and small populations, but also because it is found in marginal or ruderal habitats in communities of common weeds. Clearly no such site is likely to merit statutory protection, and colonies could be destroyed overnight by municipal tidying or other cosmetic attention. Conservation is likely to be possible only through local vigilance combined with voluntary agreements on land management.

S. bocconei is reasonably common on Guernsey and Jersey (perhaps also surviving on Sark) and is generally regarded as a probable introduction in those islands (McClintock 1984; Le Sueur 1984). Elsewhere it occurs in south-western and southern Europe, with its stronghold around the Mediterranean, extending from Spain and Portugal to Greece and eastwards to Iran. It is chiefly restricted to coastal regions.

M.J. Wigginton

362

Stachys alpina L. (Lamiaceae)
Limestone woundwort, Briwlys y Calchfaen
Status in Britain: ENDANGERED. WCA Schedule 8.
Status in Europe: Not threatened.

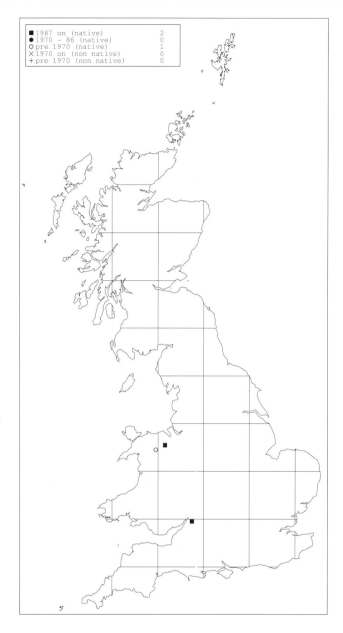

■ 1987 on (native)	2
● 1970 - 86 (native)	0
○ pre 1970 (native)	1
× 1970 on (non native)	0
+ pre 1970 (non native)	0

S. alpina is principally a plant of open woodland, glades, wood-borders, tracks and hedgebanks, growing on thin soils overlying calcareous rock, and usually in sunny and sheltered locations. A wide range of associated species include *Arrhenatherum elatius, Bromopsis ramosa, Geum urbanum, Heracleum sphondylium, Mercurialis perennis, Silene dioica, Stachys sylvatica* and *Urtica dioica*.

This species is a perennial, up to 1 m tall, flowering between June and August. The seed seems to be adapted to long periods of dormancy. It will not immediately germinate after dry storage but germination can be triggered in various ways, including chilling, perforation of the seed integuments and treatment with gibberellin (Pinfield *et al.* 1972). It can be invasive under garden conditions (J.M. Brummitt pers. comm.).

British records are from only three hectads, one in Gloucestershire and two in Denbighshire (Wilson 1927). Since 1980, *S. alpina* has occurred in small numbers in at least eight locations within a single 1 km square near Wotton-under-Edge, Gloucestershire. One small population (30-40 flowering stems in 1993) occurs by a track at the edge of a conifer plantation, and others have occurred sporadically in nearby woods. At its most regular site, in a lane-side hedgebank, numbers have ranged from a few individuals to more than a hundred plants during the past 20 years. However, this population is artificially maintained by collecting seed annually and sowing it into specially created bare patches (Taylor 1990a). In Denbighshire, one site was lost in about 1960, probably because of road-widening, but two plants were discovered in 1975 at a second site near Cilygroeslwyd (Brummitt 1981). However, both plants have now apparently gone, and the only ones now present are those that were grown *ex situ* and transplanted. In 1990, two further sites were found close by the first at Cilygroeslwyd, one holding about 14 plants in 1994 and the other only one individual. Natural populations of *S. alpina* seem to be in slow decline in Denbighshire, though with some recruitment (Evans & Ellis 1994a).

Plants usually appear following scrub or woodland clearance, or other disturbance, almost certainly from long-buried seed. It seems likely, therefore, that with appropriate management colonies could be resurrected at historic sites. Plants may persist for several years, provided that competition for light and space is not too intense. Colonies could be lost through the cessation of management or by unwitting damage: for example, a

colony in Denbighshire was adversely affected by herbicide targeted at nettles. Populations on Wildlife Trusts sites in both counties are currently being maintained by intensive management, including coppicing, herbicide treatment, rotavation, seed sowing and/or transplantation.

S. alpina occurs throughout western, central and southern Europe except the Mediterranean, northwards to Germany and Wales and eastwards to the Caucasus.

D. Evans

Stachys germanica L. (Lamiaceae)

Downy woundwort, Briwlys Tewbannog
Status in Britain: ENDANGERED. WCA Schedule 8.
Status in Europe: Not threatened.

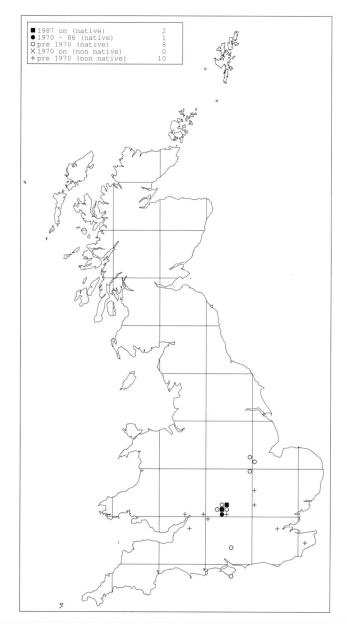

■ 1987 on (native)	2
● 1970 – 86 (native)	1
○ pre 1970 (native)	8
✕ 1970 on (non native)	0
+ pre 1970 (non native)	10

Past and present habitats of *S. germanica* include woodland edges, old hedgerows, road verges, ditches, green lanes and stony fields. Of the current four sites, all of which are in oolitic limestone country, two are in the verges of green lanes and two in game-crop fields cultivated from ancient grassland. In the green lane sites, where plants grow between a hedge of woodland origin and limestone grassland, associated species include *Bromopsis erecta, Centaurea nigra, C. scabiosa, Cirsium acaule, Clinopodium vulgare, Galium mollugo, G. verum, Helianthemum nummularium, Knautia arvensis, Mercurialis perennis, Primula veris, Stachys sylvatica* and *Viola riviniana*. In the game-crop fields, one of which is generally sown with canary-grass or buckwheat and the other invariably with thinly-sown wheat, *S. germanica* may occur not only at the edges but, when conditions are favourable, throughout the crop. These fields periodically lie fallow, and the mixture of grassland and arable species reflects the past history of these sites.

S. germanica is biennial or occasionally a short-lived perennial, with flowers normally opening from July onwards, although they can open in mid-June in hot summers (Dunn 1997). Pollination is mainly by bumblebees, and seed production is variable. Germination begins in late spring, and if conditions are favourable, continues into early autumn. Open ground is essential for germination (which takes place almost exclusively around the parent plants) and for the establishment of overwintering rosettes.

S. germanica was first recorded in 1632 by a London apothecary, Leonard Buckner, who found it "wilde in Oxfordshire in the field joyning Witney Parke, a mile from the Towne". Reliable 18th, 19th and early 20th century records show that it was probably native or well-established in North Hampshire, Oxfordshire, Northamptonshire and South Lincolnshire, though it is likely to have been more widespread in parts of southern Britain than records suggest (Marren 1988). A declining species, it is now confined to West Oxfordshire. There, Druce (1886) noted numerous locations, a number of these being in the vicinity of Brize Norton (where it grew "in the cornfields, plentifully"), Charlbury, Minster Lovell, Witney and Woodstock. Between 1983 and 1990 the number of known sites fluctuated between three and five. In 1993 there were four, following the discovery of a new site in 1991. Populations vary greatly from year to year; a minimum of one and a maximum of 461 flowering stems having been recorded at any one site during the past 12 years. The plant thrives best in warm, open situations, and is a poor competitor, disappearing as soon as the sward closes or scrub encroaches. Its characteristic erratic appearance, sometimes after long absences, is often associated with accidental or deliberate disturbance of the soil.

At all stages of growth, plants are susceptible to damage or destruction by predators, including molluscs, rabbits, and small mammals, which at one site in 1990 stripped the seeds from over 100 stems. Conservation action has been directed, where possible, towards the maintenance of open or disturbed ground, and between 1989 and 1996 such 'recovery' work led to increasing populations and enhanced seed-banks at three sites. Present sites are not considered to be under serious threat. As seed remains viable for many years, the opening up of old sites may stimulate germination. Observations on its life cycle and ecology are found in Dunn (1987, 1991, 1997).

S. germanica is widespread throughout central, western and southern Europe, reaching its northern limit in Britain. It occurs in a wide range of habitats on calcareous soils, including the edges of forests, in felling areas, hedges, meadows, rocky hills and quarries (Hegi 1964). It also occurs throughout the Mediterranean region to North Africa and eastwards to central-southern Russia.

A.J. Dunn

Taraxacum Weber (Asteraceae)
Dandelion, Beàrnan Bride, Dant y Llew

Like *Hieracium, Rubus* and *Alchemilla, Taraxacum* is a largely agamospermous genus in Britain, reproducing clonally from seed. However, unlike many members of these other genera, the majority of dandelions are opportunistic weeds that are often extremely widespread and abundant in open or grassy sites in the lowlands. Not surprisingly in a genus with such anthropogenic tendencies, many species are clearly adventive aliens that probably have their native areas in northern or eastern Europe. About 236 species of *Taraxacum* have been recorded from Britain, of which it is thought that some 100 (42%) are introduced. Only 42 species (18%) are endemic, a much smaller proportion than in the hawkweeds and brambles.

Nevertheless, dandelions do occupy semi-natural habitats, and various species specialise in such communities as fens, sand-dunes, heaths, moorland, mountain ledges and even the edges of saltmarsh. A few of the rarest and most threatened native species are listed below, together with brief notes as to their British habitat and distribution. Further information is given in Dudman & Richards (1997).

Species	British habitat and distribution
T. beeftinkii Hagendijk, van Soest & Zevenb.	In two saltmarshes in Essex, where nearly extinct. Occurs in similar sites in the Netherlands where it is also very local.
T. cenabense Sahlin.	Known only from one locality in West Sussex and another in northern France.
T. cherwellense A.J. Richards.	Restricted to a few grasslands in southern England where it is apparently very rare. Has almost disappeared from the type locality in Oxford. Endemic.
T. clovense A.J. Richards.	Known only from a single, well known and heavily-protected mountain ledge in Angus. Endemic.
T. cymbifolium H. Lind.	In Britain, known only from a single site on Ben Lawers. Widespread through the high arctic, almost absent from Scandinavia.
T. geirhildae (Beeby) Palmer & Scott.	Known for 80 years from a single site in Shetland. Endemic.
T. hygrophilum van Soest.	Restricted to a single water meadow in Kent, but known from similar sites in the Netherlands.
T. scanicum Dahlst.	Acidic species-rich grassland; native in the Breckland and the Channel Islands.
T. serpenticola A.J. Richards.	Restricted to the Unst serpentine, Shetland. Endemic.
T. tanylepis Dahlst.	Possibly restricted to a single marshy locality in Orkney. Endemic.

A.J. Richards

Tephroseris integrifolia ssp. *maritima* (Syme) Nordenstam (Asteraceae)

Senecio integrifolius ssp. *maritimus* (Syme) Chater
South Stack fleawort, Chweinllys Arfor
Status in Britain: VULNERABLE. ENDEMIC. WCA
Schedule 8.

This local endemic is known only near Holyhead, Anglesey, where it is found on mildly acidic soils (pH 5.5.-7.0) on maritime cliffs of Pre-Cambrian rock. It is found on the grassy cliff-tops, on ledges and in crevices on the steeper parts of the cliff-faces, often in precipitous, inaccessible places. Most populations are found on slopes with an aspect between south and north-west, and rarely is it found on north to north-east facing slopes (Smith 1979). Plants are thus sheltered against the extreme effects of cold winds. The most constant associates are *Agrostis stolonifera*, *Cochlearia danica*, *Festuca rubra*, *Holcus lanatus*, *Scilla verna*, *Silene uniflora* and *Tripleurospermum maritimum*, but among many others are *Anthyllis vulneraria*, *Armeria maritima*, *Daucus carota*, *Hypochaeris radicata*, *Lotus corniculatus*, *Plantago maritima*, *Potentilla erecta*, *Sedum anglicum* and *Thymus polytrichus*.

T. integrifolia ssp. *maritima* is usually a much taller plant than ssp. *integrifolia* and may exceptionally be a metre tall. Flowering is generally between early May and early July, but only 40% of the plants produce flowers in any particular season. Reproduction appears to be mainly by seed. However, there is some evidence that lateral buds in the axils of basal leaves may grow into small shoots which root and become detached from the parent plant, eventually forming such off-sets as are described for ssp. *integrifolia* (Smith 1979). On average, stems of ssp. *maritima* have more flowering heads than does ssp. *integrifolia*, producing more fruit per head and twice as much seed. Seeds germinate readily soon after ripening, at least in laboratory conditions. However, little is known of its reproductive biology.

This plant is found in many places between Porth Dafarch and South Stack (Roberts 1982), the greater part of the population occurring on Penrhyn Mawr. The total population is estimated at several thousands of plants. There appear to be few threats. Rock-climbing on the cliffs may pose a future threat, but the fact that most colonies lie within an RSPB reserve gives considerable protection from human interference.

Although it was formerly said to occur near Brough, Westmorland (Clapham *et al.* 1962), Smith (1979) concluded that it was not identical to the Holyhead plant and probably represented a second endemic subspecies, now extinct. The third subspecies (ssp. *integrifolia*) is not endemic.

R.H. Roberts and G.H. Battershall

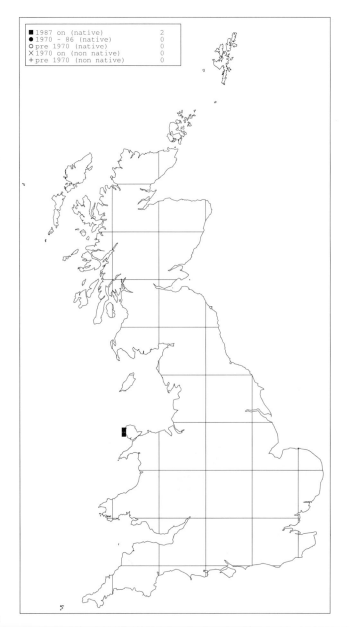

■ 1987 on (native)   2
● 1970 - 86 (native)   0
○ pre 1970 (native)   0
✕ 1970 on (non native)   0
+ pre 1970 (non native)   0

Teucrium botrys L. (Lamiaceae)
Cut-leaved germander
Status in Britain: VULNERABLE. WCA Schedule 8.
Status in Europe: Not threatened.

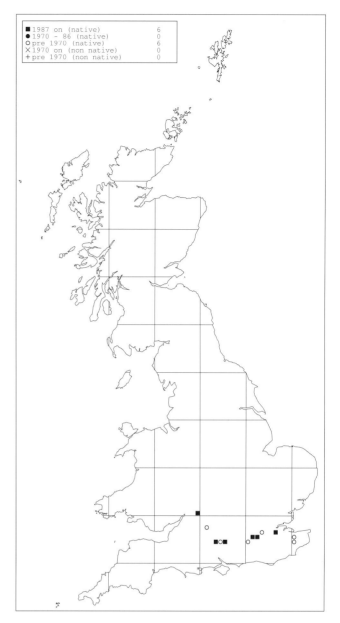

T. botrys is a plant of bare or sparsely vegetated places on chalk and limestone. It has been recorded from a variety of habitats, including open grassland, arable field margins, chalk and limestone spoil tips, a disused chalk quarry, and open fallow. A very open community is essential, as it is intolerant of shade or competition. A wide range of associated species include *Ajuga chamaepitys*, *Carex flacca*, *Cerastium pumilum*, *Chaenorrhinum minus*, *Euphorbia exigua*, *Galium parisiense*, *Gentianella amarella*, *Kickxia spuria*, *Leontodon hispidus*, *Pilosella officinarum*, *Thymus polytrichus* and *Vulpia unilateralis*.

In Britain, it is predominantly a monocarpic biennial (occasionally an annual) that reproduces exclusively by seed. Seed germinates mainly in spring and autumn, but seedlings may appear at other times of year, depending on the weather, and plants from a spring germination flower in the same year. It flowers from July to September and is pollinated by bees, though self-pollination is possible (Clapham *et al*. 1987). Open, broken ground is essential for germination. Since the seeds are relatively heavy and tend to fall within a few centimetres of the parent plant, *T. botrys* is a poor coloniser.

At its peak, *T. botrys* was recorded from 12 hectads, but since 1930 many sites have been lost and it is now extant in only six hectads, and at only six sites. It is present at Upper Halling in West Kent, the Chipstead valley and Box Hill in Surrey, Micheldever and Harewood Forest in Hampshire, and near Stroud in Gloucestershire. *T. botrys* has been extinct for more than 40 years at Uffcott Hill in Wiltshire, at Godmersham in Kent and at Selsdon in Surrey. Populations vary considerably from year to year, depending on the state of the habitat. In recent years, those in Gloucestershire, Hampshire and Surrey have numbered in the thousands in response to cultivation or other appropriate conservation management. Whilst some other populations are currently small, recovery can be rapid following soil disturbance.

In the absence of domestic livestock, habitat management has included local harrowing and cultivation, scrub cutting and the cutting of turf (Winship 1994a). The main threat to the species is the cessation of such conservation management, and agricultural intensification, which has reduced the extent of one of the Hampshire populations. Urban development has encroached on to a Surrey site. However, five of the six populations are safeguarded within SSSIs and two are managed by county Wildlife Trusts.

T. botrys is widespread in open habitats in southern, western and central Europe, extending northwards to Britain and eastwards to Poland and Romania. It is also recorded from Algeria.

H.R. Winship

Teucrium chamaedrys L. (Lamiaceae)

Wall germander, Chwerwlys y Mur
Status in Britain: ENDANGERED.
Status in Europe: Not threatened.

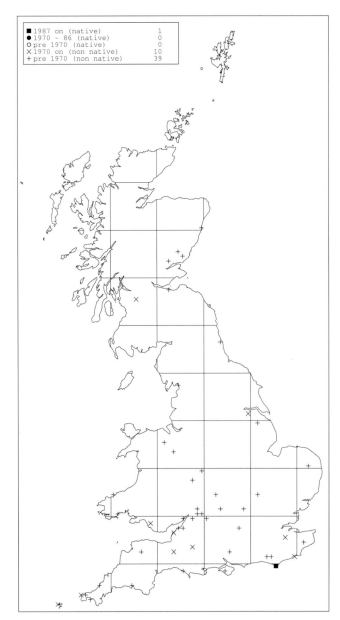

T. chamaedrys mostly occurs as a wall-denizen in Britain, sometimes long-persisting, and is known to have originated from gardens or from apothecary's plots around old settlements. The most intriguing colony is at Cuckmere Haven, East Sussex, where it grows in cliff-top chalk grassland in association with such species as *Carlina vulgaris, Euphrasia nemorosa, Koeleria macrantha* and *Thymus polytrichus.* The plants are small, almost prostrate, and scattered in the short turf.

It is a long-lived woody perennial, lacking rhizomes, and flowering between July and September. It is pollinated by bees, and though protandrous, self-pollination is possible.

T. chamaedrys is apparently now known from only seven sites in Britain, and there is strong evidence that it is an introduction at six of these. However, F. Rose (pers. comm.) considers the habitat and growth form of the Cuckmere plants to be identical to those in presumed native sites in northern France, supporting the assertion that it is native at this site (e.g. Stace 1991). The record is mapped as such, though further studies are needed to establish the exact taxonomic identity of these plants. Garden plants are often more robust and less hairy, and are probably the hybrid *T. chamaedrys* x *T. lucidum*; some or most naturalised plants might also be this hybrid (Stace 1991).

T. chamaedrys occurs throughout western, central and southern Europe, extending to eastern Asia and North Africa. Across its range, a large number of variable subspecies have been described, though with much confusion in nomenclature, particularly in North Africa and in China (R. Clement pers. comm.).

M. Briggs

Teucrium scordium L. (Lamiaceae)
Water germander
Status in Britain: VULNERABLE. WCA Schedule 8.
Status in Europe: Not threatened.

T. scordium is a lowland plant of dune-slacks, river banks, ditches and pits on moist calcareous soils. Dune-slacks now hold much the largest populations in Britain, where it grows in open communities around the damp margins of dune-slack pools. Common associates include *Agrostis stolonifera, Anagallis tenella, Hydrocotyle vulgaris, Pulicaria dysenterica, Ranunculus flammula* and *Salix repens*. Though open habitats are most favourable, it will survive for a time in closed communities dominated by grasses (even amongst *Rubus* and *Salix* bushes), but perhaps only where grasses and shrubs are kept in check by rabbit-grazing. There is some evidence that rabbits graze selectively, avoiding the *T. scordium*. In contrast, the plant seems to survive well where there is a thick moss carpet. In its other habitats, it also occurs in fairly open, semi-aquatic communities, with such species as *Carex acutiformis, Filipendula ulmaria, Iris pseudacorus, Lythrum salicaria, Mentha aquatica* and *Rumex hydrolapathum*.

T. scordium is a perennial plant with a stoloniferous creeping rootstock and erect or decumbent stems up to 50 cm high, but often much shorter. The pale pink-purple flowers appear from June to October. Flowering and seed production may be poor and erratic in dry years, and the plant requires open sites for germination and seedling growth. Colonies may be irregular in appearance, re-establishing themselves in areas bared of vegetation. Studies in the Netherlands have shown that dune-slack colonies of *T. scordium* decline during periods of drought, and years of above average rainfall are beneficial (Laan & Smant 1985).

This species has suffered a great decline, and currently occurs at only two sites, in Devon and Cambridgeshire. The only other post-1970 record is from Stallode Wash, Suffolk where it was apparently unknown between about 1830 and 1976. However, the newly-discovered colony persisted only until 1979, the site having become dominated by *Phragmites australis*. It formerly occurred in several counties from Berkshire to Yorkshire, though the East Anglian fenland appears to have been its stronghold. Sites have been lost mainly through drainage and land claim. Devon now holds by far the largest populations, several thousand plants occurring in a dozen or more dune-slacks at Braunton Burrows. The population is probably stable at the present time, colonies appearing at new sites whilst disappearing at others as a result of competition. However, a possibly falling water table and overgrowth of shrubs may threaten it there. In the Cambridgeshire pit, the colonies of *T. scordium* have declined recently because of overgrowth by more vigorous species. Ground must be cleared to provide the right

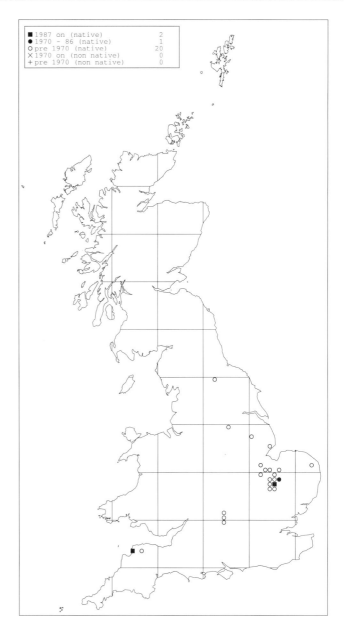

■ 1987 on (native)	2
● 1970 - 86 (native)	1
○ pre 1970 (native)	20
✕ 1970 on (non native)	0
+ pre 1970 (non native)	0

conditions for the plant to re-establish. Water quality is probably important to the plant's survival, and eutrophication appears to be particularly detrimental.

T. scordium is still abundant in some parts of western Ireland but is now extinct in the Channel Islands. It occurs in most European countries, except Iceland. However, it is declining in many, and is regarded as endangered or vulnerable at least in the Netherlands, Switzerland, the Czech Republic, Germany and Denmark. It also occurs in western Siberia and the Aral-Caspian region.

M.J. Wigginton

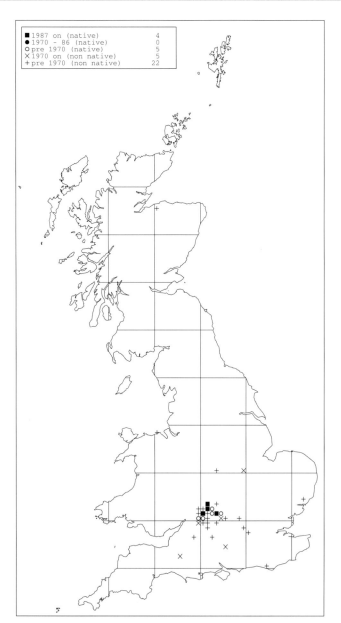

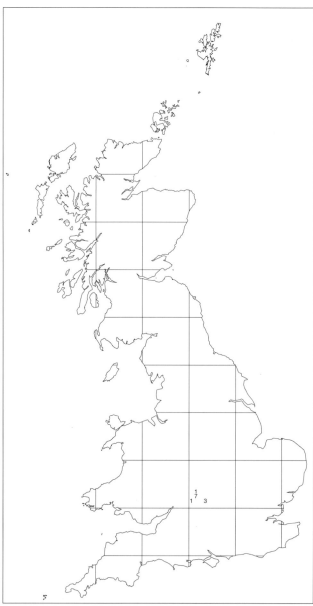

```
■ 1987 on (native)          4
● 1970 - 86 (native)        0
○ pre 1970 (native)         5
✗ 1970 on (non native)      5
+ pre 1970 (non native)    22
```

Thlaspi perfoliatum L. (Brassicaceae)

Perfoliate pennycress
Status in Britain: VULNERABLE. WCA Schedule 8.
Status in Europe: Not threatened.

Of all the species associated with limestone,
T. perfoliatum is one of the rarest. It is a plant of bare,
open, broken limestone in natural habitats and in artificial
sites such as quarries and railway banks. The only species
with which it can be said to grow regularly is *Erophila verna*,
but other associates are *Anthyllis vulneraria, Centaurea nigra,
Cerastium glomeratum, Hippocrepis comosa, Lotus corniculatus,
Myosotis arvensis, Pilosella officinarum* and *Thymus
polytrichus*.

This species is an annual. It is variable in size, but
often no more than 3 cm high when the first flowers open,
which may be as early as late March (Rich 1991). At this
time the inflorescence is more or less sessile in the leaf
rosette and individual plants are easily overlooked, but

plants subsequently become taller and much more
noticeable. Since it has no special means of seed dispersal
its ability to spread is limited, and, once established, it
must rely on continual disturbance of the soil to keep
down competitors and to maintain an open habitat. The
longevity of seed in the soil and the size of seed-banks are
not known (Baskin & Baskin 1979).

T. perfoliatum has been recorded from a total of 35
hectads in Britain and is considered to be native in nine of
them. Within these hectads, a recent evaluation has found
that it has probably been recorded in about 45 native and
37 introduced sites. Since 1986, *T. perfoliatum* has been
recorded in only nine native sites (seven in Gloucestershire
and two in Oxfordshire) and three introduced sites. It was
extant in all nine native sites in 1996 (Rich, Lambrick,
Kitchen & Kitchen in press). Populations can vary greatly
in size according to climatic and habitat conditions; a
particular site may, for instance, hold a few tens of plants
and many thousands in successive years. National totals in

1992, 1994, 1996 and 1997 were about 10,000, 4300, 71,000 and 8,700. Most of the large increase in 1996 was accounted for by 50,000 plants at a site not counted in 1994 and 1995. However, every site showed an increase in 1996, this being attributed to the long summer drought of 1995, which reduced competitors and created open ground for germination (Rich, Alder *et al.* 1996). There are few records of *T. perfoliatum* away from its main area, and these are generally of non-persistent colonies in ruderal habitats. However, colonies on railway banks in Somerset and Rutland have persisted for more than 20 years (Rich, Kitchen & Kitchen 1989).

Losses in the past have been due to habitat destruction and degradation, including agricultural improvement, herbicides, ploughing, grubbing up of hedges and invasion by scrub. The maintenance of open communities with areas of dry, open, broken soil is essential for the survival of *T. perfoliatum*. In some sites, grazing by domestic livestock or rabbits achieves this, but in the absence of grazing other conservation management is essential. At some sites it is necessary to disturb the ground more or less annually to remove potential competitors and provide suitable conditions for germination. Some populations are under threat because of the lack of appropriate management.

It occurs widely in southern Europe, becoming less common as it reaches its northern limits in Belgium and central Germany. There are scattered populations further north in southern Sweden and on the Baltic coast. It also occurs in North Africa and the Near East, and as an introduction in North America.

A.J. Showler

Thymus serpyllum L. (Lamiaceae)
Breckland thyme
Status in Britain: LOWER RISK - Near Threatened.
Status in Europe: Not threatened.

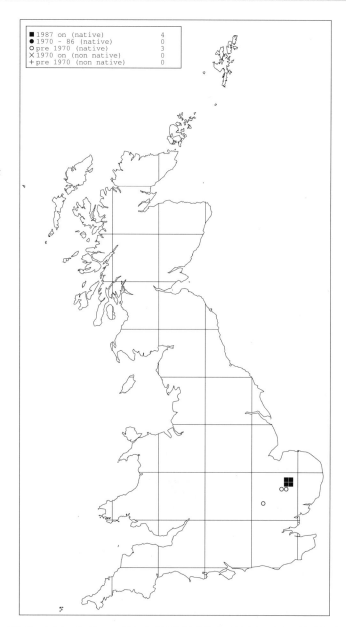

T. serpyllum is confined in Britain to the Breckland, where it is more or less restricted to dry, sandy soils overlying chalky drift or on inland dunes (Pigott 1955). It grows in areas of short, open grassland, particularly on heaths grazed by rabbits or sheep, and frequently on bare patches between rabbit-grazed bushes of *Calluna vulgaris*. It is tolerant of acidic sands and is frequently seen in association with *Filago minima, Ornithopus perpusillus, Teesdalia nudicaulis* and *Cladonia* species, together with *Festuca ovina, Galium verum, Koeleria macrantha* and *Pilosella officinarum*. It is light-demanding and intolerant of competition.

It is a slow-growing mat-forming perennial, extending by runners. Flowering is in July and August, later than *T. polytrichus*, which also has non-flowering runners, but at the same time as *T. pulegioides*, which lacks them. Reproduction is by vegetative spread and by seed. The rather woody runners root at the internodes and in suitably open habitats may form extensive and profusely flowering tracts in late summer. Identification of this species can be difficult, as plants are often stunted by drought, nutrient-deficiency and grazing (Pigott 1954).

Sites for *T. serpyllum* were lost as heathland was ploughed for cultivation or forestry. Furthermore, the decline in rabbit and sheep-grazing also reduced the areas of short, open turf. Surveys undertaken in 1991 and 1993 (Leonard 1993) have shown that heaths and roadside verges in West Suffolk support 15 stations for *T. serpyllum*. It occurs in three sites in West Norfolk but no longer occurs in Cambridgeshire. At the present time, the closely mown turf surrounding the runways at RAF Lakenheath make this its most extensive known site. The recovery of the rabbit population and increase in sheep-grazing in Breckland should ensure the creation and maintenance of the bare ground that it requires for germination and vigorous growth (Pigott 1955).

Unlike the British populations, *T. serpyllum* is variable in Europe, where several subspecies and varieties are recognised. The type subspecies ranges from England, north-east France and the Netherlands, eastwards to Germany, southern Scandinavia and north and central European Russia. Its habitats in Europe are similar to those of Breckland, sharing such species as *Herniaria glabra, Silene conica* and *Veronica spicata*.

Y. Leonard

Tordylium maximum L. (Apiaceae)
Hartwort
Status in Britain: ENDANGERED.
Status in Europe: Not threatened.

T. maximum seems to have become established along the Thames valley some time before 1670, having been recorded in Middlesex between St James and Chelsea, and in two or three places between Isleworth and Twickenham, persisting at one of the latter until at least 1837 (Kent 1975). A single plant, presumably casual, was recorded from Esher in 1871. In 1875 it was again found close to the Thames, on the verges of Fort Road in the vicinity of Tilbury Fort, South Essex, where it persisted, despite heavy predation for herbarium specimens (Crompton 1974-86), until destroyed by a pipe-laying operation in 1984. Its discovery on south-facing slopes by the Thames at Benfleet, also in South Essex, at two separate locations in 1949 and 1966 respectively (Jermyn 1974), suggests that *T. maximum* may be an overlooked native plant that firmly established itself during (or perhaps before) the climatic optimum of the sixteenth century - along the broken alluvial banks and clay foothills of the north bank of the Thames, together with other continental thermophiles, such as *Lactuca saligna* and *Vicia bithynica*.

Typical habitats for *T. maximum* are unstable, south-facing sunny banks, often at the interface between thorn scrub and grassland, in a zone frequently grazed by rabbits, on mineral soils derived from either Marine Alluvium or London Clay. Its commonest associate is *Arrhenatherum elatius*, but *Lathyrus hirsutus*, *Petroselinum segetum*, *Smyrnium olusatrum* and *Vicia bithynica* grow with it in quantity at both of its Benfleet localities.

T. maximum is an annual or biennial, or possibly a short-lived perennial when excessively grazed. When in flower, in June and July, it bears a superficial resemblance to *Torilis japonica*, although, later, the conspicuous fruit is distinctive. Seed is ready to drop by late August or September. It germinates readily, giving rise in a good year to large numbers of seedlings in the vicinity of mature plants. In hot dry summers both first and second year plants fruit and then die off or die back by July/August. Plants stay green and overwinter, however, when grazed by rabbits. There is some evidence that seed may remain viable for only a short period. Seed collected and sown the same day resulted in a large new colony at one site, but seed kept and sown the following spring at several sites failed to produce any plants, despite the large amount sown (pers. obs.).

The colony on broken ground amongst encroaching scrub by Benfleet Creek, to the west of Benfleet, produced 25 plants in 1984, but none could be found in 1995. The main, and possibly only, surviving colonies in Britain occur in Hadleigh Country Park near the top of Benfleet Downs, where two patches are known some 300 m apart, producing between them no more than 100 mature plants in most years. Fortuitously, a fire site made during scrub clearance in 1991 created the ideal conditions for *T. maximum*, and a large dense patch of at least 1,000

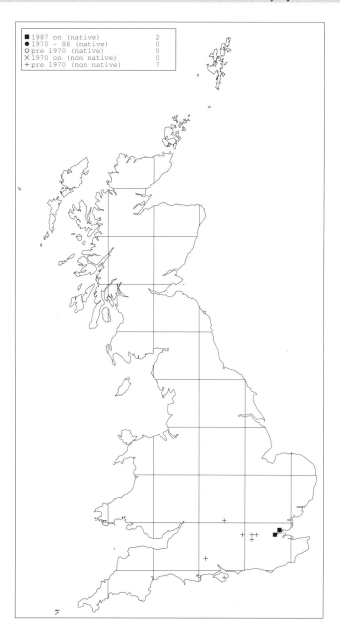

■ 1987 on (native)	2
● 1970 – 86 (native)	0
○ pre 1970 (native)	0
✕ 1970 on (non native)	0
+ pre 1970 (non native)	7

plants produced an abundance of seed in 1995 and 1996. The main threats to its survival are scrub encroachment of the sunny banks and, in recent decades, flail-mowing of the scrub margins during the growing season. Seed collected and dispersed in the area saved the colony from near extinction in 1968, and seed was again collected in August 1995 and dispersed to apparently suitable nearby scrub margins on Benfleet Downs and to a second site near Tilbury Fort.

This species is widespread in the Mediterranean region, including North Africa, extending northwards to Britain and eastwards to the Caucasus. It is regarded as native in northern France, though German and Belgian populations are believed to be associated with human activity, and it may not be native there.

K.J. Adams

Trichomanes speciosum Willd.
(Hymenophyllaceae)
Killarney fern, Raineach Chill Airne, Rhedynen Wrychog
Status in Britain: VULNERABLE. WCA Schedule 8. EC
Habitats & Species Directive, Annexes II and IV.
Status in Europe: Vulnerable.

The British habitats of *T. speciosum* include deep recesses behind cascades, other caves and holes and steep to overhanging rock faces in ravines. Their common features are rocks in constant shade and dampness, and most are situated in wooded glens or cliffs, though the wetness deriving from running water varies greatly. In Ireland a wider range of habitats is occupied, including block scree in shady corries, open mountain cliffs, and, occasionally, soil- or peat-covered banks and tree trunks. This is a warmth-loving plant, evidently sensitive to winter frost, so that nearly all its British localities are in mild oceanic districts of the far west and at low altitudes (10-215 m) though one Welsh station is at 380 m. The Scottish colonies are confined to the mild south-western coastal district and to near sea level. Substrates are acidic to mildly basic, and rock types include granite, sandstone, slate, rhyolite, dolerite and mica-schist.

Ratcliffe *et al.* (1993) recognised four main floristic groupings of associated species, but the most constant associates are bryophytes such as *Conocephalum conicum, Jubula hutchinsiae, Pellia epiphylla, Riccardia chamedryfolia, Saccogyna viticulosa* and *Thamnobryum alopecurum*. Many vascular woodland plants, including various ferns, can grow close to *T. speciosum*, though none is constant. Each colony consists of a single patch of creeping rhizome bearing from three to several hundred or even more fronds, which vary from seven to 60 cm long. Although the area covered by each colony varies from 0.5-4.0 square metres, all appear to be both long-lived and remarkably constant in size. Frond number and size have varied little over a 25-30 year period at five British colonies.

T. speciosum was first recorded in Britain in 1724 at Bingley in Yorkshire by Dr Richard Richardson (Ray 1724). After proving to be widespread in south-west Ireland during 1800-1850, it was discovered in Snowdonia around 1850, and by 1900 at least five localities were reported from North Wales. There were also 19th century records from Cornwall, the Lake District, Arran and Cowal. Few other finds of this fern in the sporophyte stage have ever been made in Britain, and its recorded distribution is in Cornwall, Cardiganshire, Merioneth, Caernarvonshire, Yorkshire, Cumberland, Westmorland, Clyde Isles, Argyll and Kintyre. Out of at least 24 separate colonies in 17 localities reported up to 1995, only 14 colonies in 10 localities are known to be extant. In 1995, a large population of more than 100 separate patches was discovered in Yorkshire, and two additional localities have also been reported from South Wales.

The mystery of the extreme scarcity of the *T. speciosum* sporophyte - given the abundance of apparently suitable habitats within its geographical range - has been given a new twist by the recent discovery that gametophyte colonies are both more widespread and more numerous in Britain (Rumsey *et al.* 1990, 1991). By 1997, colonies had been recorded in 105 hectads. These gametophytes appear to exist independently of the sporophytes, in a state of indefinitely suspended development under present conditions; so that their occurrence does not alter views on the conservation needs of the sporophytes. The scattered colonies of the mature sporophyte may represent fragmented relics of a wider distribution during warmer post-glacial conditions, now surviving under a sub-optimal climate. For these reasons, and for reasons of confidentiality, no map of the currently-known distribution of the species is included.

The biggest threat to *T. speciosum* has been from collecting, which is believed to have eradicated at least five British colonies, as well as uncounted populations in Ireland. Any alterations to the flow and chemical composition of the water or to the microclimate in the habitats of this fern, such as through tree removal, hydro-electric developments and other water extraction, commercial afforestation, mining, quarrying and pollution, are likely to be inimical to its survival. The natural random scouring of ravines by stream torrents under flood conditions can be aggravated by human interference within catchments. One small colony was lost during the intense frosts of early 1963, but probably after human activity had previously reduced its vigour. Reintroduction or re-stocking of *T. speciosum* should be limited to the precise places from which it is known to have been eradicated by human agency. Most of the extant sporophyte colonies and some of the gametophytes are within SSSIs, three of which are also NNRs.

T. speciosum is confined to Europe and Macaronesia, occurring also in Ireland, western France, Spain, Portugal, north-west Italy, Madeira, the Canary Islands and the Azores. In continental Europe it occurs mainly in shady ravines and caves, but it has colonised numerous old wells in Brittany. In the Atlantic Isles it grows especially in the cloud zone, usually in evergreen mist forests and often as an epiphyte.

D.A. Ratcliffe, H.J.B. Birks and H.H. Birks

Trifolium bocconei Savi (Fabaceae)
Twin-headed clover
Status in Britain: VULNERABLE.
Status in Europe: Not threatened. Near endemic.

In Britain, *T. bocconei* is known only from the Lizard peninsula, where it occurs on soils overlying serpentine and, rarely, on schist. Most populations occur in the species-rich and ecologically distinctive grasslands that are developed on shallow soils on south-facing valley slopes and around rock outcrops with a southerly aspect. Although moist in winter, these grasslands become severely droughted in summer, and most of them are grazed, further increasing the openness of the sward. Associates include such notable species as *Isoetes histrix*, *Juncus capitatus*, *Trifolium strictum* and *T. subterraneum*. *T. bocconei* is absent from the most exposed cliff-slopes, but small populations occur in maritime therophyte vegetation on ledges on the more sheltered areas of cliff. It can also be found on rock outcrops in areas of recently burnt *Erica vagans-Ulex europaeus* heath but soon disappears as the heather and gorse become re-established.

In common with many rare species on the Lizard, *T. bocconei* is a winter annual. Germination takes place in autumn and plants overwinter as a vegetative rosette. Flowering is usually in April and May, though it can be delayed in a cold spring. At some sites *T. bocconei* is seen only in favourable years, suggesting that a viable seed-bank exists in the soil. The terminal inflorescences of this species are most often borne in pairs, a unique feature amongst British clovers and from which its common name is derived.

The main populations occur in the cove valleys that dissect the coastal area of the Lizard and (to a lesser extent) on the upper sections of sea-cliffs. Only a few small populations occur on rock outcrops further inland, possibly because of a less mild microclimate. There are about 10 known sites, though it has not been recorded in two of them since 1970. In several of the cove valleys there are a number of small populations scattered on south-facing slopes. Further inland the populations are small and isolated: in total, fewer than 100 individuals may occur even in favourable years. Populations fluctuate in numbers, increasing in warm wet springs, particularly following years of intense summer drought when the grassland sward becomes open, but they tend to be much smaller after dry cold springs. Grazing by livestock appears to be essential for the survival of most populations.

T. bocconei is a plant of southern and western Europe, extending northwards to Britain and Jersey and eastwards to Bulgaria and Greece; it also occurs in Turkey.

J.J. Hopkins

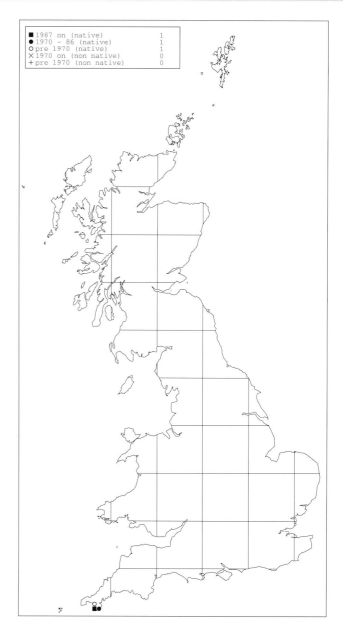

■ 1987 on (native)	1
● 1970 – 86 (native)	1
○ pre 1970 (native)	1
✕ 1970 on (non native)	0
+ pre 1970 (non native)	0

Trifolium incarnatum ssp. *molinerii* (Balbis ex Hornem.) Syme (Fabaceae)

Long-headed clover
Status in Britain: VULNERABLE.
Status in Europe: Not threatened. Endemic.

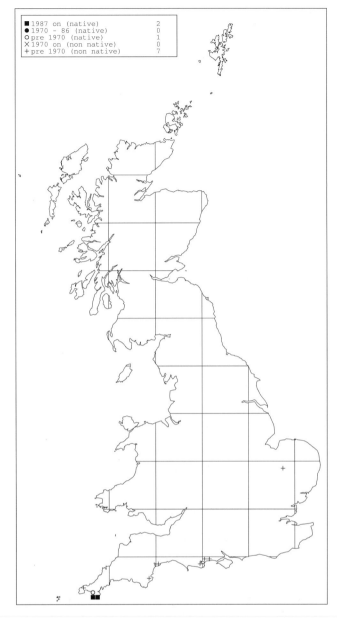

In Britain, this taxon occurs only on the Lizard peninsula and is the most restricted in range of the group of local *Trifolium* species for which that area is noted. It occurs mainly on cliff-slopes, extending inland for only about 200 m along the south-facing slopes of the cove valleys that dissect the Lizard coasts. Typically it occurs in open grassland that is severely droughted in summer. Common associates include *Anthyllis vulneraria, Armeria maritima, Bromus hordeaceus* ssp. *ferronii, Catapodium marinum, Cerastium diffusum, Dactylis glomerata, Daucus carota* ssp. *gummifer, Festuca rubra, Plantago coronopus, P. lanceolata, Sedum anglicum* and *Silene uniflora. T. incarnatum* ssp. *molinerii* is confined to hornblende and mica schists (apart from one isolated site on slates), growing on freely-draining soils derived from peri-glacial material. It is absent from serpentine (contrary to the statement in Duffey *et al.* 1974) and other igneous rocks. Most sites have a southerly aspect, indicating a requirement for a warm microclimate, as also does its restriction to frost-free maritime locations.

T. incarnatum ssp. *molinerii* is an annual of low competitive ability. Germination is in September and October, with plants overwintering as relatively conspicuous, prostrate, large-leaved rosettes. Flowering occurs in late May and early June, and with the onset of summer drought in late June the plants soon die. Abundance is affected by the climate prevailing at critical times. A cold, wet spring causes a high mortality of rosettes. In contrast, plants do particularly well in a warm spring, especially following a summer of intense drought (which kills or suppresses perennial competitors) and a mild winter. At some sites *T. incarnatum* ssp. *molinerii* may not appear every year. In the laboratory, seed has germinated after 23 years' storage, suggesting that it may remain viable for very long periods in the soil (Martin & Frost 1980).

Historically this species has been recorded from about five localities and has been seen at all of them in the past 20 years, including its re-discovery at Gunwalloe in 1985. Populations fluctuate markedly in size, mainly influenced by climatic conditions. For instance, in 1977 following an intense summer drought and a warm wet autumn and winter, more than 36,000 plants were recorded. However, this declined to about 1,600 in 1979, after wet summers that encouraged the vigorous growth of perennial competitors. Some sites may hold 10,000 or more plants in favourable years, but at others fewer than 100 may occur. Its appearance at sites holding only small populations tends to be rather erratic (Martin & Frost 1980).

The distribution and abundance of this species have remained relatively constant in the past 100 years, partly because most of its sites are on sea-cliffs that have escaped the effects of agricultural intensification and other development. At some sites, human activity, including the trampling of ground on paths and vantage points, the mowing of footpath edges, and accidental summer fires, helps to keep the community open. Cessation of grazing in the coastal valleys and on some cliff-slopes could threaten some populations, as the plant is quickly suppressed in rank grassland. However the grazed sites are currently under conservation management and populations seem secure in these under the present regimes.

T. incarnatum ssp. *molinerii* is a plant of western and southern Europe, extending eastwards from Iberia to Greece and, in western Europe, northwards to Britain and Jersey. Ssp. *incarnatum* occurs in Britain as an introduction. It was formerly much grown as a fodder crop but is now uncommon (Stace 1991).

J.J. Hopkins

Trifolium strictum L. (Fabaceae)
Upright clover, Meillionen Unionsyth
Status in Britain: VULNERABLE.
Status in Europe: Not threatened. Near endemic.

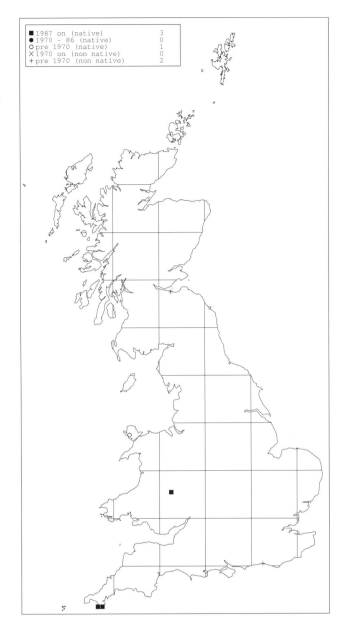

```
■ 1987 on  (native)            3
● 1970 - 86 (native)           0
○ pre 1970  (native)           1
✗ 1970 on  (non native)        0
+ pre 1970  (non native)       2
```

T. strictum is found only on the Lizard peninsula and at one site in Radnorshire. At the Lizard, it occurs in species-rich, short-grazed grassland on the upper and middle slopes of sea-cliffs, and up to about 500 m inland on the slopes of cove valleys, with some populations on large rock outcrops in enclosed and intensively managed grasslands. On a few occasions it has spread after burning into areas of open gorse scrub, though it has become excluded as closed vegetation develops. The grasslands overlie metamorphic hornblende schist, mica schist and serpentine, suggesting a wide tolerance of soil base status and nutrient availability. The sward typically contains maritime perennials, including *Armeria maritima* and *Plantago coronopus*, as well as annuals such as *Aira caryophyllea, Aphanes arvensis, Moenchia erecta, Trifolium arvense, T. incarnatum* ssp. *molinerii, T. scabrum, T. striatum* and *T. subterraneum*. On serpentine, species indicative of base enrichment, such as *Allium schoenoprasum* and *Filipendula vulgaris*, may also occur. In Radnorshire, *T. strictum* occurs in grasslands overlying basalt. Here also the vegetation is rich in annual species, including other *Trifolium* species, together with some additional associates, such as *Teesdalia nudicaulis*, that do not occur at its Lizard sites. Nearly all sites are on south-facing slopes, never on those with a northerly aspect, and in many of them the grassland is developed on very shallow soil that is severely droughted in summer.

T. strictum is an annual that germinates in September and October and overwinters as a rosette. Flowers are produced in late May and early June in most years and, with the onset of drought in late June or early July, the plants quickly die. It does not appear every year at some of its sites, indicating the presence of a viable seed-bank. After a mild winter and spring, particularly following intense summer droughts when competitive perennial species are suppressed, populations of *T. strictum* do well, but there are reduced numbers of flowering plants following a cold wet winter and spring.

In the past 20 years *T. strictum* has been observed at all the Cornish sites in which it has historically been recorded. It occurs at about nine sites, although several of these contain many disjunct populations in small patches of grassland scattered over several hectares. However at some sites, *T. strictum* is confined to an area of less than 100 square metres. In favourable years there are many thousands of plants at the Lizard. By contrast, in Radnorshire the tiny population occurs in an area of about 10 square metres, with the highest count in recent years of only 13 plants.

On cliff-slopes, droughting of the shallow soils (together with maritime exposure at the Lizard) maintains the open conditions that *T. strictum* requires, and there is some evidence that light trampling by summer tourists at some sites also helps in this. However, most populations occur in areas of cattle-grazed heath/grass and in grassland enclosures. In a few sites, where cattle-grazing has been absent or at low levels in some years, the growth of rank grassland has suppressed *T. strictum*. Since rabbit-grazing at its present levels cannot maintain the short turf, there can be little doubt that *T. strictum* would disappear from many of its sites if grazing by domestic stock ceased. Generally, the sites with their shallow soils and frequent rock outcrops are not amenable to agricultural improvement, and threats of other development are also relatively limited. Nearly all sites are protected within SSSIs or are on National Trust land, and most are under conservation management.

This species occurs widely in western and southern Europe, ranging northwards to Britain and Jersey, and from Spain and Portugal eastwards to Bulgaria and Greece. It also occurs in Turkey.

J.J. Hopkins

Trinia glauca (L.) Dumort. (Apiaceae)

Honewort, Githrog
Status in Britain: LOWER RISK - Near Threatened.
Status in Europe: Not threatened.

T. glauca is restricted in Britain to dry stony limestone sites, typically occurring in short, open, grazed turf on south-facing slopes. Characteristic associates include *Festuca ovina, Pilosella officinarum, Sanguisorba minor, Thymus polytrichus* and a number of annuals and pauciennials including *Carlina vulgaris, Centaurium erythraea* and *Euphrasia nemorosa*, together with *Homalothecium lutescens* and *Weissia* species. *T. glauca* occurs in some of its sites with two other rare plants of the western limestones, *Helianthemum apenninum* and *Koeleria vallesiana*, but is not restricted to the xeric conditions that those two species require. It is thus frequently also associated with such species as *Briza media, Carex flacca, Helictotrichon pratense* and *Scabiosa columbaria* and occasionally with *Aster linosyris* and *Potentilla neumanniana*. In turf closely grazed by rabbits, plants of *T. glauca* are no more than a few centimetres tall.

This species is unique amongst British Apiaceae in being dioecious (very rarely monoecious). Flowering is in May and June, the creamy-white flowers of the male plants being especially conspicuous. It behaves either as a biennial or a monocarpic perennial. Regeneration from seed is important in maintaining populations, although its discontinuous distribution is in part a reflection of the very exacting conditions needed for the establishment of new plants. Under grazed conditions, the plant can be a perennial, surviving in a vegetative state until the opportunity arises to flower.

T. glauca is currently known only from the Carboniferous limestone in North Somerset and Gloucestershire and the Devonian limestone in South Devon, occurring in about 20 sites in six hectads. Populations are generally stable, and records from some sites date back more than two centuries. During surveys in 1988-89 (FitzGerald 1990a; Taylor 1990f) the two largest native populations were found at Sand Point, with an estimated 10,000 plants, and at Crook Peak, with similar numbers. It was abundant at several other sites, ranging from hundreds to a few thousand plants. The large population at Goblin Combe, estimated to number over 18,000 plants in 1989, originated from six rooted plants and 40 seeds sown in 1955 (Hope-Simpson 1987). Most of its sites are protected within SSSIs. A few are grazed by sheep or cattle, and in those that are not, rabbits are especially important in maintaining short open turf. In the Avon Gorge, in the absence of grazing, it is the harsh conditions of the rocky south-facing bluffs that enable *T. glauca* to survive. This species is not immediately

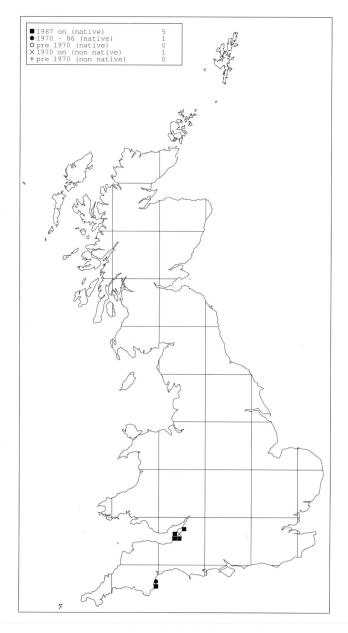

■ 1987 on (native)	5
● 1970 - 86 (native)	1
○ pre 1970 (native)	0
✕ 1970 on (non native)	1
+ pre 1970 (non native)	0

threatened, although potential threats to some populations include a reduction in grazing by domestic stock or by rabbits and excessive trampling by visitors.

T. glauca occurs throughout western, central and southern Europe, extending to south-west Asia. Its stronghold is in southern and central Europe, and it becomes rare further north, reaching its northern limit in England. It is threatened in Germany.

R.D. Porley

Tuberaria guttata (L.) Fourr. (Cistaceae)
Spotted rockrose, Cor-rosyn Rhuddfannog
Status in Britain: VULNERABLE.
Status in Europe: Not threatened. Near Endemic.

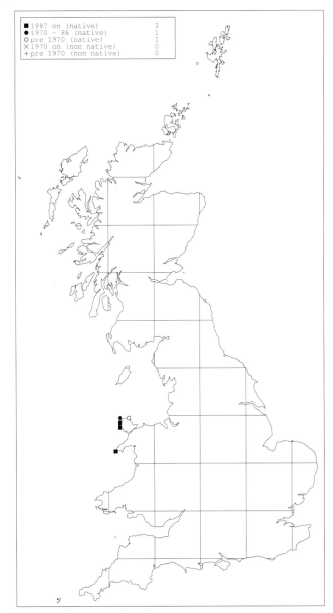

The habitat of *T. guttata* is invariably thin soil overlying hard igneous rock near the sea. It grows most abundantly at sites with plenty of bare ground, and its immediate and most dominant associates are generally *Cladonia* and *Polytrichum* species (typically *C. portentosa* and *P. juniperinum*), together with frequent *Aira praecox*, *Scilla verna* and *Sedum anglicum*. These pockets of skeletal soil are usually found within a more closed community of *Calluna vulgaris*, *Carex panicea*, *Erica cinerea*, *Festuca ovina* and *Ulex gallii*, in which *T. guttata* can, however, also occur. These associates reflect the low nutrient status and the mixture of drought and brief waterlogging to which the habitat is prone.

This species is an annual, flowering and seeding in May and June. Mature plants may be only 6 cm tall, and the flowers, which can be quite conspicuous, may last for only a few hours, dropping their petals by midday. In a warm rainy summer, *T. guttata* can complete its life cycle within a matter of weeks, but the more typical pattern is late summer germination, followed by an overwintering rosette. This species shows an adaptive plasticity in its morphology, to the extent that the dwarf maritime ecotype in Wales (and in Ireland) was formerly treated as an endemic taxon, ssp. *breweri* (e.g. Clapham *et al.* 1962). Recent work suggests, however, that these populations, at the edge of their range and long isolated, have developed a unique pattern of genetic variation, and one which is now as disjunct and broken within Wales as elsewhere (Kay & John 1995). Subspecific status for British plants is not considered appropriate, though recognition at a lower taxonomic level may be so (Stace 1991).

T. guttata is confined to North Wales. Most populations are on the west coast of Holy Island, Anglesey, between South Stack and Rhoscolyn. Away from this area, the only other known colonies are on the north-west coast of Anglesey and on the Lleyn peninsula. Tens, if not hundreds, of thousands of plants have been recorded at each of its six major sites. In some years, however, plants may appear only in their hundreds, and in widely-scattered, tiny patches. Three populations (one of which has not been refound in recent years) have never exceeded 500 plants, and at some sites in some years the plant has not appeared at all. The Lleyn population is tiny, with only about 30 plants in 1993. When not flowering, this small species is inconspicuous, and some small populations might still be overlooked on Anglesey and Lleyn.

It is not yet clear if a recent pattern of local absence and low numbers represents temporary seasonal downturns, perhaps as a consequence of poor weather conditions, or is evidence of a long-term decline. At all but one site, however, there are worrying signs that poor annual performance is due to adverse management, such

as heather burning, scrub overgrowth, fertiliser drift, and nutrient enrichment from overwintering livestock - all part of the ongoing isolation of rocky heath outcrops within a landscape of grazing 'improvement', housing development and caravan parks.

T. guttata occurs in western and south-western Ireland and in the Channel Islands (Jersey and Alderney). Elsewhere in Europe it is found in the Mediterranean region, extending northwards in western Europe to north-west Germany. It is also occurs in the Canary Islands.

For further reading on the ecology and taxonomy of this species, see Proctor (1960, 1962) and Gallego & Aparicio (1993).

A. Jones

Valerianella eriocarpa Desv. (Valerianaceae)
Hairy-fruited cornsalad, Gwylaeth yr Oen Gwlanog
Status in Britain: VULNERABLE.
Status in Europe: Not threatened.

V. eriocarpa is an annual weed of calcareous banks, walls, cliff edges and other dry open habitats. It is intolerant of competition, growing usually with a variety of other annuals, such as *Arenaria serpyllifolia*, *Bromus hordeaceus* ssp. *ferronii*, *Cerastium diffusum*, *Geranium rotundifolium*, *Sherardia arvensis*, *Valerianella carinata* and *V. dentata*. On the richer ledges of its cliff-ledge habitats in Dorset these species are accompanied by *Euphorbia portlandica*, *Hippocrepis comosa*, *Petroselinum segetum*, *Ranunculus parviflorus* and *Salvia verbenaca*. Formerly it occurred in arable sites, where it is assumed that its associates would have been annuals similar to those listed above.

V. eriocarpa is an annual, germinating in early spring and flowering in May, setting seed rapidly in dry years. In favourable conditions production of seed is very good, even from tiny (2 cm) plants. Nothing is known of the viability of its seed.

The currently accepted wisdom is that *V. eriocarpa* is an arable introduction, though perhaps an ancient one (Perring & Farrell 1983; Stace 1991), that has spread to other open habitats such as car parks and rocky cliffs and quarries. In its best-known locality, at Portland in Dorset, there are early 19th century records of it as an arable weed on the summit plateau, and it now persists in good numbers in a situation that "is in a sense artificial, being quarry spoils and footpaths, but the ground is kept open by exposure and desiccation, as much as by trampling, and very closely approximates a natural cliff edge or rocky outcrop" (FitzGerald 1990d). However, over the last three years it has been found in seven sites on bare and disturbed limestone from Lulworth to Swanage, and it is considered to be a true native in these sites. In recent times, the only other persistent colonies are on sandy walls in Cornwall, though there are other records (almost all of non-persistent colonies), north to Scotland. It appears to have disappeared completely from its arable field habitat, though records were always scattered, and this species seems to have been ephemeral in that habitat.

V. eriocarpa has a distribution centred in the Mediterranean region, extending as far east as Turkey in Europe. It is also found in the Canary Islands and is scattered northwards to Britain, France and the Channel Islands. Its European range is not quite clear because of confusion with records of *V. muricata* (which overlaps in range to the south) and possibly with those of *V. dentata* var. *mixta*. Although often described as having hairy seeds, the seeds of some British plants are glabrous.

D.A. Pearman

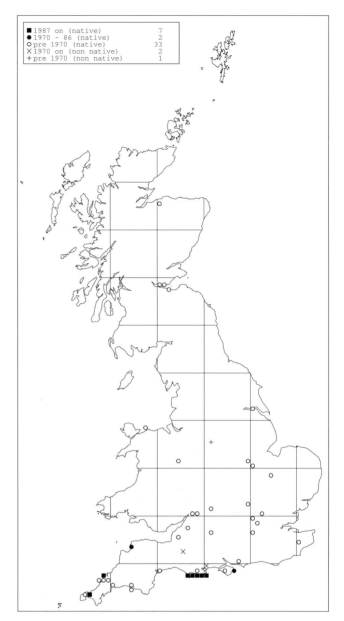

■ 1987 on (native)	7
● 1970 - 86 (native)	2
○ pre 1970 (native)	33
✕ 1970 on (non native)	2
+ pre 1970 (non native)	1

Valerianella rimosa Bast. (Valerianaceae)
Broad-fruited cornsalad, Gwylaeth yr Oen Llyfyn
Status in Britain: CRITICALLY ENDANGERED.
Status in Europe: Not threatened.

V. rimosa is a plant of arable fields and other open habitats on chalk and limestone and has usually been recorded in a species-rich community of annual weeds, including such species as *Euphorbia platyphyllos, Papaver argemone, P. hybridum, Ranunculus arvensis, Scandix pecten-veneris, Torilis arvensis* and *Valerianella dentata*. In Gloucestershire, it occurred around the edges of a limestone quarry, and in 1997 one plant was found on old over-burden together with *Bromus hordeaceus, Geranium molle, Potentilla reptans, Polygonum aviculare* and *Trifolium dubium*.

It is an annual, germinating mainly in the late autumn but also in smaller numbers in late spring (Wilson 1990). It is consequently found largely in fields cultivated in autumn. It competes very poorly with modern cereal crops that have received the usual quantities of nitrogen fertiliser: a study has shown that only 20% of *V. rimosa* plants survived in a fully fertilised cereal crop, compared with in an unfertilised one, and the number of seeds produced in the former was less than a quarter of the total in the latter (Wilson 1990). Plants flower from late June to late July (sometimes August), producing most seed by the time winter wheat and spring barley is harvested in August. However, it is likely that little seed would be produced before winter barley is harvested in southern England, typically around mid-July. Little is known about the dormancy characteristics of the seed. However, the reappearance of the plant after many years' absence (for example, between 1987 and 1997 at one of the Gloucestershire sites) suggests some seed can remain viable for a fairly long period.

V. rimosa was once quite widespread in southern England, extending north to Fife, but has suffered a sharp decline, with records from only eight hectads since 1970. It has been recorded at only four sites since 1987, one each in Somerset and Hampshire and two in Gloucestershire. The Somerset site is managed for *V. rimosa*, and this site may now be the only one in Britain where it regularly occurs. It seems to be of sporadic occurrence elsewhere, appearing only when the habitat is suitably open and disappearing when it closes. It is possible that non-fruiting plants of *V. rimosa* may have been passed over as the common *V. dentata* or *V. locusta*, since they can be differentiated reliably only by characters of the ripe fruit.

This species appears to be critically endangered in Britain, mainly because of agricultural intensification and changes in land use, including the cessation of cultivation.

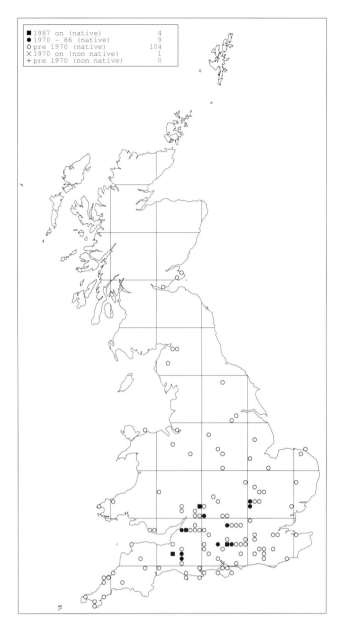

■ 1987 on (native)	4
● 1970 – 86 (native)	9
○ pre 1970 (native)	104
✕ 1970 on (non native)	1
+ pre 1970 (non native)	0

However, the main extant site, near Taunton, is an SSSI and nature reserve, and management is directed towards the conservation of *V. rimosa*. A national review of this species and a 'recovery' programme are urgently required.

V. rimosa has a mainly southern-continental distribution in Europe, but extends northwards to Ireland, Britain and Denmark and eastwards to Ukraine and the Crimea. It is threatened in the Netherlands, Germany and Switzerland.

P.J. Wilson

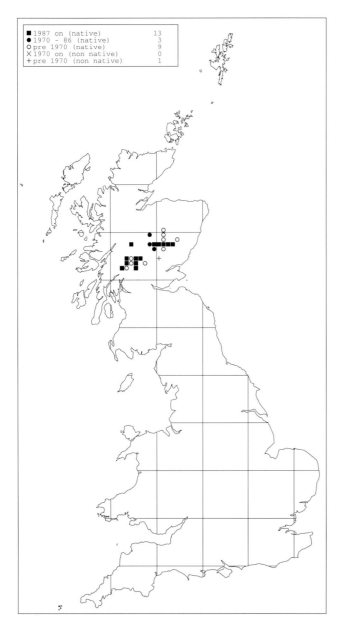

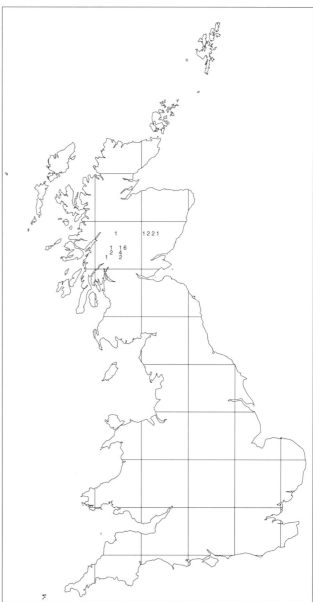

Veronica fruticans Jacquin (Scrophulariaceae)
V. saxatilis Scop.
Rock speedwell, Lus-crè na Creige
Status in Britain: LOWER RISK - Near Threatened.
Status in Europe: Not threatened.

In Britain, *V. fruticans* has a restricted and localised distribution in the Scottish Highlands, where it occurs on base-rich alpine rocks at altitudes from 500 to 1,000 m. It most typically occurs on dry, open calcareous slopes and rock ledges on crags, usually in south-facing sites and almost invariably inaccessible to grazing. It is an early coloniser of freshly exposed soil and thrives best on rather bare substrates, in ungrazed sites where there is minimal competition from other species. Associated species include *Agrostis capillaris*, *Alchemilla alpina*, *Erigeron borealis*, *Euphrasia nemorosa*, *Festuca vivipara*, *Gentiana nivalis*, *Persicaria vivipara*, *Saxifraga oppositifolia*, *Silene acaulis*, *Thymus polytrichus* and mosses such as *Ctenidium molluscum* and *Tortella tortuosa*. Where it spreads (rarely)

into the grassy areas along the base of crags, associates may include *Anthoxanthum odoratum* and *Nardus stricta*.

It is a spreading, rather woody perennial with numerous ascending, often branching, shoots. Flowering normally occurs from the end of June to the end of July but may occasionally extend into August. Flowers are visited by insects but are apparently often self-pollinated (Clapham *et al.* 1987). Flowers may last only a short time, sometimes as little as one day, and seem especially vulnerable to rain and wind. When vegetative, plants can be difficult to see and are likely to be overlooked unless searched for deliberately.

V. fruticans is seldom abundant in any locality. Most of the strongest colonies are found in the mountains of Perthshire, the main cluster of sites being on the Ben Lawers and Meall nan Tarmachan ranges, and the largest concentrations occurring on the crags of Cam Chreag and An Stuc. Away from the main area, there are scattered sites ranging from Beinn an Dothaidh in the west to the Cairngorms and Clova in the east. Of the 47 populations recorded prior to 1990, 26 have been confirmed since 1990 and eight new ones found. Ten populations have been lost, and 11 have not been verified in the 1990s, though they may be extant. Most populations are small, with fewer than 10 plants; one or two populations have more than 100, and one holds more than 1,000 plants. The national total appears to be only a little over 2,000 plants.

V. fruticans has an arctic-alpine distribution and occurs in most of the mountain ranges of western Europe. Its range extends northwards to the Arctic Circle and locally eastwards to north-west Russia, Croatia and Albania, and it also occurs in Greenland, Iceland and the Faeroe Islands. In Europe, its strongholds are in the Alps, the Pyrenees and the uplands of Norway and Sweden. A wider range of habitats in Europe include rough, stony pastures, meadows and south-facing scree slopes up to 3,000 m in altitude.

B.G. Hogarth

Veronica spicata L. ssp. *spicata*
(Scrophulariaceae)
Spiked speedwell
Status in Britain: VULNERABLE.
Status in Europe: Not threatened.

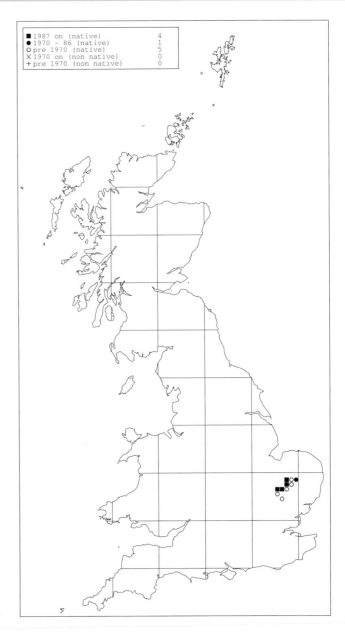

V. spicata ssp. *spicata* occurs in short, dry grasslands of the Breckland. The soils are sandy, well-drained, liable to drought, nutrient-poor and often non-calcareous and are frequently associated with peri-glacial stripes or patches (Coombe 1987b; Birkinshaw 1990a). It is generally intolerant of shade and competition and grows poorly in tall, tussocky, closed swards. Associated species include *Achillea millefolium, Agrostis capillaris, Bromopsis erecta, Festuca ovina, F. rubra, Filipendula vulgaris, Galium verum, Lotus corniculatus, Koeleria macrantha* and *Senecio jacobaea*. In lightly-grazed or mown pasture its micro-distribution is determined by tussocks of *Festuca ovina* and soil patterns (Watt 1964; Coombe 1987b).

It is a polycarpic perennial that can potentially live indefinitely by clonal growth. Abundant seed is produced, but most growth is vegetative by rhizomes and stolons, and it is quick to exploit open ground (Birkinshaw 1990a). It is evergreen and new growth starts from April. Flowering is usually from July to October, but may exceptionally be as early as April and last into November. Pollination is by various insects, mainly bumble-bees. It is self-compatible. Seeds germinate well in cultivation, and establishment in the wild is likely to be mainly in autumn, as seedlings are readily droughted. Establishment is most successful in bare soil and is progressively less so as the sward closes and becomes more coarse.

V. spicata ssp. *spicata* has declined because of the cultivation of former grazed grass heaths and as a result of afforestation, under-grazing, bracken-invasion and road improvements. It has been recorded recently at only three out of 12 native sites: at Newmarket, Cambridgeshire, near Brandon, West Suffolk, and on Weeting Heath in West Norfolk. It occurs in several parts of the Newmarket Heath race-course, in areas mown regularly to a height of about 10 cm. The plants flourish over a fairly wide area, but flowering is rather sparse (Coombe 1987b) and there are very few plants in some locations. At Weeting, management is mainly by rabbit-grazing. Five populations are protected against rabbit-grazing in enclosures that allow the plants to flower and set seed (hundreds of spikes in good years) but which are opened to rabbits between November and April. Three populations that are not enclosed may produce flowers and set seed if the grazing is not too heavy; otherwise they spread only vegetatively. The small population (30 spikes in 1991) near Brandon is hand-weeded to keep competing vegetation down. Seeds and plants were experimentally re-introduced to Cavenham Heath and West Harling Heath in 1989. Plants put into an enclosure at Cavenham survived until at least 1995, but no plants have persisted outside.

It is drought-tolerant, and in wet conditions it may be susceptible to fungal attack (Birkinshaw 1990a). It is tolerant of heavy grazing by rabbits, and by sheep at high stocking densities, but may get uprooted by rabbits burrowing and scraping. Rabbits and small mammals will nip off the inflorescences but do not eat the leaves to any extent (Watt 1971).

V. spicata is a variable species and occurs throughout most of Europe from Spain and Britain northwards to Fennoscandia and eastwards to Russia, Bulgaria and Turkey, though it is rare in the west and absent from most of the Mediterranean islands. Some authorities (e.g. Pigott & Walters 1954) regard ssp. *hybrida* as an ecotype of ssp. *spicata*, and indeed the two taxa appear to differ in only minor characteristics. *Flora Europaea* (Tutin *et al.* 1972) does not recognise ssp. *hybrida* as distinct from ssp. *spicata*, though both are retained as subspecies in British lists (e.g. Kent 1992). Ssp. *spicata* is widespread and common in continental Europe.

T.C.G. Rich

Veronica triphyllos L. (Scrophulariaceae)
Fingered speedwell
Status in Britain: ENDANGERED. WCA Schedule 8.
Status in Europe: Not threatened.

This is the rarest of the annual *Veronica* species and the
earliest to flower. It grows on open calcareous or slightly
acidic sandy soil (pH 6.4-8.0), usually at the margins of
arable fields or on soil blown from arable land (Watt 1971),
but historically it has also been recorded on sandy tracks
and waste ground, in fallow fields and gravel pits. Its
associates are mainly short-lived species, including
*Arabidopsis thaliana, Arenaria serpyllifolia, Cerastium
semidecandrum, Erophila verna, Lamium purpureum, Myosotis
ramosissima, Saxifraga tridactylites, Stellaria pallida, Veronica
persica* and *V. hederifolia.*

The seedlings of this annual species usually appear in
January and the flowers in mid-March, though it may
flower as early as February. When well-grown, the slender
stems may attain 10 cm in height and sprawl over the
ground, but they are often very much shorter. Climatic
conditions control the timing of seed ripening and
shedding; this may occur in April or as late as June. After
seed is shed, the plant quickly dies and disappears. Open
ground is essential for germination, and *V. triphyllos* is
absent where there is winter competition.

V. triphyllos has always had a very restricted
distribution in Britain but has declined drastically as
agricultural and other development have taken their toll. It
now occurs at only three sites in Breckland, one in Norfolk
and two in Suffolk. Its sole Norfolk site is now included
within a housing estate, in which turves were transplanted
to a sandy bank in order to save some of the original
population. In Suffolk, a tiny population still remains at
the single native site: one plant was recorded there in 1978
and five in 1991. *V. triphyllos* was introduced to its other
Suffolk site in 1967 - a broad strip of cultivated ground at
the edge of an arable field and which is maintained as a
reserve principally for rare *Veronica* species. *V. triphyllos* is
considered doubtfully native in Surrey, where it was once
a weed of sandy fields (Lousley 1976) and persisted in a
nursery at Byfleet probably until the early 1980s. It has
also been recorded in several other counties, from
Cornwall to Yorkshire, generally as a non-persistent casual.

Sites have been lost to development and many
populations destroyed by changes in agricultural practice,
including the application of herbicides. Conservation
management at the reserve site mainly comprises regular
disturbance or shallow cultivation of the soil, with the aim
of maintaining suitably open conditions for germination
and growth. But these cultivations have not always been

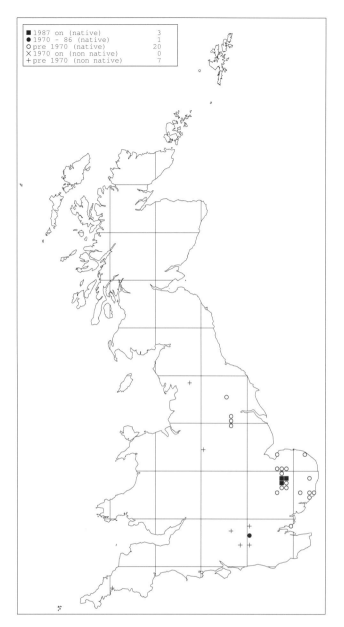

successful in the past, and germination and flowering have
been erratic (Trist 1979). Numbers vary according to the
cultivation regime, but in good years several hundred
plants can occur.

In Europe it is widespread in dry grassland, cultivated
ground and waste places from Sweden and Latvia
southwards but is rather rare in the Mediterranean region.
It is also recorded in North Africa and western Asia.

L. Farrell

Veronica verna L. (Scrophulariaceae)
Spring speedwell
Status in Britain: VULNERABLE.
Status in Europe: Not threatened.

Like the other two rare annual Breckland speedwells, *V. praecox* and *V. triphyllos*, this species grows on infertile sandy soil in sparsely vegetated habitats or open ground. However, unlike those two species, it is mainly a plant of short grassland in stony areas where there is no regular cultivation but relatively intense grazing by sheep or rabbits, which keep the habitat open. Associated species include *Aphanes arvensis, Carex arenaria, Festuca ovina, Myosotis ramosissima, Ornithopus perpusillus, Scleranthus annuus, Senecio sylvaticus* and *Teesdalia nudicaulis*.

It is an annual plant, usually germinating in spring, though some autumn germination may occur in a wet season. Flowering is in late March and April (sometimes continuing into June), the flowers opening only in bright sun. There is a good seed-bank.

It is now confined to the vicinity of Icklingham, West Suffolk, where it occurs at about 12 sites, all but one of which lie within a single hectad. It has been introduced to two County Trust reserves, where it is flourishing. Populations are small, ranging from about 50 plants down to a singleton. It formerly occurred in Norfolk, and there are old records from Devon, where it was introduced.

In Europe it grows in cultivated fields and other dry places and is widely scattered, although it is absent from the extreme north and west and from the Mediterranean region. Elsewhere, it is recorded in western Asia and Morocco.

L. Farrell

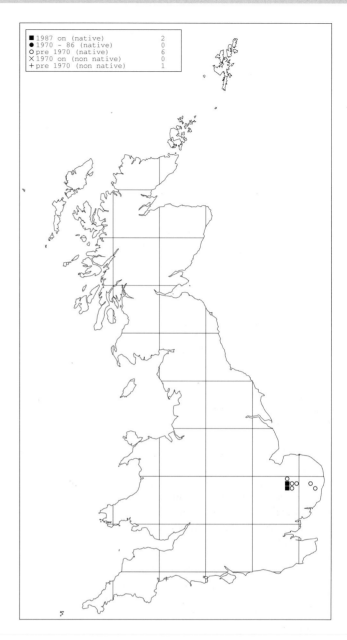

■ 1987 on (native)	2
● 1970 - 86 (native)	0
○ pre 1970 (native)	6
✕ 1970 on (non native)	0
+ pre 1970 (non native)	1

Viola canina L. ssp. *montana* (L.) Hartman (Violaceae)
Heath dog-violet
Status in Britain: ENDANGERED.
Status in Europe: Uncertain.

This rare subspecies of the widespread but local
V. canina differs from the typical ssp. *canina* in being much
taller, with suberect leafy stems to 30 cm and large stipules
up to more than 1 cm long. In habit the plant resembles
V. persicifolia but has more triangular-cordate leaves that
are thicker in texture and no wide-creeping underground
rhizome. The pale milky-white colour of the open flower
of *V. persicifolia* contrasts strongly with the blue of
V. canina, and the hybrid between these two species
(*V.* x *ritschliana*) has an intermediate pale colour.

V. canina ssp. *montana* is confined, so far as is known,
to Cambridgeshire, occurring in the NNRs of Wicken,
Woodwalton and Holme Fens. Earlier records suggest that
the plant was formerly more widespread in that county,
and it was recorded at Lakenheath Poor's Fen in West
Suffolk in 1954 (S.M. Walters pers. obs.).

There is some difficulty in assessing the *V. canina*
subspecies involved where rough grazed pasture or heath
grades into fen. Thus at Otmoor, Oxfordshire, a survey
carried out in 1993 of ground suitable for *V. persicifolia*
(which was last seen there in 1963-4) revealed only the
sterile hybrid *V.* x *ritschliana* (*V. canina* x *V. persicifolia*) in
the absence of both parents. At several Cambridgeshire
sites, herbarium specimens, referable to one or other
subspecies, confirm that *V. canina* was once more
widespread and usually occurred at the edge of fen areas
where *V. persicifolia* also grew and with which it
hybridised.

Hybridisation between *V. persicifolia* and *V. canina* was
very obvious at Woodwalton Fen, apparently from the
early days of the Nature Reserve until at least 1965. The
author's own knowledge of the Fen is mainly based on the
period 1945-1965, when the south end of the Fen showed a
bewildering array of 'fen violets', referable to *V. persicifolia*,
V. canina ssp. *montana,* and the sterile hybrid. Sterility of
interspecific violet crosses is obvious late in the season,
because fertile plants set abundant capsules from the
cleistogamous flowers produced on the elongated stems
after the open flowers, but sterile plants set no fruit. It was
these complex violet populations that were described,
amongst others, by E.M. Gregory in 1912 in her classic
monograph, *British Violets*.

In continental Europe, where *V. canina* is more common
than in Britain, it is obvious that several tall 'fen-meadow'
ecotypic variants occur, and inter-specific taxonomy is
more complicated (e.g. Røren *et al.* 1994). The use of the
name ssp. *montana* for our fenland plant, based as it is on
the Swedish plant that Linnaeus knew, is a practical

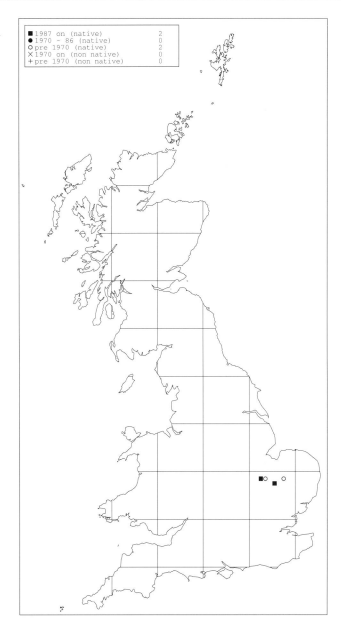

■ 1987 on (native)	2
● 1970 - 86 (native)	0
○ pre 1970 (native)	2
✕ 1970 on (non native)	0
+ pre 1970 (non native)	0

taxonomic recognition of significant variation, but
continental '*montana*' has a much wider ecological
amplitude than the British plant.

Recent investigations at Wicken Fen, primarily
concerning *V. persicifolia*, have revealed that dormant seed
is very important, not only for *V. persicifolia* but also for the
small population of *V. canina* ssp. *montana* accompanying it
(Rowell *et al.* 1982, 1983). Both taxa are clearly dependent
upon a combination of disturbance factors in their Wicken
and Woodwalton habitats that have the effect of reducing
competition and providing open habitats of damp, bare
peat for dormant seed germination.

S.M. Walters

Viola kitaibeliana Schultes (Violaceae)

Viola nana (DC.) Godron
Dwarf pansy
Status in Britain: VULNERABLE.
Status in Europe: Not threatened.

In Britain, *V. kitaibeliana* is found only in the Isles of
Scilly, where it occurs in short, open turf on sandy soil, or
on open sand, such as in eroded areas of sand-dune and
around rabbit burrows. Associated plants are mostly
dwarfed forms of common sand-dune species such as
*Anagallis arvensis, Cerastium diffusum, Erodium cicutarium,
E. maritimum, Euphorbia portlandica, Festuca rubra, Lotus
corniculatus, Myosotis ramosissima, Plantago coronopus* and
Senecio jacobaea. Formerly it was often found in arable
fields, but very few of those populations persist.

It is an annual, flowering from the end of March into
April. In natural habitats it generally withers and
disappears by the end of May, but in arable ground it may
persist as straggling late-flowered plants into June or even
early July (Lousley 1971). *V. kitaibeliana* can be readily
distinguished by its small size, although some dwarf forms
of *V. arvensis* and *V. tricolor* can be similar in habit.

Several sites for this plant have been lost since the
1950s, and it has apparently declined in numbers over the
past few years. It now occurs almost exclusively on Bryher
and Tean, with one recent record from St Martin's. Most of
its populations are very limited in extent, though
thousands of plants may appear in a patch of only a few
square metres. In some places, there may be a good seed-
bank, since plants have been known to reappear at old
sites following cultivation or other disturbance.

Most of its sites appear to be threatened to some
degree. Dunes are subject to coastal erosion and
inundation from the sea and, conversely, from the erection
of coastal defences against both of these natural events.
Locally, the digging of sand for building materials may
destroy colonies. In the past, sites have been built over and
others have been invaded by coarse vegetation.

V. kitaibeliana is found in the Channel Islands, west-
central France, southern and central Europe and eastern
Ukraine. However, preliminary studies by Kirschner have
shown that the taxon encompasses a number of
recognisable forms in Europe, and the taxonomic position
of British plants requires further investigation.

R.E. Parslow and M.J. Wigginton

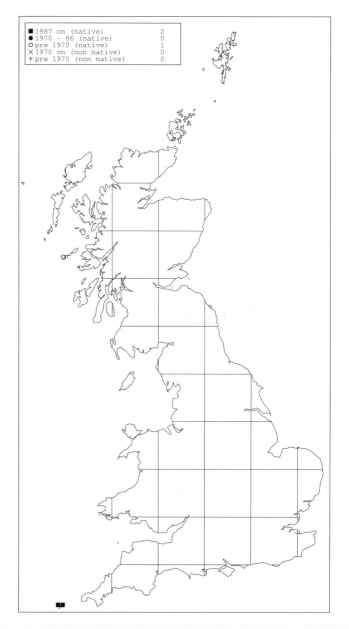

■ 1987 on (native)	2
● 1970 - 86 (native)	0
○ pre 1970 (native)	1
✕ 1970 on (non native)	0
+ pre 1970 (non native)	0

Viola persicifolia Schreb. (Violaceae)
Viola stagnina Kit.
Fen violet
Status in Britain: ENDANGERED. WCA Schedule 8.
Status in Europe: Not threatened?

V. persicifolia is a plant of wet, peaty, base-rich soils in
fens and formerly river valleys, growing in relatively open
vegetation, often with some peat. It is found in a
mixed fen community, typically including *Calamagrostis
canescens, C. epigejos, Carex panicea, Hydrocotyle vulgaris* and
Mentha aquatica. Other associates may include *Anagallis
tenella, Angelica sylvestris, Cladium mariscus, Filipendula
ulmaria* and *Luzula pallidula. V. persicifolia* seems to favour
sites that are seasonally wet (usually in winter), rather than
those that are permanently waterlogged. Plants do not
persist when the vegetation becomes overgrown and do
best where the surface of the peat is periodically disturbed,
as for example by peat-digging or grazing by cattle.

It is a creeping perennial, with narrow pointed leaves
and pale bluish-white flowers, from which the plant gets
the name 'milk violet' in some European countries.
Flowers appear in May and June. In addition to the
normal ones, the plant produces small, self-fertile
cleistogamous flowers, which can also set seed in great
abundance.

This species has been recorded from more than 20 sites
in England, though it has long been lost from most of
these. Two of the most recent losses were from Suffolk in
1968 and from the Doncaster area in 1975. It is now
confined to two NNR fenland sites in Cambridgeshire and
one in Oxfordshire. At the Cambridgeshire sites,
Woodwalton Fen and Wicken Fen (Rowell *et al.* 1982;
Rowell 1983), the sizes of the populations fluctuate
spectacularly. At Wicken Fen, several hundreds flowered
in the early 1980s, after a gap of over 60 years during
which no plants had been recorded (Rowell 1984), but in
1994 only 366 plants (and numerous seedlings) were
recorded there. No plants were seen at Woodwalton Fen in
1993 or 1994 (Wells *et al.* 1995), but in 1995 and 1996 more
than 1,000 plants were in flower, many of them in an
experimental disturbance plot set up in 1994 (T.C.E. Wells,
pers. comm.). The unhybridised plant was thought to be
extinct at Otmoor, Oxfordshire, by 1965 (Woodell 1967),
but a population of about 50 plants was discovered there
in 1997.

Seed can remain viable in the soil for long periods, and
experience in England and on the continent suggests that,
where appropriate physical conditions persist, a
population can re-establish well under suitable
management (Pullin & Woodell 1987). Few of its historical
sites still have an appropriate hydrology and nutrient
status, but it is not impossible that some populations
persist in these areas as buried seed. Many factors have
contributed to the loss of populations in the past,
including agricultural improvement, drainage, nutrient
enrichment and commercial peat-digging. In the known
remaining populations, overgrowth of the fen vegetation
and excessive waterlogging have suppressed the growth of

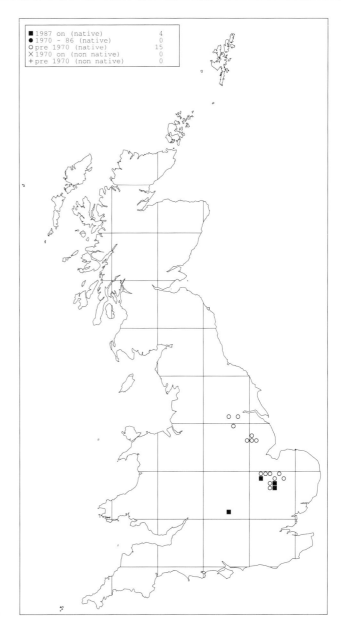

■ 1987 on (native)	4
● 1970 – 86 (native)	0
○ pre 1970 (native)	15
✕ 1970 on (non native)	0
+ pre 1970 (non native)	0

V. persicifolia. At Wicken Fen the plants grow in an area
where, in the past, peat was dug by hand. This site would
benefit from a lowering of the water level or perhaps
remodelling of the land to bring some peat above the
water surface. Part of Woodwalton Fen is grazed to
maintain a small population, but periodic disturbance,
including scrub removal, is necessary in order to produce
enough fruiting plants to replenish the seed-bank.

In Europe this species is known from similar habitats
but is frequently found in grazed vegetation, for example
in the Irish turloughs (Pullin 1986) and in the Netherlands.
It is distributed widely from Scandinavia to northern Spain
and from western Ireland to Russia but has declined
throughout western Europe with the loss of peatland
habitats. *V. persicifolia* is now rare in Ireland and either rare
or extinct in most other western European countries,
although its position is more secure in the Baltic region.

V.M. Morgan

Viola rupestris Schmidt (Violaceae)

Teesdale violet
Status in Britain: LOWER RISK - Near Threatened.
Status in Europe: Not threatened.

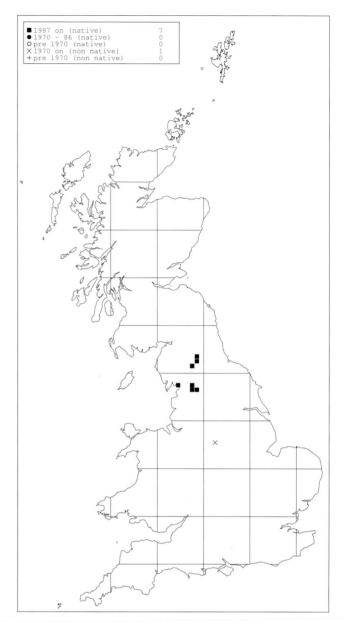

In Britain, *V. rupestris* is a small, early-flowering species of open turf on base-rich sites, occurring in two apparently distinct habitats. At Arnside Knott it grows on warm, south-facing slopes at about 140 m altitude, in tussocky *Sesleria caerulea* grassland, whilst on the cold, high limestone of the northern Pennines between 400 and 575 m it occurs in tightly-grazed *Festuca-Sesleria*. Although there are obvious differences between the Arnside site and the others, there are also many shared habitat features, and, in particular, all are likely to have remained relatively open throughout history as non-forest relict habitats. The turf is almost always open and dominated by *S. caerulea*, giving some ecological separation from *V. riviniana*, which is favoured by the more closed *Festuca ovina* swards. Typical associates in both upland and lowlands sites include *Campanula rotundifolia, Carex caryophyllea, C. ericetorum, Danthonia decumbens, Helianthemum nummularium, Lotus corniculatus, Pilosella officinarum* and *Thymus polytrichus*.

V. rupestris is a long-lived perennial. Reproduction is by vegetative spread or by seed. It flowers in May, though the incidence of flowering and fruiting is generally very low in most or all populations. For instance, in a survey of Widdybank populations in 1995, less than 0,1% of the plants were in fruit (S. Hedley *in litt.*), and at Arnside in 1993, only 3% (Robinson 1993). Fruits that are produced and survive to maturity set abundant seed late in the year. It seems uncertain whether germination is normally in autumn or spring. The Arnside population is distinct from the others in having white flowers.

This species occurs on Widdybank Fell, Co. Durham, on Long Fell and Arnside Knott, Cumbria, and in the Craven district of West Yorkshire. Widdybank holds the largest population, perhaps of the order of 10,000 plants, where they are restricted to the small areas of sugar limestone. There is no indication that the plant has declined overall, though the northernmost colony on Widdybank has been lost since 1985, probably because of erosion. In Craven, it is known in about 13 sites, some of which hold populations of fewer than 10 plants, but at least one holds more than 1,000 plants. At Arnside, the population has declined dramatically over the last 20 years, although it now appears to have stabilised at a much lower level (fewer than 100 plants).

There appears to be little overt threat to *V. rupestris* in the uplands, even where the grasslands are hard grazed by sheep. However, the possibility of attrition of certain populations through hybridisation with *V. riviniana* has frequently been raised, and observations on the Long Fell population, where it grows in dense *Festuca ovina* turf in close association with *V. riviniana*, suggest that unhybridised *V. rupestris* is rare there. Hybrids are also reported on Widdybank, where both species occur in proximity, but plants that appear to be intermediate in character do not necessarily show the vigour and patch-

density said to be typical of the hybrid, *V. x burnatii*. At Arnside, the cause of the population decline is likely to have been the gradually increasing coarseness of the sward of *Sesleria caerulea* in the absence of sufficient grazing. This a continuing threat, though short turf is maintained for the present by rabbit-grazing, together with occasional strimming or cutting.

An examination of its widespread international distribution clearly indicates that *V. rupestris* does not belong to the oceanic element of the British flora. In recognition of this, Valentine & Harvey (1961) predicted that this species might be found in the Craven district of Yorkshire, at a time when only the Arnside, Widdybank Fell and Long Fell populations were known. They also suggested that *V. rupestris* may be found in other limestone areas in Britain.

It has its stronghold in eastern Europe but can be found throughout Eurasia from England in the west to Sakhalin in the Far East. It is scarce in western Europe and is confined to montane areas in the southern parts of its range. Glabrous and pubescent forms occur in central Europe and these have been recognised as subspecies.

However, the two forms often occur together, and Kirschner & Skalicky (1989) suggest that sub-specific taxa cannot be distinguished on the basis of this character.

I. Taylor

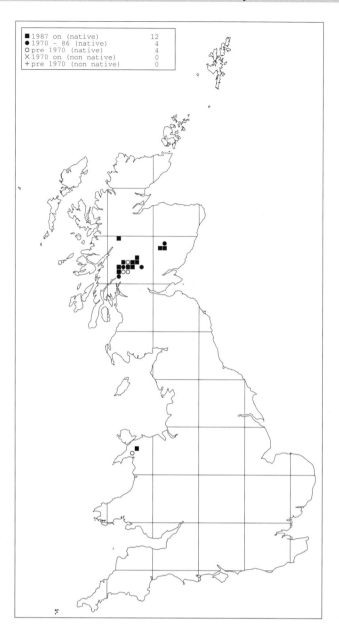

Woodsia alpina (Bolton) Gray (Woodsiaceae)

Alpine woodsia, Raineach Mhion Ailpeach, Coredynen Alpaidd
Status in Britain: LOWER RISK - Near Threatened. WCA
Schedule 8.
Status in Europe: Not threatened.

W. alpina occupies an extremely specialised habitat,
namely the steep, bare faces of strongly calcareous rocks,
within the montane zone, at an altitude of 525-915 m. It
grows on Ordovician pumice tuffs in Snowdonia,
Dalradian mica-schists and limestones in Perthshire and
Angus, hornblende schist in Angus, and Moine schist in
West Inverness. *W. alpina* is always on treeless crags in
Britain and grows where competition is minimal, rooting
in small crevices and niches of knobbly and uneven rock
surfaces with only the slightest accumulation of soil.
Aspect varies widely, though plants on south-facing rocks
can become quite desiccated in dry summers. The
immediate associates are most often rupestral lichens and
mosses, but vascular plants usually in fairly close company

are *Asplenium viride*, *A. trichomanes*, *Campanula rotundifolia*,
Cerastium alpinum, *Galium boreale*, *Poa alpina*, *Saxifraga
nivalis*, *S. oppositifolia*, *Sedum rosea* and *Silene acaulis*. In
number and size of colonies, this is the less rare of our two
Woodsia species.

Under exceptional conditions, fronds reach a length of
up to 15 cm (Page 1982). Where undisturbed, *W. alpina*
appears to maintain its populations by vegetative renewal of
long-established plants, the creeping rhizomes sending up
new crowns along their length. Monitoring between 1977
and 1994 has shown that many populations have been
remarkably stable. However, some can be dynamic, as
shown by one population in which the number of clumps
increased from eight to 10 between these years, but only
three were common to both 1977 and 1994 (Fleming 1995).

W. alpina was first reported by Knowlton from
Snowdonia in 1790 (Hyde *et al.* 1978). It is geographically
more restricted than *W. ilvensis* in Britain and is found

only in Caernarvonshire, Argyll, Perthshire, Angus and West Inverness. The hills between Glen Turret, Perthshire, and Glen Fyne, Argyll, and especially the Breadalbane range, are the headquarters of this fern, which is known from at least 14 different hills here. Even in its Breadalbane stronghold it is unaccountably scarce and is absent from many suitable-looking cliffs - a true relict from a much colder age. Eastern outliers are in Caenlochan and Glen Doll, and the northernmost locality is on a hill above the Great Glen. *W. alpina* is still on at least two cliffs in Snowdonia and was formerly reported from two others. In all, there are thus at least 19 separate populations. Some are difficult to count accurately, since they are partly or wholly inaccessible, and a few are spread over lofty precipices where even viewing through binoculars is difficult. Colonies range from a few tufts to several hundreds. The hybrid between *W. alpina* and *W. ilvensis* has not been satisfactorily confirmed from Britain, despite one or two old reports (Rickard 1972).

This species was seriously depleted in its main Snowdon locality by Victorian fern collectors, and there are large numbers of specimens from Scotland in herbaria. Even during the last 35 years, it has declined in accessible places on the Ben Lawers range and Caenlochan. Most of its sites are NNRs or SSSIs. Few of its cliffs are climbed upon, so 'gardening' by rock climbers is hardly a problem.

W. alpina is widespread in Alpine and Boreal-Arctic regions of both Old and New Worlds and appears to be a plant of calcareous rocks wherever it occurs. In Scandinavia, some populations are on rocks within the Boreal forest zone, and its habitats include large, lichen-crowned blocks in rather open pinewood.

J. Mitchell and D.A. Ratcliffe

Woodsia ilvensis (L.) R.Br. (Woodsiaceae)

Oblong woodsia, Raineach Mhion Fhad-shliosach, Coredynen Hirgul
Status in Britain: ENDANGERED. WCA Schedule 8.
Status in Europe: Not threatened.

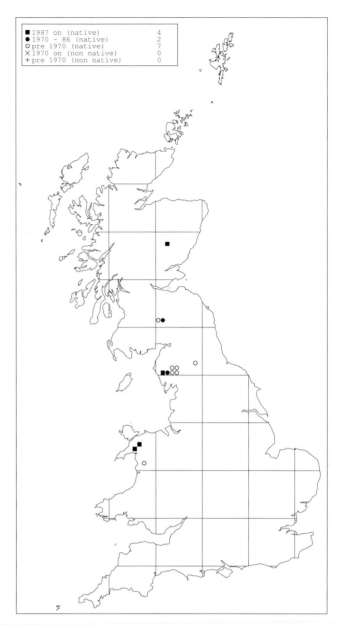

A plant of open rock habitats with freedom from competition, *W. ilvensis* grows on both basic and acidic substrates: calcareous pumice tuff in Snowdonia and hornblende schist in Clova, but non-calcareous Borrowdale Volcanic rocks in Lakeland and Silurian greywackes and shales in the Moffat Hills. Its associates on base-rich rocks suggest that these have become leached and somewhat acidified at the surface. Most colonies are on treeless crags and small outcrops, at 365-760 m, over a wide range of aspects. One colony is on cliffs and montane screes abutting ancient native pine - juniper woodland (Page 1988). Individual tufts usually root in the crevices and recesses of steep, dry rocks with virtual absence of soil, but the largest extant British colony, in the Lake District, grows on shattered, unstable rocks with earthy niches and ledges. *W. ilvensis* does not belong to any particular community, and its typical associates are other plants of dry mountain rocks, including a wide variety of rupestral bryophytes (especially *Racomitrium* species) and lichens, and vascular species such as *Alchemilla alpina, Calluna vulgaris, Cryptogramma crispa, Deschampsia flexuosa, Festuca ovina, F. vivipara, Galium saxatile, Saxifraga hypnoides, Sedum rosea, Silene uniflora* and *Thymus polytrichus*.

W. ilvensis was first discovered in Britain in Snowdonia by Lhuyd in 1690 (Hyde *et al.* 1978), though some of the early records were confused with *W. alpina*. It is the more widespread of the two species, with confirmed records from Caernarvonshire, Cumberland, Dumfriesshire, Angus, Inverness-shire (not mapped), Merioneth, Durham and Westmorland, though it is no longer found in the last three counties. It is, however, much the rarer, with only eight extant colonies. The finest remaining colony numbered 73 separate tufts in 1996 (V. Fleming *in litt.*), some of which have waxed and waned over the past 40 years. The largest of these reaches a diameter of 25 cm with up to 100 fronds during favourable periods. Growth was most luxuriant during a wet summer (1954), when many fronds reached 18 cm in length, approaching the size commonly seen outside Britain. Most other colonies fall far short of this tuft and frond size and either exist under sub-optimal conditions or have suffered some kind of genetic attenuation. A total of not more than 30 clumps are known at the other seven sites, with five sites holding three or fewer clumps. The Moffat Hills of Dumfriesshire were said once to have had many hundreds of plants, in at least five sites (Mitchell 1980), but *W. ilvensis* now survives at only two of these sites, with three clumps at one and a single clump at the other (V. Fleming pers. comm.). One of these has declined from 22 tufts in 1954 to only three in 1994, apparently through natural causes. In Wales, there are only two sites (in 1996), with a total of 12 plants.

The rarity of the plant in Britain is puzzling, given that its habitats appear not to be special and indeed are widespread. It is clearly a relict from colder late-glacial or early post-glacial times. Although sporophytes have been produced from spores in cultivation, the wild plant appears to be unable to reproduce and spread naturally in this way here and persists by the rejuvenation of long-

lived plants. Yet in Norway it is able to colonise walls and road construction rubble. It is possible that colonies in Britain are now suffering from inbreeding depression and genetic drift, and this may pose the greatest threat. This fern has been made rarer by collecting, and few British plants have suffered so severely. The only certain Pennine colony, on Falcon Clints in Upper Teesdale, was eradicated over a century ago, and only one of nine former Lakeland sites is now known to hold the fern. Some populations are vulnerable to rock falls. Most remaining colonies are within NNRs or SSSIs.

Abroad, *W. ilvensis* is a widespread Alpine and Boreal-Arctic species of both the Old and New Worlds, growing on both mildly or partially calcareous and non-calcareous rocks. In Scandinavia and North America its habitats are especially cliffs and stable screes within the Boreal forest zone, but where tree-cover is open and light intensity high.

D.A. Ratcliffe and J. Mitchell

6.3 Native species extinct in the wild

Square brackets around the year of last record (column 2) denote that the species has since been reintroduced.

Species	Year of last record in the wild in GB	Habitat	Comments/reasons for losses
Agrostemma githago	1970s?	Arable	Lost through agricultural intensification; now only casual, or deliberately planted.
Arnoseris minima	1970	Arable	Lost through agricultural intensification.
Bromus interruptus	1972	Arable	Agricultural intensification; native plants maintained in cultivation.
Bupleurum rotundifolium	1960s	Arable	Agricultural intensification; now only a rare casual; the rather more frequent casual *B. subovatum* is sometimes mistaken for it.
Carex davalliana	c. 1845	Calcareous fen	Drainage: known at only one site, in Somerset, until about 1845, possibly formerly native.
Crepis foetida	[1980]	Coastal habitats	Reintroduced to Dungeness.
Euphorbia peplis	1965	Sand/shingle beach	Still occurs in the Channel Isles.
Filago gallica	[1955]	Arable	Reintroduced to former site in Essex.
Galeopsis segetum	1980s?	Arable	Now perhaps only a rare casual.
Holosteum umbellatum	1930	Arable fields, old walls, thatched roofs, banks	Presumed native on light soils; not seen since 1930.
Hydrilla verticillata	c. 1934	Freshwater	Known only between 1915 and about 1934 in Esthwaite Water.
Neotinea maculata	1986?	Dune/heath	Isle of Man only, but now perhaps extinct.
Otanthus maritimus	1970s?	Dunes, shingle	Last known in GB in Cornwall and Sussex; reason for loss of this predominantly Mediterranean species perhaps climate change.
Pinguicula alpina	Just after 1900?	Upland bogs	Recorded from only one site; habitat degradation the presumed reason for its loss. There may be some doubt that this plant is correctly named.
Rubus arcticus	1841	Montane	Several records from the Scottish Highlands.
Sagina boydii	1878?	Montane?	Presumed to have been collected from the wild in S. Aberdeen in 1878, but not seen in the wild since, though still in cultivation.
Saxifraga rosacea ssp. *rosacea*	c. 1974	Montane	Native stock still in cultivation.
Spiranthes aestivalis	1959	Bogs	Lost because of drainage of bogs, and also from collecting (large numbers in herbaria).
Tephroseris palustris ssp. *congesta* (*Senecio congestus*)	1899	Fen ditches	Formerly local from Sussex to Yorkshire, and lost through drainage.
Trichophorum alpinum (*Scirpus hudsonianus*)	c. 1813	Bogs	Known only from one bog in Angus from 1791 to about 1813 when the habitat was dredged, then flooded.

6.4 Excluded taxa

Non-native species, introductions, casuals, escapes from cultivation, varieties and hybrids (including those between native species) and dubious taxa are excluded. The following taxa were included in the 2nd edition Red Data Book (Perring & Farrell 1983) but are treated here as non-natives.

Ajuga genevensis
Alyssum alyssoides
Anisantha tectorum
Anthoxanthum aristatum
Campanula persicifolia
Campanula rapunculus
Caucalis platycarpos
Centaurium latifolium
Crocus vernus
Cyclamen hederifolium
Elatine hydropiper
Epipactis leptochila var. *dunensis*
Equisetum ramosissimum
Euphorbia villosa
Euphrasia eurycarpa
Euphrasia rhumica
Galium fleurotii
Galium spurium
Gentianella anglica 'ssp. *cornubiensis'*

Iris spuria
Iris versicolor
Isatis tinctoria
Juncus filiformis
Juncus nodulosus
Juncus subulatus
Juncus tenuis var. *dudleyi*
Ledum groenlandicum
Leucojum vernum
Linaria supina
Matthiola incana
Narcissus obvallaris
Oenothera stricta
Orobanche minor var. *maritima*
Paeonia mascula
Polygala 'austriaca'
Rorippa austriaca
Sagina x *normaniana*
Sagittaria rigida
Silene italica
Sisymbrium irio
Spartina alterniflora
Tetragonolobus maritimus
Trifolium stellatum
Veronica praecox

7   Habitats of rare species in Britain

In Stewart *et al.* (1994), most of the Nationally Scarce species described in that volume were assigned to one of ten main habitat types, with a few species remaining unassigned. In order that direct comparisons can be made, nationally rare species have been assigned to the same habitat types in the present book. Attention is drawn to the excellent descriptions of habitats by C.D. Preston and D.A. Pearman in Stewart *et al.* (1994), and reference should be made to these for overviews of historical change, current state, threats, and land use and management. Many of the remarks in these habitat descriptions are just as appropriate to our nationally rare species.

The representation of rare species in each of the 10 habitat types is shown below under the habitat sub-heads (six species could not easily be placed and remain unassigned). Table 7 shows the numbers of rare species in each of the threat categories in each habitat, and the data are graphically displayed in Figure 3.

Table 7   Occurrence of *Threatened, Lower Risk (near threatened)* and *Data Deficient* species in habitat formations

Habitat	CR	EN	VU	LR(nt)	Percentage CR + EN	Total Threatened and LR(nt)		Nationally Scarce **	
	No. of species	No. of species	No. of species	No. of species	%	No.	%	No.	%
Montane habitats	0	3	22	14	8	39	14	38	14
Heaths, moors & bogs	0	1	5	9	7	15	5	13	5
Neutral and acid grassland	3	4	10	6	30	23	8	10	4
Calcareous grassland and rock	4	12	31	15	26	62	22	44	17
Fens and basic flushes	1	5	7	5	33	18	6	17	6
Water margins and damp mud	3	4	8	4	37	19	7	8	3
Freshwater habitats	0	0	3	2	0	5	2	13	5
Woods, scrub and hedges	5	6	11	4	42	26	9	28	11
Cultivated land	3	2	3	1	56	9	3	13	5
Coastal habitats	4	6	33	21	15	64*	22*	61	23
Unclassified habitats	1	2	2	1	50	6	2	18	7
Total	24	45	134	83	24	286	-	263	-

Key: CR = Critically Endangered; EN = Endangered; VU = Vulnerable; LR (nt) = Lower Risk (near threatened); *55 species (19%) excluding subspecies of *Limonium*; **data from Stewart *et al.* (1994).

Figure 3   Occurrence of rare species in threat categories and habitat types

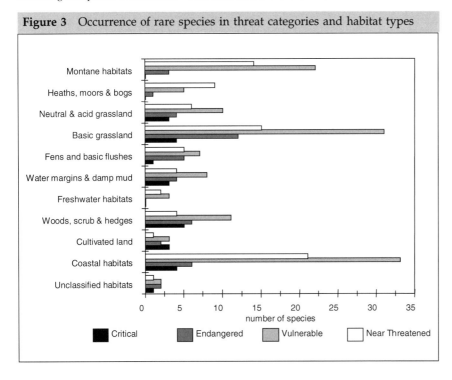

■ Critical   ■ Endangered   ■ Vulnerable   □ Near Threatened

The habitat types are, for the most part, well differentiated ecologically, though because they have not been precisely defined (e.g. the ecological and altitudinal limits of the 'montane' region), the placing of some species is a matter of interpretation. Most species occur predominantly in a single broad habitat type. A few are found with similar or different frequencies in more than one habitat type (e.g. *Cystopteris dickieana* and *Oxytropis halleri* in both montane and coastal, and *Physospermum cornubiense* in woodland, grassland and heath), but each species has been assigned to only one group in the lists and tables. The analysis of the data should, therefore, be viewed in that light.

The proportions of species in the threat categories in each habitat type may reflect the degree of threat to the habitat and the effectiveness of conservation measures.

Montane habitats

Of the 39 rare species of montane habitats, almost all are placed in the *Vulnerable* and *Lower Risk (near threatened)* categories, with only three species (8%) qualifying as *Endangered*. Of these three, *Arabis alpina* and *Homogyne alpina* qualify on the basis of their small population size, and only *Woodsia ilvensis* on the basis of significant and continuing decline. This picture is not surprising since many montane species are relict species confined to few sites, with populations that have been fairly stable over many years, in habitats that are generally less subject to man's activities than are lowland ones. However, many potential threats are mentioned in the accounts of montane species, and of these, increased grazing by domestic stock and deer, and recreational activities, may be the most immediate. Changes in weather patterns brought on by global warming are likely to have an impact on our montane flora, though the outcomes are largely unpredictable.

Arabis alpina (EN)
Arenaria norvegica ssp. *norvegica* (NT)
Artemisia norvegica (VU)
Astragalus alpinus (VU)
Athyrium flexile (VU)
Bartsia alpina (NT)
Carex atrofusca (NT)
Carex lachenalii (NT)
Carex microglochin (VU)
Carex norvegica (VU)
Carex rariflora (NT)
Cerastium fontanum ssp. *scoticum* (VU)
Cicerbita alpina (VU)
Cystopteris dickieana (VU)
Cystopteris montana (NT)
Diapensia lapponica (VU)
Erigeron borealis (VU)
Euphrasia cambrica (VU)
Euphrasia rivularis (NT)
Gentiana nivalis (VU)

Helianthemum canum ssp. *levigatum* (VU)
Homogyne alpina (EN)
Koenigia islandica (NT)
Lloydia serotina (VU)
Luzula arcuata (NT)
Lychnis alpina (VU)
Minuartia rubella (NT)
Myosotis alpestris (NT)
Oxytropis campestris (VU)
Phyllodoce caerulea (VU)
Poa flexuosa (VU)
Sagina nivalis (VU)
Salix lanata (VU)
Saxifraga cernua (VU)
Saxifraga cespitosa (VU)
Saxifraga rivularis (NT)
Veronica fruticans (NT)
Woodsia alpina (NT)
Woodsia ilvensis (EN)

Heaths, moors and bogs

Relatively few *Threatened* and *Lower Risk (near threatened)* species have been assigned to these inherently rather species-poor habitats, and of the 15 species, only five are deemed to be threatened. The small numbers of *Minuartia stricta* render it *Endangered*, though there is no evidence of any overall decline in the past 20 years

(arguably this species could have been placed in the montane group). On the other hand, there seems to have been a considerable decline in populations of *Euphrasia vigursii*. *Scheuchzeria palustris* is assigned with doubt to the *Vulnerable* category because of possible future threats from afforestation and drainage, but it might better be regarded as *Lower Risk (near threatened)*.

Diphasiastrum issleri (NT)
Erica ciliaris (NT)
Erica vagans (NT)
Eriophorum gracile (VU)
Euphrasia vigursii (VU)
Genista pilosa (NT)
Herniaria ciliolata ssp. *ciliolata* (NT)
Isoetes histrix (NT)

Juncus capitatus (NT)
Minuartia stricta (EN)
Ophioglossum lusitanicum (VU)
Ornithopus pinnatus (NT)
Saxifraga hirculus (VU)
Scheuchzeria palustris (VU)
Thymus serpyllum (NT)

Neutral and acid grassland, and acid rock

The 23 species in this group grow in a wide diversity of habitats. Eleven occur in neutral grasslands, including hay-meadows (e.g. *Alchemilla* spp.), flood-meadows (e.g. *Lythrum hyssopifolia*) and grazed pasture (e.g. *Apium repens, Lactuca saligna*), and the remainder on acid substrates. Seven species (30%) are regarded as *Critical* or *Endangered*, and of these, all but one require an open community or bare soil for maintaining their populations. Many of the sites for species in this group are not specially protected and are all too vulnerable to agricultural improvement, cultivation, or drainage. For example, the rare *Alchemilla* species depend on extensive management of their upland hay-meadows, and the survival of *Lactuca saligna* in Essex on trampling and grazing of grassy banks by cattle.

Alchemilla acutiloba (NT)
Alchemilla monticola (NT)
Alchemilla subcrenata (EN)
Apium repens (CR)
Armeria maritima ssp. *elongata* (VU)
Eryngium campestre (VU)
Festuca longifolia (VU)
Gagea bohemica (VU)
Gladiolus illyricus (NT)
Herniaria glabra (NT)
Hypericum linariifolium (NT)
Lactuca saligna (EN)

Lobelia urens (VU)
Lythrum hyssopifolia (VU)
Orobanche reticulata (NT)
Petrorhagia prolifera (CR)
Scleranthus perennis ssp. *perennis* (CR)
Scleranthus perennis ssp. *prostratus* (EN)
Scorzonera humilis (VU)
Sorbus arranensis (VU)
Sorbus pseudofennica (VU)
Tordylium maximum (EN)
Veronica verna (VU)

Calcareous grassland and basic rock

Grasslands on chalk and other limestones (and on other basic substrates, including serpentine) are often characterised by their great diversity of species, including many that are now rare. These habitats support more rare species than any other habitat formation (unless all the sub-taxa of *Limonium* are taken into account), of which 16 (25%) are *Critical* or *Endangered*. About 13 rare species are confined to the chalk and one (*Cerastium nigrescens*) to serpentine. *Carex muricata* ssp. *muricata* and *Gentianella ciliata* are declining and appear to be the most threatened of the species in this habitat group. *Sorbus leptophylla* and *S. leyana* must be placed in the *Critical* category because of their small numbers, but their populations are not in decline. The grassland species vary in their habitat requirements. Those that require a short turf or gaps in the sward are threatened in many sites by the rank growth of robust grasses (such as *Brachypodium pinnatum*) and scrub, brought about by a relaxation of grazing.

Ajuga chamaepitys (VU)
Alchemilla glaucescens (NT)
Alchemilla gracilis (VU)
Alchemilla minima (VU)
Allium sphaerocephalon (EN)
Althaea hirsuta (EN)
Anisantha madritensis (VU)
Arabis glabra (VU)
Arabis scabra (VU)
Arenaria norvegica ssp. *anglica* (VU)
Artemisia campestris (EN)
Bunium bulbocastanum (NT)
Carex filiformis (NT)
Carex muricata ssp. *muricata* (CR)
Carex ornithopoda (NT)
Cerastium brachypetalum (EN)
Cerastium nigrescens (VU)
Cirsium tuberosum (VU)
Cotoneaster integerrimus (EN)
Crepis praemorsa (EN)
Cypripedium calceolus (CR)
Dianthus armeria (VU)
Dianthus gratianopolitanus (VU)
Filago pyramidata (EN)
Gentiana verna (NT)
Gentianella ciliata (CR)
Helianthemum apenninum (NT)
Himantoglossum hircinum (VU)
Hypochaeris maculata (VU)
Koeleria vallesiana (NT)
Lychnis viscaria (VU)

Melampyrum arvense (EN)
Muscari neglectum (VU)
Ophrys fuciflora (VU)
Ophrys sphegodes (NT)
Orchis militaris (VU)
Orchis simia (VU)
Orobanche artemisiae-campestris (EN)
Orobanche caryophyllea (VU)
Orobanche purpurea (VU)
Phleum phleoides (NT)
Pilosella flagellaris ssp. *bicapitata* (VU)
Pilosella peleteriana (VU)
Potentilla fruticosa (NT)
Potentilla rupestris (VU)
Polemonium caeruleum (NT)
Polygala amarella (VU)
Rhinanthus angustifolius (VU)
Seseli libanotis (VU)
Silene otites (NT)
Sorbus anglica (NT)
Sorbus eminens (VU)
Sorbus leptophylla (EN)
Sorbus leyana (CR)
Sorbus minima (VU)
Stachys germanica (EN)
Teucrium botrys (VU)
Teucrium chamaedrys (EN)
Thlaspi perfoliatum (VU)
Trinia glauca (NT)
Veronica spicata ssp. *spicata* (VU)
Viola rupestris (NT)

Fen, marsh and lowland calcareous flushes

The 18 species in this group, of which 13 are considered threatened, occupy a range of mesotrophic and basic, mainly lowland, mire habitats. There has been a long history of drainage of lowland mires, and few large areas remain. The consequent loss of species has been particularly marked in East Anglia, where some species (e.g. *Teucrium scordium* and *Viola persicifolia*) are now highly restricted. Whilst many sites for rare fenland species seem secure within specially protected areas, the lowering of water tables resulting from the drainage of adjacent areas, or drought, is an ever present threat. *Dactylorhiza incarnata* ssp. *ochroleuca* may be one of the most threatened taxa in Britain, with few plants remaining, in one site only. *Liparis loeselii* is also characteristic of dune-slacks (which now hold its largest populations), and could have been placed in the coastal habitats group.

Calamagrostis purpurea (NT)
Calamagrostis stricta (NT)
Carex buxbaumii (VU)
Carex chordorrhiza (VU)
Carex flava (VU)
Dactylorhiza incarnata ssp. *cruenta* (EN)
Dactylorhiza incarnata ssp. *ochroleuca* (CR)
Dactylorhiza lapponica (NT)

Dryopteris cristata (NT)
Kobresia simpliciuscula (NT)
Liparis loeselii (EN)
Luzula pallidula (VU)
Ranunculus ophioglossifolius (EN)
Schoenus ferrugineus (VU)
Selinum carvifolia (VU)
Teucrium scordium (VU)
Viola canina ssp. *montana* (EN)
Viola persicifolia (EN)

Freshwater margins and damp mud

Species in this group are mainly small, often annual, species of shallow water, mud seasonally exposed at the edges of ponds and watercourses, and of damp short-grazed swards near water. Six of these species (33%) are placed in the Critical and *Endangered* categories, and most have suffered a decline. *Schoenoplectus triqueter* may now be the rarest of Britain's native plants. The lack of

traditional management of ponds has long been recognised as a threat to many species of margins (e.g. *Cyperus fuscus, Damasonium alisma, Mentha pulegium*), though conservation management together with site protection has given them a greater degree of security in recent years. *Juncus pygmaeus* seems to have undergone a severe decline in recent years, though because of the long-persistence of its seed in the soil, many of its populations can probably be revived, given appropriate management of the habitat.

Alisma gramineum (CR)
Calamagrostis scotica (VU)
Carex vulpina (VU)
Crassula aquatica (VU)
Cyperus fuscus (VU)
Damasonium alisma (EN)
Eleocharis austriaca (NT)
Eriocaulon aquaticum (NT)
Galium constrictum (NT)
Juncus pygmaeus (EN)

Leersia oryzoides (EN)
Ludwigia palustris (NT)
Mentha pulegium (VU)
Pulicaria vulgaris (VU)
Ranunculus reptans (EN)
Ranunculus tripartitus (VU)
Rumex aquaticus (VU)
Schoenoplectus triqueter (CR)
Senecio paludosus (CR)

Freshwater

Though there are few aquatic species that are nationally rare, many others have undergone a significant decline in recent years because of eutrophication, pollution, increased recreation, drainage of ditches and the canalisation of waterways. One species, *Potamogeton acutifolius*, has moved to the threatened category in the past 10 years, and if current trends continue, others may follow (e.g. *P. compressus*, which has shown a marked decline). *Najas flexilis*, regarded as rare in Perring & Farrell (1983) and Stewart *et al.* (1994), is now known in 19 hectads and is treated as *Nationally Scarce*. Only one aquatic species, *Hydrilla verticillata*, has become extinct (not seen after 1934) since botanical recording began, though it has been recorded from only one site, and the reasons for its loss are not known.

Najas marina (VU)
Potamogeton acutifolius (VU)
Potamogeton epihydrus (VU)

Potamogeton nodosus (NT)
Potamogeton rutilus (NT)

Woodland, scrub and hedges

Of the 26 rare taxa of woodlands and related habitats, 11 (42%) are considered to be especially threatened because of their extreme geographical restriction or small numbers. *Cynoglossum germanicum* has shown the greatest decline of any woodland species: it has been lost from many sites, and must be regarded as extremely threatened outside its main range in Surrey. Many light-demanding woodland species, (e.g. *Euphorbia serrulata, Phyteuma spicatum*), even if not eliminated from sites, have declined because of the cessation of coppice rotation or the lack of other management regimes that had hitherto maintained an open canopy. In such circumstances, some plants may persist for a while as a weak form and without flowering, but most will eventually succumb. However, seed of some species (*Carex depauperata, Euphorbia serrulata*) is known to have considerable longevity, and populations may often be revived after long neglect of the habitat.

Buxus sempervirens (NT)
Carex depauperata (CR)
Cephalanthera rubra (CR)
Clinopodium menthifolium (EN)
Cynoglossum germanicum (VU)

Epipactis youngiana (EN)
Epipogium aphyllum (EN)
Euphorbia hyberna (VU)
Euphorbia serrulata (VU)
Leucojum aestivum (NT)

Lithospermum purpureocaeruleum (NT)
Lonicera xylosteum (EN)
Maianthemum bifolium (VU)
Moneses uniflora (VU)
Physospermum cornubiense (VU)
Phyteuma spicatum (VU)
Polygonatum verticillatum (VU)
Pulmonaria obscura (VU)

Pyrus cordata (EN)
Sorbus bristoliensis (EN)
Sorbus domestica ((CR)
Sorbus lancastriensis (NT)
Sorbus subcuneata (VU)
Sorbus vexans (VU)
Sorbus wilmottiana (CR)
Stachys alpina (EN)

Cultivated land

Changes in farming practice have had an enormous impact on the weed flora of arable land, and the effect on species is a measure of the threat to the habitat. Many species have become scarce or rare, and seven have become extinct in the wild since 1970 (only seven other species have become extinct this century - see Section 6.3). Of the nine species placed in this habitat group, five (56%) are regarded as *Critical* or *Endangered*, the largest percentage of species in any of the habitat groups. Of these, *Galium tricornutum* may be amongst the most threatened plants in Britain. As mentioned in section 3.4, other declining species of arable fields (such as *Scandix pecten-veneris* and *Torilis arvensis*) might also have qualified as threatened, though this is uncertain because of the lack of recent data.

Adonis annua (VU)
Centaurea cyanus (EN)
Echium plantagineum (EN)
Filago gallica (CR)
Filago lutescens (VU)

Galium tricornutum (CR)
Lavatera cretica (VU)
Polycarpon tetraphyllum (NT)
Valerianella rimosa (CR)

Coastal habitats

A total of 65 rare taxa occur in coastal habitats (55 taxa if subspecies of *Limonium* are not listed separately), though only 10 (15%) are considered to be *Critical* or *Endangered*. Many coastal species are very localised geographically, or confined to very few sites, though some have been fairly stable in terms of geographical distribution and population size over many years. Several species (e.g. *Atriplex pedunculata, Crepis foetida, Corrigiola litoralis, Petrorhagia nanteuilii, Romulea columnae*) are confined to a single locality, though not all are severely threatened. Not unexpectedly, species of cliffs and cliff-top grassland are among the least threatened of our rare species, and those of low habitats (dune, shingle, foreshore) among the most threatened.

Although much of the coastal region is protected and rare species are to some degree safeguarded within protected areas, coastal habitats and species remain vulnerable to a wide range of damaging activities, including commercial development, impacts of tourism, excavation of sand and shingle, construction of hard sea-defences, and estuary barrages. The reduction or lack of grazing of coastal grassland habitats in recent years poses an additional threat to species requiring short swards. Sea-level rise will lead to the steepening of coastal profiles on soft coasts and the loss of low habitats, and increased storminess (considered a possible product of global warming) would threaten dune and shingle habitats, which support several threatened species, including *Corrigiola litoralis* and *Petrorhagia nanteuilii*.

Anthyllis vulneraria ssp. corbierei (NT)
Asparagus prostratus (VU)
Aster linosyris (NT)
Atriplex pedunculata (CR)
Bupleurum baldense (EN)
Carex recta (VU)
Centaurium scilloides (VU)
Centaurium tenuiflorum (VU)
Chenopodium chenopodioides (NT)
Chenopodium vulvaria (VU)
Coincya wrightii (VU)
Corrigiola litoralis (CR)
Corynephorus canescens (NT)
Crepis foetida (EN)
Cynodon dactylon (VU)
Cytisus scoparius ssp. maritimus (VU)
Draba aizoides (NT)
Eleocharis parvula (VU)
Euphrasia campbelliae (NT)
Euphrasia heslop-harrisonii (NT)
Euphrasia marshallii (NT)
Euphrasia rotundifolia (EN)
Gentianella uliginosa (VU)
Gnaphalium luteoalbum (CR)
Hierochloe odorata (NT)
Limonium bellidifolium (NT)
Limonium binervosum ssp. anglicum (NT)
Limonium binervosum ssp. binervosum (NT)
Limonium binervosum ssp. cantianum (VU)
Limonium binervosum ssp. mutatum (VU)
Limonium binervosum ssp. saxonicum (NT)
Limonium britannicum ssp. britannicum (NT)
Limonium britannicum ssp. celticum (NT)

Limonium britannicum ssp. coombense (NT)
Limonium britannicum ssp. transcanalis (NT)
Limonium dodartiforme (VU)
Limonium loganicum (VU)
Limonium paradoxum (VU)
Limonium parvum (VU)
Limonium procerum ssp. cambrense (VU)
Limonium procerum ssp. devoniense (VU)
Limonium recurvum ssp. portlandicum (VU)
Limonium recurvum ssp. humile (NT)
Limonium recurvum ssp. recurvum (VU)
Limonium transwallianum (VU)
Limosella australis (VU)
Matthiola sinuata (VU)
Mibora minima (NT)
Ononis reclinata (VU)
Oxytropis halleri (NT)
Petrorhagia nanteuilii (EN)
Peucedanum officinale (NT)
Polygonum maritimum (EN)
Romulea columnae (VU)
Rumex rupestris (EN)
Scirpoides holoschoenus (VU)
Spergularia bocconei (CR)
Tephroseris integrifolia ssp. maritima (VU)
Trifolium bocconei (VU)
Trifolium incarnatum ssp. molinerii (VU)
Trifolium strictum (VU)
Tuberaria guttata (VU)
Viola kitaibeliana (VU)
Valerianella eriocarpa (VU)

Ruderal and unclassified habitats

Bupleurum falcatum (CR)
Centaurea calcitrapa (VU)
Fumaria reuteri (EN)

Senecio cambrensis (NT)
Trichomanes speciosum (VU)
Veronica triphyllos (EN)

Appendix 1 The IUCN criteria for Critically Endangered, Endangered and Vulnerable species (IUCN 1994)

Critically Endangered (CR)

A taxon is *Critically Endangered* when it is facing an extremely high risk of extinction in the wild in the immediate future, as defined by any of the following criteria (A to E):

A. Population reduction in the form of either of the following:

1 an observed, estimated, inferred or suspected reduction of at least 80% over the last 10 years or three generations, whichever is the longer, based on (and specifying) any of the following:
   a. direct observation
   b. an index of abundance appropriate for the taxon
   c. a decline in area of occupancy, extent of occurrence and/or quality of habitat
   d. actual or potential levels of exploitation
   e. the effects of introduced taxa, hybridisation, pathogens, pollutants, competitors or parasites;

2 a reduction of at least 80%, projected or suspected to be met within the 10 years or three generations, whichever is the longer, based on (and specifying) any of b, c, d or e above.

B. Extent of occurrence estimated to be less than 100 km^2 or areas of occupancy estimated to be less than 10 km^2, and estimates indicating any two of the following:

1 severely fragmented or known to exist at only a single location;

2 continuing decline, observed, inferred or projected, in any of the following:
   a. extent of occurrence
   b. area of occupancy
   c. area, extent and/or quality of habitat
   d. number of locations or sub-populations
   e. number of mature individuals;

3 extreme fluctuations in any of the following:
   a. extent of occurrence
   b. area of occupancy
   c. number of locations or sub-populations
   d. number of mature individuals.

C. Population estimated to number less than 250 mature individuals and either:

1 an estimated continuing decline of at least 25% within three years or one generation, whichever is longer, or

2 a continuing decline, observed, projected, or inferred, in numbers of mature individuals and population structure, in the form of either
   a. severely fragmented (i.e. no sub-population estimated to contain more than 50 mature individuals), or
   b. all individuals are in a single sub-population.

D. Population estimated to number less than 50 mature individuals.

E. Quantitative analysis showing the probability of extinction in the wild at least 50% within 10 years or three generations, whichever is the longer.

Endangered (EN)

A taxon is *Endangered* when it is not *Critically Endangered* but is facing a very high risk of extinction in the wild in the near future, as defined by any of the following criteria (A to E):

A. Population reduction in the form of either of the following:

1 an observed, estimated, inferred or suspected reduction of at least 50% over the last 10 years or three generations, whichever is the longer, based on (and specifying) any of the following:
   a. direct observation
   b. an index of abundance appropriate for the taxon
   c. a decline in area of occupancy, extent of occurrence and/or quality of habitat
   d. actual or potential levels of exploitation
   e. the effects of introduced taxa, hybridisation, pathogens, pollutants, competitors or parasites;

2 a reduction of at least 50%, projected or suspected to be met within the next ten years or three generations, whichever is the longer, based on (and specifying) any of b, c, d, or e above.

B. Extent of occurrence estimated to be less than 5,000 km^2 or area of occupancy estimated to be less than 500 km^2, and estimates indicating any two of the following:

1 sverely fragmented or known to exist at no more than five locations;

2 continuing decline, inferred, observed or projected, in any of the following:
   a. extent of occurrence
   b. area of occupancy
   c. area, extent and/or quality of habitat
   d. number of locations or sub-populations
   e. number of mature individuals;

3 extreme fluctuations in any of the following:
   a. extent of occurrence
   b. area of occupancy
   c. number of locations or sub-populations
   d. number of mature individuals.

C. Population estimated to number less than 2,500 mature individuals and either:

1 an estimated continuing decline of at least 20% within five years or two generations, whichever is longer, or

2 a continuing decline, observed, projected or inferred, in numbers of mature individuals and population structure in the form of either:

a. severely fragmented (i.e. no sub-population estimated to contain more than 250 mature individuals)

b. all individuals are in a single sub-population.

D. Population estimated to number less than 250 mature individuals.

E. Quantitative analysis showing the probability of extinction in the wild is at least 20% within 20 years or five generations, whichever is the longer.

Vulnerable (VU)

A taxon is *Vulnerable* when it is not *Critically Endangered* or *Endangered* but is facing a high risk of extinction in the wild in the medium-term future, as defined by any of the following criteria (A to E):

A. Population reduction in the form of either of the following:

1 an observed, estimated, inferred or suspected reduction of at least 20% over the last 10 years or three generations, whichever is the longer, based on (and specifying) any of the following:
   a. direct observation;
   b. an index of abundance appropriate for the taxon;
   c. a decline in area of occupancy, extent of occurrence and/or quality of habitat;
   d. actual or potential levels of exploitation;
   e. the effects of introduced taxa, hybridisation, pathogens, pollutants, competitors or parasites;

2 a reduction of at least 20%, projected or suspected to be met within the next 10 years or three generations, whichever is the longer, based on (and specifying) any of b, c, d or e above.

B. Extent of occurrence estimated to be less than 20,000 km^2 or area of occupancy estimated to be less than 2,000 km^2, and estimates indicating any two of the following:

1 severely fragmented or known to exist at no more than ten locations;

2 continuing decline, inferred, observed or projected, in any of the following:
   a. extent of occurrence
   b. area of occupancy
   c. area, extent and/or quality of habitat
   d. number of locations or sub-populations
   e. number of mature individuals;

3 extreme fluctuations in any of the following:
   a. extent of occurrence
   b. area of occupancy
   c. number of locations or sub-populations
   d. number of mature individuals.

C. Population estimated to number less than 10,000 mature individuals and either:

1 an estimated continuing decline of at least 10% within 10 years or 3 generations, whichever is longer, or

2 a continuing decline, observed, projected, or inferred, in numbers of mature individuals and population structure in the form of either:
   a. population estimated to number less than 1,000 mature individuals; or
   b. population is characterised by an acute restriction in its area of occupancy (typically less than 100 km^2) or in the number of locations (typically less than five). Such a taxon would thus be prone to the effects of human activities (or stochastic events whose impact is increased by human activities) within a very short period of time in an unforeseeable future, and is thus capable of becoming *Critically Endangered* or even *Extinct* in a very short period.

D. Population very small or restricted in the form of either of the following:

1 population estimated to number less than 1,000 mature individuals; or

2 population is characterised by an acute restriction in its area of occupancy (typically less than 100 km^2) or in the number of locations (typically less than five). Such a taxon would thus be prone to the effects of human activities (or stochastic events whose impact is increased by human activities) within a very short period of time in an unforeseeable future, and is thus capable of becoming *Critically Endangered* or even *Extinct* in a very short period.

E. Quantitative analysis showing the probability of extinction in the wild is at least 10% within 100 years.

Definitions

Extent of occurrence

Extent of occurrence is defined as the area contained within the shortest continuous imaginary boundary that can be drawn to encompass all the known, inferred or projected sites of the present occurrence of a taxon, excluding cases of vagrancy. This measure may exclude discontinuities or disjunctions within the overall distributions of taxa (e.g. large areas of obviously unsuitable habitat) (but see 'area of occupancy'). Extent of occurrence can often be measured by a minimum convex polygon (the smallest polygon in which no internal angle exceeds 180° and which contains all the sites of occurrence).

Area of occupancy

Area of occupancy is defined as the area within its 'extent of occurrence' (see definition) that is occupied by a taxon, excluding cases of vagrancy. The measure reflects the fact that a taxon will not usually occur throughout the area of its extent of occurrence, which may, for example, contain unsuitable habitats. The area of occupancy is the smallest area essential at any stage to the survival of existing populations of a taxon (e.g. colonial nesting sites, feeding sites for migratory taxa). The size of the area of occupancy will be a function of the scale at which it is measured, and should be at a scale appropriate to relevant biological aspects of the taxon. The criteria include values in km^2, and to avoid errors in classification, the area of occupancy should be measured on grid squares (or equivalents) that are sufficiently small.

Appendix 2 Species assigned to IUCN threat categories

Critically Endangered

Species	Threat category (1997)	A	B	C	D	Perring & Farrell (1983)
Alisma gramineum	CR	A1a	B1+3d			E
Apium repens	CR		B1+2e	C2a		E
Atriplex pedunculata	CR		B1+3d			EX
Bupleurum falcatum	CR		B1+3d	C2b	D	EN
Carex depauperata	CR				D	E
Carex muricata ssp. *muricata*	CR			C2a		R
Cephalanthera rubra	CR				D	E
Corrigiola litoralis	CR	A1a			D	V
Cypripedium calceolus	CR				D	E
Dactylorhiza incarnata ssp. *ochroleuca*	CR		B1+2e	C2a	D	-
Epipogium aphyllum	CR				D	E
Filago gallica	CR				D	EX
Galium tricornutum	CR	A1a	B1+3d		D	NS
Gentianella ciliata	CR	A1a		C2b	D	-
Gnaphalium luteoalbum	CR		B1+3d			E
Petrorhagia prolifera	CR		B1+3d		D	-
Schoenoplectus triqueter	CR	A1a?			D	E
Scleranthus perennis ssp. *perennis*	CR		B1+3d		D	E
Senecio paludosus	CR				D	V
Sorbus domestica	CR				D	-
Sorbus leptophylla	CR				D	R
Sorbus leyana	CR				D	R
Sorbus wilmottiana	CR				D	R
Spergularia bocconei	CR	A1a?		C2a	D	E
Valerianella rimosa	CR	A1a?		C2?		V

Endangered

Species	Threat category (1997)	A	B	C	D	Perring & Farrell (1983)
			Criteria sub-heads			
Alchemilla subcrenata	EN	A1c			D	V
Allium sphaerocephalon	EN		B1+2e		D	V
Althaea hirsuta	EN		B1+3d			E
Arabis alpina	EN				D	R
Artemisia campestris	EN		B1+2c			E
Bupleurum baldense	EN		B1+3d			E
Centaurea cyanus	EN		B1+3d			NS
Cerastium brachypetalum	EN	A2c		C1+2a		R
Clinopodium menthifolium	EN			C2b?		V
Cotoneaster integerrimus	EN				D	E
Crepis foetida	EN				D	V
Crepis praemorsa	EN				D	-
Dactylorhiza incarnata ssp. *cruenta*	EN				D	-
Damasonium alisma	EN		B1+3d			E
Echium plantagineum	EN		B1+2d+3d			V
Epipactis youngiana	EN				D	-
Euphrasia rotundifolia	EN		B1+2c/d		D	R
Filago pyramidata	EN		B2d+3d			V
Fumaria reuteri	EN		B1+3d			E
Homogyne alpina	EN				D	V
Juncus pygmaeus	EN	A1a	B1+2d/e +3d			R
Lactuca saligna	EN	A1a	B1+?3d			E
Leersia oryzoides	EN	A1a?	B1+2d			V
Liparis loeselii	EN	A1a	B1+2c/d			E
Lonicera xylosteum	EN				D	V
Melampyrum arvense	EN		B1+3d			E
Minuartia stricta	EN		B1+3d			V
Orobanche artemisiae-campestris	EN				D	E
Petrorhagia nanteuilii	EN		B1+3d			E
Polygonum maritimum	EN			C2a?		E
Pyrus cordata	EN				D	E
Ranunculus ophioglossifolius	EN		B1+3d			E
Ranunculus reptans	EN		B1+3d			-
Rumex rupestris	EN			C2a		V
Scleranthus perennis ssp. *prostratus*	EN		B1+3d			E
Sorbus bristoliensis	EN				D	R
Stachys alpina	EN		B1+3d		D	E
Stachys germanica	EN		B1+3d			E
Teucrium chamaedrys	EN				D	R
Tordylium maximum	EN	A2c?	B1+3d			-
Veronica triphyllos	EN		B1+2d+3d		D	E
Viola canina ssp. *montana*	EN		B1+3d		D	-
Viola persicifolia	EN		B1+3d		D	E
Woodsia ilvensis	EN			C2a	D	V

Vulnerable

Species	Threat category (1997)	A	B	C	D	Perring & Farrell (1983)
			Criteria sub-heads			
Adonis annua	VU	A2?	B2d+3c			NS
Ajuga chamaepitys	VU		B2b+3d			NS
Alchemilla gracilis	VU			C1+2a	D2	V
Alchemilla minima	VU				D2	V
Anisantha madritensis	VU				D2	R
Arabis glabra	VU	A1a?	B2b+3d			NS
Arabis scabra	VU		B1+3d			V
Arenaria norvegica ssp. *anglica*	VU		B1+3d			E
Armeria maritima ssp. *elongata*	VU				D2	V
Artemisia norvegica	VU				D2	R
Asparagus prostratus	VU				D1	V
Astragalus alpinus	VU				D2	R
Athyrium flexile	VU				D1	-
Calamagrostis scotica	VU				D2	V
Carex buxbaumii	VU				D2	V
Carex chordorrhiza	VU				D2	R
Carex flava	VU				D1+2	R
Carex microglochin	VU				D2	R
Carex norvegica	VU				D1	R
Carex recta	VU				D2	R
Carex vulpina	VU	A1a/b	B2d/e			NS
Centaurea calcitrapa	VU		B1+3d			R
Centaurium scilloides	VU				D2	V
Centaurium tenuiflorum	VU				D2	V
Cerastium fontanum ssp. *scoticum*	VU				D2	-
Cerastium nigrescens	VU				D2	R
Chenopodium vulvaria	VU		B1+3d		D2	V
Cicerbita alpina	VU				D1+2	R
Cirsium tuberosum	VU	A2e				R
Coincya wrightii	VU				D2	R
Crassula aquatica	VU				D2	V
Cynodon dactylon	VU				D2	R
Cynoglossum germanicum	VU				D2	V
Cyperus fuscus	VU		B1+3d			E
Cystopteris dickieana	VU				D2	E
Cytisus scoparius ssp. *maritimus*	VU				D2	-
Dianthus armeria	VU		B2d+3d	C2a?		NS
Dianthus gratianopolitanus	VU				D2	V
Diapensia lapponica	VU				D2	V
Eleocharis parvula	VU		B1+?2c			R
Erigeron borealis	VU		B1+2e		D1	R
Eriophorum gracile	VU		B1+2c			V
Eryngium campestre	VU				D2	V
Euphorbia hyberna	VU				D2	R

Vulnerable continued

Species	Threat category (1997)	A	B	C	D	Perring & Farrell (1983)
Euphorbia serrulata	VU	A1a				R
Euphrasia cambrica	VU				D2?	R
Euphrasia vigursii	VU	A1a/c				R
Festuca longifolia	VU		B1+2d	C1?		R
Filago lutescens	VU		B2d+3d			V
Gagea bohemica	VU				D2	V
Gentiana nivalis	VU		B1+3d		D2	V
Gentianella uliginosa	VU		B1+3d	C2a		V
Helianthemum canum ssp. *levigatum*	VU				D2	R
Himantoglossum hircinum	VU		B1+3c			V
Hypochaeris maculata	VU		B1+2d			R
Lavatera cretica	VU		B2d+3c			R
Limonium binervosum ssp. *cantianum*	VU				D2	
Limonium binervosum ssp. *mutatum*	VU				D2	
Limonium dodartiforme	VU				D2	
Limonium loganicum	VU				D2	
Limonium paradoxum	VU				D2	E
Limonium parvum	VU				D2	
Limonium procerum ssp. *cambrense*	VU				D2	
Limonium procerum ssp. *devoniense*	VU				D2	
Limonium recurvum ssp. *portlandicum*	VU				D2	
Limonium recurvum ssp. *recurvum*	VU				D2	E
Limonium transwallianum	VU				D2	R
Limosella australis	VU		B1+3c/d			R
Lloydia serotina	VU				D2	V
Lobelia urens	VU		B1+3d			V
Luzula pallidula	VU				D1+2	R
Lychnis alpina	VU				D2	V
Lychnis viscaria	VU	A1a		C1+2a		R
Lythrum hyssopifolia	VU		B1+3d			V
Maianthemum bifolium	VU				D2	V
Matthiola sinuata	VU		B1+3d		D2	V
Mentha pulegium	VU			?C1		V
Moneses uniflora	VU			C1?		V
Muscari neglectum	VU			C2a?		V
Najas marina	VU		B1+3c/d			V
Ononis reclinata	VU		B1+3d			V
Ophioglossum lusitanicum	VU		B1+3d		D2	R
Ophrys fuciflora	VU				D1+2	V
Orchis militaris	VU				D2	V
Orchis simia	VU				D1+2	V
Orobanche caryophyllea	VU				D2	E
Orobanche purpurea	VU	A2			D1	V
Oxytropis campestris	VU				D2	R

Vulnerable   continued

Species	Threat category (1997)	A	B	C	D	Perring & Farrell (1983)
Phyllodoce caerulea	VU				D1	V
Physospermum cornubiense	VU	A1a				R
Phyteuma spicatum	VU	A1a	B1+2e		D1	R
Pilosella flagellaris ssp. *bicapitata*	VU				D2	-
Pilosella peleteriana	VU				D2	-
Poa flexuosa	VU				D1	R
Polygala amarella	VU	A1a				R
Polygonatum verticillatum	VU		B1+2d			E
Potamogeton acutifolius	VU	A1a/c				-
Potamogeton epihydrus	VU				D2	R
Potentilla rupestris	VU				D1	V
Pulicaria vulgaris	VU	A1a				V
Pulmonaria obscura	VU				D1+2	-
Ranunculus tripartitus	VU	A1a	B2d+3d			NS
Rhinanthus angustifolius	VU	A1c				E
Romulea columnae	VU				D2	V
Rumex aquaticus	VU				D2	R
Sagina nivalis	VU		B1+3d			R
Salix lanata	VU	A1c+2c				R
Saxifraga cernua	VU				D1	V
Saxifraga cespitosa	VU				D1	R
Saxifraga hirculus	VU	A2c	B1+2c			R
Scheuchzeria palustris	VU			C2a?		V
Schoenus ferrugineus	VU				D2	V
Scirpoides holoschoenus	VU				D2	R
Scorzonera humilis	VU				D2	V
Selinum carvifolia	VU				D2	V
Seseli libanotis	VU				D2	R
Sorbus anglica	VU				D1	R
Sorbus arranensis	VU				D1+2	R
Sorbus eminens	VU				D1+2	R
Sorbus minima	VU				D1+2	R
Sorbus pseudofennica	VU				D1+2	R
Sorbus subcuneata	VU				D1	V
Sorbus vexans	VU				D1	V
Tephroseris integrifolia ssp. *maritima*	VU				D2	-
Teucrium botrys	VU		B1+3d		D2	V
Teucrium scordium	VU	A1c?			D2	V
Thlaspi perfoliatum	VU	A1c	B1+3d			R
Trichomanes speciosum	VU			C2a?		E
Trifolium bocconei	VU		B1+3d			R
Trifolium incarnatum ssp. *molinerii*	VU		B1+3d			R
Trifolium strictum	VU		B1+3d			R
Tuberaria guttata	VU		B2d+3d	C2a?		V

Vulnerable continued

Species	Threat category (1997)	A	B	C	D	Perring & Farrell (1983)
Valerianella eriocarpa	VU		B1+3c		D2	V
Veronica spicata ssp. *spicata*	VU				D2	V
Veronica verna	VU				D2	V
Viola kitaibeliana	VU				D2	R

Lower Risk (near threatened)

Species	Threat category (1997)	Perring & Farrell (1983)
Alchemilla acutiloba	LR - nt	R
Alchemilla glaucescens	LR - nt	R
Alchemilla monticola	LR - nt	R
Anthyllis vulneraria ssp. *corbierei*	LR - nt	-
Arenaria norvegica ssp. *norvegica*	LR - nt	R
Aster linosyris	LR - nt	R
Bartsia alpina	LR - nt	R
Bunium bulbocastanum	LR - nt	R
Buxus sempervirens	LR - nt	R
Calamagrostis purpurea	LR - nt	-
Calamagrostis stricta	LR - nt	NS
Carex atrofusca	LR - nt	R
Carex filiformis	LR - nt	R
Carex lachenalii	LR - nt	R
Carex ornithopoda	LR - nt	R
Carex rariflora	LR - nt	R
Chenopodium chenopodioides	LR - nt	NS
Corynephorus canescens	LR - nt	NS
Cystopteris montana	LR - nt	NS
Dactylorhiza lapponica	LR - nt	-
Diphasiastrum issleri	LR - nt	-
Draba aizoides	LR - nt	R
Dryopteris cristata	LR - nt	V
Eleocharis austriaca	LR - nt	R
Erica ciliaris	LR - nt	R
Erica vagans	LR - nt	R
Eriocaulon aquaticum	LR - nt	R
Euphrasia campbelliae	LR - nt	R
Euphrasia heslop-harrisonii	LR - nt	R
Euphrasia marshallii	LR - nt	R
Euphrasia rivularis	LR - nt	R
Galium constrictum	LR - nt	R
Genista pilosa	LR - nt	R
Gentiana verna	LR - nt	R
Gladiolus illyricus	LR - nt	V

Lower Risk (near threatened) continued

Species	Threat category (1997)	Perring & Farrell (1983)
Helianthemum apenninum	LR - nt	R
Herniaria ciliolata ssp. *ciliolata*	LR - nt	R
Herniaria glabra	LR - nt	V
Hierochloe odorata	LR - nt	R
Hypericum linariifolium	LR - nt	V
Isoetes histrix	LR - nt	R
Juncus capitatus	LR - nt	R
Kobresia simpliciuscula	LR - nt	R
Koeleria vallesiana	LR - nt	R
Koenigia islandica	LR - nt	R
Leucojum aestivum	LR - nt	R
Limonium bellidifolium	LR - nt	R
L. binervosum ssp. *anglicum*	LR - nt	-
L. binervosum ssp. *binervosum*	LR - nt	-
L. binervosum ssp. *saxonicum*	LR - nt	-
L. britannicum ssp. *britannicum*	LR - nt	-
L. britannicum ssp. *celticum*	LR - nt	-
L. britannicum ssp. *coombense*	LR - nt	-
L. britannicum ssp. *transcanalis*	LR - nt	-
Limonium recurvum ssp. *humile*	LR - nt	-
Lithospermum purpureocaeruleum	LR - nt	R
Ludwigia palustris	LR - nt	R
Luzula arcuata	LR - nt	-
Mibora minima	LR - nt	R
Minuartia rubella	LR - nt	R
Myosotis alpestris	LR - nt	R
Ophrys sphegodes	LR - nt	V
Ornithopus pinnatus	LR - nt	R
Orobanche reticulata	LR - nt	E
Oxytropis halleri	LR - nt	R
Peucedanum officinale	LR - nt	R
Phleum phleoides	LR - nt	R
Polemonium caeruleum	LR - nt	R
Polycarpon tetraphyllum	LR - nt	R
Potamogeton nodosus	LR - nt	R

Lower Risk (near threatened) continued

Species	Threat category (1997)	Perring & Farrell (1983)
Potamogeton rutilus	LR - nt	R
Potentilla fruticosa	LR - nt	R
Saxifraga rivularis	LR - nt	R
Senecio cambrensis	LR - nt	R
Silene otites	LR - nt	R
Sorbus lancastriensis	LR - nt	R
Thymus serpyllum	LR - nt	R
Trinia glauca	LR - nt	R
Veronica fruticans	LR - nt	R
Viola rupestris	LR - nt	R
Woodsia alpina	LR - nt	R

Data Deficient

Species	Threat category (1997)	Perring & Farrell (1983)
Cochlearia atlantica	DD	-
Asplenium trichomanes ssp. *pachyrachis*	DD	-

Nationally Scarce

Species	Threat category (1997)	Perring & Farrell (1983)
Allium ampeloprasum	LR(ns)	R
Cochlearia micacea	LR(ns)	-
Fumaria occidentalis	LR(ns)	R
Gastridium ventricosum	LR(ns)	V
Gnaphalium norvegicum	LR(ns)	R
Geranium purpureum	LR(ns)	R
Limonium procerum ssp. *procerum*	LR(ns)	-
Lotus angustissimus	LR(ns)	R
Poa infirma	LR(ns)	-
Rorippa islandica	LR(ns)	-
Rosa agrestis	LR(ns)	-
Salvia pratensis	LR(ns)	V
Scrophularia scorodonia	LR(ns)	R

Extinct in the wild in Britain (year of last record in the wild in Britain)

Species	Threat category (1997)	Perring & Farrell (1983)
Agrostemma githago (1970s?)	EW	EX
Bromus interruptus (1972)	EW	EX
Bupleurum rotundifolium (1960s)	EW	EX
Euphorbia peplis (1965)	EW	EX
Sagina boydii (1878)	EW	EX
Saxifraga rosacea ssp. *rosacea* (c. 1974)	EW	EX

Extinct in Britain (year of last record in Britain)

Species	Threat category (1997)	Perring & Farrell (1983)
Arnoseris minima (1970)	EX	EX
Carex davalliana (c. 1845)	EX	EX
Crepis foetida (1980)	EX*	EX
Filago gallica (1955)	EX*	EX
Galeopsis segetum (1980s?)	EX	E
Holosteum umbellatum (1930)	EX	EX
Hydrilla verticillata (c. 1934)	EX	EX
Neotinea maculata (1986?)	EX	R
Otanthus maritimus (1970s?)	EX	EX
Pinguicula alpina (1900s)	EX	EX
Rubus arcticus (1841)	EX	EX
Spiranthes aestivalis (1959)	EX	EX
Tephroseris palustris ssp. *congestus* (1899)	EX	EX
Trichophorum alpinum (1813)	EX	EX

Key: * reintroduced

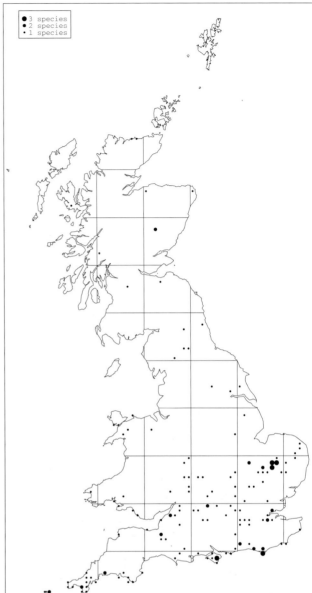

Figure A2.1 Number of Critically Endangered species recorded from 1987 onwards in each 10 km square. The species mapped are listed in Appendix 2. The smaller dot indicates 1 species per square, and the larger, 2 species per square.

Figure A2.2 Number of Endangered species recorded from 1987 onwards in each 10 km square. The species mapped are listed in Appendix 2. Dots of increasing size indicate 1, 2 and 3 species per square.

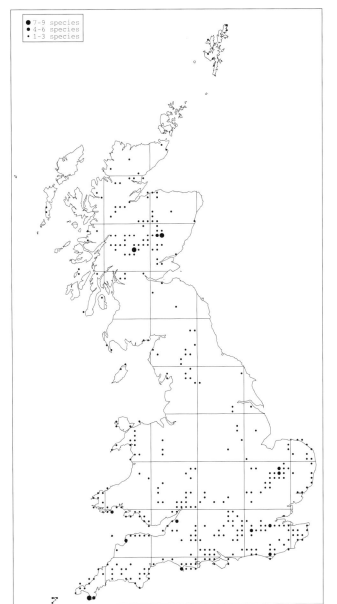

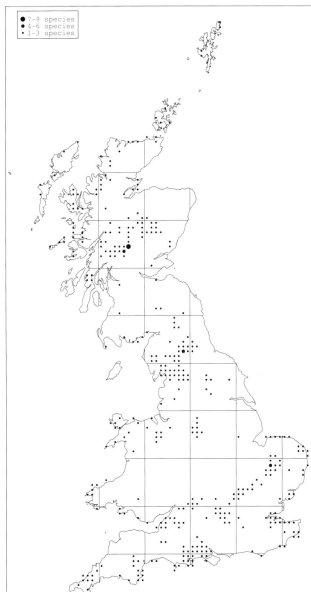

Figure A2.3 Number of Vulnerable species recorded from 1987 onwards in each 10 km square. The species mapped are listed in Appendix 2. Dots of increasing size indicate 1-3, 4-6 and 7-9 species per square.

Figure A2.4 Number of Near Threatened species recorded from 1987 onwards in each 10 km square. The species mapped are listed in Appendix 2. Dots of increasing size indicate 1-3, 4-6 and 7-9 species per square.

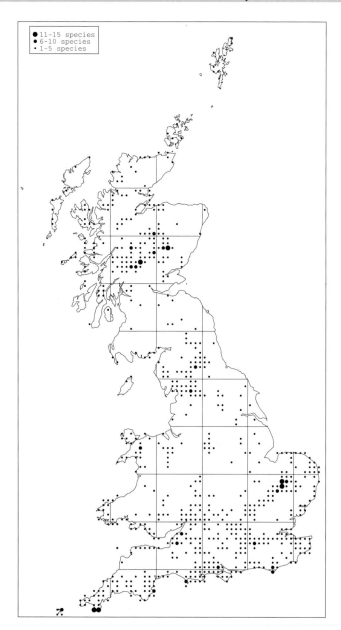

Figure A2.5 Number of all Threatened and Near Threatened species recorded from 1987 onwards in each 10 km square. The species mapped are listed in Appendix 2. Dots of increasing size indicate 1-5, 6-10 and 11-15 species per square.

Appendix 3 Species specially protected under UK and international legislation

Species	WCA * (1981)	EC Habitats Directive Annex	Bern Convention Appendix	CITES Appendix
Ajuga chamaepitys	8			
Alisma gramineum	8			
Allium sphaerocephalon	8			
Althaea hirsuta	8			
Alyssum alyssoides	8			
Apium repens	8	IIb, IVb	I	
Arabis alpina	8			
Arabis scabra	8			
Arenaria norvegica	8			
Artemisia campestris	8			
Atriplex pedunculata	8			
Bupleurum baldense	8			
Bupleurum falcatum	8			
Carex depauperata	8			
Centaurium tenuiflorum	8			
Cephalanthera rubra	8			II (C1)
Chenopodium vulvaria	8			
Cicerbita alpina	8			
Clinopodium menthifolium	8			
Coincya wrightii	8			
Corrigiola litoralis	8			
Cotoneaster integerrimus	8			
Crassula aquatica	8			
Crepis foetida	8			
Cynoglossum germanicum	8			
Cyperus fuscus	8			
Cypripedium calceolus	8	IIb, IVb	I	II (C1)
Cystopteris dickieana	8			
Dactylorhiza lapponica	8			II (C1)
Dactylorhiza incarnata ssp. *cruenta*				II (C1)
Dactylorhiza incarnata ssp. *ochroleuca*				II (C1)
Damasonium alisma	8			
Dianthus armeria	8			
Dianthus gratianopolitanus	8			
Diapensia lapponica	8			
Epipactis youngiana	8			
Epipogium aphyllum	8			II (C1)
Equisetum ramosissimum	8			
Erigeron borealis	8			
Eriophorum gracile	8			
Eryngium campestre	8			
Filago lutescens	8			
Filago pyramidata	8			

Species	WCA * (1981)	EC Habitats Directive Annex	Bern Convention Appendix	CITES Appendix
Fumaria reuteri	8			
Gagea bohemica	8			
Galanthus nivalis (if native)		Vb		II (C1)
Gentiana nivalis	8			
Gentiana verna	8			
Gentianella anglica	8	IIb, IVb	I	
Gentianella ciliata	8			
Gentianella uliginosa	8			
Gladiolus illyricus	8			
Gnaphalium luteoalbum	8			
Hieracium attenuatifolium	8			
Hieracium northroense	8			
Hieracium zetlandicum	8			
Himantoglossum hircinum	8			II (C1)
Homogyne alpina	8			
Lactuca saligna	8			
Leersia oryzoides	8			
Limosella australis	8			
Liparis loeselii	8	IIb, IVb	I	II (C1)
Lloydia serotina	8			
Luronium natans	8	IIb, IVb	I	
Lychnis alpina	8			
Lycopodium, sensu lato - all species		Vb		
Lythrum hyssopifolia	8			
Melampyrum arvense	8			
Mentha pulegium	8			
Minuartia stricta	8			
Najas flexilis	8	IIb, IVb	I	
Najas marina	8			
Ononis reclinata	8			
Ophioglossum lusitanicum	8			
Ophrys fuciflora	8			II (C1)
Ophrys sphegodes	8			II (C1)
Orchis militaris	8			II (C1)
Orchis simia	8			II (C1)
Orobanche artemisiae-campestris	8			
Orobanche caryophyllacea	8			
Orobanche reticulata	8			
Petrorhagia nanteuilii	8			
Phyllodoce caerulea	8			
Phyteuma spicatum	8			
Polygonatum verticillatum	8			
Polygonum maritimum	8			
Potentilla rupestris	8			
Pulicaria vulgaris	8			
Pyrus cordata	8			

Species	WCA * (1981)	EC Habitats Directive Annex	Bern Convention Appendix	CITES Appendix
Ranunculus ophioglossifolius	8			
Rhinanthus angustifolius	8			
Romulea columnae	8			
Ruscus aculeatus		Vb		
Rumex rupestris	8	IIb, IVb	I	
Salvia pratensis	8			
Saxifraga cernua	8			
Saxifraga cespitosa	8			
Saxifraga hirculus	8	IIb, IVb		
Schoenoplectus triqueter	8			
Scleranthus perennis	8			
Scorzonera humilis	8			
Selinum carvifolia	8			
Senecio paludosus	8			
Stachys alpina	8			
Stachys germanica	8			
Tephroseris integrifolia ssp. *maritima*	8			
Teucrium botrys	8			
Teucrium scordium	8			
Thlaspi perfoliatum	8			
Trichomanes speciosum	8	IIb, IVb	I	
Veronica spicata ssp. *spicata*	8			
Veronica triphyllos	8			
Viola persicifolia	8			
Woodsia alpina	8			
Woodsia ilvensis	8			

Key: *Wildlife & Countryside Act; EC Habitat & Species Directive, Annex Iib: designation of protected areas for plant species; Annex Ivb: special protection necessary for plant species; Annex Vb: exploitation of plant to be subject to management if necessary; Bern Convention, Appendix I: special protection for plant species; CITES, Appendix I: trade permitted only in exceptional circumstances; Appendix II: trade subject to licensing; EC regulations treat all species of orchids as if they were listed in Appendix I (category C1).

Appendix 4 Occurrence of rare species in counties and unitary authority areas

With very few exceptions, this list includes only post-1987 occurrences. It includes only those nationally scarce species for which an account is given in this book. Species have been placed in threat categories according to their national status. The numbers of species recorded in each vice-county are tabulated in Appendices 5 and 6.

Isles of Scilly

Endangered	*Echium plantagineum, Rumex rupestris*
Vulnerable	*Lavatera cretica, Ophioglossum lusitanicum, Viola kitaibeliana*
Lower Risk (near threatened)	*Ornithopus pinnatus, Polycarpon tetraphyllum*
Nationally Scarce	*Allium ampeloprasum, Fumaria occidentalis, Poa infirma, Scrophularia scorodonia*

Cornwall

Critical	*Spergularia bocconei.*
Endangered	*Echium plantagineum, Fumaria reuteri, Juncus pygmaeus, Polygonum maritimum, Pyrus cordata, Rumex rupestris*
Vulnerable	*Asparagus prostratus, Cynodon dactylon, Cytisus scoparius* ssp. *maritimus, Dianthus armeria, Euphorbia hyberna, Euphrasia vigursii, Hypochaeris maculata, Isoetes histrix, Limonium loganicum, Lobelia urens, Mentha pulegium, Physospermum cornubiense, Ranunculus tripartitus, Trichomanes speciosum, Trifolium bocconei, Trifolium incarnatum* ssp. *molinerii, Trifolium strictum, Valerianella eriocarpa*
Lower Risk (near threatened)	*Anthyllis vulneraria* ssp. *corbierei, Erica ciliaris, Erica vagans, Genista pilosa, Herniaria ciliolata* ssp. *ciliolata, Hypericum linariifolium, Juncus capitatus, Limonium britannicum* ssp. *britannicum, Limonium britannicum* ssp. *coombense, Lotus angustissimus, Polycarpon tetraphyllum*
Nationally Scarce	*Allium ampeloprasum, Fumaria occidentalis, Gastridium ventricosum, Geranium purpureum, Lotus angustissimus, Poa infirma, Scrophularia scorodonia*

Devon, Plymouth Unitary Authority & Torbay Unitary Authority.

Critical	*Corrigiola litoralis, Schoenoplectus triqueter*
Endangered	*Bupleurum baldense, Pyrus cordata, Rumex rupestri*
Vulnerable	*Coincya wrightii, Cytisus scoparius* ssp. *maritimus, Eleocharis parvula, Eryngium campestre, Euphorbia hyberna, Euphrasia vigursii, Limonium binervosum* ssp. *mutatum, Limonium procerum* ssp.

devoniense, Lobelia urens, Matthiola sinuata, Mentha pulegium, Ononis reclinata, Romulea columnae, Scirpoides holoschoenus, Sorbus anglica, Sorbus subcuneata, Sorbus vexans, Teucrium scordium, Valerianella eriocarpa

Lower Risk (near threatened)	*Aster linosyris, Cynodon dactylon, Erica ciliaris, Helianthemum apenninum, Hypericum linariifolium, Leucojum aestivum, Limonium britannicum* ssp. *coombense, Limonium britannicum* ssp. *transcanalis, Lithospermum purpureocaeruleum, Lotus angustissimus, Polycarpon tetraphyllum, Trinia glauca*
Nationally Scarce	*Gastridium ventricosum, Geranium purpureum, Lotus angustissimus, Poa infirma, Scrophularia scorodonia*

Somerset, N.W. Somerset Unitary Authority and Bath & N.E. Somerset Unitary Authority

Critical	*Carex depauperata, Sorbus wilmottiana, Valerianella rimosa*
Endangered	*Althaea hirsuta, Leersia oryzoides, Sorbus bristoliensis*
Vulnerable	*Anisantha madritensis, Arabis scabra, Cyperus fuscus, Dianthus armeria, Dianthus gratianopolitanus, Euphorbia serrulata, Himantoglossum hircinum, Scirpoides holoschoenus, Sorbus anglica, Sorbus eminens, Sorbus subcuneata, Sorbus vexans, Thlaspi perfoliatum, Valerianella eriocarpa*
Lower Risk (near threatened	*Aster linosyris, Cynodon dactylon, Helianthemum apenninum, Herniaria glabra, Koeleria vallesiana, Lithospermum purpureocaeruleum, Potamogeton nodosus, Trinia glauca*
Nationally Scarce	*Allium ampeloprasum, Gastridium ventricosum, Geranium purpureum, Rosa agrestis*

Wiltshire & Thamesdown Unitary Authority

Endangered	*Melampyrum arvense*
Vulnerable	*Adonis annua, Arabis glabra*
Lower Risk (near threatened)	*Carex filiformis, Cirsium tuberosum, Leucojum aestivum, Ophrys sphegodes, Potamogeton nodosus*
Nationally Scarce	*Rosa agrestis, Salvia pratensis*

Dorset, Bournemouth Unitary Authority & Poole Unitary Authority

Endangered	*Polygonum maritimum*
Vulnerable	*Adonis annua, Asparagus prostratus, Centaurium tenuiflorum, Chenopodium vulvaria, Himantoglossum hircinum, Limonium dodartiforme, Limonium recurvum* ssp. *portlandicum, Limonium recurvum* ssp. *recurvum, Lobelia urens,*

	Lythrum hyssopifolia, Orobanche purpurea, Pilosella peleteriana, Potamogeton acutifolius, Scorzonera humilis, Valerianella eriocarpa
Lower Risk (near threatened	Cirsium tuberosum, Erica ciliaris, Leucojum aestivum, Ludwigia palustris, Mibora minima, Ophrys sphegodes, Polycarpon tetraphyllum, Potamogeton nodosus
Nationally Scarce	Allium ampeloprasum, Gastridium ventricosum, Geranium purpureum, Rosa agrestis

Isle of Wight

Endangered	Clinopodium menthifolium, Fumaria reuteri, Melampyrum arvense, Orobanche artemisiae-campestris
Vulnerable	Orobanche purpurea, Pilosella peleteriana
Lower Risk (near threatened)	Ophrys sphegodes
Nationally Scarce	Gastridium ventricosum

Hampshire, Portsmouth Unitary Authority & Southampton Unitary Authority

Critical	Cephalanthera rubra, Valerianella rimosa
Endangered	Polygonum maritimum
Vulnerable	Adonis annua, Ajuga chamaepitys, Arabis glabra, Cyperus fuscus, Dianthus armeria, Eleocharis parvula, Eriophorum gracile, Filago lutescens, Himantoglossum hircinum, Lobelia urens, Mentha pulegium, Orobanche purpurea, Pulicaria vulgaris, Ranunculus tripartitus, Teucrium botrys
Lower Risk (near threatened)	Buxus sempervirens, Galium constrictum, Gladiolus illyricus, Lotus angustissimus, Ludwigia palustris
Nationally Scarce	Gastridium ventricosum, Geranium purpureum, Lotus angustissimus, Rosa agrestis

West Sussex

Endangered	Filago pyramidata, Leersia oryzoides, Lonicera xylosteum, Orobanche artemisiae-campestris, Petrorhagia nanteuilii
Vulnerable	Carex vulpina, Centaurea calcitrapa, Filago lutescens, Lythrum hyssopifolia, Potamogeton acutifolius, Ranunculus tripartitus
Nationally Scarce	Salvia pratensis.

East Sussex and Brighton Unitary Authority & Hove Unitary Authority

Endangered:	Althaea hirsuta, Bupleurum baldense, Lactuca saligna, Lonicera xylosteum, Polygonum maritimum, Teucrium chamaedrys
Vulnerable	Adonis annua, Carex vulpina, Centaurea calcitrapa, Centaurium scilloides, Dianthus armeria, Himantoglossum hircinum, Lobelia urens, Phyteuma spicatum, Potamogeton acutifolius, Seseli libanotis

Lower Risk (near threatened)	Chenopodium chenopodioides, Genista pilosa, Limonium binervosum ssp. binervosum, Ludwigia palustris, Ophrys sphegodes
Nationally Scarce	Rosa agrestis

Kent, and Rochester & Gillingham proposed Unitary Authority

Endangered	Althaea hirsuta, Cerastium brachypetalum, Crepis foetida, Filago pyramidata, Orobanche artemisiae-campestris
Vulnerable	Adonis annua, Ajuga chamaepitys, Carex vulpina, Centaurium scilloides, Chenopodium vulvaria, Cynoglossum germanicum, Dianthus armeria, Himantoglossum hircinum, Limonium binervosum ssp. cantianum, Ophrys fuciflora, Orchis simia, Orobanche caryophyllacea, Orobanche purpurea, Polygala amarella, Potamogeton acutifolius, Teucrium botrys
Lower Risk (near threatened	Buxus sempervirens, Chenopodium chenopodioides, Limonium binervosum ssp. binervosum, Lotus angustissimus, Ophrys sphegodes, Peucedanum officinale
Nationally Scarce	Lotus angustissimus, Salvia pratensis

Surrey

Critical	Carex depauperata
Endangered	Damasonium alisma, Filago pyramidata, Leersia oryzoides
Vulnerable	Ajuga chamaepitys, Arabis glabra, Carex vulpina, Cynoglossum germanicum, Cyperus fuscus, Dianthus armeria, Eriophorum gracile, Filago lutescens, Himantoglossum hircinum, Potamogeton acutifolius, Pulicaria vulgaris, Ranunculus tripartitus, Rhinanthus angustifolius, Teucrium botrys
Lower Risk (near threatened	Buxus sempervirens, Carex filiformis, Dryopteris cristata, Polycarpon tetraphyllum
Nationally Scarce	Rosa agrestis, Salvia pratensis

Essex & Southend Unitary Authority

Critical	Atriplex pedunculata, Bupleurum falcatum, Filago gallica
Endangered	Lactuca saligna, Melampyrum arvense, Tordylium maximum
Vulnerable	Dianthus armeria, Filago lutescens
Lower Risk (near threatened)	Chenopodium chenopodioides, Limonium binervosum ssp. saxonicum, Peucedanum officinale

Hertfordshire

Critical	Galium tricornutum
Vulnerable	Dianthus armeria, Hypochaeris maculata
Lower Risk (near threatened)	Bunium bulbocastanum, Phleum phleoides

Greater London

Vulnerable	Arabis glabra, Rhinanthus angustifolius

Berkshire

Endangered	*Althaea hirsuta*
Vulnerable	*Arabis glabra, Cyperus fuscus, Mentha pulegium*
Lower Risk (near threatened)	*Dryopteris cristata, Leucojum aestivum, Potamogeton nodosus*
Nationally Scarce	*Salvia pratensis*

Oxfordshire

Critical	*Althaea hirsuta, Apium repens, Galium tricornutum*
Endangered	*Filago pyramidata, Stachys germanica*
Vulnerable	*Adonis annua, Carex vulpina, Cynoglossum germanicum, Dianthus armeria, Himantoglossum hircinum, Lythrum hyssopifolia, Orchis militaris, Orchis simia, Thlaspi perfoliatum*
Lower Risk (near threatened)	*Carex filiformis, Leucojum aestivum*
Nationally Scarce	*Salvia pratensis*

Buckinghamshire

Critical	*Cephalanthera rubra, Epipogium aphyllum, Gentianella ciliata*
Endangered	*Damasonium alisma*
Vulnerable	*Adonis annua, Cyperus fuscus, Orchis militaris, Physospermum cornubiense*
Lower Risk (near threatened)	*Bunium bulbocastanum, Buxus sempervirens, Leucojum aestivum*
Nationally Scarce	*Salvia pratensis*

Suffolk

Endangered	*Artemisia campestris, Centaurea cyanus, Scleranthus perennis* ssp. *prostratus, Veronica triphyllos*
Vulnerable	*Ajuga chamaepitys, Arabis glabra, Festuca longifolia, Filago lutescens, Himantoglossum hircinum, Hypochaeris maculata, Muscari neglectum, Orchis militaris, Orobanche purpurea, Pulmonaria obscura, Veronica spicata* ssp. *spicata, Veronica verna*
Lower Risk (near threatened)	*Chenopodium chenopodioides, Corynephorus canescens, Herniaria glabra, Ophrys sphegodes, Peucedanum officinale, Phleum phleoides, Silene otites, Thymus serpyllum*

Norfolk

Critical	*Gnaphalium luteoalbum, Petrorhagia prolifera*
Endangered	*Artemisia campestris, Liparis loeselii, Veronica triphyllos*
Vulnerable	*Arabis glabra, Dianthus armeria, Maianthemum bifolium, Najas marina, Orobanche purpurea, Potamogeton acutifolius, Veronica spicata* ssp. *Spicata*
Lower Risk (near threatened	*Calamagrostis stricta, Corynephorus canescens, Dryopteris cristata, Herniaria glabra, Limonium bellidifolium, Limonium binervosum* ssp. *anglicum, Phleum phleoides, Silene otites, Thymus serpyllum*
Nationally Scarce	*Rosa agrestis*

Cambridgeshire

Critical	*Dactylorhiza incarnata* ssp. *ochroleuca, Galium tricornutum, Senecio paludosus*
Endangered	*Filago pyramidata, Viola canina* ssp. *montana, Viola persicifolia*
Vulnerable	*Adonis annua, Ajuga chamaepitys, Filago lutescens, Himantoglossum hircinum, Hypochaeris maculata, Luzula pallidula, Lythrum hyssopifolia, Muscari neglectum, Selinum carvifolia, Seseli libanotis, Teucrium scordium, Veronica spicata* ssp. *Spicata*
Lower Risk (near threatened)	*Bunium bulbocastanum, Herniaria glabra, Phleum phleoides, Silene otites*

Bedfordshire & Luton Unitary Authority

Critical	*Valerianella rimosa*
Endangered	*Cerastium brachypetalum, Melampyrum arvense, Petrorhagia prolifera*
Vulnerable	*Adonis annua, Ajuga chamaepitys, Hypochaeris maculata, Seseli libanotis*
Lower Risk (near threatened	*Bunium bulbocastanum*

Northamptonshire

Vulnerable	*Hypochaeris maculata*

Gloucestershire, S. Gloucestershire Unitary Authority, and City & County of Bristol Unitary Authority

Critical	*Carex muricata* ssp. *muricata, Cephalanthera rubra, Sorbus domestica, Sorbus wilmottiana*
Endangered	*Allium sphaerocephalon, Ranunculus ophioglossifolius, Sorbus bristoliensis, Stachys alpina*
Vulnerable	*Anisantha madritensis, Arabis scabra, Carex vulpina, Cynoglossum germanicum, Euphorbia serrulata, Lythrum hyssopifolia, Sorbus eminens, Teucrium botrys, Thlaspi perfoliatum*
Lower Risk (near threatened)	*Buxus sempervirens, Carex filiformis, Ophrys sphegodes, Potamogeton nodosus, Trinia glauca*
Nationally Scarce	*Salvia pratensis*
Data Deficient	*Asplenium trichomanes* ssp. *pachyrachis*

Herefordshire

Critical	*Epipogium aphyllum*
Vulnerable	*Mentha pulegium, Ranunculus tripartitus, Sorbus eminens*
Data Deficient	*Asplenium trichomanes* ssp. *pachyrachis*

Worcestershire

Critical	*Alisma gramineum*
Vulnerable	*Arabis glabra, Dianthus armeria*
Nationally Scarce	*Rosa agrestis*

Warwickshire

Vulnerable	*Dianthus armeria*

Staffordshire and Stoke on Trent Unitary Authority

Lower Risk (near threatened)	*Polemonium caeruleum*

Shropshire

Vulnerable	*Sorbus anglica*

Lincolnshire, & N. Lincolnshire Unitary Authority

Critical	*Alisma gramineum*
Vulnerable	*Armeria maritima* ssp. *elongata, Dianthus armeria, Festuca longifolia, Maianthemum bifolium, Orobanche purpurea, Rhinanthus angustifolius*
Lower Risk (near threatened)	*Herniaria glabra, Limonium binervosum* ssp. *anglicum*

Nottinghamshire

Vulnerable	*Dianthus armeria, Festuca longifolia*

Derbyshire

Vulnerable	*Arabis glabra*
Lower Risk (near threatened)	*Carex ornithopoda, Polemonium caeruleum*

Merseyside

Lower Risk (near threatened	*Corynephorus canescens*

Lancashire

Lower Risk (near threatened)	*Calamagrostis stricta, Eleocharis austriaca, Limonium britannicum* ssp. *celticum, Sorbus lancastriensis*

East Riding of Yorkshire

Vulnerable	*Maianthemum bifolium*
Lower Risk (near threatened)	*Calamagrostis stricta*

South Yorkshire

Lower Risk (near threatened)	*Calamagrostis stricta, Orobanche reticulata*

West Yorkshire

Endangered	*Epipactis youngiana*
Lower Risk (near threatened)	*Orobanche reticulata*

North Yorkshire

Critical	*Carex muricata* ssp. *muricata, Cypripedium calceolus*
Vulnerable	*Alchemilla minima, Arabis glabra, Arenaria norvegica* ssp. *anglica, Maianthemum bifolium, Polygala amarella, Saxifraga hirculus*
Lower Risk (near threatened)	*Alchemilla glaucescens, Bartsia alpina, Calamagrostis stricta, Carex ornithopoda, Eleocharis austriaca, Orobanche reticulata, Polemonium caeruleum, Viola rupestris*

Durham

Endangered	*Alchemilla subcrenata, Minuartia stricta*
Vulnerable	*Helianthemum canum* ssp. *levigatum, Maianthemum bifolium, Polygala amarella, Saxifraga hirculus*
Lower Risk (near threatened)	*Alchemilla acutiloba, Alchemilla monticola, Bartsia alpina, Gentiana verna,*

Kobresia simpliciuscula, Myosotis alpestris, Potentilla fruticosa, Viola rupestris

Northumberland

Endangered	*Epipactis youngiana*
Vulnerable	*Alchemilla gracilis*
Lower Risk (near threatened	*Alchemilla acutiloba, Eleocharis austriaca, Polemonium caeruleum*
Data Deficient	*Asplenium trichomanes* ssp. *pachyrachis*

Cumbria

Endangered	*Crepis praemorsa, Woodsia ilvensis*
Vulnerable	*Alchemilla minima, Carex flava, Hypochaeris maculata, Lychnis alpina, Orobanche purpurea, Polygala amarella, Saxifraga hirculus, Trichomanes speciosum*
Lower Risk (near threatened)	*Alchemilla glaucescens, Aster linosyris, Bartsia alpina, Calamagrostis purpurea, Carex ornithopoda, Eleocharis austriaca, Euphrasia rivularis, Gentiana verna, Limonium recurvum* ssp. *humile, Myosotis alpestris, Potentilla fruticosa, Sorbus lancastriensis, Viola rupestris*

Isle of Man

Vulnerable	*Mentha pulegium*

WALES

Monmouth, Newport Unitary Authority, Torfaen Unitary Authority, Blaenau Gwent Unitary Authority, & Caerphilly Unitary Authority (Part)

Vulnerable	*Euphorbia serrulata, Sorbus anglica, Sorbus eminens*
Nationally Scarce	*Salvia pratensis*
Data Deficient	*Asplenium trichomanes* ssp. *Pachyrachis*

Glamorgan

Bridgend

Endangered	*Liparis loeselii*
Vulnerable	*Matthiola sinuata, Scorzonera humilis*
Nationally Scarce	*Gastridium ventricosum*

Neath & Port Talbot

Vulnerable	*Eriophorum gracile, Matthiola sinuata*

Swansea

Endangered	*Liparis loeselii, Rumex rupestris*
Vulnerable	*Asparagus prostratus, Dianthus armeria, Eriophorum gracile, Gentianella uliginosa, Matthiola sinuata, Ononis reclinata*
Lower Risk (near threatened)	*Aster linosyris, Draba aizoides, Lithospermum purpureocaeruleum, Mibora minima*
Nationally Scarce	*Gastridium ventricosum*

Vale of Glamorgan

Lower Risk (near threatened)	*Cirsium tuberosum, Lithospermum purpureocaeruleum*

Powys

Critical	*Sorbus leptophylla, Sorbus leyana, Scleranthus perennis* ssp. *perennis.*
Vulnerable	*Gagea bohemica, Lychnis viscaria, Mentha pulegium, Pilosella peleteriana, Potentilla rupestris, Sorbus anglica, Sorbus minima, Trifolium strictum*
Lower Risk (near threatened)	*Genista pilosa*

Carmarthenshire

Vulnerable	*Gentianella uliginosa*

Pembrokeshire

Vulnerable	*Anisantha madritensis, Asparagus prostratus, Centaurium scilloides, Cytisus scoparius* ssp. *maritimus, Gentianella uliginosa, Limonium paradoxum, Limonium parvum, Limonium procerum* ssp. *cambrense, Limonium transwallianum, Ononis reclinata, Orobanche purpurea, Ranunculus tripartitus*
Lower Risk (near threatened)	*Aster linosyris, Genista pilosa, Limonium britannicum* ssp. *transcanalis*

Caernarvonshire & Merionethshire

Endangered	*Woodsia ilvensis*
Vulnerable	*Cytisus scoparius* ssp. *maritimus, Eleocharis parvula, Eriophorum gracile, Euphrasia cambrica, Lloydia serotina, Sorbus eminens, Trichomanes speciosum, Tuberaria guttata*
Lower Risk (near threatened	*Euphrasia rivularis, Genista pilosa, Hypericum linariifolium, Limonium britannicum* ssp. *celticum, Limosella australis, Saxifraga cespitosa, Woodsia alpina*
Nationally Scarce	*Allium ampeloprasum.*

Aberconwy & Colwyn

Endangered	*Cotoneaster integerrimus*
Vulnerable	*Euphrasia cambrica, Hypochaeris maculata, Lloydia serotina*
Lower Risk (near threatened)	*Aster linosyris, Senecio cambrensis*
Nationally Scarce	*Rosa agrestis*

Denbighshire

Endangered	*Stachys alpina*
Vulnerable	*Sorbus anglica*
Lower Risk (near threatened)	*Lithospermum purpureocaeruleum*

Wrexham

Critical	*Carex muricata* ssp. *muricata*
Lower Risk (near threatened)	*Senecio cambrensis*

Flintshire

Vulnerable	*Dianthus armeria*
Lower Risk (near threatened)	*Senecio cambrensis*

Anglesey

Endangered:	*Rumex rupestris*
Vulnerable	*Tephroseris integrifolia* ssp. *maritima, Tuberaria guttata*
Lower Risk (near threatened)	*Anthyllis vulneraria* ssp. *corbierei, Juncus capitatus, Limonium britannicum* ssp. *celticum, Mibora minima*

SCOTLAND
Dumfries & Galloway

Endangered	*Woodsia ilvensis*
Vulnerable	*Lychnis viscaria, Ononis reclinata*
Lower Risk (near threatened)	*Hierochloe odorata, Limonium recurvum* ssp. *humile, Oxytropis halleri*

East Ayrshire

Lower Risk (near threatened)	*Calamagrostis stricta*

North Ayrshire

Vulnerable	*Sorbus arranensis, Sorbus pseudofennica, Trichomanes speciosum*
Data Deficient	*Cochlearia atlantica*

Renfrewshire

Lower Risk (near threatened)	*Hierochloe odorata*

City of Glasgow

Endangered	*Epipactis youngiana*

The Borders

Vulnerable	*Lychnis viscaria*
Lower Risk (near threatened):	*Calamagrostis stricta, Eleocharis austriaca, Hierochloe odorata*

Midlothian

Endangered	*Epipactis youngiana*

City of Edinburgh

Vulnerable	*Lychnis viscaria*

West Lothian

Vulnerable	*Saxifraga hirculus*

Falkirk

Endangered	*Epipactis youngiana*

Fife

Vulnerable	*Lychnis viscaria*

Argyll & Bute

Endangered	*Ranunculus reptans*
Vulnerable	*Athyrium flexile, Carex buxbaumii, Gentianella uliginosa, Oxytropis campestris, Rumex aquaticus, Trichomanes speciosum*
Lower Risk (near threatened)	*Arenaria norvegica* ssp. *norvegica, Bartsia alpina, Calamagrostis purpurea, Cystopteris montana, Dactylorhiza*

lapponica, Eriocaulon aquaticum, Euphrasia heslop-harrisonii, Kobresia simpliciuscula, Koenigia islandica, Oxytropis halleri, Potamogeton rutilus, Veronica fruticans, Woodsia alpina

Nationally Scarce   *Cochlearia micacea*

Stirling

Vulnerable	*Lychnis viscaria, Poa flexuosa, Rumex aquaticus, Sagina nivalis, Salix lanata*
Lower Risk (near threatened)	*Bartsia alpina, Cystopteris montana, Kobresia simpliciuscula, Veronica fruticans*

Perthshire & Kinross

Vulnerable	*Astragalus alpinus, Carex microglochin, Carex norvegica, Cystopteris dickieana, Erigeron borealis, Gentiana nivalis, Lychnis viscaria, Oxytropis campestris, Phyllodoce caerulea, Polygonatum verticillatum, Sagina nivalis, Salix lanata, Saxifraga cernua, Scheuchzeria palustris, Schoenus ferrugineus*
Lower Risk (near threatened)	*Bartsia alpina, Calamagrostis purpurea, Carex atrofusca, Carex rariflora, Cystopteris montana, Hierochloe odorata, Kobresia simpliciuscula, Minuartia rubella, Myosotis alpestris, Oxytropis halleri, Saxifraga rivularis, Veronica fruticans, Woodsia alpina*
Nationally Scarce	*Cochlearia micacea*

Angus

Endangered	*Homogyne alpina, Woodsia ilvensis*
Vulnerable	*Astragalus alpinus, Athyrium flexile, Carex norvegica, Cerastium fontanum ssp. scoticum, Cicerbita alpina, Erigeron borealis, Gentiana nivalis, Lychnis alpina, Oxytropis campestris, Rhinanthus angustifolius, Salix lanata*
Lower Risk (near threatened)	*Calamagrostis purpurea, Calamagrostis stricta, Carex rariflora, Cystopteris montana, Veronica fruticans, Woodsia alpina*
Nationally Scarce	*Cochlearia micacea, Gnaphalium norvegicum*

City of Aberdeen

Vulnerable	*Cystopteris dickieana*

Aberdeenshire

Endangered	*Ranunculus reptans*
Vulnerable	*Astragalus alpinus, Athyrium flexile, Carex norvegica, Cicerbita alpina, Erigeron borealis, Poa flexuosa, Salix lanata, Saxifraga cespitosa, Saxifraga hirculus*
Lower Risk (near threatened)	*Calamagrostis purpurea, Carex lachenalii, Carex rariflora, Luzula arcuata, Minuartia rubella, Saxifraga rivularis, Woodsia alpina*
Nationally Scarce	*Cochlearia micacea, Gnaphalium norvegicum*

Moray

Vulnerable	*Moneses uniflora, Poa flexuosa, Saxifraga cespitosa, Saxifraga hirculus*
Lower Risk (near threatened)	*Carex lachenalii, Carex rariflora, Luzula arcuata, Saxifraga rivularis*

Highland (East Inverness-shire)

Endangered	*Woodsia ilvensis*
Vulnerable	*Athyrium flexile, Carex buxbaumii, Carex chordorrhiza, Cystopteris dickieana, Moneses uniflora, Phyllodoce caerulea, Poa flexuosa, Salix lanata*
Lower Risk (near threatened)	*Carex lachenalii, Carex rariflora, Carex recta, Cystopteris montana, Luzula arcuata, Saxifraga rivularis, Veronica fruticans*
Nationally Scarce	*Cochlearia micacea, Gnaphalium norvegicum*

Highland (West Inverness-shire)

Vulnerable	*Athyrium flexile, Carex buxbaumii, Crassula aquatica, Diapensia lapponica, Phyllodoce caerulea, Poa flexuosa, Salix lanata, Saxifraga cernua, Saxifraga cespitosa, Scheuchzeria palustris*
Lower Risk (near threatened)	*Arenaria norvegica ssp. norvegica, Carex atrofusca, Carex lachenalii, Carex rariflora, Cystopteris montana, Dactylorhiza lapponica, Eriocaulon aquaticum, Euphrasia heslop-harrisonii, Luzula arcuata, Minuartia rubella, Saxifraga rivularis, Veronica fruticans, Woodsia alpina*
Nationally Scarce	*Gnaphalium norvegicum*
Data Deficient	*Cochlearia atlantica.*

Highland (Skye)

Endangered	*Arabis alpina*
Lower Risk (near threatened)	*Arenaria norvegica ssp. norvegica, Dactylorhiza lapponica, Eriocaulon aquaticum, Euphrasia heslop-harrisonii, Euphrasia marshallii, Koenigia islandica*
Data Deficient	*Cochlearia atlantica*

Highland (Ross & Cromarty)

Endangered	*Dactylorhiza incarnata ssp. Cruenta*
Vulnerable	*Artemisia norvegica, Carex recta, Poa flexuosa, Saxifraga cespitosa*
Lower Risk (near threatened)	*Alchemilla glaucescens, Dactylorhiza lapponica, Luzula arcuata, Oxytropis halleri, Potamogeton rutilus, Saxifraga rivularis*
Nationally Scarce	*Cochlearia micacea, Gnaphalium norvegicum*

Highland (Sutherland)

Endangered	*Euphrasia rotundifolia*
Vulnerable	*Carex chordorrhiza, Carex recta, Moneses uniflora, Potentilla rupestris*
Lower Risk (near threatened)	*Alchemilla glaucescens, Arenaria norvegica ssp. norvegica, Dactylorhiza lapponica, Euphrasia marshallii, Minuartia rubella, Oxytropis halleri, Potamogeton rutilus*

422

Highland (Caithness)

Vulnerable	*Calamagrostis scotica, Carex recta*
Lower Risk (near threatened)	*Anthyllis vulneraria* ssp. *corbierei, Calamagrostis stricta, Hierochloe odorata*

Western Isles

Vulnerable	*Potamogeton epihydrus*
Lower Risk (near threatened)	*Dactylorhiza lapponica, Euphrasia campbelliae, Euphrasia heslop-harrisonii, Hierochloe odorata, Potamogeton rutilus*
Data Deficient	*Cochlearia atlantica*

Orkney Islands

Lower Risk (near threatened	*Euphrasia heslop-harrisonii, Euphrasia marshallii, Hierochloe odorata*

Shetland Islands

Vulnerable	*Cerastium nigrescens, Pilosella flagellaris* ssp. *Bicapitata*
Lower Risk (near threatened)	*Arenaria norvegica* ssp. *norvegica, Euphrasia heslop-harrisonii, Euphrasia marshallii, Potamogeton rutilus*

Appendix 5 Records of Threatened and Near Threatened species in vice-counties

This table shows the numbers of Threatened and Near Threatened species that have been recorded in each vice-county in three date classes. Also given is the total number of Threatened and Near Threatened species that have been recorded in each vice-county since 1970. It should be noted that the figures in a date class may mask changes that have taken place in more than one.

Nmbers of taxa [excluding introductions]

v.c. no.	Vice-county	Critical Pre-1970	Critical 1970-1986	Critical 1987-1997	Critical Total 1970-1997	Endangered Pre-1970	Endangered 1970-1986	Endangered 1987-1997	Endangered Total 1970-1997	Vulnerable Pre-1970	Vulnerable 1970-1986	Vulnerable 1987-1997	Vulnerable Total 1970-1997	Near Threatened Pre-1970	Near Threatened 1970-1986	Near Threatened 1987-1997	Near Threatened Total 1970-1997
1	West Cornwall	3		1	1		1	6	7	3	2	15	17	3	1	8	9
2	East Cornwall	3		1	1	3		2	2	7		6	6		2	3	5
3	South Devon	2		2	2	3		3	3	5	4	9	13	1	1	7	8
4	North Devon	2				4		1	1	6	2	9	11	1			1
5	South Somerset	2		1	1			3	3	5	1	2	3			1	1
6	North Somerset	1	1	2	3	3		4	4	2	1	9	1			6	6
7	North Wiltshire	2						2	2	6	1	2	3	1		3	3
8	South Wiltshire	3				1		1	1	3	2	2	4			3	3
9	Dorset	3	1		1	4	1	1	2	9	1	14	14	1		7	7
10	Isle of Wight	2				3		6	6	8	1	2	3	1		7	7
11	South Hampshire	2				7		1	1	6		8	8	2		1	1
12	North Hampshire	1		2	2	3	2		2	4	2	7	9	1	2	3	5
13	West Sussex	4				3	1	2	2	6		7	7			1	1
14	East Sussex	2				3		6	7	5	2	7	1	3			
15	East Kent	5		1	1	5	2	6	6	8	4	9	13	1	2	2	4
16	West Kent	6				1	2	2	4	1		8	8	1		5	5
17	Surrey	4		1	1	1	3	3	5	6		13	13	4		1	1
18	South Essex	1	1	1	2	3		3	5	6	1	1	1	1		3	3
19	North Essex	1	1	1	2	1	2	3	3	7		1	1			1	1
20	Hertfordshire	2		1	1	3		1	2	11	2	2	4	1		2	3
21	Middlesex	3				5			1	9	1	2	3			2	2
22	Berkshire	1	2		2	1		3	3	6	2	4	6			3	3
23	Oxfordshire	1	2	1	3	1	1	3	4	4	5	5	1			2	2
24	Buckinghamshire	2	1	2	3	2		2	2	9	1	3	4	2		4	4
25	East Suffolk	2	1		1	4		2	2	6	1	3	4	1		3	5
26	West Suffolk	4		1	2	4	1	4	5	5	2	9	11	1	2	6	7
27	East Norfolk	4		2	3	4		2	2	1	2	4	4	1	1	2	4
28	West Norfolk	4	1	2	3	7		2	2	2	2	4	6		2	8	9
29	Cambridgeshire	2	1	3	4	3	4	4	4	5	1	9	1		1	4	4
30	Bedfordshire		2	1	3	2		2	2	7	1	3	4	2		1	2
31	Huntingdonshire	3			3	2	2	1	3	4	2	1	3				

No.	County	1	2	3	4	5	6	7	8	9	10	11
32	Northamptonshire	2					1	1	1		1	1
33	East Gloucestershire	1	2			1	2	3	3		1	1
34	West Gloucestershire	1	4	1		5	5	8	7		4	4
35	Monmouthshire	1	1		1	1	1	3	3		1	
36	Herefordshire	3		1		1	1	2	4			
37	Worcestershire	1	1		1	1	1	2	8		1	1
38	Warwickshire	2	2			1	1	1	7			
39	Staffordshire	1				1	1	1	5		1	1
40	Shropshire	1	1		1		2	1	2		1	1
41	Glamorgan	2	1	2		2		8	4		4	4
42	Breconshire	1	2	2	1		3	3	1		1	1
43	Radnorshire		1	1	1			5	1	1		
44	Carmarthenshire	2			1	1	2	2	2		1	
45	Pembrokeshire	1	1			1	2	11	13	1	2	3
46	Cardiganshire			1			1	1	2		1	1
47	Montgomeryshire		1					4	4			
48	Merionethshire		2	1			1	4	4	2	1	1
49	Caernarvonshire	1			3		1	11	12		5	5
50	Denbighshire	1		1				1	1	1	2	2
51	Flintshire	1						1	1		1	1
52	Anglesey	1	1		1	6	2	2		1	3	4
53	South Lincolnshire	4	1		1	5	2	2	5	2	1	1
54	North Lincolnshire	4			1	6	6	5		2		2
55	Leicestershire	2			2	9						
56	Nottinghamshire	1			1	6	2	2		1		
57	Derbyshire	3			1	3	1	2	1		2	2
58	Cheshire				1	5				1	2	2
59	South Lancashire	1			1	1					3	3
60	West Lancashire					1				1	3	4
61	South-east Yorkshire	2				3	1	1		1	1	2
62	North-east Yorkshire	3				3	1	2		1	1	1
63	South-west Yorkshire	2	4			4	1	2		1	1	1
64	Mid-west Yorkshire	1	1		1	5	1	4		1	8	8
65	North-west Yorkshire	2	1		1	4	2	4		2	6	9
66	Durham	2	1	3		1		2			8	8

Vice-county	v.c. no.	Critical				Endangered				Vulnerable				Near Threatened			
		Pre-1970	1970-1986	1987-1997	Total 1970-1997	Pre-1970	1970-1986	1987-1997	Total 1970-1997	Pre-1970	1970-1986	1987-1997	Total 1970-1997	Pre-1970	1970-1986	1987-1997	Total 1970-1997
South Northumberland	67	1				1		1	1	3		1	1			2	2
North Northumberland	68	1					1	1	1	1						1	1
Westmorland with North Lancashire	69	1				3		1	1	1		3	3	2	1	12	13
Cumberland	70	2				1	1	1	2	2		2	2		1	6	7
Isle of Man	71					1						1	1				
Dumfriesshire	72						1		1							1	1
Kirkcudbrightshire	73											1	1			1	1
Wigtownshire	74						1		1			1	1		2	1	3
Ayrshire	75					1										1	1
Renfrewshire	76					1	1		2						1	1	2
Lanarkshire	77					1	1		2					1			
Peeblesshire	78					1				1				1			
Selkirkshire	79										1	1	1			3	3
Roxburghshire	80	1				2	1		1	2		1	1		2		2
Berwickshire	81	1				1				1							
East Lothian	82	1				1				1							
Midlothian	83	1				1		1	1	1		1	1			2	2
West Lothian	84																1
Fifeshire	85	2				1				1		1	1			1	1
Stirlingshire	86											2	2				
West Perthshire	87					1				1		1	1	1	1	2	3
Mid Perthshire	88					1				2	1	12	13	3	3	9	12
East Perthshire	89					1				2	1	7	8	2	1	5	6
Angus	90					1		2	2	2		11	11	3	1	5	6
Kincardineshire	91					1				1		1	1	1			
South Aberdeenshire	92					1				3	2	5	7	2	3	6	9
North Aberdeenshire	93					1		1	1	1						1	1
Banffshire	94					1				2		2	2	1	2	3	5
Moray	95					1		1	1	1		1	1				
East Inverness-shire	96					1	1		1	1	1	8	9		2	6	8
West Inverness-shire	97	1						1	1	1	1	9	1		3	11	13
Argyll Main	98					2		1	1	2		4	4	1		9	9

No.	Vice-county							Total
99	Dunbartonshire	1			1	1	1	
100	Clyde Isles	1		1	2	2	2	
101	Kintyre	1			1	1	1	1
102	South Ebudes	1		1	1	1	1	1
103	Mid Ebudes		2				4	4
104	North Ebudes	1	1		2	1	5	6
105	West Ross	1	1	1	3	3	4	6
106	East Ross	1	1	1	3	3	4	4
107	East Sutherland	1		1	2	3	1	
108	West Sutherland	1	1	1	1	1	2	8
109	Caithness	1			2	2	3	3
110	Outer Hebrides		3		1	1	5	5
111	Orkney Islands		1	1	2	3	3	3
112	Shetland Islands	1		1	2	2	2	4

The number of threatened and near-threatened taxa last recorded in each vice-county during these periods. e.g. in North Somerset, 1 species has not been recorded since 1970, and 2 species not since 1986.

429

Appendix 6a Occurrence of Threatened and Near Threatened species in grid squares

The following table shows the numbers of hectads (10 x 10 km squares) and 1 km x 1 km squares in which each taxon has been recorded during each period. For many species the figures provide a reliable measure of change, indicating, for example, a real increase in the occurrence of *Allium babingtonii* and a real decrease in the occurrence of *Dianthus armeria* and *Thlaspi perfoliatum*. However, some interpretation of the figures is needed. For instance, for some species a change in the number of records from one period to another may only reflect greater or lesser recording effort during particular periods. There is, for instance, often a dip in the number of records in the period 1970-1986, this being very often an artifact of recording effort.

The *Ratio* column compares the number of 1 km x 1 km squares in which the taxon occurred between 1970 and 1997 with the number of hectads in which it has been recorded between 1970 and 1997 (i.e. it divides the number of 1970-1997 one km squares from which records are known by the number of 1970-1997 hectads from which records are known). The ratio gives a measure of local frequency: the higher the ratio, the more frequent the taxon, even if only in a very limited geographical range.

Species	10 km x 10 km squares				1 km x 1 km squares			Ratio
	Pre-1970	1970-1986	1987-1997	Total 1970-1997	1970-1986	1987-1997	Total 1970-1997	
Adonis annua	62	12	10	16	15	11	22	1.38
Ajuga chamaepitys	37	18	11	20	18	25	37	1.85
Alchemilla acutiloba	13	7	9	11	12	15	22	2.00
Alchemilla glaucescens	14	6	12	12	5	24	28	2.33
Alchemilla gracilis	0	2	2	2	3	3	3	1.50
Alchemilla minima	2	2	3	3	2	10	11	3.67
Alchemilla monticola	9	2	3	3	2	6	8	2.67
Alchemilla subcrenata	2	1	1	1	2	2	3	3.00
Alisma gramineum	2	4	2	4	4	4	7	1.75
Allium ampeloprasum var. *ampeloprasum*	3	2	3	3	2	4	4	1.33
Allium ampeloprasum var. *babingtonii*	18	18	35	38	38	116	136	3.58
Allium sphaerocephalon	1	1	1	1	2	2	2	2.00
Althaea hirsuta	1	3	5	5	4	5	6	1.20
Anisantha madritensis	2	2	2	2	3	6	6	3.00
Anthyllis vulneraria ssp. *corbierei*	5	4	1	5	3	1	5	1.00
Apium repens	4	2	1	2	2	1	2	1.00
Arabis alpina	1	1	1	1	1	1	1	1.00
Arabis glabra	128	25	14	32	30	20	44	1.38
Arabis scabra	1	1	1	1	3	3	4	4.00
Arenaria norvegica ssp. *anglica*	2	2	2	2	9	10	10	5.00
Arenaria norvegica ssp. *norvegica*	8	11	8	11	15	15	22	2.00
Armeria maritima ssp. *elongata*	6	1	1	1	1	1	1	1.00
Artemisia campestris	11	3	4	4	8	7	10	2.50
Artemisia norvegica	3	3	3	3	6	3	6	2.00
Asparagus officinalis ssp. *prostratus*	10	5	10	10	9	16	17	1.70
Asplenium trichomanes ssp. *pachyrachis*	0	1	6	6	1	9	9	1.50
Aster linosyris	8	7	7	7	10	11	14	2.00
Astragalus alpinus	3	4	4	4	8	8	11	2.75
Athyrium flexile	11	5	6	8	6	6	11	1.38
Atriplex pedunculata	12	0	1	1	0	1	1	1.00
Bartsia alpina	14	12	11	13	32	35	49	3.77

Species	10 km x 10 km squares				1 km x 1 km squares			Ratio
	Pre-1970	*1970-1986*	*1987-1997*	*Total 1970-1997*	*1970-1986*	*1987-1997*	*Total 1970-1997*	
Bunium bulbocastanum	13	11	11	12	40	33	46	3.83
Bupleurum baldense	3	2	2	2	2	2	2	1.00
Bupleurum falcatum	1	1	0	1	1	0	1	1.00
Buxus sempervirens	6	3	5	5	5	18	18	3.60
Calamagrostis purpurea	3	5	6	9	5	6	9	1.00
Calamagrostis scotica	1	1	1	1	1	2	2	2.00
Calamagrostis stricta	19	14	8	15	14	14	25	1.67
Carex atrofusca	5	5	4	5	12	11	16	3.20
Carex buxbaumii	2	3	3	3	4	5	5	1.67
Carex chordorrhiza	1	3	3	3	7	3	7	2.33
Carex depauperata	7	1	2	2	1	3	3	1.50
Carex filiformis	9	11	10	12	13	17	21	1.75
Carex flava	1	1	1	1	1	1	1	1.00
Carex lachenalii	7	5	3	5	10	12	17	3.40
Carex microglochin	1	1	1	1	8	8	9	9.00
Carex muricata ssp. *muricata*	3	4	5	5	4	5	5	1.00
Carex norvegica	4	5	5	5	7	6	8	1.60
Carex ornithopoda	11	11	13	13	44	33	55	4.23
Carex rariflora	11	16	14	18	55	34	68	3.78
Carex recta	2	3	3	3	8	7	11	3.67
Carex vulpina	21	12	8	13	17	12	22	1.69
Centaurea calcitrapa	6	4	5	5	4	9	9	1.80
Centaurea cyanus	408	88	43	125	92	46	136	1.09
Centaurium scilloides	3	1	1	1	4	2	4	4.00
Centaurium tenuiflorum	2	1	1	1	4	5	5	5.00
Cephalanthera rubra	8	4	3	5	5	3	6	1.20
Cerastium brachypetalum	0	2	2	2	5	5	6	3.00
Cerastium fontanum ssp. *scoticum*	0	1	1	1	1	1	1	1.00
Cerastium nigrescens	2	2	2	2	2	2	2	1.00
Chenopodium chenopodioides	29	10	11	13	4	58	62	4.77
Chenopodium vulvaria	95	3	2	4	3	2	5	1.25
Cicerbita alpina	4	4	4	4	5	4	5	1.25
Cirsium tuberosum	11	13	11	14	37	29	45	3.21
Clinopodium menthifolium	1	1	1	1	1	1	1	1.00
Cochlearia micacea	21	6	15	15	6	35	37	2.47
Coincya wrightii	1	1	1	1	5	1	5	5.00
Corrigiola litoralis	2	1	1	1	1	2	2	2.00
Corynephorus canescens	23	7	10	12	12	14	21	1.75
Cotoneaster integerrimus	1	1	1	1	2	6	6	6.00
Crassula aquatica	2	1	1	1	2	5	5	5.00
Crepis foetida	13	1	1	1	2	1	3	3.00
Crepis praemorsa	0	0	1	1	0	1	1	1.00
Cynodon dactylon	2	2	2	2	2	3	5	2.50
Cynoglossum germanicum	29	4	3	4	9	15	16	4.00
Cyperus fuscus	9	5	6	6	6	9	10	1.67
Cypripedium calceolus	20	1	1	1	1	1	2	2.00

Species	10 km x 10 km squares				1 km x 1 km squares			Ratio
	Pre-1970	1970-1986	1987-1997	Total 1970-1997	1970-1986	1987-1997	Total 1970-1997	
Cystopteris dickieana	2	1	4	4	1	5	5	1.25
Cystopteris montana	19	13	9	14	24	16	29	2.07
Cytisus scoparius ssp. *maritimus*	13	5	5	8	4	4	8	1.00
Dactylorhiza incarnata ssp. *cruenta*	0	1	0	1	1	0	1	1.00
Dactylorhiza incarnata ssp. *ochroleuca*	2	0	1	1	0	1	1	1.00
Dactylorhiza lapponica	2	4	15	15	4	22	23	1.53
Damasonium alisma	50	3	5	6	5	6	8	1.33
Dianthus armeria	157	24	19	37	25	19	40	1.08
Dianthus gratianopolitanus	2	2	2	2	6	4	9	4.50
Diapensia lapponica	1	1	1	1	1	0	1	1.00
Diphasiastrum complanatum	7	1	3	3	1	3	3	1.00
Draba aizoides	3	2	2	2	13	10	15	7.50
Dryopteris cristata	18	9	6	11	29	26	34	3.09
Echium plantagineum	7	3	3	3	3	6	6	2.00
Eleocharis austriaca	8	8	7	10	19	7	24	2.40
Eleocharis parvula	11	6	7	7	9	13	15	2.14
Epipactis youngiana	0	1	5	6	1	7	8	1.33
Epipogium aphyllum	5	2	0	2	5	0	5	2.50
Erica ciliaris	12	9	13	14	75	81	98	7.00
Erica vagans	7	5	4	5	69	41	75	15.00
Erigeron borealis	7	4	4	4	13	11	14	3.50
Eriocaulon aquaticum	7	8	9	9	28	37	42	4.67
Eriophorum gracile	12	4	6	7	4	8	9	1.29
Eryngium campestre	2	3	3	3	3	4	4	1.33
Euphorbia hyberna	2	2	2	2	2	7	8	4.00
Euphorbia serrulata	6	6	8	9	11	20	23	2.56
Euphrasia cambrica	4	2	3	3	2	7	9	3.00
Euphrasia campbelliae	6	3	2	4	3	3	5	1.25
Euphrasia heslop-harrisonii	11	3	8	10	3	10	12	1.20
Euphrasia marshallii	26	14	3	15	19	2	20	1.33
Euphrasia rivularis	10	7	5	8	4	15	19	2.38
Euphrasia rotundifolia	3	1	2	2	1	1	2	1.00
Euphrasia vigursii	24	12	7	15	25	12	32	2.13
Festuca longifolia	2	4	5	5	13	12	18	3.60
Filago gallica	11	1	1	2	1	0	2	1.00
Filago lutescens	69	11	12	16	10	16	21	1.31
Filago pyramidata	99	12	9	14	15	15	24	1.71
Fumaria occidentalis	22	25	19	28	66	50	105	3.75
Fumaria reuteri	8	3	2	3	3	2	3	1.00
Gagea bohemica	1	1	1	1	1	1	1	1.00
Galium constrictum	5	8	8	10	35	39	51	5.10
Galium tricornutum	274	6	2	7	6	2	7	1.00
Gastridium ventricosum	72	11	22	25	26	51	65	2.60
Genista pilosa	16	11	11	12	39	38	59	4.92
Gentiana nivalis	4	3	3	3	7	9	11	3.67
Gentiana verna	4	3	4	4	40	40	55	13.75

Species	10 km x 10 km squares				1 km x 1 km squares			Ratio
	Pre-1970	1970-1986	1987-1997	Total 1970-1997	1970-1986	1987-1997	Total 1970-1997	
Gentianella ciliata	2	0	1	1	0	1	1	1.00
Gentianella uliginosa	6	7	7	8	15	10	20	2.50
Geranium purpureum	21	15	23	26	22	54	64	2.46
Gladiolus illyricus	9	5	6	6	32	34	41	6.83
Gnaphalium luteoalbum	6	1	2	2	1	3	3	1.50
Gnaphalium norvegicum	12	8	16	16	9	27	31	1.94
Helianthemum apenninum	4	4	4	4	8	7	10	2.50
Helianthemum canum ssp. *levigatum*	1	1	1	1	2	2	2	2.00
Herniaria ciliolata	4	3	2	3	30	6	30	10.00
Herniaria glabra	12	7	9	11	29	30	42	3.82
Hierochloe odorata	9	13	11	17	19	16	27	1.59
Himantoglossum hircinum	99	17	15	23	20	20	28	1.22
Homogyne alpina	1	1	1	1	1	1	1	1.00
Hypericum linariifolium	14	8	9	9	13	23	23	2.56
Hypochaeris maculata	14	9	9	9	14	15	19	2.11
Isoetes histrix	2	3	3	3	16	11	18	6.00
Juncus capitatus	7	4	4	5	17	15	24	4.80
Juncus pygmaeus	2	3	3	3	15	5	17	5.67
Kobresia simpliciuscula	5	10	11	14	30	24	46	3.29
Koeleria vallesiana	3	3	4	4	8	13	16	4.00
Koenigia islandica	5	4	5	5	5	19	21	4.20
Lactuca saligna	23	5	2	5	12	3	12	2.40
Lavatera cretica	4	5	4	5	7	8	10	2.00
Leersia oryzoides	18	5	5	5	7	7	8	1.60
Leucojum aestivum	17	14	14	17	43	36	62	3.65
Limonium bellidifolium	10	5	5	5	25	14	28	5.60
Limonium binervosum	2	16	3	16	25	5	27	1.69
Limonium britannicum	6	17	5	21	25	7	31	1.48
Limonium dodartiforme	1	3	4	4	3	4	4	1.00
Limonium loganicum	0	1	0	1	4	0	4	4.00
Limonium paradoxum	0	1	1	1	1	1	1	1.00
Limonium parvum	0	1	1	1	1	1	1	1.00
Limonium procerum	0	3	0	3	4	0	4	1.33
Limonium recurvum	1	6	3	6	10	6	11	1.83
Limonium transwallianum	1	1	1	1	2	2	2	2.00
Limosella australis	4	3	4	5	8	7	10	2.00
Liparis loeselii	25	9	6	9	13	16	20	2.22
Lithospermum purpureocaeruleum	21	13	13	14	20	32	37	2.64
Lloydia serotina	2	2	2	2	6	7	8	4.00
Lobelia urens	9	8	6	8	10	6	10	1.25
Lonicera xylosteum	2	2	2	2	5	7	7	3.50
Lotus angustissimus	36	17	19	25	20	36	46	1.84
Ludwigia palustris	8	4	5	5	12	24	24	4.80
Luzula arcuata	19	12	7	15	15	15	28	1.87
Luzula pallidula	2	2	2	2	2	2	3	1.50
Lychnis alpina	2	1	2	2	1	2	2	1.00

Species	10 km x 10 km squares				1 km x 1 km squares			Ratio
	Pre-1970	*1970-1986*	*1987-1997*	*Total 1970-1997*	*1970-1986*	*1987-1997*	*Total 1970-1997*	
Lychnis viscaria	22	13	14	15	26	22	30	2.00
Lythrum hyssopifolia	42	2	6	7	9	16	17	2.43
Maianthemum bifolium	5	3	3	3	4	3	4	1.33
Matthiola sinuata	11	4	7	7	11	17	22	3.14
Melampyrum arvense	28	4	4	6	5	5	8	1.33
Mentha pulegium	150	13	13	20	20	20	32	1.60
Mibora minima	4	5	6	6	8	13	16	2.67
Minuartia rubella	6	3	5	5	9	11	11	2.20
Minuartia stricta	1	1	1	1	2	2	2	2.00
Moneses uniflora	19	8	12	13	7	17	22	1.69
Muscari neglectum	11	12	8	12	19	17	24	2.00
Myosotis alpestris	6	5	5	5	14	12	16	3.20
Najas marina	4	3	2	3	12	12	17	5.67
Ononis reclinata	7	6	8	9	7	13	15	1.67
Ophioglossum lusitanicum	1	1	1	1	1	1	1	1.00
Ophrys fuciflora	5	4	4	4	11	9	13	3.25
Ophrys sphegodes	48	12	15	17	37	48	59	3.47
Orchis militaris	14	4	3	4	4	3	4	1.00
Orchis simia	6	4	2	4	4	2	4	1.00
Ornithopus pinnatus	4	3	3	3	6	6	10	3.33
Orobanche artemisiae-campestris	5	3	2	3	4	4	5	1.67
Orobanche caryophyllacea	4	5	4	5	7	6	8	1.60
Orobanche purpurea	14	10	13	16	22	20	33	2.06
Orobanche reticulata	4	6	6	6	11	22	23	3.83
Oxytropis campestris	3	3	3	3	5	6	7	2.33
Oxytropis halleri	13	10	9	10	23	22	33	3.30
Petrorhagia nanteuilii	5	1	1	1	3	3	4	4.00
Petrorhagia prolifera	0	1	2	2	1	2	2	1.00
Peucedanum officinale	6	6	7	8	32	34	42	5.25
Phleum phleoides	16	10	9	12	26	31	42	3.50
Phyllodoce caerulea	2	3	3	3	12	5	12	4.00
Physospermum cornubiense	9	4	6	8	12	20	27	3.38
Phyteuma spicatum	6	3	4	4	13	12	16	4.00
Pilosella flagellaris ssp. *bicapitata*	1	0	3	3	0	6	6	2.00
Pilosella peleteriana	7	1	6	6	1	12	13	2.17
Poa flexuosa	4	6	4	8	9	7	13	1.63
Poa infirma	6	10	31	34	13	54	63	1.85
Polemonium caeruleum	14	13	13	14	25	22	30	2.14
Polycarpon tetraphyllum	12	8	9	10	12	26	30	3.00
Polygala amarella	13	12	10	15	20	17	25	1.67
Polygonatum verticillatum	10	5	8	8	6	9	10	1.25
Polygonum maritimum	10	1	6	6	1	7	7	1.17
Potamogeton acutifolius	27	10	10	13	36	37	53	4.08
Potamogeton epihydrus	2	1	2	2	2	3	3	1.50
Potamogeton nodosus	13	7	9	9	35	45	55	6.11
Potamogeton rutilus	7	8	9	12	10	14	20	1.67

Species	10 km x 10 km squares				1 km x 1 km squares			Ratio
	Pre-1970	1970-1986	1987-1997	Total 1970-1997	1970-1986	1987-1997	Total 1970-1997	
Potentilla fruticosa	5	7	5	7	18	14	22	3.14
Potentilla rupestris	3	4	4	4	5	4	5	1.25
Pulicaria vulgaris	56	8	9	10	14	14	18	1.80
Pulmonaria obscura	0	0	1	1	0	3	3	3.00
Pyrus cordata	6	2	3	3	2	4	4	1.33
Ranunculus ophioglossifolius	4	2	2	2	2	2	3	1.50
Ranunculus reptans	2	0	2	2	0	2	2	1.00
Ranunculus tripartitus	66	17	10	20	32	16	42	2.10
Rhinanthus angustifolius	38	6	9	10	10	20	23	2.30
Romulea columnae	2	1	1	1	2	1	2	2.00
Rorippa islandica sensu stricto	8	13	23	30	19	37	51	1.70
Rosa agrestis	26	5	16	21	5	20	25	1.19
Rumex aquaticus	2	3	3	3	11	14	15	5.00
Rumex rupestris	33	15	15	18	26	26	35	1.94
Sagina nivalis	3	4	2	4	14	10	14	3.50
Salix lanata	11	11	10	13	12	15	20	1.54
Salvia pratensis	31	21	19	24	32	36	47	1.96
Saxifraga cernua	4	3	3	3	5	4	5	1.67
Saxifraga cespitosa	9	9	6	10	15	8	18	1.80
Saxifraga hirculus	19	7	6	7	18	17	23	3.29
Saxifraga rivularis	16	12	14	17	26	23	35	2.06
Scheuchzeria palustris	9	4	5	6	16	19	29	4.83
Schoenoplectus triqueter	7	1	1	1	3	1	3	3.00
Schoenus ferrugineus	2	2	3	3	5	6	6	2.00
Scirpoides holoschoenus	3	2	2	2	4	5	7	3.50
Scleranthus perennis ssp. perennis	1	1	1	1	1	1	1	1.00
Scleranthus perennis ssp. prostratus	5	2	3	3	5	4	6	2.00
Scorzonera humilis	3	1	2	2	1	3	3	1.50
Scrophularia scorodonia	22	16	32	33	38	118	130	3.94
Selinum carvifolia	4	2	2	2	5	4	5	2.50
Senecio cambrensis	3	11	7	11	15	17	26	2.36
Senecio paludosus	5	1	1	1	1	1	1	1.00
Seseli libanotis	4	4	3	4	8	6	8	2.00
Silene otites	14	7	5	7	33	22	37	5.29
Sorbus anglica	12	11	10	13	19	25	34	2.62
Sorbus arranensis	1	1	1	1	5	7	8	8.00
Sorbus bristoliensis	0	1	1	1	4	2	5	5.00
Sorbus domestica	0	0	4	4	0	4	4	1.00
Sorbus eminens	4	4	6	6	7	9	13	2.17
Sorbus lancastriensis	4	6	7	9	33	24	51	5.67
Sorbus leptophylla	3	3	4	4	4	3	5	1.25
Sorbus leyana	2	2	2	2	3	3	4	2.00
Sorbus minima	1	1	1	1	5	3	5	5.00
Sorbus pseudofennica	1	1	1	1	3	3	4	4.00
Sorbus subcuneata	3	4	4	4	12	11	17	4.25
Sorbus vexans	2	2	3	3	3	5	8	2.67

Species	10 km x 10 km squares				1 km x 1 km squares			Ratio
	Pre-1970	1970-1986	1987-1997	Total 1970-1997	1970-1986	1987-1997	Total 1970-1997	
Sorbus wilmottiana	1	1	1	1	2	2	3	3.00
Spergularia bocconei	6	4	4	4	2	5	6	1.50
Stachys alpina	2	2	2	2	2	2	2	1.00
Stachys germanica	9	2	2	3	7	5	9	3.00
Tephroseris integrifolia ssp. *maritima*	1	1	2	2	0	2	2	1.00
Teucrium botrys	12	6	6	6	10	9	12	2.00
Teucrium chamaedrys	1	1	1	1	1	1	1	1.00
Teucrium scordium	17	3	2	3	6	4	8	2.67
Thlaspi perfoliatum	9	3	4	4	11	12	16	4.00
Thymus serpyllum	5	4	4	4	20	17	29	7.25
Tordylium maximum	2	1	2	2	3	3	4	2.00
Trifolium bocconei	3	2	1	2	7	4	7	3.50
Trifolium incarnatum ssp. *molinerii*	3	2	2	2	4	4	5	2.50
Trifolium strictum	4	3	3	3	13	6	14	4.67
Trinia glauca	5	6	5	6	14	18	22	3.67
Tuberaria guttata	4	4	3	4	8	9	10	2.50
Valerianella eriocarpa	22	2	7	9	3	11	14	1.56
Valerianella rimosa	61	11	4	13	10	4	13	1.00
Veronica fruticans	18	14	13	16	22	24	30	1.88
Veronica spicata ssp. *spicata*	7	3	4	5	4	7	8	1.60
Veronica triphyllos	10	3	3	4	5	3	6	1.50
Veronica verna	4	2	2	2	19	13	24	12.00
Viola canina ssp. *montana*	3	2	2	2	2	3	4	2.00
Viola kitaibeliana	3	2	2	2	6	4	8	4.00
Viola persicifolia	18	3	4	4	5	5	7	1.75
Viola rupestris	4	6	7	7	10	15	17	2.43
Woodsia alpina	15	15	12	16	29	19	40	2.50
Woodsia ilvensis	9	6	4	6	6	2	7	1.17

Appendix 6b Occurrence of Threatened and Near Threatened species in grid squares, arranged by numbers of 1 km x 1 km squares with records between 1987-1997

This table lists species recorded in more than 20 1 km squares between 1987 and 1997. The *Ratio* column compares the number of 1 km x 1 km squares in which the taxon occurred between 1970-1997 with the number of hectads (10 km x 10 km squares) in which it has been recorded between 1970-1997 (i.e. it divides the number of 1970-1997 one km squares from which records are known by the number of 1970-1997 hectads from which records are known). The ratio gives a measure of local frequency:

the higher the ratio, the more frequent the taxon, even if only in a very limited geographical range.

This table highlights those species that are locally frequent within their restricted range and which, though qualifying as Threatened or Near Threatened under the hectad threshold used in this book, might not do so if a 1 km threshold were adopted. Under such a scheme, at least the two species heading the list would be prime candidates for the Nationally Scarce category.

Species	*10 km x 10 km squares*				*1 km x 1 km squares*			*Ratio*
	Pre-1970	*1970-1986*	*1987-1997*	*Total 1970-1997*	*1970-1986*	*1987-1997*	*Total 1970-1997*	
Scrophularia scorodonia	22	16	32	33	38	118	130	3.94
Allium ampeloprasum var. *babingtonii*	18	18	35	38	38	116	136	3.58
Erica ciliaris	12	9	13	14	75	81	98	7.00
Chenopodium chenopodioides	29	10	11	13	4	58	62	4.77
Geranium purpureum	21	15	23	26	22	54	64	2.46
Poa infirma	6	10	31	34	13	54	63	1.85
Gastridium ventricosum	72	11	22	25	26	51	65	2.60
Fumaria occidentalis	22	25	19	28	66	50	105	3.75
Ophrys sphegodes	48	12	15	17	37	48	59	3.47
Centaurea cyanus	408	88	43	125	92	46	136	1.09
Potamogeton nodosus	13	7	9	9	35	45	55	6.11
Erica vagans	7	5	4	5	69	41	75	15.00
Gentiana verna	4	3	4	4	40	40	55	13.75
Galium constrictum	5	8	8	10	35	39	51	5.10
Genista pilosa	16	11	11	12	39	38	59	4.92
Eriocaulon aquaticum	7	8	9	9	28	37	42	4.67
Potamogeton acutifolius	27	10	10	13	36	37	53	4.08
Rorippa islandica sensu stricto	8	13	23	30	19	37	51	1.70
Leucojum aestivum	17	14	14	17	43	36	62	3.65
Lotus angustissimus	36	17	19	25	20	36	46	1.84
Salvia pratensis	31	21	19	24	32	36	47	1.96
Bartsia alpina	14	12	11	13	32	35	49	3.77
Cochlearia micacea	21	6	15	15	6	35	37	2.47
Carex rariflora	11	16	14	18	55	34	68	3.78
Gladiolus illyricus	9	5	6	6	32	34	41	6.83
Peucedanum officinale	6	6	7	8	32	34	42	5.25
Bunium bulbocastanum	13	11	11	12	40	33	46	3.83
Carex ornithopoda	11	11	13	13	44	33	55	4.23
Lithospermum purpureocaeruleum	21	13	13	14	20	32	37	2.64
Phleum phleoides	16	10	9	12	26	31	42	3.50

Species	10 km x 10 km squares				1 km x 1 km squares			Ratio
	Pre-1970	*1970-1986*	*1987-1997*	*Total 1970-1997*	*1970-1986*	*1987-1997*	*Total 1970-1997*	
Herniaria glabra	12	7	9	11	29	30	42	3.82
Cirsium tuberosum	11	13	11	14	37	29	45	3.21
Gnaphalium norvegicum	12	8	16	16	9	27	31	1.94
Dryopteris cristata	18	9	6	11	29	26	34	3.09
Polycarpon tetraphyllum	12	8	9	10	12	26	30	3.00
Rumex rupestris	33	15	15	18	26	26	35	1.94
Ajuga chamaepitys	37	18	11	20	18	25	37	1.85
Sorbus anglica	12	11	10	13	19	25	34	2.62
Alchemilla glaucescens	14	6	12	12	5	24	28	2.33
Kobresia simpliciuscula	5	10	11	14	30	24	46	3.29
Ludwigia palustris	8	4	5	5	12	24	24	4.80
Sorbus lancastriensis	4	6	7	9	33	24	51	5.67
Veronica fruticans	18	14	13	16	22	24	30	1.88
Hypericum linariifolium	14	8	9	9	13	23	23	2.56
Saxifraga rivularis	16	12	14	17	26	23	35	2.06
Dactylorhiza lapponica	2	4	15	15	4	22	23	1.53
Lychnis viscaria	22	13	14	15	26	22	30	2.00
Orobanche reticulata	4	6	6	6	11	22	23	3.83
Oxytropis halleri	13	10	9	10	23	22	33	3.30
Polemonium caeruleum	14	13	13	14	25	22	30	2.14
Silene otites	14	7	5	7	33	22	37	5.29

Acknowledgements

This Red Data Book is the result of co-operation between botanists, conservationists and other naturalists, both amateur and professional, from a wide diversity of backgrounds, all of whom gave freely of their time and knowledge. We are grateful to all who contributed to it, and if any names have been inadvertently omitted, then this is an oversight on the part of the editor. We are most indebted to the officers and members of the BSBI, without whose help this project would not have been possible. Special thanks are due to D.A. Pearman, who gave unstintingly of his time throughout the project's duration. The other members of the BSBI steering group, D. McCosh and Dr F.H. Perring, contributed individually, as also did Dr C.D. Preston, in addition to his contribution at Monks Wood. I am especially grateful to the following BSBI vice-county recorders who gave invaluable help checking and amending data sheets, adding new records and occasionally undertaking fieldwork:

Mrs P.P. Abbott, Dr K.J. Adams, G.H. Ballantyne, Mrs M. Barron, G.H. Battershall, B.D. Batty, Mrs G. Beckett, P.M. Benoit, C.R. Boon, Dr H.J.M. Bowen, R.P. Bowman, M.E. Braithwaite, Lady Anne Brewis, Mrs M. Briggs, A.L. Bull, Miss E.R. Bullard, J.K. Butler, A.O. Chater, A.R. Church, Mrs P. Copson, Dr R.W.M. Corner, Dr E.M. Crackles, Mrs G. Crompton, J.J. Day, J.R. Edelsten, Mrs P.A. Evans, S.B. Evans, T.G. Evans, Mrs K.M. Fallowfield, B.R.W. Fowler, Dr C.N. French, Dr L.S. Garrad, Mrs G.M. Gent, Rev. G.G. Graham, D.E. Green, I.P. Green, Mrs J.A. Green, P.R. Green, E.F. Greenwood, Dr R.L. Gulliver, Dr G. Halliday, P.A. Harmes, Prof. D.M. Henderson, Mrs B.G. Hogarth, Dr D.R. Humphreys, Miss A.M. Hutchison, Mrs E.M. Hyde, Miss E.H. Jackson, T.J. James, M.B. Jeeves, P.A. Jones, G.M. Kay, Dr Q.O.N. Kay, D.H. Kent, H.J. Killick, M.A.R. & Mrs C. Kitchen, Dr P. Macpherson, R. Maycock, D.J. McCosh, D. McKean, T.F. Medd, Miss R.J. Murphy, Mrs C.W. Murray, Dr R.J. Pankhurst, Mrs R.E. Parslow, E.G. Philp, M. Porter, R.D. Pryce, R.H. Roberts, H.A. Salzen, M.N. Sanford, W. Scott, B. Shepard, F.W. Simpson, Mrs J.E. Smith, R. Smith, Dr R.A.H. Smith, L.M. Spalton, Mrs E.W. Stewart, Mrs O.M. Stewart, A. McG. Stirling, Prof. G.A. Swan, N.W. Taylor, Dr R.E. Thomas, B.H. Thompson, Mrs S.E. Thomson, Prof. I.C. Trueman, W.H. Tucker, Mrs M. Wainwright, K. Watson, Dr D.W. Welch, Mrs I. Weston, G.T.D. Wilmore, Dr A. Wilmott, D.C. Wood, P.C.H. Wortham, Dr G. Wynne,

The following people wrote or contributed to species accounts:

Miss F. Abraham, Dr K.J. Adams, Dr P.A. Ashton, G.H. Battershall, Mrs P.M. Batty, Mrs G. Beckett, C.R. Birkinshaw, Mrs M. Briggs, J.K. Butler, A.J. Byfield, D.A. Callaghan, Dr P.D. Carey, C. Chatters, A.R. Church, J.M. Church, Dr D.E. Coombe, N.R. Cowie, Dr J.H.S. Cox, Mrs G. Crompton, C.S. Crook, Miss J.M. Dinsdale, Miss A.J. Dunn, D. Evans, S.B. Evans, T.G. Evans, Miss L. Farrell, Dr B.W. Ferry, M.J.Y. Foley, J.E. Gaffney,

Dr C. Gibson, Dr A.J. Gray, I.P. Green, Dr A.D.R. Hare, P.A. Harmes, N.G. Hodgetts, Mrs B.G. Hogarth, Dr J.J. Hopkins, Miss A. Horsfall, A. Jackson, A.C. Jermy, A. Jones, Miss B. Jones, Dr M. Jones, P.S. Jones, Dr R.F. John, Miss D. Junghanns, Dr Q.O.N. Kay, G.C.B. Kennison, M.P. King, M.A.R. & Mrs C. Kitchen, S.J. Leach, Mrs Y. Leonard, P.S. Lusby, D.K. Mardon, P.R. Marren, R.W. Marriot, Mrs W. McCarthy, Mrs E.J.McDonnell, Miss H. McHaffie, Miss H.M. Meredith, J. Mitchell, Ms V.M. Morgan, Dr J.L. Muddeman, Miss R.J. Murphy, A. Newton, Mrs R.E. Parslow, A.G. Payne, D.A. Pearman, Dr A. Pickering, Prof. C.D. Pigott, Dr C. Pope, R.D. Porley, M.S. Porter, Dr C.D. Preston, Rev. A.L. Primavesi, Dr D.A. Ratcliffe, Dr T.C.G. Rich, Dr A.J. Richards, M.H. Rickard, M. Rix, R.H. Roberts, Dr F. Rose, R. Rose, G.P. Rothero, F.J. Roberts, F.J. Rumsey, M.N. Sanford, Miss R. Scott, Dr A.J. Showler, Dr A.J. Silverside, Dr F.M. Slater, Dr R.A.H. Smith, D. Soden, M.J. Southam, N.F. Stewart, Mrs O.M. Stewart, J. Stokes, D.A. Stone, C. Sydes, I. Taylor, Dr P.E. Taylor, R.E. Thomas, P.J.O. Trist, P.F. Ulf-Hansen, J.C. Vogel, R.M. Walls, Dr S.M. Walters, T.C.E. Wells, Dr B.D. Wheeler, Dr B.R. Wheeler, Prof. W.J. Whittington, Miss L. Wilkinson, Dr P.J. Wilson, Miss H.R. Winship, J. Woodman, J. Wright, M. Wright, Dr M.B. Wyse Jackson.

Many others, including members of voluntary societies and staff of statutory country agencies and non-government organisations, supported the project in various ways - by providing species records, reports and the results of research, commenting on text or and carrying out fieldwork:

Mrs P.P. Abbott, R. Abbot, R. Archer, Dr J. Backshall, Dr S.G. Ball, T. Barfield, B.D. Batty, Miss F. Bayley, S. Bell, Miss T. Bennett, P.M. Benoit, K. Betteridge, C.R. Birkinshaw, Mrs E. Birse, D.G. Boddington, A. Bolton, R.P. Bowman, R. Bradford, A.J.P. Branson, Dr A. Brenchley, Mrs M. Briggs, P. Cashman, Dr D. Chamberlain, A.O. Chater, C. Chatters, J. Clitheroe, R.L. Cole, J. Conaghan, Miss A.P. Conolly, R.J. Cooke, Dr D.E. Coombe, Mrs P. Copson, P. Corkhill, Dr N.R. Cowie, Prof. R.M. Crawford, Mrs G. Crompton, Q.C.B. Cronk, C.S. Crook, M. Cuddy, R.L. D'Ayala, C. Doarks, P. Duncan, M.J. Edgington, R.G. Ellis, Dr G.R. Else, Mrs C. Evans, S.B. Evans, Miss L. Farrell, Dr J. Fenton, R.A. Finch, R.S.R. Fitter, Lady R. FitzGerald, Dr V. Fleming, M.J.Y. Foley, I. Forrest, Dr W. Foyt, Mrs J.E. Gaffney, J. Gallaher, P. Gateley, Mrs C. Geddes, Dr H. Gillett, Prof. C.H. Gimingham, Dr D.R. Glendinning, Rev. G.G. & Mrs P.S. Graham, D.E. Green, I.P. Green, Mrs J.A. Green, P. Green, Dr R.J. Gulliver, P.A. Harmes, R. Harold, Dr A.D. Headley, K.A. Hearn, S. Hedley, J. Hickling, D.D. Hobson, N.G. Hodgetts, Mrs B.G. Hogarth, Dr J.J. Hopkins, Miss A. Horsfall, C. Hurford, Miss A.M. Hutchinson, P. Jepson, A.C. Jermy, Miss D. Junghanns, Dr R.F. John, A. Jones, Dr P.S. Jones, Dr M. Jones, Dr Q.O.N. Kay, G.C.B. Kennison, Dr R. Key, J. Killick, M.P. King, M.A.R. & Mrs C. Kitchen, P. Lambley, Dr C.A. Lambrick, I. Lakin, Mrs F. Le Sueur, S.J. Leach,

Mrs Y. Leonard, J. Love, C.J. Lowe, P.S. Lusby,
Dr P. Macpherson, R. McBeath, Mrs W. McCarthy,
D. McClintock, D. McCosh, Mrs E.J. McDonnell,
P. McSweeney, D.K. Mardon, Miss H.M. Meredith,
Dr C. Miles, Mrs D.J. Millward, J. Mitchell,
Ms V.M. Morgan, Miss R.J. Murphy, Mrs C.W. Murray,
Mrs M.A. Palmer, S. Parker, Mrs R.E. Parslow, A.G. Payne,
D.A. Pearman, Dr F.H. Perring, B. Phillips, Prof. C.D. Pigott,
Mrs J. Pitt, C. Pope, R.D. Porley, Dr C.D. Preston,
Dr M.C.F. Proctor, C. Pulteney, Miss D. Ramsay,
Dr D.A. Ratcliffe, Dr T.C.G. Rich, F.J. Roberts, R.H. Roberts,
Dr F. Rose, Dr S.L. Rothera, G. Rothero, M. Scott,
Miss R. Scott, J. Shackles, B. Shepard, Dr D.A. Sheppard,
Dr A.J. Showler, N. Sibbett, M.B.E. Simpson, L. Slack,
I.M. Slater, K. Slater, M. Smallbones, Mrs J. Smart,
A.E. Smith, Mrs J.E. Smith, L.M. Spalton, Prof. C.A. Stace,
Miss H.E. Stace, J. Stacey, E. Steer, R.A. Stevens,
N.F. Stewart, Mrs O.M. Stewart, N.J.H. Stuart,
C. Studholme, A. Takagi-Arigho, I. Taylor, Dr P. Taylor,
Dr R. Thomas, K. Walsh, R. Weaving, S. Webb, D.A. Wells,
Mrs H. Whetter, Mrs S.J. Whild, Miss L. Wilkinson,
M.A. Wilkinson, P. Williams, Dr P.J. Wilson,
Miss H.R. Winship, D.C. Wood, Dr S.J.R. Woodell,
A. Woodhall, R.G. Woods, P.C.H. Wortham, Miss J. Wright,
B. Yates.

I am grateful to Dr D. Stevens, Countryside Council
for Wales, for contributing the section on genetic
aspects, to N.G. Hodgetts for the chapter on
Conservation, Mrs M.A. Palmer and Dr I. McLean for
overall management of the project at JNCC, and to the
other members of the country agencies' steering
committee: Miss L. Farrell, Dr C. Sydes and A. Jones.
I also thank members of the editorial panel, and others,
who read the draft text and made many valuable
comments - Miss L. Farrell, Mrs M.A. Palmer,
Dr C.D. Preston, Dr D.A. Ratcliffe and, most especially,
S.J. Leach and D.A. Pearman, who read and commented
on two drafts.

The maps were produced by the Biological Records
Centre, Monks Wood. I extend thanks to P.T. Harding,
Head of the BRC, for support, and to Dr C.D. Preston,
H.R. Arnold and Mrs J.M. Croft for processing records and
overseeing the production of the species distribution maps.
Thanks also go to Ms S.S. Kaznowska, JNCC, for proof-
reading the whole document.

Finally, our grateful thanks to Professor A.J. Willis and
D.A. Pearman for reviewing the complete document prior
to publication.

Bibliography

Abbott, P. 1995. Orobanche reticulata *in Yorkshire in 1995*. Wakefield, English Nature. (Recovery Programme Report.)

Abbott, P. 1996. Orobanche reticulata *in Yorkshire, 1996*. Unpublished Report to Leeds City Council. [Includes descriptions and maps of all known populations.]

Abbott, R.J., Noltie, H.J., & Ingram, R. 1983. The origin and distribution of *S. cambrensis* Rosser in Edinburgh. *Transactions of the Botanical Society of Edinburgh, 44*: 103-106.

Adams, K.J. 1994. *Survey of the* Lactuca saligna *populations at Vange and Fobbing Marshes, 1993/4*. Colchester, unpublished Report to English Nature.

Akeroyd, J.R. 1975. *The possible allopolyploid origin of* Petrorhagia nanteuilii. BSc Thesis, University of St. Andrews.

Akeroyd, J.R., & Beckett, G. 1995. *Petrorhagia prolifera* (L.) P.W.Ball & Heywood (Caryophyllaceae), an overlooked native species in Eastern England. *Watsonia, 20*: 405-407.

Akeroyd, J.R., & Clarke, K. 1993. *Dianthus armeria* L. new to Ireland, and other rare plants in West Cork. *Watsonia, 19*: 185-193.

Aldasoro, J.J., Aedo, C., & Muñoz Garmendia, F. 1996. The genus *Pyrus* L. (Rosaceae) in south-west Europe and North Africa. *Botanical Journal of the Linnean Society, 121*: 148-158.

Allen, B., & Woods, P. 1993. *Wild Orchids of Scotland*. Edinburgh, HMSO.

Anon. 1875. *Gentiana pneumonanthe* in Bucks. *Journal of Botany, London, 6*: 295.

Archer, D. 1987. An investigation into the effect of ammonia addition and water aeration on the growth of *Potamogeton nodosus* Poir. Peterborough, unpublished report to the Nature Conservancy Council.

Archer Briggs, T.R. 1880. *Flora of Plymouth*. London, van Voorst.

Arnold, F.H. 1887. *Flora of Sussex*. London, Simpkin, Marshall, Hamilton, Kent & Co.

Ashton, P.A. 1990. Multiple origins of *Senecio cambrensis* Rosser, and related evolutionary studies in British *Senecio*. PhD Thesis, University of St Andrews.

Ashton, P.A., & Abbott, R.J. 1992. Multiple origins and genetic diversity in the newly arisen allopolploid species, *Senecio cambrensis* Rosser (Compositae). *Heredity, 68*: 25-32.

Babcock, E.B. 1947. The Genus *Crepis*. Part 1. The taxonomy, plylogeny, distribution and evolution of *Crepis*. *University of California Publications in Botany, 21*: 1-198.

Babington, C.C. 1863. On the discovery of *Gladiolus illyricus* as a British plant. *Journal of Botany, 1*: 97-98.

Baker, H.G. 1955. *Geranium purpureum* Vill. and *G. robertianum* L. in the British Flora: 1. *Geranium purpureum*. *Watsonia, 3*: 160-167.

Baker, J.G. 1906. *North Yorkshire: studies of its botany, geology, climate and physical geography*. 2nd ed. London, A. Brown & Sons.

Baldock, D., & Rich, T.C.G. 1996. *Tower mustard* (Arabis glabra) *in Surrey*. London, Plantlife. (Back from the Brink Project Report, No. 69.)

Ball, P.W., & Heywood, V.H. 1962. The taxonomic separation of the cytological races of *Kohlrauschia prolifera* (L.) Kunth *sensu lato*. *Watsonia, 5*: 113-116.

Ball, P.W., & Heywood, V.H. 1964. A revision of the genus *Petrorhagia*. *Bulletin of the British Museum (Natural History), Botany, 3*: 121-172.

Baskin, J.M., & Baskin, C.C. 1979. The ecological life cycle of *Thlaspi perfoliatum* and a comparison with published studies on *Thlaspi arvense*. *Weed Research, 19*: 285-292.

Bateman, R.M. 1980. *Cephalanthera rubra* - Red Helleborine. *The Orchid Society of Great Britain Journal, 1(1)*.

Bateman, R.M., & Denholm, I. 1983. *Dactylorhiza incarnata* (L.) Soó subsp. *ochloleuca* (Boll.) P.F. Hunt & Summerhayes. *Watsonia, 14*: 410-411.

Batty, B.D., & Batty, P.M. 1978. *Saxifraga cernua* study report, 1978. London, unpublished report to the National Trust for Scotland and the Nature Conservancy Council.

Batty, P.M., Batty, B.D., & Miller, G.M. 1984. Population size and reproduction of *Gentiana nivalis* L. at Ben Lawers NNR. *Transactions of the Botanical Society of Edinburgh, 44*: 269-280.

Beckett, G. 1992. Childing pink, a Norfolk plant. *Botanical Society of the British Isles News, 62*: 20-21.

Beckett, G. 1993. *Rare plant survey of the Norfolk Breckland, 1993*. Peterborough, unpublished report to English Nature.

Bennett, A. 1899. *Senecio paludosus* and *S. palustris* in East Anglia. *Transactions of the Norfolk and Norwich Naturalists' Society, 6*: 457-462.

Bennett, A. 1908. *Potamogeton pensylvanicus* in England. *Naturalist, Hull (1908)*: 10-11.

Bennett, A. 1905. Distribution of *Sonchus palustris* L. and *Atriplex pedunculata* L. in England. *Transactions of the Norfolk and Norwich Naturalists' Society, 8*: 35-43.

Benum, P. 1958. *The flora of Troms Fylke*. Tromso Museums Skrifter.

Bevis, J., Kettel, R., & Shepard, B. 1978. *Flora of the Isle of Wight*. Newport, Isle of Wight Natural History and Archaeology Society.

Bignal, E. 1980. The endemic whitebeams of North Arran. *Glasgow Naturalist, 20*: 50-64.

Birkinshaw, C.R. 1990a. The biology of *Veronica spicata* subspecies *spicata* and its re-introduction to West Harling Heath and Cavenham Heath. *Nature Conservancy Council, CSD Report*, No. 1,149.

Birkinshaw, C.R. 1990b. The biology of *Leersia oryzoides* and its re-establishment on the Basingstoke canal, Woking, and the Oberwater, New Forest. *Nature Conservancy Council, CSD Report*, No. 1,151.

Birkinshaw, C.R. 1990c. The ecology of *Carex depauperata* and its reinforcement at Cheddar Wood, Somerset. *Nature Conservancy Council, CSD Report*, No. 1,152.

Birkinshaw, C.R. 1990d. The biology of *Artemisia campestris*, and its introduction to High Lodge pit, Mildenhall,

Suffolk. *Nature Conservancy Council, CSD Report*, No. 1,153.

Birkinshaw, C.R. 1990e. A report of the re-introduction of *Buplerum falcatum* to Norton Heath, Essex. *Nature Conservancy Council, CSD Report*, No. 1,154.

Birkinshaw, C.R. 1991. The habitat preferences of *Carex depauperata* Curtis ex With. *Bulletin of the British Ecological Society*, 22: 26-31.

Birkinshaw, C.R. 1994. Aspects of the ecology and conservation of *Damasonium alisma* Miller in Western Europe. *Watsonia*, 20: 33-99.

Birkinshaw, C.R., & Sanford, M.N. 1996. *Pulmonaria obscura* Dumort. (Boraginaceae) in Suffolk. *Watsonia*, 21: 169-178.

Birse, E.M. 1997. Creeping spearwort, *Ranunculus reptans* L., at the Loch of Strathbeg. *Botanical Society of the British Isles News*, 74: 17-19.

Björkqvist, I. 1967. Studies in *Alisma* L. I. Distribution, variation and germination. *Opera Botanica*, 17: 1-128.

Blab, J., Nowek, E., Trautmann, W., & Sukopp, H. 1984. *Rote Liste der Gefährdeten Tiere und Pflanzen in der Bundesrepublick Deutschland*. Greven.

Blackstock T.H., & Jones, R.A. 1997. *Juncus capitatus* Weigel (Juncaceae) redisocovered near its orginal locality in Anglesey (v.c. 52). *Watsonia*, 21: 277-278.

Blakemore, J. 1979. Rare plant survey of Durham. Unpublished report to the Nature Conservancy Council.

Blakemore, J. 1980. Rare plant survey of Northumberland. Unpublished report to the Nature Conservancy Council.

Blakemore, J. 1981. Rare plant survey of North Yorkshire. Unpublished report to the Nature Conservancy Council.

Böcher, T.W. 1963. Experimental and cytological studies on plant species. VIII. Racial differentiation in amphi-atlantic *Viscaria alpina*. *Biologiske Skrifter udg. af det Kongelige Danske Videnskabernes Selskab*, 11: 1-33.

Böcher, T.W., & Larsen, K. 1958. Secondary polyploid and ecological differentiation in *Sarothamnus scoparius*. *New Phytologist*, 57: 311-317.

Boddington, D.G. 1995. *The relationship of* Carex microglochin *Wahlenberg to soil chemistry and hydrology in the Ben Lawers range*. MSc Thesis, University College of North Wales.

Bolliger, M. 1978. Die *Pulmonaria obscura-officinalis*-Gruppe in der Schweiz. *Berichte Schweiz Botanischer Gesellschaft*, 88: 30-62.

Bowen, H.J.M., & Dymond, J.A. 1955. Strontium and barium in plants and soils. *Proceedings of the Royal Society, ser. B*, 144: 355-368.

Bradshaw, M.E. 1962. The distribution and status of five species of the *Alchemilla vulgaris* L. aggregate in Upper Teesdale. *Journal of Ecology*, 50: 681-706.

Bradshaw, M.E. 1964. Studies on *Alchemilla filicaulis* Bus. *sensu lato* and *A. minima* Walters. III. *Alchemilla minima*. *Watsonia*, 6: 76-81.

Branwell, A.E. 1872. Short notes and queries. *Phyteuma spicatum*. *Journal of Botany*, 10: 307-308.

Bratton, J.H. 1991. *British Red Data Books: 3. Invertebrates other than insects*. Peterborough, Joint Nature Conservation Committee.

Brett, O. 1955. Cytotaxonomy of the genus *Cerastium*. 1. Cytology. *New Phytologist*, 54: 138-148.

Brewis, A., Bowman, P., & Rose, F. 1996. *The Flora of Hampshire*. Eastleigh, The Hampshire and Isle of Wight Trust Limited.

Briggs, M. 1983. Perennial centaury in the lawn. *Botanical Society of the British Isles News*, 33: 9.

Briggs, M., & Maurice, P. 1993. *Ludwigia palustris* - first record in Sussex since 1876. *Botanical Society of the British Isles News*, 64: 13-14.

Brightmore, D. 1968. Biological Flora of the British Isles. *Lobelia urens* L. *Journal of Ecology*, 56: 613-620.

Britten, J. 1879. *Gentiana pneumonanthe* in Bucks. *Journal of Botany, London*, 8: 44.

Bromfield, W.A. 1843. Notice of a new British *Calamintha* discovered on the Isle of Wight. *The Phytologist*, 177: 768-770.

Brookes, B.S. 1981. The discovery, extermination, translocation and eventual survival of *Schoenus ferrugineus* in Britain. *In: The biological aspects of rare plant conservation*, ed. by H. Synge, 421-428. London, Wiley & Sons.

Brooks, R.R., Trow, J.M., & Bolviken, B. 1979. Biogeochemical anomalies in Fennoscandia: a study of copper, lead and nickel levels in *Melandrium dioicum* and *Viscaria alpina*. *Journal of Geochemical Exploration*, 11: 73-87.

Bruederle, L.P., & Jensen, U. 1991. Genetic differentation of *Carex flava* and *Carex viridula* in West Europe (Cyperaceae). *Systematic Botany*, 16: 41-49.

Brummitt, J.M. 1981. *Stachys alpina* in Denbighshire. *Botanical Society of the British Isles Welsh Bulletin*, 34: 7-9.

Bucknall, C. 1897. *Stachys alpina* in Britain. *Journal of Botany, London*, 35: 380-381.

Butcher, R.W. 1921. A new British flowering plant *Tillaea aquatica* L. *Naturalist, Hull (1921)*: 369-370.

Byfield, A.J. 1991. Classic British wildlife sites - the Lizard Peninsula. *British Wildlife*, 3: 92-105.

Byfield, A.J. 1992. *The status and ecology of* Eleocharis parvula *in Britain (excluding Ireland)*. Unpublished report to English Nature - South Region, Newbury.

Callaghan, D.A. 1996. The conservation status of *Lythrum hyssopifolia* L. in the British Isles. *Watsonia*, 21: 179-186.

Carlsson, D. 1991. Kransrams - vastsvensk fagningsblomma [Clearing flower - a species favoured by burning in western Sweden]. *Svensk Botanisk Tidskrift*, 85: 81-86.

Carver, R.J.I. 1990. *The status of childing pink* Petrorhagia nanteuilii *at Pagham Harbour National Nature Reserve and Site of Special Scientific Interest (and other closely associated sites)*. A report to West Sussex County Council and the Nature Conservancy Council.

Cassidi, M.D. 1980. Status of the Lundy cabbage *Rhynchosinapis wrightii*. *Lundy Field Society Report*, 31: 64-67.

Catling, P.M. 1980. Rain-assisted autogamy in *Liparis loeselii* (L.) L.C.M. Rich. (Orchidaceae). *Bulletin of the Torrey Botanical Club*, 107: 525-529.

Cauwood, D. 1981. *Salvia pratensis: an ecological survey*. BSc Thesis, University College of Wales, Aberystwyth.

Chapman, S.B. 1975. The distribution and composition of hybrid populations of *Erica ciliaris* L. and *Erica tetralix* L. in Dorset. *Journal of Ecology*, 63: 809-824.

Chapman, S.B., & Rose, R.J. 1994. Changes in the distribution of *Erica ciliaris* L. and *E.* x *watsonii* Benth. in Dorset, 1963-1987, *Watsonia, 20*: 89-95.

Chater, A.O., & Akeroyd, J.R. 1993. *Polycarpon* Loefl. ex L. *In*: *Flora Europaea, Volume 1, 2nd ed.*, ed. by T.G. Tutin *et al*. Cambridge, Cambridge University Press.

Chater, A.O., & Rich, T.C.G. 1995. *Rorippa islandica* (Oeder ex Murray) Borbás (Brassicaceae) in Wales. *Watsonia, 20*: 229-238.

Chatters, C. 1991. The status of *Pulicaria vulgaris* Gaertner in Britain in 1990. *Watsonia, 18*: 405-406.

Church, J.M., Coppins, B.J., Gilbert, O.L., James, P.W., & Stewart, N.F. 1996. *Red Data Books of Britain and Ireland: lichens. Vol. 1: Britain.* Peterborough, Joint Nature Conservation Committee.

Clapham, A.R. 1978. *Upper Teesdale. The area and its natural history.* London, Collins.

Clapham, A.R., Tutin, T.G., & Moore, D.M. 1987. *Flora of the British Isles*. 3rd ed. Cambridge, Cambridge University Press.

Clapham, A.R., Tutin, T.G., & Warburg, E.F. 1952. *Flora of the British Isles.* Cambridge, Cambridge University Press.

Clapham, A.R., Tutin, T.G., & Warburg, E.F. 1962. *Flora of the British Isles.* 2nd ed. Cambridge, Cambridge University Press.

Clark, W.D. 1943. Pondweeds from North Uist (v.c. 110) with a special consideration of *Potamogeton rutilus* Wolfg. and a new hybrid. *Proceedings of the University of Durham Philosophical Society, 10*: 368-373.

Clarke, S. 1993. *A survey of* Cochlearia micacea *on Ben Lawers NNR 1993.* Killin, National Trust for Scotland.

Clement, E.J. 1997. *Ludwigia x muellertii* Hort. new to Britain. *Botanical Society of the British Isles News, 77*: 54.

Clement, E.J., & Foster, M.C. 1994. *Alien plants of the British Isles.* London, Botanical Society of the British Isles.

Coker, P.D. 1962. Biological Flora of the British Isles. *Corrigiola litoralis* L. *Journal of Ecology, 50*: 833-840.

Coker, P.D. 1969. The site of *Arenaria norvegica* ssp. *norvegica* in Morvern, Argyll. *Transactions of the Botanical Society of Edinburgh, 40*: 557-564.

Coker, P.D., & Coker, A.M. 1973. Biological Flora of the British Isles. *Phyllodoce caerulea* (L.) Bab. *Journal of Ecology, 61*: 901-913.

Cook, C.D.K. 1966. A monographic study of *Ranunculus* subgenus *Batrachium* (DC) A. Gray. *Mitteilungen der Botanischen Staatssammlung, München, 6*: 47-237.

Cook, C.D.K. 1983. Aquatic plants endemic to Europe and the Mediterranean. *Botanische Jahrbücher, 103*: 539-582.

Cook, C.D.K. 1988. New and noteworthy plants from the northern Italian ricefields. *Berichte Schweizerischen Botanischen Gesellschaft, 83*: 54-65.

Coombe, D.E. 1987a. Longevity of spores of some terrestrial *Isoetes* species. *Botanical Society of the British Isles News, 45*: 38.

Coombe, D.E. 1987b. Spiked speedwell, soil stripes and polygons, and the vanishing chalk heaths of Cambridgeshire. *Nature in Cambridgeshire, 29*: 26-37.

Coombe, D.E. 1992. *Ophioglossum lusitanicum*: 31 years in cultivation. *Botanical Society of the British Isles News, 60*: 62.

Coombe, D.E., & Frost, L.C. 1956a. The heaths of the Cornish serpentine. *Journal of Ecology, 44*: 226-256.

Coombe, D.E., & Frost, L.C. 1965b. The nature and origin of the soils over the Cornish serpentine. *Journal of Ecology, 44*: 605-615.

Corner, R.W.M. 1975. *Eleocharis austriaca* Hayek new to Scotland. *Watsonia, 10*: 411-412.

Corner, R.W.M., & Roberts, F.J. 1989. *Carex ornithopoda* Willd. in Cumbria. *Watsonia, 17*: 437-438.

Council of Europe. 1991. *Convention on the Conservation of European Wildlife and Natural Habitats.* Scientific Group of Experts.

Countryman, W.D. 1968. *Alisma gramineum* in Vermont. *Rhodora, 70*: 577-579.

Cowie, N.R., & Sydes, C. 1995. *Status, distribution, ecology and management of Lapland marsh-orchid* Dactylorhiza lapponica. Edinburgh, Scottish Natural Heritage. (Review, No. 42.)

Cowie, N.R., & Sydes, C. 1995. *Status, distribution, ecology & management of brown bog-rush* Schoenus ferrugineus. Edinburgh, Scottish Natural Heritage. (Review No. 43.)

Cox, J.H.S. 1997. Hampshire purslane found in Dorset. *Recording Dorset, 7*: 20-21.

Crackles, F.E. 1990. *Flora of the East Riding of Yorkshire.* Hull, Hull University Press & Humberside County Council.

Crackles, F.E. 1994. *Calamagrostis stricta* (Timm) Koeler, *C. canescens* (Wigg.) Roth and their hybrids in S.E. Yorks, v.c. 61, northern England. *Watsonia, 20*: 51-60.

Crackles, F.E. 1995. A graphical analysis of the characters of *Calamagrostis stricta* (Timm.) Koeler, *C. canescens* (Wigg.) Roth and their hybrid populations in S.E. Yorks, v.c. 61, northern England. *Watsonia, 20*: 397-404.

Crins, W.J., & Ball, P.W. 1989. Taxonomy of the *Carex flava* complex (Cyperaceae) in North America and northern Eurasia. II: Taxonomic treatment. *Canadian Journal of Botany, 67*: 1048-1065.

Crompton, G. 1974-1986. rare plant survey of eastern England. *Nature Conservancy Council, CSD Report*, No. 1059 [Bedfordshire, Hertfordshire, Northamptonshire, Leicestershire, Nottinghamshire, Lincolnshire, Humberside, Cambridgeshire, Suffolk, Norfolk, Essex].

Cuddy, M. 1988. Purple viper's bugloss (*Echium plantagineum*) at St Just. *Botanical Cornwall Newsletter, 2*: 5-6.

Cuddy, M. 1991. *Flora of St Just in Penwith.* Cornwall Institute of Cornish Studies. Redruth, Cornish Biological Records Unit.

Cullen, J. 1986. *Anthyllis* in the British Isles. *Notes of the Royal Botanic Garden, Edinburgh, 43(2)*: 277-281.

Cunningham, M.H., & Kenneth, A.G. 1979. *The Flora of Kintyre.* East Ardsley, EP Publishing Limited.

Curtis,T.G.F., & McGough, H.N. 1988. *The Irish Red Data Book. 1. Vascular plants.* Dublin, Wildlife Service Ireland.

Curwen, E.C., & Ross-Williamson, R.P. 1931. The date of Clissbury Camp. *Antiquaries Journal, 11*: 31.

Dalby, D.H., & Rich, T.C.G. 1995. *The history, taxonomy, distribution and ecology of mountain scurvy-grass* (Cochlearia micacea *Marshall*). Back from the Brink Project Report, No. 42. (Contractor: Plantlife.) Edinburgh, Scottish Natural Heritage.

Dandy, J.E., & Taylor, G. 1938. Studies of British Potamogetons. - III. *Potamogeton rutilus* in Britain. *Journal of Botany, 76*: 239-241.

Dandy, J.E., & Taylor, G. 1939. Studies of British Potamogetons. - IV. The identity of *Potamogeton drucei*. *Journal of Botany, 77*: 56-62.

Daniels, R.E. 1990. Variability in three marginal populations of *Lobelia urens* L. in Britain. *Vegetatio, 90*: 63-71.

Daniels, R.E., McDonnell, E.J., & Moy, I.L. 1996. *Species Recovery Programme - shore dock (Rumex rupestris Le Gall). Second Report.* Peterborough, English Nature (Contract Report.)

Daniels, R.E., Raybould, A.F., & Farkas, J.M. 1996. Conserving genetic variation in British populations of *Lobelia urens*. *Biological Conservation, 79*: 15-22.

Davey, F.H. 1907. *Euphrasia vigursii* sp. n. *Journal of Botany, 45*: 17-20, Plate 486.

Davey, F.H. 1909. *Flora of Cornwall: being an account of the flowering plants and ferns found in the county of Cornwall including the Scilly Isles.* Penrhyn, Chegwidden.

David, R.W. 1979. Another British locality for *Carex muricata* L. *sensu stricto*. *Watsonia, 12*: 335.

David, R.W. 1980a. The distribution of *Carex ornithopoda* Willd. in Britain. *Watsonia, 13*: 53-54.

David, R.W. 1980b. The distribution of *Carex rariflora* (Wahlenb.) Sm. in Britain. *Watsonia, 13*: 124-125.

David, R.W. 1981. *Carex ornithopoda* Willd. east of the Pennines. *Watsonia, 13*: 321.

David, R.W. 1983. The distribution of *Carex tomentosa* (*C. filiformis* auct.) in Britain. *Watsonia, 14*: 412-414.

David, R.W. 1993. *Carex filiformis* L. Downy-fruited sedge and *Carex humilis* Leyss. dwarf sedge. In: *The Wiltshire Flora*, ed. by B. Gillam. Newbury, Pisces Publications.

David, R.W., & Kelcey, J.G. 1975. *Carex muricata* L. *sensu* Nelmes and *Carex bullockiana* Nelmes. *Watsonia, 10*: 412-414.

Davies, D., & Jones, A. 1995. *Welsh names of plants*. Cardiff, National Museum of Wales, Department of Botany.

Davies, E.W. 1956. The ecology and distribution of *Carex flava* and its allies in the British Isles. *Botanisk Notiser, 109*: 50-74.

Davies, T.D., Abrahams, P.W., Tranter, M., Blackwood, I., Brimblecombe, P., & Vincent, C.E. 1984. Black acidic snow in the Scottish Highlands. *Nature, 312*: 58-61.

Davis, P.H. 1967. *Flora of Turkey and the East Aegean Islands, Vol. 2.* Edinburgh, Edinburgh University Press.

Davis, P.H. 1975. *Flora of Turkey and the East Aegean Islands, Vol. 5.* Edinburgh, Edinburgh University Press.

Davis, T.A.W., & Evans, S.B. 1980. Irregular times of flowering of *Ononis reclinata* L. *Watsonia, 13*: 125-126.

Day, R.T., & Scott, P.J. 1985. The biology of *Diapensia lapponica* in Newfoundland. *The Canadian Field Naturalist, 98*: 425-439.

De Lemos, M. 1992. *Sand crocus monitoring of Dawlish Warren Local Nature Reserve 1982-1992.* Report for Teignbridge District Council.

Delvosalle, L., Dermet, F., Lambinon, J., & Lawalree, A. 1969. *Plantes rares, disparues ou menacées de disparition en Belgique: l'appauvrissement de la flora indigène.* Ministère de l'Agriculture, Service des Réserves Naturelles Dominales et de la Conservation de la Nature.

Doarks, C. 1993. Proposal for a Species Recovery Project, *Liparis loeselii* (L.) Richard. Norwich, unpublished report to English Nature.

Doe, J. 1993. *Badgeworth Nature Reserve Site Management Plan, 1993-1997.* Gloucester, Gloucestershire Wildlife Trust.

Dony, J.G. 1953. *Flora of Bedfordshire.* Luton, The Corporation of Luton Museum and Art Gallery.

Dowlan, C.M., & Ho, T.N. 1995. *Gentianella ciliata* (L.) Borkh. in Wiltshire (v.c. 8). *Watsonia, 20*: 279.

Dring, M.J., & Frost, L.C. 1971. Studies of *Ranunculus ophioglossifolius* in relation to its conservation at the Badgeworth Nature Reserve, Gloucestershire, England. *Biological Conservation, 4*: 48-56.

Druce, G.C. 1886. *The Flora of Oxfordshire.* Oxford & London, Parker & Co.

Druce, G.C. 1906. Notes on the flora of the Channel Islands. *Journal of Botany, 45*: 419-428.

Druce, G.C. 1922. *Tillaea aquatica* L. *Report of the Botanical Society and Exchange Club of the British Isles, 6*: 281-282.

Druce, G.C. 1926. *The Flora of Buckinghamshire.* Arbroath, T. Buncle.

Druce, G.C. 1927. *Flora of Oxfordshire.* Oxford, Clarendon Press.

Druce, G.C. 1929. Short Notes: *Scorzonera humilis* L. *Journal of Botany, 67*: 26-27.

Dudman, A.A., & Richards, A.J. 1997. *Dandelions of Great Britian and Ireland.* B.S.B.I. Handbook No. 9. London, Botanical Society of the British Isles.

Duffey, E., Morris, M.G., Sheaill, J., Ward, L.K., Wells, D.A., & Wells, T.C.E. 1974. *Grassland ecology and wildlife management.* London, Chapman & Hall.

Duffield, J.C.H. 1979. Observations on the survival and flowering of some late spider orchids *Ophrys fuciflora* on the North Downs in East Kent, with particular reference to the colonies on the Wye National Nature Reserve. Report No. 25. Wye, Nature Conservancy Council.

Dunn, A.J. 1987. Observations on *Stachys germanica* L. at a new site in Oxfordshire. *Watsonia, 16*: 430-431.

Dunn, A.J. 1991. Further observations on *Stachys germanica*. *Watsonia, 18*: 359-367.

Dunn, A.J. 1997. Biological Flora of the British Isles. *Stachys germanica* L. *Journal of Ecology, 85*: 531-539.

Eddy, A., Welch, D., & Rawes, M. 1969. The vegetation of the Moor House National Nature Reserve in the northern Pennines, England. *Vegetatio, 16*: 239-284.

Edees, E.S., & Newton, A.L. 1988. *Brambles of the British Isles.* London, Ray Society.

Elkington, T.T. 1963. Biological Flora of the British Isles. *Gentiana verna* L. *Journal of Ecology, 51*: 755-761.

Elkington, T.T. 1964. Biological Flora of the British Isles. *Myosotis alpestris* F.W.Schmidt. *Journal of Ecology, 52*: 709-722.

Elkington, T.T. 1972. Variation in *Gentiana verna* L. *New Phytologist, 71*: 1203-1211.

Elkington, T.T., & Woodell, S.R.J. 1963. Biological Flora of the British Isles. *Potentilla fruticosa* L. *Journal of Ecology, 51*: 769-781.

Ellis, R.G. 1983. *Flowering plants of Wales.* Cardiff, National Museum of Wales.

Ellstrand, N.C., & Elam, D.R. 1993. Population genetics consequences of small population size: implications for plant conservation. *Annual Review of Ecology and Systematics, 24:* 217-42.

Ennos, R.A., Cowie, R.R., Legg, C.J., & Sydes, C. (1997). Which measures of genetic variation are relevant in plant conservation? A case study of *Primula scotica. In: The role of genetics in conserving small populations,* ed. by T. Tew *et al.* Peterborough, Joint Nature Conservation Committee.

Evans, D. 1982. *Distribution and soil preferences of* Carex microglochin *and* Sagina intermedia *in the Ben Lawers area.* BSc Thesis, University of Stirling.

Evans, D., & Ellis, M. 1994a. *The status of* Stachys alpina *in Clwyd, 1993 & 1994.* Bangor, Countryside Council for Wales. (Species & Monitoring Report, No. 94/2/7.)

Evans, D., & Ellis, M. 1994b. Lithospermum purpureocaeruleum *monitoring.* Bangor, Countryside Council for Wales. (Species & Monitoring Report, No. 94/2/5.)

Everett, S. 1988. *Rare plant survey for southern England.* Peterborough, Nature Conservancy Council. (Contract Surveys, No. 23.) [Hampshire, Wiltshire, Oxfordshire, Buckinghamshire, Berkshire.]

Everett, S. 1993. *Cirsium tuberosum* (L.) All. tuberous thistle. *In: The Wiltshire Flora,* ed. by B. Gillam, 83-90. Newbury, Pisces Publications.

Falk, D.A., & Holsinger, K.E., *eds.* 1991. *Genetics and conservation of rare plants.* New York, Oxford University Press.

Farmer, A.M., & Spence, D.H.N. 1986. The growth strategies and distribution of isoetids in the Scottish Freshwater lochs. *Aquatic Botany, 26:* 247-258.

Farrell, L. 1979. The distribution of *Leucojum aestivum* in the British Isles. *Watsonia, 12:* 325-332.

Farrell, L. 1983. The status of *Eriocaulon aquaticum* in Western Scotland. Unpublished Report, Nature Conservancy Council, South-West (Scotland) region.

Farrell, L. 1985. Biological Flora of the British Isles. *Orchis militaris* L. *Journal of Ecology, 73:* 1041-1053.

Farrell, L. 1991. Population changes and management of *Orchis militaris* at two sites in England. *In: Population Ecology of Terrestrial Orchids,* ed. by T.C.E. Wells & J.H. Williams, 63-68. The Hague, SPB Academic Publishing.

Faulkner, J.S. 1972. Chromosome studies on *Carex* section *Acutae* in NW Europe. *Botanical Journal of the Linnean Society, 65:* 271-301.

Fearn, G.M. 1975. Variation of *P. amarella* Crantz in Britain. *Watsonia, 10:* 371-383.

Ferguson, I.K., & Ferguson, L.F. 1971. *Eriocaulon aquaticum* (Hill) Druce - on Ardnamurchan (Westerness, v.c. 97). *Watsonia, 8:* 400.

Fernald, M.L. 1932. The linear-leaved North American species of *Potamogeton,* section *Axillares. Memoirs of the American Academy of Arts and Sciences, 17:* 1-183.

Ferris, C., Callow, R.S., & Gray, A.J. 1992. Mixed first and second division restitution in male meiosis of *Hierochloe odorata* (L.) Beauv. (Holy Grass). *Heredity, 69:* 21-31

Ferry, B.W. 1995. *Species Recovery Programme - stinking hawksbeard (*Crepis foetida). Peterborough, English Nature. (Contract Report.)

Field, M.H. 1994. The status of *Bupleurum falcatum* L. (Apiaceae) in the British flora. *Watsonia, 20:* 115-117.

Findlay, D.C., Colborne, G.J.N., Cope, D.W., Harrod, T.R., Hogan, D.V., & Staines, S.J. 1984. *Soils and their use in South West England.* Harpenden, Soil Survey of England and Wales.

Fitt, G. 1844. Remarks on some species of *Chenopodium. Phytologist, 1:* 1136-1138.

Fitter, R.S.R. 1945. *London's Natural History.* London, Collins.

FitzGerald, R. 1988a. *Rare plant survey of South-East England, 1985-1987. Surrey.* Peterborough, Nature Conservancy Council. (Contract Surveys, No. 32.)

FitzGerald, R. 1988b. *Rare plant survey of South-East England, 1985-87. East Sussex.* Peterborough, Nature Conservancy Council. (Contract Surveys, No. 32.)

FitzGerald, R. 1988c. *Rare plant survey of South-East England, 1985-87. West Sussex.* Peterborough, Nature Conservancy Council. (Contract Surveys, No. 32.)

FitzGerald, R. 1988d. *Rare plant survey of South-East England, 1985-87. West Kent.* Peterborough: Nature Conservancy Council. (Contract Surveys, No. 32.)

FitzGerald, R. 1988e. *Rare Plant survey of South-East England, 1985-87. East Kent.* Peterborough, Nature Conservancy Council. (Contract Surveys, No. 32.)

FitzGerald, R. 1990a. Rare plant survey of South-West England. Vol. 1. Somerset. *Nature Conservancy Council, CSD Report,* No. 1058.

FitzGerald, R. 1990b. Rare plant survey of South-West England. Vol. 2. Devon. *Nature Conservancy Council, CSD Report,* No. 1059.

FitzGerald, R. 1990c. Rare plant survey of South-West England. Vol. 3. Cornwall. *Nature Conservancy Council, CSD Report,* No. 1060.

FitzGerald, R. 1990d. Rare plant survey of South-West England. Vol. 4. Dorset. *Nature Conservancy Council, CSD Report,* No. 1061.

FitzGerald, R. 1993. *Leucojum vernum* L. *In: The Wiltshire Flora,* ed. by B. Gillam. Newbury, Pisces Publications.

FitzGerald, R., Field, G.D., & Chatters, C. 1997. The status of *Pulicaria vulgaris* Gaertner in Britain in 1995. *Watsonia, 21:* 279-280.

Fleming, V. 1995. Woodsias in Scotland. *Pteridologist, 2(6):* 287-288.

Foley, M.J.Y. 1993. *Orobanche reticulata* Wallr. populations in Yorkshire (north-east England). *Watsonia, 19:* 247-257.

Friesner, J. 1994. A study into the development of *Physospermum cornubiense.* Unpublished report, University of Plymouth.

Frost, L.C. 1981. The study of *Ranunculus ophioglossifolius* and its successful conservation at the Badgeworth Nature Reserve, Gloucestershire. *In: The biological aspects of rare plant conservation,* ed. by H. Synge. Chichester, John Wiley & Sons Ltd.

Frost, L.C., Houston, L., Lovatt, C.M., & Beckett, A. 1991. *Allium sphaerocephalon* L. and introduced *A. carinatum* L., *A. roseum* L. and *Nectaroscordium siculum* (Ucria) Lindley on St Vincent Rocks, Avon Gorge, Bristol. *Watsonia, 18:* 381-385

Frost, L.C., Hughes, M.G.B., Nichols, C., & Lawman, J.M. 1982. *A total population estimate of the land quillwort*

(Isoetes histrix) *at the Lizard District and recommendations for its conservation*. Bristol, University of Bristol.

Fryer, J., & Hylmö, B. 1994. The native British *Cotoneaster - Great Orme berry - renamed*. *Watsonia, 20*: 61-63.

Fuller, R.M. 1987. The changing extent and conservation interest of lowland grasslands in England and Wales: a review of grassland surveys 1930-1984. *Biological Conservation, 40*: 281-300.

Gallego, M.J., & Aparicio, A. 1993. Karyological study in the genus *Tuberaria* Sect. *Scorpioides* (Cistacae) - taxonomy and evolutionary references. *Plant Sytematics and Evolution, 184*: 11-25.

Gauslaa, Y. 1984. Heat resistance and energy budget in different Scandinavian plants. *Holarctic Ecology, 7*: 1-78.

Geddes, C. 1994. *Monitoring of rare montane vascular plants on Ben Lawers NNR and Caenlochan NNR*. Edinburgh, Scottish Natural Heritage. (Contract Report.)

Geddes, C. 1996. *Monitoring of rare montane vascular plants on Ben Lawers NNR and Caenlochan NNR*. Edinbugh, Scottish Natural Heritage (Review, No. 44.)

Géhu, J.M. 1969. La persistance de l'*Obione pedunculata* en baie du Mont-Saint-Michel et en quelques points du littoral du nord de la France. Sa signification biologique. *Le Monde des Plantes*, No. 359: 1-4.

Géhu, J.M., & Meslin, R. 1968. Sur la répartition et l'écologie d'*Haliminone pedunculata* (L.) Aell. (Dicotyledoneae, Chenopodiaceae) en France. *Bulletin du Laboratoire maritime de Dinard, 1(1)*: 116-136.

Gerard, J. 1597. *Herball, or General History of Plants* [edition amended by Thomas Johnson, of 1633]. London.

Gibbons, E.J., & Lousley, J.E. 1958. An inland *Armeria* overlooked in Britain. *Watsonia, 4*: 125-135.

Gibbs, P.E., & Gornall, R.J. 1976. A biosystematic study of the creeping spearworts at Loch Leven, Kinross. *New Phytologist, 77*: 777-785.

Gibson, C. 1991. Annual sea-purslane at Foulness, Essex. *Sanctuary, 20*: 9.

Gill, J.J.B., & Walker, S. 1971. Studies in *Cytisus scoparius* (L.) Link with particular emphasis on the prostrate forms. *Watsonia, 8*: 345-356.

Gillam, B., ed. 1993. *The Wiltshire Flora*. Newbury, Pisces Publications.

Giller, K.E., & Wheeler, B.D. 1988. Acidification and succession in a flood-plain mire in Broadland, Norfolk. *Journal of Ecology, 76*: 849-866.

Gilli, A. 1966. Orobanchaceae. *In: Illustrierte Flora von Mitteleuropa*, ed. by G. Hegi. München.

Glover, B.J., & Abbot, R.J. 1995. Low genetic diversity in the Scottish endemic *Primula scotica* Hook. *New Phytotogist, 129*: 147-153.

Godfree, J.S. 1979. *Saxifraga cernua* L. A report on seed production. *Botanical Society of the British Isles News, 23*: 27.

Godwin, H. 1975. *The History of the British flora*. 2nd ed. Cambridge, Cambridge University Press.

Good, R. 1936. On the distribution of the lizard orchid (*Himantoglossum hircinum* Koch). *New Phytologist, 35*: 142-170.

Good, R. 1984. *A concise flora of Dorset*. Dorset Natural History and Archaeology Society.

Goodwillie, R. 1995. Additions to the Irish range of *Rorippa islandica* (Oeder ex Murray) Borbás. *Irish Naturalists' Journal, 25*: 57-59.

Gornall, R.J. 1987. Notes on a hybrid spearwort *Ranunculus flammula* x *R. reptans*. *Watsonia, 16*: 383-388.

Graham, G.G. 1988. *The flora and vegetation of County Durham*. The Durham Flora Committee and the Durham County Conservation Trust.

Graham, G.G., & Graham, P.S. 1993a. Gentiana verna *survey in Durham*. Peterborough, Joint Nature Conservation Committee.

Graham, G.G., & Graham, P.S. 1993b. Potentilla fruticosa *survey in Durham*. Peterborough, Joint Nature Conservation Committee.

Graham, G.G., & Primavesi, A.L. 1993. *Roses of Great Britain and Ireland*. B.S.B.I. Handbook No. 7. London, Botanical Society of the British Isles.

Grant Roger, J. 1952. *Diapensia lapponica* in Scotland. *Transactions of the Botanical Society of Scotland, 36*: 34-37

Grassly, N.C., Harris, S.A., & Cronk, Q.C.B. 1996. British *Apium repens* (Jacq.) Lag. (Apiaceae) status assessed using random amplified polymorphic DNA (RAPD). *Watsonia, 21*: 103-111.

Gravett, T. 1994. *Cotoneaster integerrimus* - a step further away from extinction. *Botanical Society of the British Isles Welsh Bulletin, 57*: 4.

Gray, A.J. 1997. Genecology, the genetic system and the conservation genetics of British grasses. *In: The role of genetics in conserving small populations*, ed. by T. Tew *et al*. Peterborough, Joint Nature Conservation Committee.

Gregory, E.S. 1912. *British violets*. Cambridge, Heffers.

Griffiths, M.E., & Proctor, M.C.F. 1956. Biological Flora of the British Isles. *Helianthemum canum* (L.) Baumg. *Journal of Ecology, 44*: 677-682.

Grime J.P., Hodgson J.G., & Hunt R. 1987. *Comparative plant ecology*. London, Unwin Hyman.

Grimshaw, S. 1991. *The ecology of* Selinum carvifolia. Diploma in Field Biology. University of London.

Grose, D. 1957. *The Flora of Wiltshire*. Devizes, Wiltshire Natural History and Archaeological Society.

Grubb, P.J. 1976. A theoretical background to the conservation of ecologically distinct groups of annuals and biennials in the chalk grassland ecosystem. *Biological Conservation, 10*: 53-76.

Haggett, G.M. 1952. Observations on a colony of *Anepia irregularis* in Norfolk. *Entomologist, 80*: 36-38

Hall, P.C. 1980. *Sussex plant atlas*. Brighton, Booth Museum of Natural History.

Halliday, G.H. 1960. *Taxonomic and ecological studies in the* Arenaria ciliata *and* Minuartia verna *complexes*. PhD Thesis, University of Cambridge.

Halliday, G.H. 1990. *Crepis praemorsa* (L.) Tausch, new to western Europe. *Watsonia, 18*: 85-86.

Hampton, M. 1996. *Sorbus domestica* L. Comparative morphology and habitats. *Botanical Society of the British Isles News, 73*: 32-37.

Hampton, M., & Kay, Q.O.N. 1995. *Sorbus domestica* L., new to Wales and the British Isles. *Watsonia, 20*: 381-386.

Hamrick, J.L., Dogt, M.J.W., Murawski, D.A., & Loveless, M.D. 1991. Correlations between species traits and allozyme diversity: implications for conservation biology. *In: Genetics and conservation of rare plants*, ed. by

D.A. Falk & K.E. Holsinger, 75-86. New York, Oxford University Press.

Hansen, K. 1976. Ecological studies in Danish heath vegetation. *Dansk Botanik Archiv, 31*: 1-118.

Hansson, S. 1994. En studie av lappnycklar, *Dactylorhiza lapponica*. *Svensk Botanisk Tidskrift, 88*: 17-28.

Haraldsen, K.B., & Wesenberg, J. 1993. Population genetic analysis of an amphi-Atlantic species: *Lychnis alpina* (Caryophyllaceae). *Nordic Journal of Botany, 13*: 377-387.

Hare, A.D.R. 1986. *The ecology and conservation of* Lactuca saligna *L. and* Pulicaria vulgaris *Gaertn. in Great Britain.* PhD Thesis, Queen Mary College, University of London.

Hare, A.D.R. 1990. Lesser fleabane - a plant of seasonal hollows. *British Wildlife, 2*: 77-79.

Harmes, P.A., & Spiers, A. 1993. *Polygonum maritimum* L. in East Sussex (v.c. 14). *Watsonia, 19*: 271-273.

Harris, S.A., & Ingram, R. 1992. Molecular systematics of the genus *Senecio* L. 1. Hybridisation in a British polyploid complex. *Heredity, 69*: 1-10.

Haskins, L.E. 1978. *The vegetational history of south-east Dorset.* PhD Thesis, University of Southampton.

Hawkes, J.G., & Phipps, J.B. 1954. *Scorzonera humilis* in Warwickshire. *Proceedings of the Botanical Society of the British Isles, 1*: 152-153.

Hedberg, K.O. 1992. Taxonomic differentiation in *Saxifraga hirculus* L. (Saxifragaceae) - a circumpolar Arctic-Boreal species of central Asian origin. *Botanical Journal of the Linnean Society, 109*: 377-393.

Hegi, G. 1964. *Gesamtbeschreibung von* Stachys germanica L. (*Stachys germanica* L.: Verbreitung und soziologischer Anschlub in Deutschland, zusammengefabt von Dominik D. Schmidt, Aigsburg).

Heslop-Harrison, J.W. 1949. Potamogetons in the Scottish Western Isles, with some remarks on the general natural history of the species. *Transactions of the Botanical Society of Edinburgh, 35*: 1-25.

Heslop-Harrison, J.W. 1950. A pondweed, new to the European flora, from the Scottish Western Isles, with some remarks on the phytogeography of the island group. *Phyton, Horn, 2*: 104-109.

Heslop-Harrison, J.W. 1953. The North American and Lusitanian elements in the flora of the British Isles. *In*: *The changing flora of Britain*, ed. by J.E. Lousley, 105-123. Oxford, Botanical Society of the British Isles.

HMSO. 1994. *Biodiversity. The UK Action Plan.* Cm 2428. London, HMSO.

Hogarth, B.G. 1987. *A survey of* Minuartia rubella *on Ben Lawers NNR, 1987.* Edinburgh, unpublished report to the National Trust for Scotland and the Nature Conservancy Council.

Holland, S.C. 1977. *Badgeworth Nature Reserve Handbook: site of* Ranunculus ophioglossifolius *Vill. (Adder's-tongue Spearwort).* Gloucester, Gloucestershire Wildlife Trust.

Holland, S. C., Caddick, H. M., & Dudley-Smith, D.S. 1986. *Supplement to the Flora of Gloucestershire.* Bristol, Grenfell Publications.

Hope-Simpson, J.F. 1987. Cautionary Tale. III. *Botanical Society of the British Isles News, 47*: 22-23.

Horn, P.C. 1994. *Cerastium brachypetalum* decline in Bedfordshire. *Botanical Society of the British Isles News, 65*: 18-21.

Hubbard, C.E. 1984. *Grasses, a guide to their structure, identification, uses and distribution in the British Isles.* 3rd ed. Harmondsworth, Penguin Books Ltd.

Hull, P., & Smart, G.J.B. 1984. Variation in two *Sorbus* species endemic to the Isle of Arran, Scotland. *Annals of Botany, 53*: 641-648.

Hultén, E. 1954. *Artemisia norvegica* Fr. and its allies. *Nytt magasin for botanikk, 3*: 63-82.

Hultén, E. 1962. The circumpolar plants, Vol. 1, Vascular cryptogams, conifers, monocotyledons. *Kungliga Svenska vetenskapakademiens handlingar, ser. 8 (5)*: 186.

Hultén, E., & Fries, M. 1986. *Atlas of north European vascular plants: north of the Tropic of Cancer. Vols I-III.* Königstein, Koeltz Scientific Books.

Hutchings, M.J. 1987a. The population biology of the early spider orchid, *Ophrys sphegodes* Mill. I. A demographic study from 1975 to 1984. *Journal of Ecology, 75*: 711-727.

Hutchings, M.J. 1987b. The population biology of the early spider orchid, *Ophrys sphegodes* Mill. II. Temporal patterns in behaviour. *Journal of Ecology, 75*: 729-742.

Huxley, A. 1967. *Mountain flowers.* Poole, Blandford Press.

Hyde, H.A., Wade, A.E., & Harrison, S.G. 1978. *Welsh ferns, clubmosses, quillworts and horsetails.* 6th ed. Cardiff, National Museum of Wales.

Idle, E.T. 1968. *Rumex aquaticus* L. at Loch Lomondside. *Transactions of the Botanical Society of Edinburgh, 40*: 445-449.

Idle, E.T. 1975. The vegetation of Loch Libo and its management. *Western Naturalist, 4*: 58-63.

Ietswaart, J.H., & Schoorl, J.W. 1985. Fructification in Dutch *Maianthemum* populations. *Acta Botanica Neerlandica, 34*: 381-391.

Ingram, J.S.I., & Ingram, G.I.C. 1981. A soil analysis of the Wye Downs National Nature Reserve, Kent. *Transactions of the Kent Field Club, 8*: 146-148.

Ingram, R., & Noltie, H.J. 1995. Biological Flora of the British Isles. *Senecio cambrensis* Rosser. *Journal of Ecology, 83*: 537-546.

Ingrouille, M.J., & Stace, C.A. 1986. The *Limonium binervosum* aggregate (Plumbaginaceae) in the British Isles. *Botanical Journal of the Linnean Society, 92*: 177-217.

Irmisch, T. 1853. *Beiträge zur Biologie und Morphologie der Orchideen.* Leipzig, Ambrosias Abel.

Irmisch, T. 1863. Beiträge zur vergleichend Morphologie der Pflanzen. *Lloydia serotina. Botanishe Zeitschrift, 21*: 161-164, 169-173.

Irving, R.A. 1984. Notes on the distribution of the Lundy cabbage (*Rhynchosinapis wrightii*). *Lundy Field Society Report, 35*: 25-27.

IUCN. 1994. *IUCN Red List Categories.* Prepared by the IUCN Species Surivial Commission. As approved by the 40th Meeting of the IUCN Council, Gland, Switzerland, The World Conservation Union.

Ivimey-Cook, R.B. 1963. *Hypericum linariifolium* Vahl. *Journal of Ecology, 51*: 727-732.

Ivimey-Cook, R.B. 1984. *Atlas of the Devon Flora: flowering plants and ferns.* Torquay, Devonshire Association.

Jackson, A. 1995. The Plymouth pear - the recovery programme for one of Britain's rarest trees. *British Wildlife, 6*: 273-278.

Jackson, A., Erry, B., & Culham, A. (1997). Genetic aspects of the species recovery programme for the Plymouth pear (*Pyrus cordata* Desv.). *In: The role of genetics in conserving small populations,* ed. by T. Tew *et al.* Peterborough, Joint Nature Conservation Committee.

Jackson, A.B. 1913. *Maianthemum bifolium* Schmidt in England. *Journal of Botany, 51*: 202-208.

Jackson, A.B., & Domin, K. 1908. *Scirpus lacustris* x *triqueter* (*S. carinatus* Sm.). *Report of the Botanical Exchange Club, 2*: 314-316.

Jackson, M.J. 1981. *A chronological record of aquatic macrophytes found in each of the Norfolk Broads.* Report to the Nature Conservancy Council, Norwich.

Jalas, J., & Sell, P.D. 1967. *Cerastium fontanum* Baumg. *In:* Taxonomic and nomenclatural notes on the British flora, ed. by P.D. Sell. *Watsonia, 6*: 291-318.

Jalas, J., & Suominen, J., eds. 1972. *Atlas Florae Europaeae. Vol.1. Pteridophyta (Psilotaceae to Azollaceae).* Helsinki, Committee for Mapping the Flora of Europe and Societas Biologica Fennica Vanamo.

Jalas, J., & Suominen, J., eds. 1980. *Atlas Florae Europaeae. Vol. 5. Chenopodiaceae to Basellaceae.* Helsinki, Committee for Mapping the Flora of Europe and Societas Biologica Fennica Vanamo.

Jalas, J., & Suominen, J., eds. 1986. *Atlas Flora Europaeae. Volume7. Caryophyllaceae (Silenoideae).* Helsinki, Committee for Mapping the Flora of Europe and Societas Biologica Fennica Vanamo.

Jalas, J., & Suominen, J., eds. 1989. *Atlas Flora Europaeae. Volume 8. Nymphaeaceae to Ranuculaceae.* Helsinki, Committee for Mapping the Flora of Europe and Societas Biologica Fennica Vanamo.

Jalas, J., & Suominen, J., eds. 1994. *Atlas Flora Europaeae. Volume 10. Cruciferae (Sisymbrium to Aubrieta).* Helsinki, Committee for Mapping the Flora of Europe and Societas Biologica Fennica Vanamo.

Jarvis, S.C. 1971. *A study of the vegetation of dolerite outcrops and screes at Craig Breidden SSSI, Montgomeryshire, with special reference to plant-soil and of certain other rare species.* PhD Thesis, University of Lancaster.

Jarvis, S.C. 1974. Soil factors affecting the distribution of plant communities on the cliffs of Craig Breidden, Montgomeryshire. *Journal of Ecology, 62*: 721-733.

Jermy, A.C. 1989. The history of *Diphasiatrum issleri* (Lycopodiaceae) in Britain and a review of its taxonomic status. *Fern Gazette, 13*: 257-265.

Jermy, A.C., Arnold, H.R., Farrell, L., & Perring, F.H., eds. 1978. *Atlas of ferns of the British Isles.* London, Botanical Society of the British Isles and British Pteridological Society.

Jermy, A.C., & Camus, J. 1991. *The illustrated field guide to ferns and allied plants of the British Isles.* London, Natural History Museum.

Jermy, A.C., Chater, A.O., & David, R.W. 1982. *Sedges of the British Isles.* B.S.B.I. Handbook No. 1. 2nd ed. London, Botanical Society of the British Isles.

Jermyn, S.T. 1974. *Flora of Essex.* Colchester, Essex Naturalists' Trust Ltd.

Jessen, K., Andersen, S.T., & Farrington, A. 1959. The interglacial deposit near Gort, Co. Galway, Ireland. *Proceedings of the Royal Irish Academy, 60B*: 2-77.

John, R.F. 1992. *Genetic variation, reproductive biology and conservation in isolated populations of rare plant species.* PhD Thesis, University of Wales, Swansea.

Johns, O. 1982. Botanical notes from Alderney. *Transactions de la Société Guernesiaise, 21*: 146-147.

Johnston, W.R., & Proctor, J. 1980. Ecological studies on Meikle Kilbrannoch serpentines. *Transactions of the Botanical Society of Edinburgh, 43*: 207-215.

Jones, A. 1991. Welsh mudwort? *Botanical Society of the British Isles Welsh Bulletin, 52*: 6-8.

Jones, A. 1992. *Rare plant monitoring in Wales, 1991.* Bangor, Countryside Council for Wales. (Survey and Monitoring Report, No. 91/2/-6.)

Jones, A. 1993. *Rare plant monitoring in Wales, 1992.* Bangor, Countryside Council for Wales. (Survey and Monitoring Report, No. 92/2/17.)

Jones, A. 1994. *Rare plant monitoring in Wales, 1993.* Bangor, Countryside Council for Wales. (Survey and Monitoring Report, No 93/2/4.)

Jones, B. In prep. *Population biology and genetic conservation in the Snowdon Lily,* Lloydia serotina. PhD Thesis, University of Wales, Bangor.

Jones, P.M.B. 1978. *Seed viability in* Ranunculus ophioglossifolius *Vill.* BSc Thesis, University of Bristol.

Jones, P.S., Kay, Q.O.N., & Jones, A. 1995. The decline of rare plant species and community types in the sand dune systems of South Wales. *In: Directions in European coastal management,* ed. by M.G. Healy & J.P. Doody, 547-555. Cardigan, Samara Publishing Limited.

Jonsell, B. 1968. Studies in north-west European species of *Rorippa* s. str. *Symbolae botanicae upsaliensis, 19*: 1-222.

Kawano, S., Ihara, M., & Suzuki, M. 1986. Biosystematic studies on *Maianthemum* (Liliaceae-Polygonatae) II. Geography and ecological life history. *Japanese Journal of Botany, 20*: 35-65.

Kay, G.M. 1996. *Mentha pulegium* in grass seed. *Botanical Society of the British Isles News, 72*: 46.

Kay, Q.O.N. 1993. Genetic differences between populations of rare plants - implications for recovery programmes. *Botanical Society of the British Isles News, 64*: 54-56.

Kay, Q.O.N. 1996. The conservation of *Rumex rupestris* (shore dock) in Wales. Past, present and possible future sites and habitats of *Rumex rupestris* in South and West Wales. Bangor, Countryside Council for Wales.

Kay, Q.O.N. 1997. Review of the taxonomy, biology, geographical distribution and European conservation status of *Asparagus prostratus* Dumort. (*A. officinalis* subsp. *prostratus*), sea asparagus. Cardiff, Countryside Council for Wales. (South Area Report.)

Kay, Q.O.N., & Harrison, J. 1970. Biological Flora of the British Isles. *Draba aizoides. Journal of Ecology, 58*: 877-888.

Kay, Q.O.N., & John, R. 1994. *Population genetics and demographic ecology of some scarce and declining vascular plants of Welsh lowland grassland and related habitats.* Bangor, Countryside Council for Wales. (Science Report No. 93.)

Kay, Q.O.N., & John, R. 1995. *The conservation of scarce and declining plant species in lowland Wales: population genetics,* demographic ecology and recommendations for future conservation in 32 species of lowland grassland and related habitats. Bangor, Countryside Council for Wales. (Science Report No. 110.)

Kay, Q.O.N., & John, R. (1997). Patterns of variation in relation to the conservation of rare and declining plant species. *In: The role of genetics in conserving small populations,* ed. by T. Tew *et al.* Peterborough, Joint Nature Conservation Committee.

Kay, Q.O.N., John, R., & Jones, R.A. In prep. Reproductive biology, genetic variation and conservation of *Luronium natans* (L.) Raf. in Britain and Ireland.

Kenneth, A.G., Lowe, M. R., & Tennant, D. J. 1988. *Dactylorhiza lapponica* (Laest. ex Hartman) Soó in Scotland. *Watsonia,* 17: 37-41.

Kenneth, A.G., & Tennant, D.J. 1984. *Dactylorhiza incarnata* (L.) subsp. *cruenta* (O.F.Müll.) P.D.Sell in Scotland. *Watsonia,* 15: 11-14.

Kenneth, A.G., & Tennant, D.J. 1987. Further notes on *Dactylorhiza incarnata* (L.) subsp. *cruenta* (O.F.Müll.) P.D. Sell in Scotland. *Watsonia,* 16: 332-334.

Kennison, G.C.B. 1993. *Aquatic macrophyte surveys of the Norfolk Broads, 1992.* Norwich, unpublished report to the Broads Authority.

Kent, D.H. 1975. *Historical Flora of Middlesex.* London, Ray Society.

Kent, D.H. 1992. *List of vascular plants of the British Isles.* London, Botanical Society of the British Isles.

King M.P. 1989. *An investigation into the ecology and current status of shore dock in Devon and Cornwall.* MSc dissertation, University College London.

Kirschner, J. 1995. Allozyme analysis of *Luzula* Sect. *Luzula* (Juncaceae) in Ireland: evidence of the origin of tetraploids. *Folia Geobotanica et Phytotaxonomica,* 30: 283-290.

Kirschner, J., & Rich, T.C.G. 1993. A note on *Luzula* Section *Luzula* (Juncaceae) in Ireland, with special reference to *Luzula pallidula. Irish Naturalists' Journal,* 24: 297-298.

Kirschner, J., & Skalicky, V. 1989. Notes on *Viola* in the New Flora of the Czech Lands. *Preslia (Praha),* 61: 315-319.

Kitchen M., Kitchen, C., & Rich, T.C.G. 1995. *Report on adder-tongue spearwort* (Ranunculus ophioglossifolius) *at Inglestone common in 1995.* Back from the Brink Project Report, No. 64. London, Plantlife.

Klástersky, I. 1968. *Rosa* L. *In: Flora Europaea, Volume 2,* ed. by T.G. Tutin *et al.* Cambridge, Cambridge University Press.

Klisphuis, E.K., Heringa, J., & Hogeweg, P. 1986. Cytotaxonomic studies on *Galium palustre* L. Morphological differentiation of diploids, tetraploids and octoploids. *Acta Botanica Neerlandica,* 35: 383-392. [*Galium constrictum* included in study.]

Knipe, P.R. 1988. *Gentianella ciliata* (L.) Borkh. in Buckinghamshire. *Watsonia,* 17: 94-95.

Kreutz, C.A.J. 1995. *Orobanche. The European broomrape species. 1. Central and northern Europe.* Limburg: Stichting Natuurpublicaties.

Kwak, M.M. 1978. Pollination, hybridisation and ethological isolation of *Rhinanthus minor* and *R. serotinus* (*Rhinanthoideae: Scrophulariaceae*) by bumblebees (*Bombus* Latr.). *Taxon,* 27: 145-158.

Laan, D. van der, & Smant, W. 1985. Waarnemingen aan *Teucrium scordium* L. in het duingebied van Voorne. *Gorteria,* 12: 255-267.

Lang, D. 1989. *A guide to the wild orchids of Great Britain and Ireland.* Oxford, Oxford University Press.

Lang, H.J. 1977. *Ononis reclinata* in v.c. 74 (Wigtown). *Botanical Society of the British Isles News,* 17: 29.

Leach, S.J. 1988. Rediscovery of *Haliminone pedunculata* (L.) Aellen in Britain. *Watsonia,* 17: 170-171.

Leach, S.J. 1995. Lotus angustissimus. *A survey of selected sites in Devon and Cornwall.* Peterborough, Joint Nature Conservation Committee.

Leach, S.J., Cox, J.H.S., & Porley, R.D. 1994. *A preliminary survey of* Lotus angustissimus *in Devon and Cornwall, 1991-1993.* Taunton, English Nature.

Leadley, E.A., & Heywood, V.H. 1990. The biology and systematics of the genus *Coincya* Porta & Rigo ex Rouy (Cruciferae). *Botanical Journal of the Linnean Society,* 102: 313-398.

Lees. E. 1850. On the botanical features of the Great Orme's Head, with notices of some plants observed in other parts of North Wales during the summer of 1849. *Phytologist,* 3: 869-881.

Legg, C.J., Cowie, N.R., & Hamilton, A. 1995. *Experimental investigation of the response of the string sedge* Carex chordorrhiza *to changes in water depth in summer.* Edinburgh, Scottish Natural Heritage Research. (Survey and Monitoring Report, No. 41.)

Leonard, Y. 1993. *Survey of* Phleum phleoides, Silene otites *and* Thymus serpyllum *in Breckland, 1993.* Peterborough, Joint Nature Conservation Committee.

Leonard, Y., & Leonard, D. 1991. *Rare plant survey of the Suffolk Breckland, 1991.* Peterborough, English Nature. (Contract Surveys.)

Lesica, P., & Allendorf, F.W. 1995. When are peripheral populations valuable for conservation? *Conservation Biology:* 9: 753-760.

Le Sueur, F. 1984. *Flora of Jersey.* Société Jersiaise.

Ley, A. 1887. *Potentilla rupestris* in Radnorshire. *Journal of Botany,* 25: 28.

Libbey, R.P., & Swann, E.L. 1973. *Alisma gramineum* Lej.: a new county record. *Nature in Cambridgeshire,* 16: 39-41.

Lindop, M. 1996. *Cypripedium* conservation report 1995. *Botanical Society of the British Isles News,* 72: 45.

Lock, L., & Wilson, P.J. 1996. A botanical audit of arable farmland in SW England. Produced in association with the Game Conservancy and Plantlife.

Loeschke, V., Tomiuk, J., & Jain, S.K., *eds.* 1994. *Conservation genetics.* Basel, Birkhäuser Verlag.

Lousley, J.E. 1931. *Schoenoplectus* group of the genus *Scirpus* in Britain. *Journal of Botany,* 69: 151-163.

Lousley, J.E. 1939. *Rumex aquaticus* L. as a British plant. *Journal of Botany,* 77: 149-152.

Lousley, J.E. 1944a. *Potamogeton nodosus* Poir. *Report of the Botanical Society and Exchange Club of the British Isles,* 12: 507.

Lousley, J.E. 1944b. Notes on British Rumices II. *Report of the Botanical Exchange Club of the British Isles,* 12: 547-585.

Lousley, J.E. 1950. The habitats and distribution of *Gentiana uliginosa* Willd. *Watsonia,* 1: 279-282.

Lousley, J.E. 1957. *Alisma gramineum* in Britain. *Proceedings of the Botanical Society of the British Isles,* 2: 346-353.

Lousley, J.E. 1971. *Flora of the Isles of Scilly.* Newton Abbot, David and Charles.

Lousley, J.E. 1976. *Flora of Surrey.* Newton Abbot, David and Charles.

Lovatt, C.M. 1981. The history, ecology and status of *Gastridium ventricosum* (Gouan) Schinz & Thell. in the Avon Gorge, Bristol. *Watsonia, 13:* 287-298.

Lovatt, C.M. 1982. *The history, ecology and status of the rare plants and the vegetation of the Avon Gorge, Bristol.* PhD Thesis, University of Bristol.

Löve, Á., & Löve, D. 1958. The American element in the flora of the British Isles. *Botaniska Notiser for 1958:* 376-388.

Lusby, P., & MacDonald, I. 1995. Calamagrostis scotica, *species dossier.* Edinburgh Scottish Natural Heritage.

Lusby, P., & Wright, J. 1996. *Scottish wild plants. Their history, ecology and conservation.* Edinburgh, Royal Botanic Garden Edinburgh, and The Stationery Office.

Mace, G.M., & Lande, R. 1991. Assessing extinction threats: toward a reevaluation of IUCN threatened species categories. *Conservation Biology, 5:* 148-157.

Mahon, A., & Pearman, D.A., *eds.* 1993. *Endangered wildlfe in Dorset. The County Red Data Book.* Dorset Environmental Records Centre.

Mardon, D.K. 1980. Observations of *Gentiana nivalis* plants on Ben Lawers, 1980. Unpublished report to the National Trust for Scotland and Nature Conservancy Council.

Mardon, D.K. 1984. Destruction of plants in colonies of alpine gentian *Gentiana nivalis* L. on Ben Lawers NNR, Perthshire. Unpublished report to the National Trust for Scotland and Nature Conservancy Council.

Mardon, D.K. 1985. A further study of the destruction of plants in colonies of alpine gentian *Gentiana nivalis* L. on Ben Lawers NNR, Perthshire. Unpublished report to the National Trust for Scotland and Nature Conservancy Council.

Mardon, D.K. 1992. Monitoring *Minuartia rubella* on Ben Lawers NNR, 1992. Unpublished report to the National Trust for Scotland and Scottish Natural Heritage.

Margetts, L.J., & David, R.W. 1981. *Review of the Cornish Flora 1980.* Redruth, Institute of Cornish Studies.

Margetts, L.J., & Spurgin, K.L. 1991. *The Cornish Flora: supplement 1980-1990.* St Ives, Tendrine Press.

Marren, P.R. 1971. The Lundy cabbage. *Lundy Field Society Report, 22:* 27-31.

Marren, P.R. 1984. The history of Dickie's fern in Kincardineshire. *Pteridologist, 1:* 27-32.

Marren, P.R. 1988. The past and present distribution of *Stachys germanica* L. in Britain. *Watsonia, 17:* 59-68

Marren, P.R., Payne, A.G., & Randall, R.E. 1986. The past and present status of *Cicerbita alpina* (L.) Wallr. in Britain. *Watsonia, 16:* 131-142.

Marren, P.R., & Rich, T.C.G. 1993. Back from the brink. *British Wildlife, 4:* 299-301.

Marquand, E.D. 1901. *Flora of Guernsey and the Lesser Channel Islands.* London, Dulau & Co.

Marshall, J.K. 1967. *Corynephorus cansecens* (L.) Beauv. Biological Flora of the British Isles. *Journal of Ecology, 55:* 207-220.

Martin, M.H., & Frost, L.C. 1980. Autecological studies of *Trifolium molinerii* at the Lizard Peninsula, Cornwall. *New Phytologist, 86:* 329-344.

Matthews, J.R. 1955. *Origin and distribution of the British Flora.* London, Hutchinson.

Matthies, D. 1991. *Die Populationsbiologie der annuellen Hemiparasiten* Melampyrum arvense, M. cristatum *und* M. nemorosum *(Scrophulariaceae).* PhD Thesis, Ruhr-Universität, Bochum.

Mattocks, A.R., & Pigott, C.D. 1990. Pyrrolizidine alkaloids from *Cynoglossum germanicum. Phytochemistry, 29:* 2871-2872.

McBeath, R.J.D. 1967. *Phyllodoce caerulea* (L.) Bab. on Ben Alder (v.c. 97). *Transactions of the Botanical Society of Edinburgh, 40:* 335-336.

McClintock, D. 1968. *Eriocaulon aquaticum* (E. septangulare) in Ardnamurchan. *Proceedings of the Botanical Society of the British Isles, 7:* 509.

McClintock, D. 1984. *The wild flowers of Guernsey.* London, Collins.

McDonnell, E.J. 1995a. *The status of toadflax-leaved St John's-wort* (Hypericum linariifolium *Vahl) in Britain in 1994.* Back from the Brink Project Report, No. 40. London, Plantlife.

McDonnell, E.J. 1995b. *The status of shore dock* (Rumex rupestris *Le Gall) in Britain in 1994.* Back from the Brink Project Report, No. 41. London, Plantlife.

McVean, D.N., & Ratcliffe, D.A. 1962. *Plant communities of the Scottish Highlands.* Monographs of the Nature Conservancy No.1. London, HMSO.

Mennema, J., Quené-Boterenbrood, A.J., & Plate, C.L. 1980. *Atlas of the Netherlands Flora. 1. Extinct and very rare species.* The Hague, Dr W. Junk bv Publishers.

Meredith, H.M. 1994. *Scrophularia scorodonia* L. (balm-leaved figwort) - an enigma. *Botanical Cornwall, 6:* 21-36.

Merrett, C. 1666. *Pinax rerum naturalium Britannicum.* London.

Meikle, R.D. 1984. *Willows and poplars of Great Britain and Ireland.* B.S.B.I. Handbook No. 4. London, Botanical Society of the British Isles.

Meusel, H., Jäger E., Rauschert S., & Weinert, E. 1978. *Vergleichende Chorologie der zentraleuropäischen Flora. Vol.* 2. Jena, Gustav Fischer.

Miller, G.R., Geddes, C., & Mardon, D.K. 1994. *Responses of the alpine gentian* (Gentiana nivalis) *and other montane species to protection from grazing.* Edinburgh, Scottish Natural Heritage (SNH Review, No. 32).

Mitchell, J. 1980. Historical notes on *Woodsia ilvensis* in the Moffat Hills, Southern Scotland. *Fern Gazette, 12:* 65-68.

Mitchell, J. 1992. Further notes on the Reverend John Stuart's contribution to the discovery of Britain's mountain flowers. *Glasgow Naturalist, 22:* 103-106.

Molau, U., & Prentice, H.C. 1992. Reproductive system and population structure in three arctic *Saxifraga* species. *Journal of Ecology, 80:* 149-161.

Morgan, V.M. 1987a. Rare plant survey of Wales: Clywd. *Nature Conservancy Council, CSD Report,* No. 888.

Morgan, V.M. 1987b. Rare plant survey of Wales. West Gwynedd. *Nature Conservancy Council, CSD Report,* No. 889.

Morgan, V.M. 1988a. Rare plant survey of Wales. East Gwynedd. *Nature Conservancy Council, CSD Report,* No. 890.

Morgan, V.M. 1988b. Rare plant survey of Wales. Brecknock. *Nature Conservancy Council, CSD Report*, No. 955.

Morgan, V.M. 1989a. Rare plant survey of Wales. Montgomery. *Nature Conservancy Council, CSD Report*, No. 956.

Morgan, V.M. 1989b. Rare plant survey of Wales. Ceredigion. *Nature Conservancy Council, CSD Report*, No. 957.

Morgan, V.M. 1989c. Rare plant survey of Wales. Radnor. *Nature Conservancy Council, CSD Report*, No. 958.

Morgan, V.M. 1989d. Rare plant survey of Wales. Carmarthen & Dinefwr. *Nature Conservancy Council, CSD Report*, No. 959.

Morgan, V.M. 1989e. Rare plant survey of Wales. Pembroke & Preseli. *Nature Conservancy Council, CSD Report*, No. 960.

Morgan, V.M. 1989f. Rare plant survey of Wales. Mid- & South Glamorgan. *Nature Conservancy Council, CSD Report*, No. 961.

Morgan, V.M. 1989g. Rare plant survey of Wales. West Glamorgan and Llanelli. *Nature Conservancy Council, CSD Report*, No. 962.

Morkved, B., & Nilssen, A.C., eds. 1993. *Plant life.* University of Tromso, Tromso Museum.

Morris, M. 1980. *Cotoneaster integerrimus* - a conservation exercise. *Nature in Wales, 17*: 19-22.

Morton, J.K. 1955. Chromosome studies on *Sarothamnus scoparius* (L.) Wimmer and its subspecies *prostratus* (Bailey) Tutin. *New Phytologist, 54*: 68-70.

Muddeman, J.L. 1989. *Microenvironment and distribution of* Hypericum linariifolium *(Flax-leaved St John's-wort).* Thesis. University of Exeter.

Murphy, R.J. 1991. *National Trust, Boscregan Farm, West Penwith, Arable weed survey.* Redruth, Cornish Biological Records Unit.

Nagy, L., & Proctor, J. 1996. The demography of *Lychnis alpina* L. on the Meikle Kilrannoch Ultramafic Site. *Botanical Journal of Scotland, 48*: 155-166.

Nieland, R. 1994. *Reproductive ecology of* Dactylorhiza lapponica *in Scotland.* Edinburgh, Scottish Natural Heritage. (Contract Report)

Nelson, E.C. 1994. Historical data from specimens in the herbarium, National Botanic Gardens, Glasnevin, Dublin (DBN), especially on *Cypripedium calceolus*. *Botanical Society of the British Isles News, 67*: 21-22.

Nelson, E.C. 1977. The discovery in 1810 and subsequent history of *Phyllodoce caerulea* (L.) Bab. in Scotland. *The Western Naturalist, 6*: 45-72.

Nelson, E.C., & Coker, P.D. 1974. Ecology and status of *Erica vagans* in County Fermanagh, Ireland. *Botanical Journal of the Linnean Society, 69*: 153-195.

Nilsson, L.A. 1983. Mimesis of bellflower (*Campanula*) and the red helleborine orchid *Cephalanthera rubra. Nature, 305*: 799-800.

Norfolk Wildlife Trust, 1996. *Liparis loeselii*: report on its autecology and management. Norwich, unpublished report to English Nature from the Norfolk Wildlife Trust.

Oberdorfer, E. 1978. *Süddeutsche Pflanzengesellschaften.* Teil II. Stuttgart, Gustav Fischer.

Oberdorfer, E. 1979. *Exkursionsflora für Süddeutschland.* Aufl. Stuttgart.

Odin, N. 1990. *Report of management work for* C. vulraria *in Suffolk.* Colchester, unpublished report to the Nature Conservancy Council.

Oesau, A. 1975. Untersuchungen zur Keimung und Entwicklung des Wurzelsystems in der Gattung *Melampyrum* L. (Scrophulariaceae). *Beiträge zur Biologie der Pflanzen, 51*: 121-147.

O'Leary, M. 1995. Survey of *Selinum carvifolia.* Peterborough, unpublished report to English Nature.

Olivier, L., Galland, J.-P., & Maurin, H. 1995. *Livre Rouge de la Flore Menaceé de France.* Paris, Museum National d'Histoire Naturelle, Ministère de l'Environnement.

O'Mahony, T. 1976. *Carex depauperata* Curt. in N.E. Cork (H5), a sedge new to Ireland. *Irish Naturalists Journal, 18*: 296-298.

Otley College. 1990. *Lactuca saligna* L. A survey of the population on Vange and Fobbing marshes SSSI in 1990. Colchester, unpublished report to the Nature Conservancy Council.

Ouborg, N.J., & Treuren, R.van. 1994. The significance of genetic erosion in the process of extinction. IV. Inbreeding load and heterosis in relation to population size in the mint *Salvia pratensis. Evolution, 48*: 996-1008.

Ouborg, N.J., & Treuren, R.van. 1995. Variation in fitness-related characters among small and large populations of *Salvia pratensis. Journal of Ecology; 83*: 369-380.

Ouborg, N.J., & Treuren, R.van. 1997. Inbreeding depression, environmental stochasticity and population extinction in plants. *In: The role of genetics in conserving small populations,* ed. by T. Tew *et al.* Peterborough, Joint Nature Conservation Committee.

Ouborg, N.J., Treuren, R.van, & Damme, J.M.M.van. 1991. The significance of genetic erosion in the process of extinction. II. Morphological variation and fitness components in populations of varying size of *Salvia pratensis* L. and *Scabiosa columbaria* L. *Oecologia, 86*: 359-367.

Øvstedal, D.O., & Mjaavatten, G. 1992. A multivariate comparison between three N.W. European populations of *Artemisia norvegica* (Asteraceae) by means of chemometric and morphometric data. *Plant Systematics and Evolution, 181*: 2-32.

Owen, M., Atkinson-Willes, G.L., & Salmon, D.G. 1986. *Wildfowl in Great Britain.* 2nd ed. Cambridge, Cambridge University Press.

Page, C.N. 1982. *The ferns of Britain and Ireland.* Cambridge, Cambridge University Press.

Page, C.N. 1988. *Ferns.* London, Collins. (New Naturalist, No. 74.)

Page, S.E., & Rieley, J.O. 1985. The ecology and distribution of *Carex chordorrhiza* L. fil. *Watsonia, 15*: 253-259.

Palmer, J.R. 1994. *Cerastium brachypetalum* - Status in W. Kent. *Botanical Society of the British Isles News, 65*: 21-22.

Pankhurst, R.J., & Preston, C.D. 1996. Checklist of Scotland's Native Flora. *In: Scottish plants for Scottish gardens,* ed. by Jill, Duchess of Hamilton. Edinburgh, The Stationery Office.

Parkinson, J. 1640. *Theatrum Botanicum.*

Parslow, R.E. 1994. *Poa infirma* in the Isles of Scilly. *Botanical Society of the British Isles News, 67*: 22.

Parslow, R.E, & Colston, A. 1994. *The current status of Rumex rupestris in the Isles of Scilly.* Peterborough, English Nature. (Report for Species Recovery Programme.)

Payne, A.G. 1981a. Erigeron borealis - *its distribution on the Meall nan Tarmachan and Ben Lawers hills.* Unpublished report to the National Trust for Scotland and the Nature Conservancy Council.

Payne, A.G. 1981b. Minuartia rubella - *its distribution on the Meall nan Tarmachan and Ben Lawers hills.* Unpublished report to the National Trust for Scotland and the Nature Conservancy Council.

Payne, A.G. 1981c. *The status of* Sagina intermedia *Fenzl in Great Britain, 1981.* Unpublished report to the National Trust for Scotland and the Nature Conservancy Council.

Payne, A.G., & Geddes, C. 1980. *Survey of* Carex microglochin, *1980.* Unpublished report to the Nature Conservancy Council.

Pearman, D.A. 1997. Presidential address: towards a new definition of rare and scarce species. *Watsonia, 21:* 225-245.

Pemadasa, M.A., & Lovell, P.A. 1974a. Factors affecting the distribution of some annuals in the dune system at Aberffraw, Anglesey. *Journal of Ecology, 62:* 403-416.

Pemadasa, M.A., & Lovell, P.A. 1974b. Factors controlling the flowering time of some dune annuals. *Journal of Ecology, 62:* 869-880.

Perring, F.H. 1963. The Irish problem. *Proceedings of the Bournemouth Natural Science Society, 52:* 36-48.

Perring, F.H. 1996. A bridge too far - the non-Irish element in the British flora. *Watsonia, 21:* 15-51.

Perring, F.H., & Farrell, L. 1977. *British Red Data Book. 1. Vascular plants.* 1st. ed. Lincoln, Royal Society for Nature Conservation.

Perring, F.H., & Farrell, L. 1983. *British Red Data Book. 1. Vascular plants.* 2nd. ed. Lincoln, Royal Society for Nature Conservation.

Perring, F.H., & Walters, M., eds. 1962. *Atlas of the British Flora.* 1st ed. London, Thomas Nelson & Sons.

Perring, F.H., & Walters, M., eds. 1982. *Atlas of the British Flora.* 3rd ed. London, Botanical Society of the British Isles.

Philbrick, C.T. 1983. Aspects of floral biology in three species of *Potamogeton* (pondweeds). *Michigan Botanist, 23:* 35-38.

Phillips, R.W. 1905. *The Flora of Llandudno and its neighbourhood.* National Union of Teachers' Conference Handbook.

Philp, E.G. 1982. *Atlas of the Kent flora.* Kent Field Club.

Pigott, C.D. 1954. Species delimitation and racial divergence on British *Thymus. New Phytologist, 53:* 470-495.

Pigott, C.D. 1955. Biological Flora of the British Isles. *Thymus serpyllum* Linn., *emend* Mill. subsp. *serpyllum. Journal of Ecology, 43:* 379-382.

Pigott, C.D. 1956. The vegetation of Upper Teesdale in the North Pennines. *Journal of Ecology, 44:* 545-586.

Pigott, C.D. 1958. Biological Flora of the British Isles. *Polemonium caeruleum* L. *Journal of Ecology, 46:* 507-525.

Pigott, C.D. 1987. The Whites, Box Hill, Surrey. *In: Rare vascular plant survey of south-east England: Surrey,* by R. FitzGerald. Peterborough, Nature Conservancy Council.

Pigott, C.D. 1988. The reintroduction of *Cirsium tuberosum* (L.) All. in Cambridgeshire. *Watsonia, 17:* 149-152.

Pigott, C.D., & Walters, S.M. 1953. Is the box-tree a native of England? *In: The changing Flora of Britain,* ed. by J.E. Lousley. Arbroath, T. Buncle & Co. Ltd.

Pigott, C.D., & Walters, S.M. 1954. On the interpretation of the discontinous distributions shown by certain British species of open habitats. *Journal of Ecology, 42:* 95-116.

Pinfield, N.J., Martin, M.H., & Stobart, A.K. 1972. The control of germination in *Stachys alpina* L. *New Phytologist, 99:* 99-104.

Porley, R.D. 1996. Foliicolous *Metzgeria fruticulosa* on box leaves in the Chiltern Hills, England. *Journal of Bryology, 19:* 188-189.

Preston, C.D. 1986. An additional criterion for assessing native status. *Watsonia, 16:* 83.

Preston, C.D. 1988. The *Potamogeton* L. taxa described by Alfred Fryer. *Watsonia, 17:* 23-35.

Preston, C.D. 1989. The ephemeral pools of south Cambridgeshire. *Nature in Cambridgeshire, 31:* 2-11.

Preston, C.D., & Croft, J.M. 1997. *Aquatic plants in Britain and Ireland.* Colchester, Harley Books.

Preston, C.D., & Pearman, D. 1991. A second extant Dorset locality for *Gastridium ventricosum. Proceedings of the Dorset Natural History and Archaeological Society, 113:* 207-210.

Preston, C.D., & Whitehouse, H.L.K. 1986. The habitat of *Lythrum hyssopifolia* L. in Cambridgeshire, its only surviving English locality. *Biological Conservation, 35:* 41-62.

Prince, S.D., & Hare, A.D.R. 1981. *Lactuca saligna* and *Pulicaria vulgaris* in Britain. *In: The biological aspects of rare plant conservation,* ed. by H. Synge. Chichester, Wiley & Co.

Pring, M. E. 1961. Biological Flora of the British Isles. *Arabis stricta* Huds. *(A. scabra* All.). *Journal of Ecology, 49:* 431-437.

Pritchard, N.M. 1959. *Gentianella* in Britain. *Watsonia, 4:* 169-92.

Pritchard, N.M. 1971. Where have all the gentians gone? *Transactions of the Botanical Society of Edinburgh, 41:* 279-91.

Pritchard, N.M., & Tutin, T.G. 1972. *Gentianella. In: Flora Europaea, Volume 3,* ed. by T.G. Tutin *et al.* Cambridge, Cambridge University Press.

Proctor, J., Bartlem, K., Carter, S.P., Dare, D.A., Jarvis, S.B., & Slingsby, D.R. 1991. Vegetation and soils of the Meikle Kilrannoch ultramafic sites. *Botanical Journal of Scotland, 46:* 47-64.

Proctor, J., & Johnston, W.R. 1977. *Lychnis alpina* L. in Britain. *Watsonia, 11:* 199-204.

Proctor, M.C.F. 1956. Biological Flora of the British Isles. *Helianthemum apenninum* (L.) Mill. *Journal of Ecology, 44:* 688-692.

Proctor, M.C.F. 1958. Ecological and historical factors in the distribution of British *Helianthemum* species. *Journal of Ecology, 46:* 349-71.

Proctor, M.C.F. 1960. Biological Flora of the British Isles. *Tuberaria guttata* (L.) Fourreau. *Journal of Ecology, 48:* 243-253.

Proctor, M.C.F. 1962. The British forms of *Tuberaria guttata* (L.) Fourreau. *Watsonia, 5:* 236-249.

Proctor, M.C.F., & Groenhof, A.C. 1992. Peroxidase isoenzyme and morphological variation in *Sorbus* L. in South Wales and adjoining areas, with particular reference to *S. porrigentiformis* E.F.Warb. *Watsonia, 19*: 21-37.

Proctor, M.E. 1985. *Survey of rare plants in Devon*. Taunton, Nature Conservancy Council.

Proctor, M.E., Proctor, M.C.F., & Groenhof, A.C. 1989. Evidence from peroxidase polymorphism on the taxonomy and reproduction of some *Sorbus* populations in south-west England. *New Phytologist, 112*: 569-575.

Pugsley, H.W. 1924. *Gentiana uliginosa* Willd. in Britain. *Journal of Botany, London, 62*: 193-196

Pugsley, H.W. 1926. The British *Orobanche* list. *Journal of Botany, London, 64:* 16-19.

Pugsley, H.W. 1930. A revision of the British *Euphrasia*. *Journal of the Linnean Society of London (Botany), 48*: 467-544.

Pugsley, H.W. 1931. A further new *Limonium* in Britain. *Journal of Botany, 69*: 44-47.

Pugsley, H.W. 1948. A prodromus of the British *Hieracia*. *Journal of the Linnean Society of London (Botany), 54*: 1-356, Plates 1-17.

Pullin, A.S. 1986. The status, habitat, and species association of the fen violet *Viola persicifolia* in western Ireland. *British Ecological Society Bulletin, 17(1)*: 15-19.

Pullin, A.S., & Woodell, S.R.J. 1987. Response of the fen violet, *Viola persicifolia* Schreber, to different management regimes at Woodwalton Fen National Nature Reserve, Cambridgeshire, England. *Biological Conservation, 41*: 203-217.

Pulteney, R. 1757. An account of more rare English plants observed in Leicester. *Philosophical Transactions, 49*: 156.

Pykälä, J., & Toivenen, H. 1994. Taxonomy of the *Carex flava* complex (Cyperaceae) in Finland. *Nordic Journal of Botany, 14*: 173-191.

Raatikainen, M. 1990. Oravanmarjan, *Maianthemum bifolium* (L.) F.W.Schmidt marja- ja siemensadosta. *Memoranda Societatis pro Fauna et Flora Fennica, 66*: 75-80.

Randall, R.E. 1974. *Rorippa islandica* (Oeder) Borbás *sensu stricto* in the British Isles. *Watsonia, 10*: 80-82.

Randall, R.E., & Thornton, G. 1996. Biological Flora of the British Isles. *Peucedanum officinale* L. *Journal of Ecology, 84*: 475-485.

Ratcliffe, D.A. 1960. The mountain flora of Lakeland. *Proceedings of the Botanical Society of the British Isles, 4*: 1-25.

Ratcliffe, D.A. 1962. *Potentilla rupestris* in East Sutherland. *Proceedings of the Botanical Society of the British Isles, 4*: 473-501.

Ratcliffe, D.A. 1977. *A nature conservation review*. 2 vols. Cambridge, Cambridge University Press..

Ratcliffe, D.A., Birks, H.J.B., & Birks, H.H. 1993. The ecology and conservation of the Killarney fern *Trichomanes speciosum* Willd. in Britain and Ireland. *Biological Conservation, 66*: 231-247.

Raven, J.H., & Walters, S.M. 1956. *Mountain flowers*. London, Collins. (New Naturalist, No. 33.)

Ray, J. 1660. *Catalogus plantarum circa Cantabrigiam nascentium*. Cambridge.

Ray, J. 1677. *Catalogus plantarum Angliae*. London. 2nd ed.

Ray, J. 1724. *Synopsis methodica stirpium Britannicarum*. London. 3rd ed.

Raymond, M., & Kycyniak, J. 1948. Six additions to the adventitious flora of Quebec. *Rhodora, 50*: 176-180.

Rich, T.C.G. 1991. *Crucifers of Great Britain and Ireland*. B.S.B.I. Handbook No. 6. London, Botanical Society of the British Isles.

Rich, T.C.G. 1993a. *Hairy mallow* Althaea hirsuta *L. at Cleeve Hill, Watchet, Somerset*. Back from the Brink Project Report, No. 3. London, Plantlife.

Rich, T.C.G. 1993b. *Adder's-tongue spearwort* Ranunculus ophioglossifolius *at Inglestone Common, Avon*. Back from the Brink Project Report, No. 13. London, Plantlife.

Rich, T.C.G. 1994a. *Narrow-leaved cudweed (Filago gallica) in Britain. January 1994*. Back from the Brink Project Report No. 22. Plantlife, London.

Rich, T.C.G. 1994b. *Red-tipped cudweed (Filago lutescens) in Britain*. Back from the Brink Project Report, No. 28. London, Plantlife.

Rich, T.C.G. 1995a. *Narrow-leaved cudweed (Filago gallica). Report for 1994 and re-introduction at Berechurch. April 1995*. Back from the Brink Project Report, No. 53. London, Plantlife.

Rich T.C.G. 1995b. *Broad-leaved cudweed (Filago pyramidata) in Britain*. Back from the Brink Project Report, No. 29. London, Plantlife.

Rich T.C.G. 1995c. *Broad-leaved cudweed (Filago pyramidata) in Britain in 1995*. Back from the Brink Project Report, No. 65. London, Plantlife.

Rich, T.C.G. 1995d. *The status of meadow clary (Salvia pratensis L.) in Britain in 1994*. Back from the Brink Project Report, No. 44. London, Plantlife.

Rich, T.C.G. 1995e. *The status of brown galingale (Cyperus fuscus L.) in Britain in 1995*. Back from the Brink Project Report, No. 63. London, Plantlife.

Rich, T.C.G. 1996a. *Red-tipped cudweed (Filago lutescens) in Britain in 1995*. Back from the Brink Project Report, No. 68. London, Plantlife.

Rich, T.C.G. 1996b. *Meadow clary (Salvia pratensis L.) in 1995*. Back from the Brink Project Report, No. 72. London, Plantlife.

Rich, T.C.G. 1996c. *Broad-leaved cudweed (Filago pyramidata) in Britain in 1996*. Back from the Brink Project Report, No. 83. London, Plantlife.

Rich, T.C.G. 1996d. Is *Gentianella uliginosa* (Willd.) Boerner (Gentianaceae) present in England? *Watsonia, 21*: 208-209.

Rich, T.C.G., Alder, J., McVeigh, A., & Showler, A. 1994. *Starfruit (Damasonium alisma) survey 1994*. Back from the Brink Project Report, No. 32. London, Plantlife.

Rich, T.C.G., Alder, J., McVeigh, A., Showler, A., & Sinnadurai, P. 1995. *The star goes out ... - starfruit (Damasonium alisma) was not seen in 1995*. Back from the Brink Project Report, No. 66. London, Plantlife.

Rich, T.C.G., Alder, J., Lambrick, C., McVeigh, A., & Smith, P. 1996. *Cotswold pennycress (Thlaspi perfoliatum) in Britain in 1996*. Back from the Brink Project Report, No. 81. London, Plantlife

Rich, T.C.G., & Baecker, M. 1986. The distribution of *Sorbus lancastriensis* E.F.Warburg. *Watsonia, 16*: 83-85.

Rich, T.C.G., & Baecker, M. 1992. Additional records of *Sorbus lancastriensis* E.F. Warburg (Rosaceae). *Watsonia,* 19: 139-140.

Rich, T.C.G., & Dalby, K. 1996. The status and distribution of mountain scurvy-grass (*Cochlearia micacea* Marshall) in Scotland, with ecological notes. *Botanical Journal of Scotland, 48:* 187-198.

Rich, T.C.G., & Davis, R. 1996. *Adder's-tongue spearwort* (Ranunculus ophioglossifolius) *at Inglestone Common in 1996.* Back from the Brink Project Report, No. 79. London, Plantlife.

Rich, T.C.G., & Fairbrother, A. 1995. *How starved wood-sedge* (Carex depauperata) *got its name?* Back from the Brink Project Report, No. 57. London, Plantlife.

Rich, T.C.G., Gibson, C., & Marsden, M. 1995. *Narrow-leaved cudweed* (Filago gallica) *is back from the brink ... November 1995.* Back from the Brink Project Report, No. 67. London, Plantlife.

Rich, T.C.G., Houston, L., & Martin, M. 1996. Rare plants in the Avon Gorge contaminated with heavy metals. *Botanical Society of the British Isles News, 72:* 36-37.

Rich, T.G.C., Kitchen, M.A.R., & Kitchen, C. 1989. *Thlaspi perfoliatum* L. (Cruciferae) in the British Isles: distribution. *Watsonia, 17:* 401-407.

Rich, T.C.G., Lambrick, C.R., Kitchen, C., & Kitchen M.A.R. In press. Conserving Britain's biodiversity: *Thlaspi perfoliatum* L. (Brassicaceae), Cotswold Pennycress.

Rich, T.C.G., Lambrick, C.R., & McNab, C. In prep. The status of *Salvia pratensis* L. (Lamiaceae), meadow clary, in Britain 1994-1996.

Rich, T.C.G., & Palmer, J.R. 1994. *Grey mouse-ear* (Cerastium brachypetalum *Pers.*) *under threat from the Channel Tunnel Rail Link.* Back from the Brink Project Report, No. 30. London, Plantlife.

Rich, T.C.G., & Rose, F. 1995. *Tower mustard* (Arabis glabra) *in Hampshire in 1995.* Back from the Brink Project Report, No. 61. London, Plantlife.

Rich, T.C.G., & Ulf-Hansen, P.F. 1994. *The status of hairy mallow* (Althaea hirsuta) *in Britain in 1994.* Back from the Brink Project Report, No. 35. London, Plantlife.

Rich, T.C.G., Ulf-Hansen, P.F., & Goddard, E. 1996. *Hairy mallow* (Althaea hirsuta) *in Britain in 1996.* Back from the Brink Project Report, No. 77. London, Plantlife.

Richards, A.J. 1975. *Sorbus* L. *In: Hybridization and the Flora of the British Isles,* ed. by C.A. Stace, 233-238. London, Academic Press.

Richards, A.J., & Porter, A.F. 1982. On the identity of a Northumberland *Epipactis. Watsonia, 14:* 121-128.

Richardson, I.B.K. 1980. *Gagea. In: Flora Europaea, Volume 5,* ed. by T.G. Tutin *et al.,* 26-28. Cambridge, Cambridge University Press.

Rickard, M.H. 1972. The distribution of *Woodsia ilvensis* and *W. alpina* in Britain. *British Fern Gazette, 10:* 269-280.

Rickard, M.H. 1989. Two spleenworts new to Britain - *Asplenium trichomanes* subsp. *pachyrachis* and *Asplenium trichomanes* nothosubsp. *staufferi. Pteridologist, 1:* 244-248.

Riddelsdell, H.J., Hedley, G.W., & Price, W.R. 1948. *Flora of Gloucestershire.* Cheltenham, Cotswold Naturalists' Field Club.

Rix, E.M., & Woods, R.G. 1981. *Gagea bohemica* (Zauschner) J.A. & J.H. Schultes in the British Isles and a general review of the *G. bohemica* species complex. *Watsonia, 13:* 265 -270.

Roberts, R.H. 1982. *The flowering plants and ferns of Anglesey.* Cardiff, National Museum of Wales.

Roberts, R.H., & Day, P. 1987. *Allium ampeloprasum* L. in Anglesey. *Watsonia, 16:* 335-336.

Robinson, N. 1993. Arnside Knott SSSI: *Viola rupestris.* Rare Plant Species Monitoring Status Report, 1993. Blackwell, English Nature.

Rodwell, J. 1991. *British Plant Communities. Vol. 2. Mires and heaths.* Cambridge, Cambridge University Press.

Roe, R.G.B. 1981. *Flora of Somerset.* Taunton, Somerset Archaeological and Natural History Society.

Røren, V., Stabbetorp, O., & Borgen, L. 1994. Hybridisation between *Viola canina* and *V. persicifolia* in Norway. *Nordic Journal of Botany, 14:* 165-172.

Rose, F. 1960. *Cynoglossum germanicum. Transactions of the Kent Field Club, 5:* 7.

Rose, F. 1998. *Gentianella uliginosa* (Willd.) Boerner (Gentianaceae) found in Colonsay (v.c. 102), new to Scotland. *Watsonia, 22:* 114-116.

Rose, F., & Brewis, A. 1988. *Cephalanthera rubra* (L.) Rich. in Hampshire. *Watsonia, 17:* 176-177.

Rose, R., Bannister, P., & Chapman, S.B. 1996. Biological Flora of the British Isles. *Erica ciliaris* L. *Journal of Ecology, 84:* 617-628.

Rosser, E.M. 1955. A new British species of *Senecio. Watsonia, 3:* 228-232.

Ross-Williamson, R.P. 1930. Excavations in Whitehawk neolithic camp, near Brighton. *Sussex Archaeological Collections, 71:* 82.

Rowell, T.A. 1983. The fen violet at Wicken Fen. *Nature in Cambridgeshire, 26:* 62-65.

Rowell, T.A. 1984. Further discoveries of the fen violet (*Viola persicifolia* Schreber) at Wicken Fen, Cambridgeshire. *Watsonia, 15:* 122-123.

Rowell, T.A., Walters, S.M., & Harvey, H.J. 1982. The re-discovery of the fen violet at Wicken Fen, Cambs. *Watsonia, 14:* 183-4.

Rowell, T.A., Walters, S.M., & Harvey, H.J. 1983. The fen violet at Wicken Fen. *Nature in Cambridgeshire, 26:* 62-65.

Rumsey, F.J., Sheffield, E., & Farrar, D.R. 1990. British filmy fern gametophytes. *Pteridologist, 2:* 40-42.

Rumsey, F.J., Headley, A.D., Farrar, D.R., & Sheffield, E. 1991. The Killarney fern (*Trichomanes speciosum*) in Yorkshire. *Naturalist, 116:* 41-3.

Russell, R.V. 1993. The late spider orchid *Ophrys fuciflora* on the Wye National Nature Reserve, Kent 1965-1993, with notes on early records at Wye and other East Kent sites. Wye, English Nature.

Ryves, T.B., Clement, E.J., & Foster, M.C. 1996. *Alien grasses of the British Isles.* London, Botanical Society of the British Isles.

Salisbury, E.J. 1961. *Weeds and aliens.* London, Collins. (New Naturalist, No. 43.)

Salisbury, E.J. 1970. The pioneer vegetation of exposed muds and its biological features. *Philosophical Transactions of the Royal Society, 259B:* 207-255.

Salisbury, E.J. 1972. *Ludwigia palustris* (L.) Ell. in England with special reference to its dispersal and germination. *Watsonia, 9:* 33-37.

Salmon, C.E. 1903. Notes on *Limonium*. *Journal of Botany*, *41*: 64-74.

Scampion, B.R. 1993. *The population dynamics of Jersey cudweed* Gnaphalium luteoalbum *L. on the Holkham National Nature Reserve, North Norfolk*. Peterborough: English Nature. (Contract Report.)

Scampion, B. 1994. *Swanton Novers Flora Report*. Peterborough, unpublished report to English Nature.

Schmid, B. 1982. Karyology and hybridisation in the *Carex flava* complex in Switzerland. *Feddes Repertorium, 93*: 23-59.

Schmid, B. 1983. Notes on the nomenclature and taxonomy of the *Carex flava* group in Europe. *Watsonia, 14*: 309-319.

Schmid, B. 1986. Colonising plants with persistent seeds and persistent seedlings (*Carex flava* group). *Botanica Helvetica, 96*: 19-26.

Scott, A. 1968. *Pilosella flagellaris* (Willd.) Sell & C.West subsp. *bicapitata* Sell & C.West - in Zetland. *Proceedings of the Botanical Society of the British Isles, 7*: 192-193.

Scott, A. 1989. *The ecology and conservation of* Salvia pratensis *L*. MSc Thesis, University College, London University.

Scott, A., & Palmer, R. 1978. *The flowering plants and ferns of the Shetland Islands*. Lerwick, The Shetland Times Ltd.

Scruby, M., Grove, S., & Alexander, K. 1992. *Report of National Trust Biological Survey: Brixham to Kingswear*. Cirencester, The National Trust.

Sell, P.D., & Murrell, G. 1996. *Flora of Great Britain and Ireland, Volume 5*. Cambridge, Cambridge University Press.

Shirt, D.B. 1987. *British Red Data Books: 2. Insects*. Peterborough, Joint Nature Conservation Committee.

Showler, A. 1994. *An account of the re-appearance of starfruit* (Damasonium alisma) *at Downley Common and Naphill Common and a Report for 1989-1993*. Back from the Brink Project Report, No. 23. London, Plantlife.

Silverside, A.J. 1991a. A guide to eyebrights (*Euphrasia*) IV. On northern coasts and western hills. *Wild Flower Magazine, 420*: 29-33.

Silverside, A.J. 1991b. A guide to eyebrights (*Euphrasia*) V. Local endemics and concluding remarks. *Wild Flower Magazine, 421*: 32-36.

Simpson, F.W. 1982. *Simpson's Flora of Suffolk*. Ipswich, Suffolk Naturalists' Society.

Sirjaev, G. 1932. Generio *Ononis* L. revisio critica. *Beiheft Botanisches Zentralblatt, 49*: 381-665.

Slack, A., & Dickson, J.H. 1959. A further note on the limestone flora of Ben Sgulaird. *Glasgow Naturalist, 18*: 106-108.

Slater, F.M. 1990. Biological Flora of the British Isles. *Gagea bohemica* (Zauschner) J.A. & J.H. Schultes. *Journal of Ecology, 78*: 535-546.

Sledge, W.A. 1949. The distribution and ecology of *Scheuchzeria palustris*. *Watsonia, 1*: 24-35.

Sledge, W.A. 1975. Disappearance of *Tillaea aquatica* L. on Adel. *Naturalist, Hull (1975)*: 149.

Slingsby, D.R., Carter, S.P., & Kendal, J. 1993. *The status of populations of rare plant taxa on the Keen of Hamar over the period 1978 to 1993*. Aberdeen, Scottish Natural Heritage.

Smiles, S. 1878. *Robert Dick, geologist and botanist*. London, John Murray.

Smith, A. 1986. Endangered species of disturbed habitats. *Nature Conservancy Council, CSD Report*, No. 644.

Smith, A.R., & Tutin, T.G. 1968. *Euphorbia. In: Flora Europaea, Volume 2*, ed. by T.G. Tutin *et al.* Cambidge, Cambridge University Press.

Smith, J.E., & Sowerby, J. 1812. *English Botany*. London, Taylor & Co.

Smith, R.A.H. 1980. *Schoenus ferrugineus* L. - two native localities in Perthshire. *Watsonia, 13*: 128-129.

Smith, U.K. 1979. Biological Flora of the British Isles. *Senecio integrifolius* (L.) Clairv. *Journal of Ecology, 67*: 1109-1124.

Soler, A. 1983. Revision de la especies do *Fumaria* de la Peninsula Iberica e Islas Baleares. *Lagascalia, 11*: 141-228.

Sowerby, J.E. 1834. *English botany*. London.

Sowter, A.J. 1971. *Crassula aquatica* (L.) Schönl. - new to Scotland. *Watsonia, 8*: 294.

Sowter, A.J., Sowter, M.M., & Webster, M. McC. 1972. *Crassula aquatica* (L.) Schönl. in v.c. 97 - further observations. *Watsonia 9*: 140.

Spencer, D.F., & Anderson, L.W.J. 1987. Influence of photoperiod on growth, pigment composition and vegetative propagule formation for *Potamogeton nodosus* Poir. and *Potamogeton pectinatus* L. *Aquatic Botany, 28*: 103-112.

Stace, C.A. 1975. *Hybridisation and the flora of the British Isles*. London: Academic Press.

Stace, C.A. 1991. *New Flora of the British Isles*. Cambridge, Cambridge University Press.

Stace, C.A. 1997. *New Flora of the British Isles*. 2nd ed. Cambridge, Cambridge University Press.

Stace, C.A., & Ingrouille, M. 1986. The *Limonium binervosum* aggregate (Plumbaginaceae) in the British Isles. *Botanical Journal of the Linnean Society, 92*: 177-217.

Staines, S.J. 1984. *Soils in Cornwall III. Sheet SW61/62/71/72 (The Lizard). Soil Survey Record. No. 79*. Harpenden, Soil Survey of England and Wales.

Staples, M.J.C. 1970/71. A history of box in the British Isles. *The Boxwood Bulletin, 10*: 19-23, 34-37, 54-60.

Stearn, W.T. 1987. *Allium ampeloprasum* L. var. *babingtonii* (Borrer) Syme - II. *Botanical Society of the British Isles News, 46*: 20.

Steven, G., & Dickson, J.H. 1991. The vegetational history of Glen Diomhan, North Arran, site of endemic whitebeams, *Sorbus arranensis* Hedl. and *S. pseudofennica* E.F.Warb. *New Phytologist, 117*: 501-506.

Stevens, D.P., & Blackstock, T.H. (1997). Genetics and nature conservation in Britain: the role of the statutory country agencies. *In: The role of genetics in conserving small populations*, ed. by T. Tew *et al.* Peterborough, Joint Nature Conservation Committee.

Stewart, A., Pearman, D.A., & Preston, C.D. 1994. *Scarce plants in Britain*. Peterborough, Joint Nature Conservation Committee.

Stewart, N.F., & Church, J.M. 1992. *Red Data Books of Britain and Ireland: stoneworts*. Peterborough, Joint Nature Conservation Committee.

Stewart, N.F., & Church, J.M. In prep. *Red Data Books of Britain and Ireland: mosses and liverworts*. Peterborough, Joint Nature Conservation Committee.

Stokes, J. 1987. *The ecology of the wild gladiolus* (Gladiolus illyricus) *in the New Forest, Hampshire*. MSc dissertation, University College, London.

Stone, D.A. (Unpublished). The population biology and demography of *Ophrys fuciflora* (Crantz) Moench & Reichenbach, the late spider orchid.

Summerhayes, V.S. 1951. *Wild orchids of Britain*. London, Collins. (New Naturalist, No. 19.)

Summerhayes, V.S. 1968. *Wild orchids of Britain*. 2nd ed. London, Collins. (New Naturalist, No. 19.)

Svensson, R., & Wigren, M. 1986. A survey of the history, biology and preservation of some retreating synanthropic plants. *Symbolae botanicae upsaliensis, 25*: 1-74.

Swan, G.A., & Walters, S.M. 1988. *Alchemilla gracilis* Opiz, a species new to the British Flora. *Watsonia, 17*: 133-138.

Swann, E.L. 1971. *Maianthemum bifolium* (L.) Schmidt - its status in the British Isles. *Watsonia, 8*: 295-297.

Swann, E.L. 1975. *Supplement to the Flora of Norfolk*. Norwich, F. Crowe & Sons.

Sydes, C., & MacKintosh, E.J. 1990. *Interim report on monitoring* Phyllodoce caerulea. Edinburgh, Nature Conservancy Council.

Takagi-Arigho, R. 1994. *Poa infirma* - Flourishing? ... or fleeing? *Botanical Society of the British Isles, 65*: 14-18.

Taylor, E., & Rich, T.C.G. 1997. *Gentianella ciliata* in Surrey. *Botanical Society of the British Isles News, 74*: 20-21.

Taylor, I. 1987a. *Survey of nationally rare plant species in N.W. England. Vol. 1. Cumbria*. Peterborough, Nature Conservancy Council.

Taylor, I. 1987b. *Survey of nationally rare plant species in N.W. England. Vol. 2. Lancashire*. Peterborough, Nature Conservancy Council.

Taylor, I. 1987c. *Survey of nationally rare plant species in N.W. England. Vol. 3. Greater Manchester & Merseyside*. Peterborough, Nature Conservancy Council.

Taylor, I. 1987d. *Survey of nationally rare plant species in N.W. England. Vol. 4. South & West Yorkshire*. Peterborough, Nature Conservancy Council.

Taylor, I. 1990a. Rare plant survey of the West Midlands. Vol. 1. Gloucestershire. *Nature Conservancy Council, CSD Report*, No. 1063.

Taylor, I. 1990b. Rare plant survey of the West Midlands. Vol. 2. Derbyshire. *Nature Conservancy Council, CSD Report*, No. 1064.

Taylor, I. 1990c. Rare plant survey of the West Midlands. Vol. 3. Cheshire, Shropshire & Staffordshire. *Nature Conservancy Council, CSD Report*, No. 1065.

Taylor, I. 1990d. Rare plant survey of the West Midlands. Vol. 4. Warwickshire & the West Midlands. *Nature Conservancy Council, CSD Report*, No. 1066.

Taylor, I. 1990e. Rare plant survey of the West Midlands. Vol. 5. Hereford & Worcester. *Nature Conservancy Council, CSD Report*, No. 1067.

Taylor, I. 1990f. Rare plant survey of South-West England. Vol. 5. Avon. *Nature Conservancy Council, CSD Report*, No. 1062.

Tennant, D.J. 1996. *Cystopteris dickieana* R. Sim in the central and eastern Scottish Highlands. *Watsonia, 21*: 135-139.

Teppner, H., Ehrendorfer, F., & Puff, C. 1976. Karyosystematic notes of the *Galium palustre* group (Rubiaceae). *Taxon, 25*: 95-97.

Thompson, H.S. 1928. *Arabis scabra* All. in Somerset. *Journal of Botany, 66*: 152.

Thompson, K., Bakker, J., & Bekker, R. 1997. *The soil seed banks of North West Europe: methodology, density and longevity*. Cambridge, Cambridge University Press.

Thompson, K., Band, S.G., & Hodgson, J.G. 1993. Seed size and shape predict persistence in soil. *Functional Ecology, 7*: 236-241.

Thornton, G. 1990. *An environmental Flora of* Peucedanum officinale. Unpublished dissertation, University of Cambridge Board of Extra-mural Studies.

Tiffany, W.N. 1972. Snow cover and the *Diapensia lapponica* habitat in the White Mountains, New Hampshire. *Rhodora, 74*: 358-377.

Toase, S. 1992. *A semi-quantitative assessment of the seed bank of* Ranunculus ophioglossifolius *in Britain*. Undergraduate dissertation, Royal Holloway and New Bedford College, London.

Tonkin, B. 1993. *Pyrus cordata - Species Recovery Programme*. Truro, unpublished report on the Truro populations 1992/3, English Nature.

Townsend, F. 1904. *The Flora of Hampshire*. London, Lowell Reeve.

Treuren, R. van, Bihlsma, R., Delden, W. van, & Ouborg, N.J. 1991. The significance of genetic erosion in the process of extinction, I. Genetic differentiation in *Salvia pratensis* and *Scabiosa columbaria* in relation to population size. *Heredity, 66*: 181-189.

Triest, L. 1988. A revision of the genus *Najas* (Najadaceae) in the Old World. *Memoires Academie Royale Science D'Outre-Mer, Classe des Sciences Naturelles et Médicales, n.s. 22*: 1-172.

Trist, P.J.O. 1979. *An ecological flora of Breckland*. Wakefield, EP Publishing Ltd.

Trist, P.J.O. 1983. The past and present status of *Gastridium ventricosum* (Gouan) Schinz & Thell, as an arable colonist in Britain. *Watsonia, 14*: 257-261.

Trist, P.J.O. 1986. The distribution, ecology, history and status of *Gastridium ventricosum* (Gouan) Schinz & Thell in the British Isles. *Watsonia, 16*: 43-54.

Trist, P.J.O. 1993. *Corynephorus canescens* (L.) Beauv. (Poaceae) on the west coast of Scotland. *Watsonia, 19*: 192-193.

Tubbs, C.R. 1986. *The New Forest*. London, Collins. (New Naturalist, No. 73.)

Turner, S.R. 1988. *Potentilla rupestris* in Wales. *Botanical Society of the British Isles Welsh Bulletin, 46*: 15-20.

Turpin, P.G. 1984. The heather species, hybrids and varieties of the Lizard district of Cornwall. *Cornish Studies, 10 (for 1982)*: 5-17.

Tutin, T.G. 1953. Natural factors contributing to a change in our flora. *In: The changing flora of Britain*, ed. by J.E. Lousley, 19-25. London, Botanical Society of the British Isles.

Tutin, T.G. 1976. *Lobelia. In: Flora Europaea, Volume 3*, ed. by T.G. Tutin *et al*. Cambridge, Cambridge University Press.

Tutin, T.G. 1980. *Umbellifers of the British Isles*. B.S.B.I. Handbook No. 2. London, Botanical Society of the British Isles.

Tutin, T.G., Heywood, V.H., Burges, N.A., Moore, D.M., Valentine, D.H., Walters, S.M., & Webb, D.A., *eds*. 1972. *Flora Europaea. Volume 3. Diapensiaceae to Myoporaceae*. Cambridge, Cambridge University Press.

Ulf-Hansen, P.F. 1994. *Althaea hirsuta* monitoring at Aller Hill SSSI, Somerset. Taunton, English Nature.

Uphof, J.C.Th. 1959. A review of the genus *Gagea* Salisb. *Plant Life, 15*: 151-161.

Valentine, D.H., & Harvey, M.J. 1961. *Viola rupestris* Schmidt in Britain. *Proceedings of the Botanical Society of the British Isles, 4*: 129-135.

Vuille, F.L. 1987. Reproductive biology of the genus *Damasonium* (Alismataceae). *Plant Systematics and Evolution, 157*: 63-71.

Waite, S., & Hutchings, M.J. 1991. The effects of different management regimes on the population dynamics of *Ophyrs sphegodes*: analysis and description using matrix models. *In: Population Ecology of Terrestrial Orchids*, ed. by T.C.E. Wells & J.H. Willems. The Hague, SPB Publishing.

Walker, K. 1995. *The ecology and conservation of the Yorkshire sandwort* Arenaria norvegica *ssp.* anglica *Halliday, Caryophyllaceae*. MSc dissertation in Conservation. University College, London.

Walter, K. 1993. *Species composition and vegetation structure of plant communities within the* Cypripedium calceolus *site*. Unpublished report to English Nature.

Walters, S.M. 1949. *Alchemilla vulgaris* L. agg. in Britain. *Watsonia, 1*: 6-18.

Walters, S.M. 1952. *Alchemilla subcrenata* Buser in Britain. *Watsonia, 2*: 277-278.

Walters, S.M. 1956. *Selinum carvifolia* in Britain. *Proceedings of the Botanical Society of the British Isles, 2*: 119-122.

Walters, S.M. 1963. *Eleocharis austriaca* Hayek, a species new to the British Isles. *Watsonia, 5*: 329-335.

Walters, S.M. 1970. Dwarf variants of *Alchemilla* L. *Fragmenta floristica et geobotanica, 16*: 91-98.

Walters, S.M. 1974. The rediscovery of *Senecio paludosus* in Britain. *Watsonia, 10*: 49-54.

Walters, S.M. 1980. *Eleocharis*. *In: Flora Europaea, Volume 5*, ed. by T.G. Tutin *et al*. Cambridge, Cambridge University Press.

Walters, S.M. 1986. *Alchemilla*: a challenge to biosystematics. *Symbolae Botanicae Upsaliensis, 27*: 193-198.

Warburg, E.F. 1957. Some new names in the British flora [*Sorbus*]. *Watsonia, 4*: 43-46.

Warburg, E.F., & Kárpáti, Z.E. 1968. *Sorbus* L. *In: Flora Europaea, Volume 2*, ed. by T.G. Tutin *et al*., 67-71. Cambridge, Cambridge University Press.

Watkinson, A.R., & Davy, A.J. 1985. Population biology of salt marsh and sand dune annuals. *Vegetatio, 62*: 487-497.

Watson, J. 1994. *A survey of* Cochlearia micacea *on Ben Lawers NNR 1994*. Killin, unpublished report to the National Trust for Scotland.

Watt, A.S. 1964. The community and the individual. *Journal of Ecology, 52*: 203-211.

Watt, A.S. 1971. Rare species in Breckland: their management for survival. *Journal of Applied Ecology, 8*: 593-609.

Webb, D.A. 1950. Biological Flora of the British Isles. *Saxifraga caespitosa* L. *Journal of Ecology, 38*: 194-197.

Webb, D.A. 1985. What are the criteria for presuming native status? *Watsonia, 15*: 231-236.

Webb, D.A., & Gornall, R.J. 1989. *Saxifrages of Europe*. London, Christopher Helm.

Webb, D.A. & Scannell, M.J.P. 1983. *Flora of Connemara and the Burren*. Cambridge, Royal Dublin Society & Cambridge University Press.

Webster, S.D. 1990. Three natural hybrids in *Ranunculus* L. subgenus *Batrachium* (DC) A. Gray. *Watsonia, 18*: 139-146.

Welch, D. 1970. *Saxifraga hirculus* L. in north-east Scotland. *Transactions of the Botanical Society of Edinburgh, 41*: 27-30.

Welch, D. 1992. *Survey of* Saxifraga hirculus *in N.E. Scotland, and a review of information relevent to its conservation*. (Contractor: ITE, Banchory.) Edinburgh, Scottish Natural Heritage.

Welch, D. 1995. Habitat preferences and status of *Saxifraga hirculus* in North-East Scotland. *Botanical Journal of Scotland, 48*: 177-186.

Wells, T.C.E. 1967. *Dianthus armeria* L. at Woodwalton Fen, Hunts. *Proceedings of the Botanical Society of the British Isles, 6*: 337-342.

Wells, T.C.E. 1976. Biological Flora of the British Isles. *Hypochoeris maculata* L. *Journal of Ecology, 64*: 757-774.

Wells, T.C.E., Preston, C.D., Mountford, J.O., & Croft, J.M. 1995. *Species recovery programme: fen violet* (Viola persicifolia Schreber) *2nd progress*. Unpublished report to English Nature and the Natural Environmental Research Council.

Westhoff, V., & Den Held, A.J. 1969. *Planten Gemeenschappen in Nederland*. Zutphen.

Weston, R. 1994. *Althaea hirsuta* in v.c. 54. *Botanical Society of the British Isles News, 66*: 18.

Wheeler, B.D. 1978. The wetland plant communities of the River Ant valley, Norfolk. *Transactions of the Norfolk & Norwich Naturalists' Society, 24*: 153-187.

Wheeler, B.D. 1993. Botanical diversity in British mires. *Biodiversity and Conservation, 2*: 490-512.

Wheeler, B.D., Brookes, B.S., & Smith, R.A.H. 1983. An ecological study of *Schoenus ferrugineus* L. in Scotland. *Watsonia, 14*: 149-256.

Wheeler, B.D., & Shaw, S.C. 1987. Comparative survey of habitat conditions and management characteristics of herbaceous rich-fen vegetation types. *Nature Conservancy Council, CSD Report*, No. 764.

White, F.B.W. 1898. *The Flora of Perthshire*. Edinburgh, Perthshire Society of Natural Science.

White, J.W. 1912. *Flora of Bristol*. Bristol, John Wright.

Wilkinson, L. 1997. *Asparagus officinalis* subspecies *prostratus*. Action Plan Update Report. Cardiff, Countryside Council for Wales.

Wilkinson, M.J., & Stace, C.A. 1991. A new taxonomic treatment of the *Festuca ovina* L. aggregate (Poaceae) in the British Isles. *Botanical Journal of the Linnean Society, 106*: 347-397.

Wilks, H.M. 1960. The re-discovery of *Orchis simia* Lam. in Kent. *Transactions of the Kent Field Club, 1*: 50-55.

Wilks, H.M. 1966. *The monkey orchid in East Kent.* Handbook of the Society for the Promotion of Nature Reserves.

Willems, J.H. 1982. Establishment and development of a population of *Orchis simia* Lamk. in the Netherlands, 1972-1981. *New Phytologist, 91*: 757-765.

Willems, J.H. & Bik, L. 1991. Long-term dynamics in a population of *Orchis simia* in the Netherlands. *In: Population ecology of terrestrial orchids,* ed. by T.C.E. Wells & J.H. Willems, 33-45. The Hague, SPB Academic Publishing.

Williams, P. 1996. *A survey of the ditch flora in the North Kent Marshes SSSIs, 1995.* Peterborough, English Nature. (Research Report, No. 167.) [Species distribution maps in Appendices 2i and 2ii.]

Willis, A.J. 1967. A new locality for *Liparis loeselii.* Proceedings of the Botanical Society of the British Isles, 6: 352-353.

Willis A.J. 1985. Plant diversity and change in a species-rich dune system. *Transactions of the Botanical Society of Edinburgh, 44*: 291-308.

Wilmott, A.J. 1918. *Erythraea scilloides* in Pembrokeshire. *Journal of Botany, 56*: 321-323.

Wilson, A. 1927. *Stachys alpina* in North Wales. *North Western Naturalist, 2*: 181-182.

Wilson, G.B., Whittington, W.J., & Humphries R.N. 1995. Biological Flora of the British Isles. *Potentilla rupestris* L. (*Potentilla corsica* Sieber ex Lehm). *Journal of Ecology, 83*: 335-343.

Wilson, G.B., Wright, J., Lusby, P., Whittington, W.J., & Humphries, R.N. 1995. Biological Flora of the British Isles. *Lychnis viscaria* L. *Journal of Ecology, 83*: 1039-1051.

Wilson P.J. 1990. *The ecology and conservation of rare arable weed species and communities.* PhD Thesis, University of Southampton.

Wilson, P.J. 1993. The ecology and conservation of field cow-wheat (*Melampyrum arvense*). Eastleigh, Hampshire and Isle of Wight Wildlife Trust.

Winship, H.R. 1994a. *The conservation of cut-leaved germander* Teucrium botrys *L.* Eastleigh, Hampshire and the Isles of Wight Wildlife Trust and English Nature Species Recovery Programme.

Winship, H.R. 1994b. *The conservation of slender cottongrass* Eriophorum gracile *Koch ex Roth in England.* Peterborough, English Nature, and Eastleigh, Hampshire and the Isles of Wight Wildlife Trust.

Winship, H.R. 1994c. *The conservation of wood calamint* Clinopodium menthifolium *(Host) Stace on the Isle of Wight.* Peterborough, English Nature, and Eastleigh, Hampshire and the Isles of Wight Wildlife Trust.

Wolton, R.J., & Trowbridge, B.J. 1990. *The occurrence of acidic, wet oceanic grasslands (Rhos Pasture) in Brittany, France.* Taunton, unpublished report to the Nature Conservancy Council.

Wolton, R.J., & Trowbridge, B.J. 1992. *The occurrence of Rhos Pasture in Galicia, Spain.* Taunton, unpublished report to English Nature.

Wood, D. 1989. *A vegetation survey of the Insh Marshes RSPB Reserve, 1988-1989.* Sandy, unpublished report to the Royal Society for the Protection of Birds.

Woodell, S.R.J. 1967. *Viola stagnina* in Oxfordshire. *Proceedings of the Botanical Society of the British Isles, 6*: 32-36.

Woodhead, N. 1951. Biological Flora of the British Isles. *Lloydia serotina* (L.) Rchb. *Journal of Ecology, 39*: 198-203.

Woods R.G. 1993. *Flora of Radnorshire.* Cardiff, National Museum of Wales.

Wright, J., Averis, B., & Lusby, P. 1993. Scottish Rare Plant Project. *Polygonatum verticillatum* (L.) All. Edinburgh, Royal Botanic Garden, and Scottish Natural Heritage.

Wright, J., & Lusby, P. 1993. Scottish Rare Plant Project. *Lychnis viscaria* L. Edinburgh, Royal Botanic Garden, and Scottish Natural Heritage.

Wright, J., & Lusby, P. 1994. Scottish Rare Plant Project. *Lychnis alpina* Action Plan. Part 1. Current status and distribution. Edinburgh, Royal Botanic Garden, and Scottish Natural Heritage.

Wyse Jackson, M.B. 1992. Taxonomic notes on *Cerastium fontanum* Baumg. *emend.* Jalas (Caryophyllaceae) in Europe. *Botanical Journal of the Linnean Society, 109*: 325-328.

Yeo, P.F. 1956. Hybridisation between diploid and tetraploid species of *Euphrasia. Watsonia, 3*: 253-269.

Yeo, P.F. 1966. The breeding relationships of some European *Euphrasiae. Watsonia, 6*: 216-245.

Yeo, P.F. 1978. A taxonomic revision of *Euphrasia* in Europe. *Botanical Journal of the Linnean Society, 77*: 223-334.

Youngson, J. 1986. *A study of the survival of the rare species* Gentianella uliginosa *in its natural habitats.* Undergraduate Thesis, Wolverhampton Polytechnic.

Localities mentioned in the text

The following list includes all localities mentioned in the accounts of species, together with the vice-county and the Ordnance grid square in which they occur.

Abbots Ripton, Cambridgeshire, TL27
Aberffraw, Anglesey, SH36
Abhainn Beag, Isle of Arran, NR94
Abingdon, Berkshire, SU49
Adel Dam, Mid-W. Yorkshire, SE24
Afon Teifi, Cardiganshire, SN
Ainsdale, S. Lancashire, SD21
Alderfen Broad, E. Norfolk, TG31
Alderminster, Warwickshire, SP24
Ale Water, Selkirkshire, NT31
All Hallows Marshes, W. Kent, TQ87
Allerthorpe Common, S.E. Yorkshire, SE74
Altnaharra, W. Sutherland, NC53
Amberley Wild Brooks, W. Sussex, TQ01
An Stuc, Mid-Perthshire, NN64
Ancaster, S. Lincolnshire, SK94
Andrew's Wood, S. Devon, SX75
Aonach Beag, W. Inverness, NN47
Aonach Mor, W. Inverness, NN17
Ardnamurchan, Main Argyll, NM46,56
Arisaig, W. Inverness, NM68
Arishmell, Dorset, SY88
Arnside Knott, Westmorland, SD47
Arran, NR, NS
Arthur's Seat, Edinburgh, NT27
Arundel Park, W. Sussex, TQ00
Avon Gorge, N. Somerset/W. Gloucestershire, ST57
Axbridge, N. Somerset, ST45

Babraham, Cambridgeshire, TL54
Badgeworth, E. Gloucestershire, SO92
Balmaha, Dunbartonshire/Stirlingshire, NS49
Bangor, Caernarvonshire, SH57
Barmouth, Merioneth, SH61
Barton Broad, E. Norfolk, TG32
Batson Creek, S. Devon, SX73
Beachy Head, E. Sussex, TV59
Beauly R., E. Inverness, NH54
Beer Head, S. Devon, SY28
Beinn a'Chreachain, Main Argyll, NN34
Beinn Alligin, W. Ross, NG86
Beinn an Dothaidh, Main Argyll, NN34
Beinn Chonzie, Mid-Perthshire, NN73
Beinn Dearg, W. Ross, NH28
Beinn Sgulaird, Main Argyll, NN04
Bellingham, S. Northumberland, NY88
Ben Alder, W. Inverness, NN57
Ben Eibhinn, W. Inverness, NN47
Ben Heasgarnich, Mid-Perthshire, NN33,43
Ben Hope, W. Sutherland, NC44
Ben Lawers, Mid-Perthshire, NN54,64
Ben Lui (Beinn Laoigh), Perthshire, NN22
Ben Nevis, W. Inverness, NN17
Ben Vrackie, E. Perthshire, NN96
Benfleet Downs, S. Essex, TQ78

Berechurch, N. Essex, TL92
Berrow Dunes, N. Somerset, ST25
Berry Head, S. Devon, SX95
Bettyhill, W. Sutherland, NC76
Billacombe, S. Devon, SX55
Bingley, Mid-W. Yorks., SE13
Binsey, Oxfordshire, SP40
Bishopsteignton, S. Devon, SX87
Bixley Heath, Suffolk, TM24
Black Head, W. Cornwall, SW71
Blackfleet Broad, E. Norfolk, TG42
Blackpool Moss, Roxburghshire, NT52
Blackwater, N. Hampshire, SU85
Blair Atholl, Perthshire, NN86
Blakeney, E. & W. Norfolk, TG04
Bolt Head, S. Devon, SX63
Bovey Basin, S. Devon, SX
Box Hill, Surrey, TQ15
Boxley, E. Kent, TQ75
Boxwell, W. Gloucestershire, ST89
Braemar, S. Aberdeenshire, NO18
Bramshott, S. Hants, SU83
Brandon, W. Suffolk, TL48
Braunton Burrows, N. Devon, SS43
Breadalbane Mts., Perthshire, NN
Brean Down, N. Somerset, ST25
Breckland, W. Norfolk/W. Suffolk/Cambs., TF, TL
Bridge of Orchy, Main Argyll, NN23
Bridgewater Canal, S. Somerset, ST33
Brigend, Glamorgan, SS88
Brighton, E. Sussex, TQ30
Brixham, S. Devon, SX95
Brize Norton, Oxfordshire, SP20
Bulstrode Park, Buckinghamshire, SU98
Byfleet, Surrey, TQ06

Cader Idris, Merioneth, SH71
Caenlochan, E. Perthshire/Angus, NO17
Cairngorms, NH, NJ, NN, NO
Cairntoul, S. Aberdeenshire, NN99
Calder & Hebble Navigation, Mid-W. Yorkshire, SE02, 12
Caldicot, Monmouthshire, ST48
Callington, E. Cornwall, SX36
Cam Chreag, Mid-Perthshire, NN33
Camber Castle, E. Sussex, TQ91
Canisp, W. Sutherland, NC11, 21
Cape Cornwall, W. Cornwall, SW33
Cas Troggy, Monmouthshire, ST49
Castle Carew, Pembrokeshire, SN00
Castle Hedingham, N. Essex, TL73
Cauldron Snout, Durham, NY82
Cavenham Heath, W. Suffolk, TL77
Cefn-y-Bedd, Denbighshire, SJ35
Chadlington, Oxfordshire, SP32
Charlbury, Oxfordshire, SP32
Cheddar Gorge, N. Somerset, ST45
Chepstow, Monmouthshire, ST59
Cherry Hinton, Cambridgeshire, TL45
Chesil Beach, Dorset, SY58

Cheviot Hills, Northumberland, NT
Chilterns, SP, SU
Chippenham Fen, Cambridgeshire, TL66
Chipstead Valley, Surrey, TQ25
Chirk, Denbighshire, SJ23
Christchurch, S. Hampshire, SZ29
Cilygroeslwyd, Denbighshire, SJ15
Cliffe Marshes, W. Kent, TQ77
Cligga Head, W. Cornwall, SW75
Clitheroe, S. Lancashire, SD74
Clova Mts., Angus, NO37
Cockrah Wood, N.E. Yorkshire, SE98
Cockshoot Broad, E. Norfolk, TG31
Coire an Lochan, E. Inverness, NH90
Coire Fee, Angus, NO27
Coll, Mid Ebudes, NM15, 25, 26
Colne Point, N. Essex, TM22
Colonsay, S. Ebudes, NR39, 49
Colwyn Bay, Denbighshire, SH87
Coniston Old Man, Westmorland, SD29
Cotswolds, SO, SP, ST
Coverack, W. Cornwall, SW71
Cow Green reservoir, Durham/Westmorland, NY73, 82, 83
Cowal, Main Argyll, NN, NR, NS
Craeg Meagaidh, W. Inverness, NN48
Craig Breidden, Montgomery, SJ21
Craig-y-Castell, Breconshire, SO11
Craig-y-Cilau, Breconshire, SO11
Craig-y-Rhiwarth, Breconshire, SN81
Cronkley Fell, N.W. Yorkshire, NY82
Crook Peak, N. Somerset, ST35
Crowle Moors, N. Lincolnshire, SE71
Crymlyn Bog, Glamorgan, SS69, 79
Crymlyn Burrows, Glamorgan, SS79
Cuckmere Haven, E. Sussex, TV59
Cuillin Mts., N. Ebudes, NG42
Culbone, S. Somerset, SS84
Culford, W. Suffolk, TL87
Cwm Claisfer, Breconshire, SO11
Cwm Clydach, Breconshire, SO21
Cwm Idwal, Caernarvonshire, SH65

Dalmellington, Ayr, NS40
Dart Valley, S. Devon, SX
Dartmoor, Devon, SS,SX
Dartmouth, S. Devon, SX85
Dawlish Warren, S. Devon, SX97
Devil's Elbow, E. Perthshire, NO17
Devil's Point, S. Devon, SX45
Dirleton, E. Lothian, NT48
Drumochter Hills, E. Perth./E. Inverness, NN57, 67, 68
Dun Ban, Kintyre, NR51
Dungeness, E. Kent, TR01
Durlston, Dorset, SZ07

East Haven, Angus, NO53
East Portlemouth, S. Devon, SX73
Eastbourne, E. Sussex, TV59
Eaves Wood, W. Lancashire, SD47
Eden Valley, Cumberland, NY
Eggleston, Durham, NY92, NZ02
Eglwyseg, Denbighshire, SJ24

Eigg, N. Ebudes, NM48, 49
Ellenborough Park, N. Somerset, ST36
Ellesborough, Buckinghamshire, SP80
Ely, Cambridgeshire, TL58
Epping, S. Essex, TQ49
Esthwaite Water, Westmorland, SD39
Exeter, S. Devon, SX99

Fairfield, Cumberland, NY31
Farthing Downs, Surrey, TQ25, 35
Faversham, E. Kent, TR06
Ffrith, Flintshire, SJ25
Flat Holm, Glamorgan, ST36
Flimston, Pembrokeshire, SR99
Flimwell, E. Sussex, TQ73
Fobbing, S. Essex, TQ78
Foula, Shetland, HT93, 94
Freiston, S. Lincolnshire, TF34
Friston Wood, E. Sussex, TQ50
Frogmore Creek, S. Devon, SX74
Frys Hill, N. Somerset, ST45
Fulbourne, Cambridgeshire, TL45
Fulsby Wood, N. Lincolnshire, TF26

Gannel estuary, W. Cornwall, SW86
Geal Charn, W. Inverness, NN47
Gew Graze, W. Cornwall, SW61
Gibraltar Point, N. Lincolnshire, TF55
Giltar Point, Pembrokeshire, SS19
Glas Maol, E. Perthshire/Angus, NO17
Gleann na Ciche, E. Inverness, NH11
Glen Affric, E. Inverness, NH22
Glen Callater, S. Aberdeenshire, NO18, 28
Glen Canness, Angus, NO27
Glen Clova, Angus, NO37
Glen Coe, Main Argyll, NN15
Glen Diomhan, Isle of Arran, NR94
Glen Doll, Angus, NN17,27
Glen Easan Biorach, Isle of Arran, NR94
Glen Einich, E. Inverness, NH90
Glen Feshie, E. Inverness, NN89
Glen Fyne, Argyll, NN
Glen Iorsa, Isle of Arran, NR94
Glen Prosen, Angus, NO26, 27
Glen Quoich, S. Aberdeenshire, NO09, 19
Glen Turret, Perthshire, NN73
Goblin Combe, N. Somerset, ST46
Godalming, Surrey, SU94
Godmersham, E. Kent, TR05
Godrevy Point, W. Cornwall, SW54
Gog Magog Hills, Cambridgeshire, TL45
Golden Cap, Dorset, SY49
Goonhilly Down, W. Cornwall, SW71
Gordale, Mid-W. Yorkshire, SD96
Goring, Oxfordshire, SU67
Gower coast, Glamorgan, SS48
Gragareth, W. Lancashire, SD68
Grantham, S. Lincolnshire, SK93
Great Glen, NH, NM, NN
Great Orme, Caernarvonshire, SH78
Gunwalloe, W. Cornwall, SW62
Gwennap Head, W. Cornwall, SW32

Mull of Kintyre, Argyll, NR51
Mynydd Llangattock, Breconshire, SO11

Nance Wood, W. Cornwall, SW64
Nanjizal, W. Cornwall, SW32
Neroche Forest, S. Somerset, ST21
Ness, Cheshire, SJ37
New Forest, S. Hampshire, SU,SZ
Newmarket, W. Suffolk, TL66
Newport, Monmouthshire, ST38
Newport, Pembrokeshire, SN04
Newquay, W. Cornwall, SW86
Newtown, Isle of Wight, SZ49
Norfolk Broads, TG
North Downs, TQ,TR
North Queensferry, Fife, NT18
Norton Heath, S. Essex, TL60

Oak Mere, Cheshire, SJ56
Ochil Hills, W. Perthshire, NS89
Ongar, S. Essex, TL60
Orton, Westmorland, NY60
Otmoor, Oxfordshire, SP51

Pagham, W. Sussex, SZ89
Pakenham, W. Suffolk, TL96
Par, E. Cornwall, SX05
Pegwell Bay, E. Kent, TR36
Pellitras Point, W. Cornwall, SW32
Penrhyn Mawr, Anglesey, SH28
Penlee Point, E. Cornwall, SX44
Pennard Castle, Glamorgan, SS58
Pentire Point, E. Cornwall, SW98
Penzance, W. Cornwall, SW43
Pillar, Cumberland, NY11
Pinsla Downs, E. Cornwall, SX16
Pitton, S. Wiltshire, SU23
Plymouth, S. Devon, SX45, 55
Polruan, E. Cornwall, SX15
Poolbank, Westmorland, SD48
Poole, Dorset, SZ08, 09
Port Meadow, Oxfordshire, SP40
Porth Chapel (St Leven), W. Cornwall, SW32
Porth Dafarch, Anglesey, SH28
Porthgwarra, W. Cornwall, SW32
Porthmelgan Bay, Pembrokeshire, SM72
Portland, Dorset, SY66, 67, 77
Pound End, E. Norfolk, TG31
Prawle Point, S. Devon, SX73
Pulla Croos, W. Cornwall, SW73
Pwllheli, Caernarvonshire, SH33

River Ant, E. Norfolk, TG31,32
River Arun, W. Sussex, TQ
River Avon, Somerset/Gloucestershire, ST
River Calder, Mid-W. Yorkshire, SE02,12
River Dwyryd, Merioneth, SH63
River Dysynni, Merioneth, SH50
River Endrick, Dunbartonshire, NS48
River Fowey, E. Cornwall, SX15
River Glaslyn, Caernarvonshire/Merioneth, SH53
River Glen, Lincolnshire, TF22

River Irthing, Cumberland/Northmberland NY66, 67
River Loddon, Berkshire, SU76, 77
River Medway, E. Kent, TQ76
River Naver, W. Sutherland, NC53
River Ribble, Lancashire/Yorkshire, SD
River Severn, SO, ST
River Sheil, W. Inverness, NM66
River Spey, E. Inverness, NH70
River Thames, SN, SO, ST
River Thurne, E. Norfolk, TG41, 42
River Tyne, Cumberland/Northumberland, NY, NZ
River Ure, Mid-W. Yorkshire, SE37
River Wharfe, Yorkshire, SD, SE
River Wye, Monmouthshire/Radnorshire, SN, SO, ST
Rannoch Moor, Argyll/Perthshire, NN25, 34, 44, 45
Ravenshall, Kirkudbrightshire, NX55
Reading, Berkshire, SU67, 77
Reay, Caithness, NC96
Reculver, E. Kent, TR26
Reigate Heath, Surrey, TQ25
Rhoscolyn, Anglesey, SH27
Rhostyllen, Denbighshire, SJ34
Ribblehead, Mid-W. Yorkshire, SD77
Richmond, N.W. Yorkshire, NZ10
Ridge, Dorset, SY98
Ringwood Chase, Shropshire, SO47
Rochdale Canal, S. Lancashire, *c.* SD91
Romaldkirk, N.W. Yorkshire, NZ02
Rona, N. Ebudes, NG65, 66
Ronas Voe, Shetland, HU28
Ross Prioiry, Dunbartonshire, NS48
Rothamsted, Hertfordshire, TL11
Roudsea Wood, Westmorland, SD38
Rougholme Point, Westmorland, SD37
Ruabon, Denbighshire, SJ24
Rum, N. Ebudes, NG30, 40, NM39, 49
Rye, E. Sussex, TQ91

Saddle Point, Pembrokeshire, SR99
Salterhebble Bridge, Mid-W. Yorkshire, SE02
Samson, Isles of Scilly, SV81
Sand Point, N. Somerset, ST36
Sandwich, E. Kent, TR35
Sandymouth, E. Cornwall, SS21
Sawston Hall Meadows, Cambridgeshire, TL44
Scabbacombe, S. Devon, SX95
Selsdon, Surrey, TQ36
Selsey, W. Sussex, SZ89
Sgurr Alasdair, N. Ebudes, NG42
Shap, Westmorland, NY51
Sheerwater, Surrey, TQ06
Shillinglee, W. Sussex, SU93
Shirehampton, W. Gloucestershire, ST57
Shoeburyness, S. Essex, TQ98
Shute Shelve, N. Somerset, ST45
Skipwith Common, S.E. Yorkshire, SE63
Skye, N. Ebudes, NG
Slapton Ley, S. Devon, SX84
Sleaford, S. Lincolnshire, TF04
Slimbridge, W. Gloucestershire, SO70
Small Mouth, Dorset, SY67
Snailwell Meadows, Cambridgeshire, TL66

South Downs, Sussex/Kent, SU, TQ
South Hams, S. Devon, SX
South Harris, Outer Hebrides, NG08, 09
South Stack, Anglesey, SH28
South Tyne, S. Northumberland, NY84, 85
Southampton, S. Hampshire, SU31, 41
Southsea, Denbighshire, SJ35
Southwick, Kirkudbrightshire, NX95
Southwold, E. Suffolk, TM47
Sow of Atholl, E. Perthshire, NN67
Spalford, Nottinghamshire, SK86
St Agnes, Isles of Scilly, SV80
St Alban's Head, Dorset, SH97
St Aldhelm's Head, Dorset, SY97
St David's Head, Pembrokeshire, SM72
St Helen's, Isles of Scilly, SV91
St Just, W. Cornwall, SW33
St Martin's, Isles of Scilly, SV91
St Mary's, Isles of Scilly, SV91
St Michael's Mount, W. Cornwall, SW53
Stackpole, Pembrokeshire, SR99
Stainforth, N.W. Yorkshire, SD86
Stallode Wash, W. Suffolk, TL68
Stanner Rocks, Radnorshire, SO25
Start Point, S. Devon, SX83
Stathfarrar, E. Inverness, NH33
Steep Holm, N. Somerset, ST26
Storrington, W. Sussex, TQ01
Stour Valley, Dorset, ST81
Strath Nethy, E. Inverness, NJ00
Strathy Point, W. Sutherland, NC86
Stroud, E. Gloucestershire, SO90
Strumble Head, Pembrokeshire, SM84
Strumpshaw, E. Norfolk, TG30
Swanage, Dorset, SZ07
Swanton Novers Wood, E. Norfolk, TG03
Symonds Yat, Monmouthshire, SO51

Taff Valley, Breconshire, SO01
Tamar Valley, Cornwall/Devon, SX
Tankerton Cliffs, E. Kent, TR16
Teesdale, NY, NZ
Teign valley, S. Devon, SX78, 79, 88
Tenby, Pembrokeshire, SS09
The Storr, Isle of Skye, NG45
Thetford, W. Norfolk/W. Suffolk, TL88
Thurso, Caithness, ND16
Tilbury Fort, S. Essex, TQ67
Tima Water, Selkirkshire, NT21
Torksey, N. Lincolnshire, SK87
Torquay, S. Devon, SX96

Torridon Mountains, Sutherland, NC, NG, NH
Tremadog, Caernarvonshire, SH54
Trentishoe, N. Devon, SS64
Tresco, Isles of Scilly, SV81
Truro, W. Cornwall, SW84
Tuddenham, W. Suffolk, TL77
Twistleton Glen, Mid-W. Yorkshire, SD67

Uffcott Hill, N. Wiltshire, SU17
Ullswater, Westmorland/Cumberland, NY31, 41, 42
Unst, Shetland, HP60
Uphill, N. Somerset, ST35
Upper Halling, W. Kent, TQ66
Upton Broad, E. Norfolk, TG31

Walberswick, E. Suffolk, TM47
Waltham Brooks, W. Sussex, TQ01
Walton Backwaters, N. Essex, TM22
Wandsworth, Middlesex, TQ27
Warbleton Priory, E. Sussex, TQ51
Wastwater, Cumberland, NY10
Waveney Valley, Suffolk, TL, TM
Weardale, Durham, NY, NZ
Weeting Heath, W. Norfolk, TL78
Wendover, Buckinghamshire, SP80
West Burrafirth, Shetland, HU25
West Harling Heath, W. Norfolk, TL98
West Penwith, W. Cornwall, SW
Weston-Super-Mare, N. Somerset, ST36
Westwell Gorse, Oxfordshire, SP21
Westwood Great Pool, Worcestershire, SO86
Whernside, Mid-/N.W. Yorkshire, SD77, 78
White Ness, Shetland, HU34
White Nothe, Dorset, SY78
White Peak, Derbyshire, SK17
Whiteford, Glamorgan, SS49
Wick River, Caithness, ND35
Wicken Fen, Cambridgeshire, TL57
Widdybank Fell, Durham, NY82, 83
Wilmington, E. Sussex, TQ50
Wilsford, S. Lincolnshire, TF04
Witherslack, Westmorland, SD48
Witney, Oxfordshire, SP30
Woodchester Park, S. Gloucestershire, SO80
Woodstock, Oxfordshire, SP41
Woodwalton Fen, Huntingdonshire, TL28
Wotton-under-Edge, S. Gloucestershire, ST79
Wrexham, Denbighshire, SJ35
Wye Downs, E. Kent, TR04
Wymington, Bedfordshire/Northamptonshire, SP96
Wyre Forest, Worcestershire/Shropshire, SO77
Wytham, Oxfordshire, SP40